Noise and Vibration
Control

Noise and Vibration Control

Edited by
LEO L. BERANEK
Chief Scientist, Bolt Beranek and Newman Inc.
Lecturer, Massachusetts Institute of Technology

McGRAW-HILL BOOK COMPANY

New York   St. Louis   San Francisco   Düsseldorf
Johannesburg   Kuala Lumpur   London   Mexico
Montreal   New Delhi   Panama   Rio de Janeiro
Singapore   Sydney   Toronto

This book was set in Linofilm Baskerville by Quinn & Boden
Company, Inc., and printed on permanent paper and bound by
The Maple Press Company. The editors were William G. Salo,
Jr., and Frank Purcell. The designer was Naomi Auerbach.
Stephen J. Boldish supervised production.

Contents

Preface

This book is written for the engineer who wishes to solve a noise control problem. The material is graded in technical level, starting with the fundamentals of sound in air, scales that are used to measure sound, and the measurement and analysis of noise and vibration; and followed by methods of controlling noise and vibration and the so-called criteria for noise control. The text is a successor to *Noise Reduction* (McGraw-Hill, 1960), though only three of its eighteen chapters have the same authors as in the 1960 book.

Much of the material included in the present volume was developed for Special Summer Programs on Noise and Vibration Reduction offered at the Massachusetts Institute of Technology through the years since 1953. Other material was taken from the general acoustical literature and from the research and consultation practice of the staff of Bolt Beranek and Newman Inc.

I wish to express my warm appreciation to my co-authors both outside and inside Bolt Beranek and Newman Inc. for their unstinting effort toward producing authoritative and broadly applicable chapters. Parenthetically, those authors whose affiliations are not designated on the chapters are members of the BBN staff. I wish especially to thank Tony F. W. Embleton who assisted me at two recent Special Summer Programs at M.I.T., and Robert W. Young who made detailed comments on four chapters. I am indebted to O. L. Angevine, Jr., Peter K. Baade, James H. Botsford, Robert A. Heath, Ralph Huntley, Alan H. Marsh, Helmut A. Mueller, T. D. Northwood, Douglas W. Robinson, Hale J. Sabine, S. S. Stevens, Dean G. Thomas, Clayton H. Allen, Erich K. Bender, Warren E. Blazier, James J. Coles, Charles W. Dietrich, Ira Dyer, Parker W. Hirtle, Robert M. Hoover, George W. Kamperman, David H. Kaye, Edward M. Kerwin, Jr., Ronald L. McKay, Laymon N. Miller, Robert B. Newman, Denis U. Noiseux, and Bill G. Watters for their help and criticism during the preparation of the book.

I am particularly grateful to Elizabeth M. Donnelly, whose organizational talents lightened many tasks and to Sandra N. Hale who prepared the typescript with skill and care. I wish also to thank Suzanne Wadoski and Clare Twardzik for preparing the many illustrations.

Leo L. Beranek

Introduction

Virtually every problem in noise and vibration control involves a system composed of three basic elements: a source, a path, and a receiver. Before a solution to a complex noise problem can be designed, the dominant source of the noise must be known, the characteristics of the significant transmission paths must be understood, and a criterion for the level of noise considered permissible or desirable in this situation must be available.

These three elements of the noise problem do not necessarily act independently. The sound power that is radiated depends on the environment surrounding the source. For example, a machine may radiate more sound if it is placed in the corner of a room rather than in some other location. A speaker raises or lowers his voice depending on the size and reverberation characteristics of the room in which he is talking. The path of the sound may be affected by the acoustical details of the source and the receiver, as well as by their heights above the ground. A listener's judgement of noisiness may depend on whether he is working with his hands, concentrating on a creative task, conversing, listening to music, or trying to sleep.

His attitude toward noise may be influenced not only by the nature of the path and the spectrum of the noise, but also by economic or psychological factors such as a bonus for working hours spent in noise, or the fear of audiological or financial consequences. All these considerations emphasize that each noise problem involves a complex system of interacting elements.

Noise "control" does not always involve reduction of the unwanted sound. In the modern open-plan office building, a background "noise" with carefully controlled tonal quality and loudness may be introduced into the space through concealed sources in order to "mask" or cover unwanted noises made by the occupants and their typewriters and calculators, the elevators, and the street traffic. Thus noisiness, the annoyance caused by noise, can sometimes be reduced by adding more "noise."

Solving a noise control problem usually involves a tradeoff. The cost of protecting equally the hearing of every worker in a manufacturing plant may be so prohibitive that a higher risk of damage would have to be accepted for some of the personnel. The solution might consist of a combination of partial noise-reduction measures, the institution of an audiological testing program to select those with sensitive ears who should be transferred to other jobs, and a plan to compensate the remaining few who might suffer some hearing loss.

Noise Control at the Source. A noise source is created by the motion of a solid, liquid, or gas. A solid source may be quieted if its mode of operation is changed so that it moves less, for example, by reducing the forces that cause motions and by strengthening, damping, or isolating all or parts of the structure. Liquid and gaseous sources may be quieted by eliminating turbulence, reducing flow velocity, smoothing flow, and attenuating pressure pulsations. Control at the source by planning while the product is in the design stage is often the most effective and least expensive of control measures.

Noise Control in the Path. Most corrective measures for an existing noise control problem utilize changes in the path. Consequently, a large part of the present book is devoted to that subject. Involved are control of sound propagation out of doors, in rooms, in structures, and in ducts. Solutions include barriers, porous materials, plugs, caulking, bracing, mufflers, enclosures, vibration damping, and vibration isolation.

The Demands of the Receiver. The level to which a noise should be reduced to be acceptable to human receivers often requires judgment on the part of the engineer and the owner of a building or machine. A criterion for noise control for listeners depends on whether the goal is to preserve hearing, to create space where conversation is easy, or to provide comfort in the home, at work, or in transportation vehicles. The first two goals for noise control are fairly accurately quantifiable, but comfort may depend on mental attitudes, which may be elusive or which may change on short notice.

Noise and Vibration Control

Chapter One

The Behavior
of Sound Waves

PETER A. FRANKEN

1.1 The Nature of Sound

Sound is a disturbance that propagates through an elastic material at a
speed characteristic of that medium. This brief sentence, when converted into quantitative terms, contains a large amount of scientific information that constitutes the basis of the science of acoustics. When
applying the science of sound to the practical task of noise and vibration
control, to which this book is restricted, we do not need to reveal a detailed physical understanding of sound waves. Rather we shall draw directly on the results of fundamental studies as necessary, and refer the
interested reader to basic literature for further details.

The simple definition of sound above suggests that sound can be sensed
by the measurement of some physical quantity in the medium that is disturbed from its equilibrium value. The physical quantity that is generally of interest is *sound pressure*, the incremental variation in pressure
above and below atmospheric pressure, which, in turn, is normally about
14.7 lb/in.² or about 1.013×10^5 N/m² in metric units at sea level.*

---

* See Appendix C for conversion factors.

1

Sound pressures are extremely small.   For normal speech, they average about 0.1 N/m² (1 dyn/cm²) above and below atmospheric pressure at a distance of a meter from a talker.

In order for there to be wave motion in a material medium, the medium must exhibit two properties, inertia and elasticity.   Inertia is the property that permits one element of the medium to transfer momentum to adjacent elements.   It is related to the density of the medium, that is to say, the mass of an element.   Elasticity is the property that produces a force on a displaced element, tending to return it to its equilibrium position.   In this chapter, we will be concerned only with sound waves in the air.   In later chapters, we shall study sound in structural elements such as beams, plates, and walls.

How do we know that air has the properties of inertia (mass) and elasticity?   Atmospheric pressure results from the weight of an atmospheric column of air.   This is an indirect indication that air has weight and therefore mass. We could get a more direct indication by weighing a box containing air and then evacuating the box and weighing it again.   We would find that it would take about 13.6 ft³ of air at sea level (0.751 m Hg) and a temperature of 22°C (71.6°F) to weigh 1 lb.   This corresponds to a density of about 1.18 kg/m³ in metric units.

We can show by a number of means that air has elasticity.   For example, if we take a basketball in its uninflated state and drop it on the floor, it thuds to a halt without rebounding.   Inflating the ball with air imparts a resiliency to it which causes it to rebound when dropped.   Even more simply, suppose while preparing to inflate the ball we had held a finger over the air outlet of our hand pump so that no air could come out.   Pushing on the handle to compress the trapped air in the pump produces the same sensation as pushing on a spring.   By knowing the physical dimensions of the pump and the amount of force we apply to the handle, we can determine the "spring constant" or compressibility of the gas, thus giving us the elasticity of the air quantitatively.

1.2   Basic Properties of Waves

We can gain insight into the nature of wave motion by looking at a very simple case, that of a single pulse of sound traveling down a tube, which constrains it from spreading. Figure 1.1*a* is a "spatial snapshot" of such a pulse.   This figure presents the value of the disturbance, in this case, of the pressure deviation from atmospheric pressure, plotted as a function of the spatial coordinate x, at a time $t = 0$.   The waveshape is a rectangular pulse of value A between the spatial coordinates $x = a$ and $x = b$. Elsewhere the value is zero.   Let us assume that this wave pulse propagates in the positive x direction at a speed of c units per second.   What will the spatial snapshot look like after d seconds have elapsed? The wavefront at coordinate b in Fig. 1.1a will move cd units in the positive x direction and will then be located at $b + cd$. Similarly, the wavetail at coordinate a in Fig. 1.1a will move to $a + cd$ in d seconds. The snapshot for d seconds is shown in Fig. 1.1b. Clearly, the pulse has the same width as before. Also, since there is no spreading of the wave, nor any energy loss in the medium, it must have the same value A as before.

It is not difficult to find an analytic form that expresses this idea of propagation without distortion in the positive x direction. The discussion in the previous paragraph shows that this form must permit the value at $x = a + cd$, $t = d$, to be exactly equal to the value at $x = a$, $t = 0$. Any function F that combines the spatial and temporal variables in the form $F(x - ct)$ satisfies this requirement. For the values $x = a$, $t = 0$, such a function would have the value $F(a)$, while for the values $x = a + cd$, $t = d$, the function takes the form $F(a + cd - cd) = F(a)$, exactly as required. We may emphasize the generality of this result. *Any* function $F(x - ct)$ which combines the spatial and temporal variables in the form $x - ct$ represents a wave propagating in the positive x direction at speed c. Examples of such functions are $A \sin k(x - ct)$, $A[(x - ct)^3 - 3(x - ct)^2]$, and $Ae^{jk(x-ct)}$.

We may use a pair of spatial snapshots similar to those in Fig. 1.1 to examine the case of a wave moving in the *negative* x direction. For this case we will find that the general analytic form must be any function $G(x + ct)$ which combines the spatial and temporal variables in the form $x + ct$.

We now know the general functional form that a wave traveling in the positive or negative x direction must assume. We must now examine the important physical properties of a sound wave, so that we may incorporate these properties in our functional form.

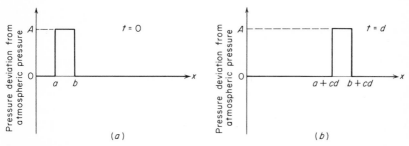

Fig. 1.1 Graphs showing the sound pressure for a single pulse of sound traveling down a tube, which constrains it from spreading. The waveshape is a rectangular pulse of amplitude A which propagates in the positive x direction (to the right) at a speed of c units per second: (a) the waveshape at $t = 0$ sec, and (b) the waveshape at $t = d$ sec.

Frequency. The subjective pitch of a simple sound is determined primarily by the number of times per second at which the sound-pressure disturbance oscillates between positive and negative values. The physical measure of this oscillation rate is called *frequency* and is denoted by the symbol f. The unit of frequency is the cycle per second (cps), which by international agreement is called hertz (Hz), after the man who first studied electromagnetic waves. The range of normal adult hearing extends approximately from 20 to 16,000 Hz. The ear is most sensitive, that is, the threshold of audibility is lowest, for sounds around 3,000 Hz.

Period. The reciprocal of the frequency f is the *period T*. It is the time required for one complete cycle. Thus, the period of a 1,000-Hz wave is 0.001 sec.

Wavelength. The *wavelength* of sound is the distance between analogous points on two successive waves. It is denoted by the Greek letter λ and is equal to the ratio of the speed of sound to the frequency of the sound, so that

$$\lambda = \frac{c}{f} = cT \tag{1.1}$$

In a very general way, we may distinguish between noises that consist of periodic sounds, that repeat regularly, and aperiodic sounds, that fluctuate randomly. The simplest periodic sound is a pure tone. A pure-tone sound is a pressure disturbance that fluctuates sinusoidally at a fixed frequency. Rotating machines, such as turbines and compressors, usually produce noise that is predominantly a set of pure-tone sounds. The set is often harmonically related: each of the *harmonic* sounds has a frequency that is an integral multiple of the lowest or fundamental frequency. For example, the fundamental frequency (first harmonic) of the noise from a turbine under certain operation conditions might be 3,000 Hz. The frequencies of the higher *harmonics* would then be 6,000 Hz, 9,000 Hz, and so on. There is a basic mathematical tool, called Fourier analysis, that can be used to subdivide any periodic signal into a series of pure-tone signals, harmonically related. Thus, any sound that repeats regularly can be subdivided into a series of harmonics, each with a particular amplitude.

The noises of a dishwasher, an air diffuser, and a rocket are examples of *aperiodic sounds*. An aperiodic sound cannot be subdivided into a set of harmonically related pure-tone sounds. It can, however, be described in terms of an infinitely large number of pure tones, of different frequencies, spaced an infinitesimal distance apart, and with different amplitudes.

It is mathematically convenient for us to study the behavior of pure-tone sounds first. We can then infer the behavior of more complex sounds.

1.3   Free Progressive Waves

Let us visualize a simple situation that will permit us to generate and study a pure-tone (sinusoidal) sound wave.   In Fig. 1.2 a long tube is shown containing air, with a movable piston at one end.   When the piston is moved back and forth by the mechanical arrangement shown, a plot of its position as a function of time is a sinusoidal function.   This to-and-fro motion of the piston causes the air molecules adjacent to it alternately to be crowded together, or compressed, and a little later to be moved apart, or rarefied.   This action of alternate compression and rarefaction moves down the tube owing to the elasticity and inertia of the medium.   The wave thus generated is sinusoidal and has a frequency equal to the number of times per second at which the piston moves back and forth.   The strength of the wave is determined by the magnitude of the displacement of the piston.

The three waves at the top of Fig. 1.2 show that the amplitude of the traveling wave is unchanged as the wave propagates to the right, and that the time delays between the same part of the wave at the plane at $x = 0$ and at the planes at $x = x_1$ and $x = 2x_1 = x_2$ are x_1/c sec and $2x_1/c$ sec respectively.

Speed of Sound.   If we varied the properties of the gas through which the sound was traveling, we would find that the square of the speed of sound varied directly with the equilibrium gas pressure p_s and inversely with the equilibrium gas density ρ.   The constant involved in the expres-

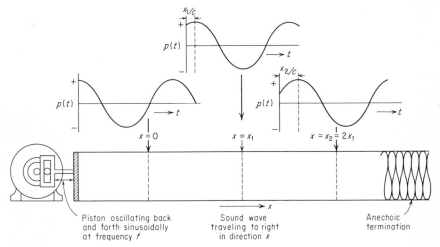

Fig. 1.2   Plane-wave propagation.   A plane wave generated by the piston at the left of the tube travels to the right and is absorbed by the anechoic termination.   The three waves at the top give the variation in sound pressure with time at the three points indicated, $x = 0$, $x = x_1$, and $x = x_2 = 2x_1$.

sion for the speed of sound is the ratio of the specific heat of the gas at constant pressure to the specific heat at constant volume. For air at most temperatures with which we deal, this ratio is 1.4. Thus the speed of sound in air is given by the equation

$$c = \sqrt{\frac{1.4 p_s}{\rho}} \qquad \text{m/sec or ft/sec} \tag{1.2}$$

If we assume that air behaves as an ideal gas, we can show that the speed of sound is dependent only on the absolute temperature of the air. This assumption is quite reasonable for most temperatures and densities with which we deal. The equations for the speed of sound become

$$c = 20.05 \sqrt{T} \qquad \text{m/sec} \tag{1.3}$$

$$c = 49.03 \sqrt{R} \qquad \text{ft/sec} \tag{1.4}$$

where R is the absolute temperature in degrees Rankine, that is, 459.7° plus the temperature in degrees Fahrenheit, and T is the absolute temperature in degrees Kelvin, that is, 273.2° plus the temperature in degrees centigrade.

Example 1.1   Determine the speed of sound at 70°F (21.1°C) in both English and metric units.

SOLUTION: The Rankine temperature is 459.7 + 70 = 529.7°R and the Kelvin temperature is 273.2 + 21.1 = 294.3°K. The speed of sound is then about

$$c = 49 \ \sqrt{R} = 49 \ \sqrt{530} = 1,128 \ \text{ft/sec}$$

$$c = 20.05 \ \sqrt{T} = 20.05 \ \sqrt{294.3} = 344 \ \text{m/sec}$$

In discussing the piston-tube experiment of Fig. 1.2, we must assume either that the tube is infinitely long or else that it has a nonreflecting (anechoic) termination so that no part of the energy in the sound wave will be reflected from the far end of the tube. The sound wave started by the piston then progresses down the tube without interacting with the side walls or the far end of the tube. All properties of the wave can then be described in terms of the distance down the tube from the piston and the action of the piston itself. This form of wave is known as a *one-dimensional, plane, free progressive* wave. The one-dimensional aspect relates to the specification of the parameters in terms of a single distance, the planar aspect to the fact that the wavefronts are parallel to each other, and the free progressive aspect to the advancement of the wave without interference from other objects or changes in the medium. In actuality, the wave is not truly free in that it is bounded by the walls of the tube and there will be some frictional losses at the boundaries.

For practical purposes, however, the motion is essentially the same as if the tube and the piston were infinite in cross section.

Sound Pressure and Particle Velocity. We have noted that at any point along the tube we can measure the varying disturbance from equilibrium pressure, the *sound pressure p*. In the mks system, its units are newtons per square meter, or N/m². Also, we can measure a varying *particle velocity u*, associated with the to-and-fro motions of the air molecules, which always occur along a line parallel to the direction of propagation. Its units are meters per second, m/sec.

Intensity. A free progressive sound wave transmits energy. The usual way in which the sound-energy propagation is described is in terms of *intensity I*, defined as the energy that flows through a unit area in a unit time. The unit for intensity is watts per square meter. In terms of the parameters describing a *free progressive sound wave* in which p and u are in phase, the *average* intensity in the direction of the wave propagation is the time average of the product of the sound pressure and the particle velocity measured *in the direction of the wave propagation*, expressed as

$$I = \overline{pu} \qquad \text{watts/m}^2 \tag{1.5}$$

1.3.1 One-dimensional Plane Wave We can now examine the quantitative behavior of a one-dimensional sound wave. Consider the piston-tube experiment (see Fig. 1.2) with an anechoic (echo-free) termination at the right of the tube, so that the tube contains a sinusoidal sound wave traveling in the positive x direction (toward the right). We have some freedom in our choice of time and space origins, and so let us select these origins so that the sound pressure at the plane $x = 0$ and at the time $t = 0$ has its maximum value of P_R. Figure 1.3 is a series of spatial snapshots representing the sound pressure as a function of distance, at four different instants of time. Each of these snapshots shows the sound pressure over a spatial extent of one wavelength λ. The vertical lines that separate the distance between $x = 0$ and $x = \lambda$ into 20 intervals help us to observe the sound-pressure behavior at any fixed point.

First, let us look at the uppermost plot of Fig. 1.3. Plot *a* represents the sound-pressure snapshot at the time $t = 0$. Because we are dealing with a periodic wave, it also represents the snapshots for times $t = T, 2T, 3T, \ldots, nT$. The value of the sound pressure at each position is represented by the location of the dots. As we required earlier, the maximum value $+P_R$ exists at the location $x = 0$ for this first snapshot. Again, because the wave is periodic in space, the same maximum value must occur at values of $x = \lambda, 2\lambda, 3\lambda, \ldots$

Plot *b* is the snapshot that describes the sound pressure a quarter of a period ($T/4$ sec) after a.   In other words, the wave in a has moved to the right a distance equal to $\lambda/4$ to become the wave in b.   Plots c and d represent the sound-pressure snapshots after two more successive intervals of $T/4$ sec.   After T sec has elapsed, the wave has traveled a full wavelength λ to the right, and the sound-pressure distribution has returned to the snapshot of a.   Then the entire sequence of snapshots repeats itself as the wave continues to travel to the right.   To convince yourself that the wave is traveling to the right, allow your eyes to jump successively from a to b to c to d.

We can now combine our knowledge of the behavior of sound waves with the snapshots of Fig. 1.3 to obtain an analytical expression for a one-dimensional sound wave.   Our earlier discussion of Fig. 1.1 showed that our expression must have the functional form $F(x - ct)$.   Figure 1.3 shows us that this function must be a sinusoid.   From trigonometry we know that a general expression must then have the form $C \cos [k(x - ct)]$

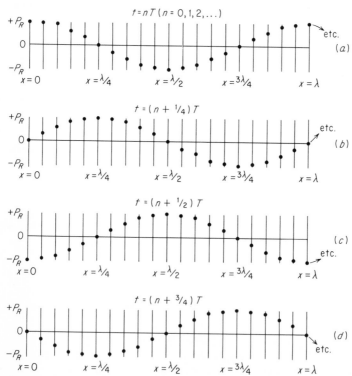

Fig. 1.3   Graphs showing the sound pressure in a plane free-progressive forward-traveling wave at 20 places in space at four instants of time t. The wave is produced by a source at the left and travels to the right with a speed c. The length of time it takes a wave to travel a distance equal to a wavelength is called the period T.   Forward-traveling wave: $p(x,t) = P_R \cos [k(x - ct)]$; $k = 2\pi/\lambda = 2\pi/(cT) = \omega/c$.

$+ S \sin [k(x - ct)]$ where C and S are unknown amplitudes and k is an unknown parameter whose meaning will become clear shortly. We have already required that the sound pressure have its maximum value P_R for $x = 0$, $t = 0$. Our expression will satisfy this requirement if we set $C = P_R$ and $S = 0$, so that

$$p(x,t) = P_R \cos [k(x - ct)] \qquad \text{N/m}^2 \tag{1.6}$$

Wavenumber. The cosine function is periodic and repeats its value every time the argument increases 2π radians (360°). From the definition of wavelength λ, we can write this periodicity condition as

$$\cos [k(x + \lambda - ct)] = \cos [k(x - ct) + 2\pi] \tag{1.7}$$

so that

$$k\lambda = 2\pi \qquad k = \frac{2\pi}{\lambda} \qquad \text{radians/m} \tag{1.8}$$

We thus see the meaning of the parameter k. It is called the *wavenumber*.

Combining Eqs. (1.1) and (1.8), we can obtain an alternate expression for k

$$k = \frac{2\pi}{c/f} = \frac{2\pi f}{c} = \frac{\omega}{c} \qquad \text{radians/m} \tag{1.9}$$

where the quantity $2\pi f$ is defined as the *circular frequency* ω, so that

$$\omega = 2\pi f = kc \qquad \text{radians/sec} \tag{1.10}$$

Thus the argument in the trigonometric function may also be written in any one of the following ways

$$k(x - ct) = \frac{2\pi}{\lambda} (x - ct) = 2\pi f \left(\frac{x}{c} - t \right) = 2\pi \left(\frac{x}{\lambda} - \frac{t}{T} \right) = \frac{2\pi x}{\lambda} - 2\pi f t$$

$$= kx - \omega t \tag{1.11}$$

In Fig. 1.3 we have looked at the spatial behavior of a traveling sound wave, at certain fixed instants of time. From Eq. (1.7) the equations for the snapshots in Fig. 1.3 are given by

(a) $t = nT$: $p = P_R \cos \dfrac{2\pi x}{\lambda}$

(b) $t = (n + \frac{1}{4})T$: $p = P_R \cos \left(\dfrac{2\pi x}{\lambda} - \dfrac{\pi}{2} \right)$

(c) $t = (n + \frac{1}{2})T$: $p = P_R \cos \left(\dfrac{2\pi x}{\lambda} - \pi \right)$

(d) $t = (n + \frac{3}{4})T$: $p = P_R \cos \left(\dfrac{2\pi x}{\lambda} - \dfrac{3\pi}{2} \right)$

where $n = 0, 1, 2, 3, \ldots$.

We could equally well look at the temporal behavior at fixed points, for any particular time t.

$$x = 0: \quad p = P_R \cos 2\pi ft$$

$$x = \lambda/4: \quad p = P_R \cos \left[2\pi f \left(t - \frac{\lambda}{4c} \right) \right] = \cos \left[2\pi f \left(t - \frac{T}{4} \right) \right]$$

$$= \cos \left(2\pi ft - \frac{\pi}{2} \right)$$

$$x = \lambda/2: \quad p = P_R \cos \left[2\pi f \left(t - \frac{\lambda}{2c} \right) \right] = \cos (2\pi ft - \pi)$$

$$x = 3\lambda/4: \quad p = P_R \cos \left[2\pi f \left(t - \frac{3\lambda}{4c} \right) \right] = \cos \left(2\pi ft - \frac{3\pi}{2} \right)$$

where we have used the trigonometric identity $\cos (-A) = \cos A$. We see that the temporal behavior at $x = \lambda/4$ is the same as the behavior at $x = 0$, except that it is delayed by a time of $T/4$ sec. Another way to express this time delay is to say that a shift in phase (the argument of the trigonometric function) of $\pi/2$ radians (90°) has occurred relative to $t = 0$. Similarly, at $x = \lambda/2$ the behavior is the same as at $x = 0$, except for a time delay of $T/2$ sec. This corresponds to a phase shift of π radians (180°), relative to $t = 0$.

Consider the piston-tube experiment with the sound source and anechoic termination interchanged, so that the tube contains a sinusoidal sound wave traveling in the negative x direction (toward the left). The corresponding sequence of spatial snapshots for this backward-traveling wave is shown in Fig. 1.4. Earlier in the chapter we stated that the general form for a wave moving in the negative x direction is any function $G(x + ct)$ which combines the spatial and temporal variables in the form $x + ct$. We can use this functional form to describe the wave shown in Fig. 1.4, and we obtain the result for the backward-traveling wave

$$p(x,t) = P_L \cos [k(x + ct)] \qquad \text{N/m}^2 \qquad (1.12)$$

where P_L is the pressure amplitude.

A comparison of Eqs. (1.6) and (1.12) emphasizes that the sign relation between the space and time variables determines the direction in which the wave propagates. If the two variables are of different sign, the wave propagates in the positive spatial direction; if the two variables are of the same sign, the wave propagates in the negative spatial direction. This fact is confirmed by scanning Fig. 1.4 from a to d. The wave can only be made to "travel" from right to left as one's eyes jump from one plot to the next.

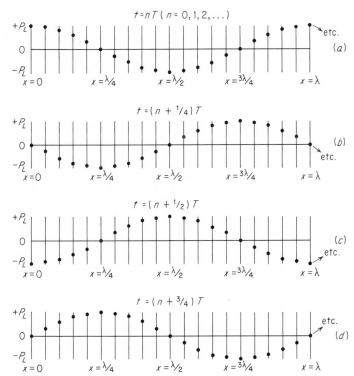

Fig. 1.4 Graphs showing the sound pressure in a plane free-progressive backward-traveling wave at 20 places in space at 4 instants of time t. The wave is produced by a source at the right (or by being reflected from a boundary) and travels to the left with a speed c. The period T is defined the same as for Fig. 1.3. Backward-traveling wave: $p(x,t) = P_L \cos [k(x + ct)]$; $k = 2\pi/\lambda = 2\pi/cT = \omega/c$.

RMS Sound Pressure. Most common sounds consist of a rapid, irregular series of positive pressure disturbances (compressions) and negative pressure disturbances (rarefactions) measured from the equilibrium pressure value. If we were to measure the mean value of the sound-pressure disturbance, we would find that it would be zero, because there are as many positive compressions as negative rarefactions. Thus, the mean value of sound pressure is not a useful measure. We must look for a measure that permits the effects of the rarefactions to be added to (rather than subtracted from) the effects of the compressions.

One such measure is the root-mean-square (rms) sound pressure p_{rms}. The rms sound pressure is obtained by squaring the value of sound-pressure disturbance at each instant of time. The squared values are then added and averaged over the sample time. The rms sound pressure is the square root of this time average. Since the squaring operation converts all the negative sound pressures to positive squared values, the rms sound pressure is a useful nonzero measure of the magnitude of the sound wave. (The rms value is also called the *effective* value.)

Example 1.2   What is the rms sound pressure of a sinusoidal traveling sound wave?

SOLUTION: From Eq. (1.6) we choose one form for a sinusoidal traveling sound wave, $p = P_R \cos [k(x - ct)]$. The rms sound pressure p_{rms} is then

$$p_{rms} = \sqrt{\overline{p^2}} = \sqrt{\overline{P_R^2 \cos^2 [k(x - ct)]}} \qquad \text{N/m}^2$$

where the overline indicates the time average. Because the time average of the cosine squared is $\frac{1}{2}$, we obtain $p_{rms} = P_R/\sqrt{2}$. Thus the rms value of a sinusoidal wave equals $1/\sqrt{2}$ or 0.707 times its amplitude.

Particle Velocity.   So far, our discussion has been in terms of the sound pressure. At each instant of time there is also a particle velocity associated with the sound wave. The relation between sound pressure p and particle velocity u is obtained from the statement of momentum conservation which expresses the motion of a particle in terms of the forces acting on it. For free-traveling plane sound waves such as are shown in Figs. 1.3 and 1.4, we find that

$$u = \frac{p}{\rho c} \qquad \text{m/sec} \qquad (1.13)$$

where $\rho =$ density of air, which for normal room temperature of $T = 22°C$ (or 71.6°F) and atmospheric pressure of $p_s = 0.751$ m (29.6 in.) of Hg equals 1.18 kg/m³

$c =$ speed of sound (see Eqs. 1.3 and 1.4), which at normal temperature of 22°C equals 344.4 m/sec

$\rho c = 406$ mks rayls (N-sec/m³) at normal room temperature and pressure

For the forward-traveling plane sound wave, we find from Eqs. (1.6) and (1.13) that

$$u = \frac{p}{\rho c} = \frac{P_R}{\rho c} \cos [k(x - ct)] \qquad \text{m/sec} \qquad (1.14)$$

The average intensity, that is, the average energy that flows through a unit area in unit time, was given in Eq. (1.5) as

$$I = \overline{pu} \qquad \text{watts/m}^2 \qquad (1.15)$$

where the overline indicates the time average of the product. Because the plane-wave behavior is the same at all locations, except for time delays (phase shifts), we can perform this time average at any location, and it is convenient to choose the location $x = 0$. Substitution of Eqs. (1.6) and (1.14) into Eq. (1.15) yields the instantaneous intensity (remembering that for a forward-traveling plane wave P_R is equal to P_{max}, the amplitude or maximum value of the sound pressure)

$$I = \frac{P_R^2}{\rho c} \cos^2 2\pi f t = \frac{P_{max}^2}{\rho c} \left(\tfrac{1}{2} + \tfrac{1}{2}\cos 4\pi f t\right) \qquad (1.16)$$

Because the time average of the cosine is zero, the average intensity is

$$I = \left(\frac{P_{max}}{\sqrt{2}}\right)^2 \frac{1}{\rho c} \qquad \text{watts/m}^2 \qquad (1.17)$$

Taking cognizance of the fact that the root-mean-square (rms) value of a sinusoidal wave equals $1/\sqrt{2}$ times its peak value, we obtain for a free progressive wave

$$I = \frac{p_{rms}^2}{\rho c} \qquad \text{watts/m}^2 \qquad (1.18)$$

Table 1.1 gives the units of common acoustic quantities in three different systems. In general, the mks system, also called SI (Système International), will be used in this book because with it acoustic powers are automatically expressed in watts, consistent with electrical powers. The current ISO international acoustical standards, to which the United States is a party, also prescribe the SI system.

We are often interested in converting cgs or fps units of power (ergs/

TABLE 1.1   Units of Common Acoustic Quantities in Consistent Systems of Units

Quantity	Units		
	meter-kilo-gram-second (mks) system (SI †)	centimeter-gram-second (cgs) system	foot-pound-second (fps) * system
Mass...........................	kilogram (kg)	gram (g)	slug (lb/32.2)
Density	kg/m³	g/cm³	slug/ft³
Displacement..................	meter (m)	centimeter (cm)	foot (ft)
Speed and velocity..........	m/sec	cm/sec	ft/sec
Force...........................	newton (N)	dyne (dyn)	pound (lb)
Pressure.......................	N/m²	dyn/cm²	lb/ft²
Power..........................	watt	ergs/sec	ft-lb/sec
Intensity.......................	watts/m²	ergs/sec-m²	lb/sec-ft²

* The English system of units is discussed in detail in Appendix B.
† SI = Système International.

sec or ft-lb/sec) into watts. This is done by multiplying the number of ergs/sec by 10^{-7} or by multiplying the number of ft-lb/sec by 0.1383. Other conversion factors are given in Appendix C.

Example 1.3 Consider the case of a sinusoidal, free progressive plane wave having a peak sound pressure of 1 N/m². What is the intensity?
SOLUTION: The effective sound pressure would be $1/\sqrt{2} = 0.707$ N/m². If ρc is taken as 406 rayls (mks units), which holds approximately at room temperature near sea level, the intensity in the direction of the wave is given by

$$I = \frac{(0.707)^2}{406} = 1.23 \times 10^{-3} \text{ watt/m}^2$$

The analogy between the expressions for acoustic power flow and electrical power flow is apparent if one substitutes electromotive force (in volts) for sound pressure. The expression $I = p^2/\rho c$, where I is acoustic intensity in watts per square meter, is analogous to $P = E^2/R$, where P is electrical power in watts, E is in volts, and R is resistance in ohms. If we identify ρc as a kind of acoustic resistance, we have analogous expressions for power.

1.3.2 One-dimensional Spherical Wave   Up to this point we have been concerned only with plane sound waves, such as can be produced in the piston-tube experiment of Fig. 1.2. If we remove the tube, the sound wavefronts will no longer be plane. There is another important one-dimensional traveling wave that we should consider: the spherical wave. This kind of wave is important when there are no reflecting surfaces near the sound source or receiver. In order to gain some idea of the nature of a spherical wave, assume that a balloonlike sound source is pulsating sinusoidally about some equilibrium size, that is, the balloon surface is contracting and expanding radially.

As shown in Fig. 1.5, the motion of the balloon surface produces a sound wave that propagates away from the balloon. The speed of propagation is the same as for the plane wave produced in the piston-tube

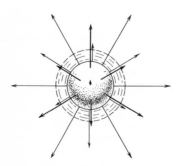

Fig. 1.5 Generation of a one-dimensional spherical wave. A balloonlike surface pulsates uniformly about some equilibrium position and produces a sound wave that propagates radially away from the balloon.

experiment, and the same relations between frequency and wavelength hold. Although the wave is now moving outward along a spherical wavefront, it is still one-dimensional, because all the parameters of the wave can be related to one variable, the radial distance r of the wavelength from the center of the balloon.

The equation for the sound pressure associated with a free progressive, spherical-traveling wave is

$$p = \frac{A}{r} \cos \left[k(r - ct) \right] \qquad \text{N/m}^2 \qquad (1.19)$$

where A is an amplitude factor with dimension N/m. The argument of the trigonometric function $k(r - ct)$ has the familiar form that indicates that this wave is traveling in the positive r direction. However, unlike the plane traveling wave where the sound-pressure amplitude remained independent of distance, in the spherical wave the amplitude is inversely proportional to the radial distance r.

Sound Intensity. An important difference between plane waves and spherical waves becomes apparent when we consider the intensity of the sound wave at any given point. If we assume that our source is imparting *energy* to the sound wave at a constant rate, ideally this *power* flow will be transmitted by the wave without diminution. In the plane wave, intensity, i.e., power per unit area, remains constant regardless of how far we move from the source, because the area of the wavefront is constant (the tube has constant cross section). In the spherical wave, however, the area of the wavefront increases as the wave progresses farther and farther from the source. In fact, the area increases with the square of the distance from the source, since the surface area of a sphere is equal to 4π times the square of its radius. In the spherical wave, therefore, with a source of constant rate of energy production (i.e., constant power), the outward-directed intensity decreases as the square of the distance from the source increases, because the product of the intensity and the area through which it passes is equal to the power transmitted by the wave. This is the origin of the "inverse square law" for intensity. In mathematical terms, for a spherical wave

$$I \text{ (at radius } r) = \frac{W}{4\pi r^2} = \frac{p_{\text{rms}}^2(\text{at } r)}{\rho c} \qquad \text{watts/m}^2 \qquad (1.20)$$

where W = total acoustic power radiated by the source of sound, watts

Example 1.4 Consider a sinusoidal, free progressive spherical wave with a sound-pressure amplitude of 1 N/m² at a distance r equal to 1 m from a source of constant power. Find the intensity at 1 m and the power radiated by the source. Also find the intensity, the rms value, and the amplitude of the sound pressure at 10 m.

SOLUTION: The effective (rms) sound pressure is 0.707 N/m². If we take ρc to be 406 rayls, the intensity at 1 m from the center of the source is given by

$$I = \frac{(0.707)^2}{406} = 1.23 \times 10^{-3} \text{ watt/m}^2$$

The total power radiated by the source is given by

$$W = 4\pi r^2 I = 4\pi \times 1.23 \times 10^{-3} = 1.546 \times 10^{-2} \text{ watt}$$

At $r = 10$ m, since W remains constant,

$$I \text{ (at } r = 10 \text{ m)} = \frac{1.546 \times 10^{-2}}{4\pi(10)^2} = 1.23 \times 10^{-5} \text{ watt/m}^2$$

$$p_{rms}(\text{at } r = 10 \text{ m}) = \sqrt{1.23 \times 10^{-5} \times 406} = 0.0707 \text{ N/m}^2$$

$$P_{max}(\text{at } r = 10 \text{ m}) = \sqrt{2} \times 0.0707 = 0.1 \text{ N/m}^2$$

We see that for 10 times the distance, the sound pressure has decreased by a factor of 10 and the sound intensity has decreased by a factor of 100. Compare this example with Example 1.3 for the free progressive plane wave.

1.3.3   One-dimensional Cylindrical Wave

In a manner similar to the formation of plane and spherical waves, we can generate waves of an even more complex nature. An example of another one-dimensional, free progressive wave can be obtained by considering a very long cylinder, say a straight hose filled with air. Suppose that we cause the surface of the hose to pulsate uniformly in a radial direction. A sound wave will now propagate outward from the cylinder. This wave is in a sense mid-

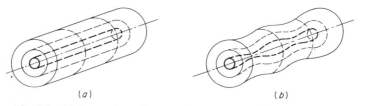

(a)   (b)

Fig. 1.6 (a) Generation of a one-dimensional cylindrical wave. A long cylindrical surface pulsates uniformly about some equilibrium position and produces a sound wave that propagates radially away from the cylinder. (b) Generation of a two-dimensional cylindrical wave. A long cylindrical surface pulsates nonuniformly about some equilibrium position and produces a sound wave that propagates away from the cylinder in a complex manner.

way between a plane wave and a spherical wave. At any two points that are the same distance from the cylinder the wave will have the same properties, but at two points having different distances from the cylinder the wave will be different. We can show that the intensity at any distance from the source is inversely proportional to the first power of the radius, rather than the second power as in the case of the spherical wave. This type of wave is called a one-dimensional cylindrical wave.

As an example of a two-dimensional wave, suppose that, in addition to the uniform radial motion of the cylinder, we imposed another motion along the axis of the cylinder itself. The resulting sound wave radiated from the cylinder would now depend not only on the radial distance from the source, but also on the distance along the axis of the source.

An approximate picture of the two waves described above is indicated by the sketches in Fig. 1.6. The action of the waves is indicated by a succession of surfaces called wavefronts. The wavefronts describe surfaces having equal phase. This implies that the sound pressure at each surface bears the same basic relation to the sound pressures at each of the other surfaces of the same phase. That is, if the sound pressure is at a maximum at one surface, it will also be at a maximum at each of the other surfaces having the same phase.

1.4   Interference and Resonance

The sound pressure and incremental density in a sound wave are generally very small in comparison with the equilibrium values on which they are superposed. This is certainly true for speech and music waves. As a result it is possible in such acoustical situations to determine the effect of two sound waves in the same space by simple linear addition of the effects of each sound wave separately. This is a statement of the principle of *superposition*.

In the previous section we presented a series of spatial snapshots for plane sound waves traveling to the right (Fig. 1.3) and to the left (Fig. 1.4). According to the principle of superposition, the effect of the sum of these two waves will be the sum of their effects, which we can see graphically by adding Figs. 1.3 and 1.4. The result is shown in Fig. 1.7, where we have set the amplitude of the forward-traveling wave P_R equal to the amplitude of the backward-traveling wave P_L.

The interference of the two waves has produced a surprising change. No longer does the sound pressure at one place occur to the right or to the left of that place at the next instant. The wave no longer travels; it is a standing wave. We see that at each point in space the sound pressure varies sinusoidally with time, except at the points $x = \lambda/4$ and $3\lambda/4$, where the pressure is always zero. The maximum value of the pressure variation at different points is different, being greatest at $x = 0$, $x = \lambda/2$, and $x = \lambda$. The sound pressures at the points between the points $x = \lambda/4$ and $x = 3\lambda/4$ always vary together, i.e., increase or decrease in phase. At the same times the sound pressures for the points to the left of $x = \lambda/4$ and to the right of $x = 3\lambda/4$ decrease or increase together (in phase). Thus all pressures are in time phase in the standing wave, but there is a space difference of phase of 180° between the sound pressures at the points at $x = 0$ and $x = \lambda/2$.

Remembering that $P_R = P_L = P$, that is, that the amplitude of the wave traveling to the right is equal to the amplitude of the wave traveling to the left, we find that the sum of the two waves is

$$p(x,t) = P \cos [k(x - ct)] + P \cos [k(x + ct)]$$

$$= 2P(\cos kx)(\cos 2\pi ft) \qquad N/m^2 \qquad (1.21)$$

From Eqs. (1.6), (1.12), and (1.21) we see very clearly the differences between a standing and a traveling wave. In a traveling wave distance x and time t occur as a sum or difference in the argument of the cosine. Hence, for the traveling wave, by adjusting both time and distance (according to the speed of sound) in the argument of the cosine, we can al-

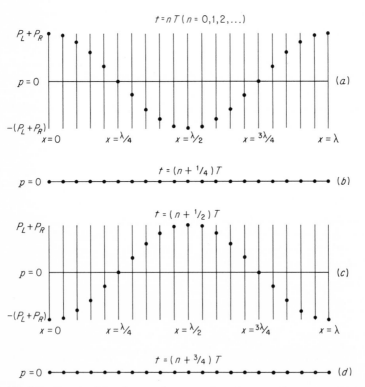

Fig. 1.7 Graphs giving the sound pressure in a plane standing wave at 20 places in space at 4 instants of time t. The wave is produced by two sources equal in strength at the right and left of the page. Standing wave: $p(x,t) = 2P(\cos kx)(\cos 2\pi ft)$; where $k = 2\pi/\lambda = 2\pi/cT$, $T = 1/f$ [see also Eq. (1.21)].

ways keep the argument and thus the magnitude of the cosine the same. In Eq. (1.21) distance and time no longer appear together in the argument of a single cosine. So the same sound pressure cannot occur at an adjacent point in space at a later time.

Standing waves will exist in any regular enclosure. In a rectangular room, for example, three classes of standing waves may exist (see Chap. 8). One class includes all waves that are perpendicular to one pair of opposing walls, i.e., that travel at grazing incidence to two pairs of walls, the $(n_x,0,0)$, $(0,n_y,0)$, $(0,0,n_z)$ modes of vibration. A second class travels at grazing incidence to only one pair of walls, the $(n_x,n_y,0)$, $(n_x,0,n_z)$, $(0,n_y,n_z)$ modes of vibration. A third class involves all walls at oblique angles of incidence, the (n_x,n_y,n_z) modes of vibration. Each free-standing wave in an acoustic space is called a normal mode of vibration, or simply a resonance. The frequencies at which resonant standing waves can exist are related to the separation between the reflecting surfaces. For example, the lowest frequency for a resonant standing wave in a one-dimensional system consisting of two rigid parallel walls is given by

$$f = \frac{c}{2d} \qquad (1.22)$$

where f = lowest frequency for resonant standing wave, Hz
c = speed of sound, m/sec
d = distance separating the two reflecting surfaces, m

Resonant standing waves can also exist at every integral multiple of this frequency. That is to say,

$$f = \frac{nc}{2d} \qquad (1.23)$$

where n is an integer: 1, 2, 3, . . .

Example 1.5 If two reflecting surfaces are located 3m (10 ft) apart at the two ends of a closed cylindrical tube, what is the lowest frequency at which a resonant standing wave can exist between the surfaces? Assume the speed of sound to be 344 m/sec (1,120 ft/sec).
SOLUTION:

$$f = \frac{344}{2 \times 3} = 57 \text{ Hz}$$

Higher-frequency resonant standing waves can exist at resonance frequencies of $2 \times 57 = 114$, $3 \times 57 = 171$, $4 \times 57 = 228$ Hz, etc.

Similar relations can be developed for any enclosure that permits the formation of standing waves. Some examples will be presented in detail in Chap. 8.

1.5 Complex Waves

At the end of Sec. 1.2, we said we would confine our study initially to pure-tone sounds, with the understanding that we could create more complex sounds by combining series of pure-tone sounds. In Sec. 1.3.1 we introduced the rms measure of sound pressure and showed that, in general, the power transmitted by the sound wave is related to the square of this measure. It will therefore be of interest for us to determine how the rms pressure of a sound wave containing two components is related to the rms pressure of each of the individual components.

Let the instantaneous pressures of two components at some fixed point in space (usually chosen as $x = 0$) be

$$p_1(t) = P_1 \cos (\omega_1 t + \theta_1) \qquad (1.24)$$

$$p_2(t) = P_2 \cos (\omega_2 t + \theta_2) \qquad (1.25)$$

where ω_1 and ω_2 are their angular frequencies and θ_1 and θ_2 are their respective phases. The instantaneous pressure sum is then

$$p(t) = p_1(t) + p_2(t) = P_1 \cos (\omega_1 t + \theta_1) + P_2 \cos (\omega_2 t + \theta_2) \quad (1.26)$$

To obtain the rms value of this sum, the electrical circuitry in an ideal sound-level meter squares $p(t)$, takes the time average, and then displays a value corresponding to the square root of this quantity. We perform this operation on Eq. (1.26) as follows:

$$[p(t)]^2 = \frac{P_1^2 + P_2^2}{2} + \frac{P_1^2 \cos 2(\omega_1 t + \theta_1)}{2} + \frac{P_2^2 \cos 2(\omega_2 t + \theta_2)}{2}$$

$$+ 2P_1 P_2 \cos (\omega_1 t + \theta_1) \cos (\omega_2 t + \theta_2) \qquad (1.27)$$

The time average of each of the single cosines is zero. It can be shown mathematically that the time average of the cosine product is also zero, when the frequencies of the two components are *not* identical. Thus for the mean-square pressure of two pure tones of different frequencies

$$p^2 = \left(\frac{P_1}{\sqrt{2}}\right)^2 + \left(\frac{P_2}{\sqrt{2}}\right)^2 \qquad (\omega_1 \neq \omega_2) \qquad (1.28)$$

The rms sound pressure of the first tone is $P_1/\sqrt{2}$, and the rms sound pressure of the second tone is $P_2/\sqrt{2}$, because the rms value of a sine wave is $1/\sqrt{2}$ times its amplitude. So, for a combination of two tones having different frequencies,

$$p = \sqrt{p_1^2 + p_2^2} \qquad (\omega_1 \neq \omega_2) \qquad \text{N/m}^2 \qquad (1.29)$$

where p, p_1, and p_2 are rms magnitudes. By extension, for n tones of different frequencies,

$$p = \sqrt{p_1^2 + p_2^2 + p_3^2 + \cdots + p_n^2} \tag{1.30}$$

Under the special condition that $\omega_1 = \omega_2 = \omega$, that is, the two pure tones have the same frequency, p_1 and p_2 combine to produce a third sine wave of the same frequency but of different amplitude and phase, that is,

$$\begin{aligned} p(t) &= p_1(t) + p_2(t) \\ &= \sqrt{P_1^2 + P_2^2 + 2P_1P_2 \cos(\theta_1 - \theta_2)} \cos(\omega t + \theta_3) \\ &= P_3 \cos(\omega t + \theta_3) \qquad \text{N/m}^2 \end{aligned} \tag{1.31}$$

where θ_3 is the new phase angle. The rms pressure is

$$p = \frac{P_3}{\sqrt{2}} = \sqrt{p_1^2 + p_2^2 + 2p_1p_2 \cos(\theta_1 - \theta_2)} \qquad \text{N/m}^2 \tag{1.32}$$

where p, p_1, and p_2 are rms values. Comparison of Eqs. (1.29) and (1.32) reveals the importance of phase when combining two sine waves of the same frequency. If the phase difference $\theta_1 - \theta_2$ is zero, the two waves are said to be in phase, and the total pressure is at its maximum. If $\theta_1 - \theta_2 = \pm 180°$, the total pressure is at its minimum.

If one wishes to find the rms sound pressure of a number of waves all of which have different frequencies except, say, two, these two are added together according to Eq. (1.32) to obtain a new rms pressure. Then this pressure and the pressure of the remainder of the components are summed according to Eq. (1.30). This point will be reviewed in the discussion on levels in the following chapter.

Appendix to Chapter One

IMPEDANCE AND ADMITTANCE

Reference to Eqs. (1.6) and (1.14) reveals that the magnitudes of sound pressure and particle velocity are directly proportional to each other. Also, in the special case of plane-wave sound propagation, the time dependence of sound pressure is exactly the same as the time dependence of particle velocity, and at any point in the wave there is no phase difference between the two quantities. Thus, in a plane sound wave the ratio of sound pressure to particle velocity at all instants of time is a constant equal to ρc.

In general, however, for linear (small-signal) acoustical phenomena in the steady state, there is a difference in the time functions of sound pressure and particle velocity, leading to a phase difference of one relative to the other. Thus, at any point the particle velocity may lead or lag the sound pressure. In many situations, both the ratio of the magnitudes and the relative phase may be functions of frequency.

In several of the chapters that follow, it is convenient in acoustical design to avoid separate consideration of steady-state sound pressure and steady-state particle velocity (or other quantities that may be derived from them, such as force and volume velocity) and instead to deal with either one and with their complex ratio, as defined below.

Complex Notation   The foundations for the designation of steady-state signals with the same frequency but different phases by complex notation are expressed in the identities

$$|A| \cos (\omega t + \theta_1) \equiv \text{Re } \bar{A}e^{j\omega t} \tag{A.1.1}$$

where $|A| \equiv$ amplitude of the cosine function
$\quad \theta_1 \equiv$ phase shift at time $t = 0$
$\quad \text{Re} \equiv$ means "the real part of"
$\quad\quad j \equiv \sqrt{-1}$

$$\bar{A} \equiv A_{\text{Re}} + jA_{\text{Im}} = |A|e^{j\theta_1} \tag{A.1.2}$$

$$|A| \equiv \sqrt{A_{\text{Re}}^2 + A_{\text{Im}}^2} \tag{A.1.3}$$

$$\theta_1 \equiv \tan^{-1} (A_{\text{Im}}/A_{\text{Re}}) \tag{A.1.4}$$

We note also that

$$e^{j\theta} \equiv \cos \theta + j \sin \theta \tag{A.1.5}$$

so that

$$A_{\text{Re}} \equiv |A| \cos \theta \tag{A.1.6}$$

$$A_{\text{Im}} \equiv |A| \sin \theta \tag{A.1.7}$$

These equations say that a cosinusoidal time-varying function, given by the left-hand side of Eq. (A.1.1), can be represented by the real-axis projection of a vector of magnitude $|A|$ given by Eq. (A.1.3), rotating at a rate ω radians per second. The angle θ_1 is the angle of the vector (in radians) relative to the positive real axis at the instant of time $t = 0$.

We might, therefore, express a time-varying steady-state sound pressure or force by

$$\bar{A}e^{j\omega t} = |A|e^{j\omega t}e^{j\theta_1} \tag{A.1.8}$$

Also, a time-varying steady-state velocity or volume velocity might be expressed by

$$\bar{B}e^{j\omega t} = |B|e^{j\omega t}e^{j\theta_2} \tag{A.1.9}$$

Definitions of Complex Impedance

Complex Impedance Z. In general, complex impedance is defined as

$$\bar{Z} \equiv \frac{\bar{A}}{\bar{B}} = \frac{|A|e^{j(\omega t+\theta_1)}}{|B|e^{j(\omega t+\theta_2)}} = \frac{|A|e^{j\theta_1}}{|B|e^{j\theta_2}} = |Z|e^{j\theta} \tag{A.1.10}$$

and

$$\bar{Z} \equiv R + jX = \sqrt{R^2 + X^2}\ e^{j\theta} = |Z|e^{j\theta} \tag{A.1.11}$$

where $\bar{Z}$ = complex impedance as given above
  A = steady-state pressure or force
  B = steady-state velocity or volume velocity
  $|Z|$ = magnitude of the complex impedance
  $\theta = \theta_1 - \theta_2$ = phase angle between the time functions A and B
  R,X = real and imaginary parts, respectively, of the complex impedance $\bar{Z}$
  $R = \mathrm{Re}\ \bar{Z}$ = resistance
  $X = \mathrm{Im}\ \bar{Z}$ = reactance

Often, $|A|$ and $|B|$ are taken to be the root-mean-square (rms) values of the phenomena they represent, although, if so taken, a factor of $\sqrt{2}$ must be added to both sides of Eq. (A.1.1) to make them correct in a physical sense. Whether amplitudes or rms values are used makes no difference in the impedance ratio.

Complex impedances are of several types, according to the quantities involved in the ratios. Types common in acoustics are given below.

Acoustic Impedance Z_A. The acoustic impedance at a given surface is defined as the complex ratio of sound pressure averaged over the surface to volume velocity through it. The surface may be either a hypothetical surface in an acoustic medium or the moving surface of a mechanical device. The unit is N-sec/m⁵, also called the mks acoustic ohm. That is,

$$Z_A = \frac{p}{u} \qquad \text{N-sec/m}^5 \text{ (mks acoustic ohms)} \tag{A.1.12}$$

Specific Acoustic Impedance Z_s. The specific acoustic impedance is the complex ratio of the sound pressure at a point of an acoustic medium or mechanical device to the particle velocity at that point. The unit is N-sec/m³, also called the mks rayl. That is,

$$Z_s = \frac{p}{u} \qquad \text{N-sec/m}^3 \text{ (mks rayls)} \tag{A.1.13}$$

Mechanical Impedance Z_M. The mechanical impedance is the complex ratio of the force acting on a specified area of an acoustic medium or mechanical device to the resulting linear velocity through or of that area, respectively. The unit is the N-sec/m, also called the mks mechanical ohm. That is,

$$Z_M = \frac{f}{u} \qquad \text{N-sec/m (mks mechanical ohms)} \qquad (A.1.14)$$

Characteristic Resistance ρc. The characteristic resistance is the ratio of the sound pressure at a given point to the particle velocity at that point in a free, plane, progressive sound wave. It is equal to the product of the density of the medium times the speed of sound in the medium (ρc). It is analogous to the characteristic impedance of an infinitely long, dissipationless transmission line. The unit is the N-sec/m³, also called the mks rayl. In the solution of problems in this book we shall assume for air that $\rho c = 406$ mks rayls which is valid for a temperature of 22°C (71.6°F) and a barometric pressure of 0.751 m (29.6 in.) Hg.

Normal Specific Acoustic Impedance Z_{sn}. At the boundary between air and a denser medium (such as a porous acoustic material) we find a further definition necessary, as follows:

When an alternating sound pressure p is produced at the surface of an acoustic material, an alternating velocity u of the air particles is produced through the surface. The to-and-fro motions of the air particles may be at any angle relative to the surface. The angle depends both on the angle of incidence of the sound wave and on the nature of the acoustic material. For example, if the material is porous and has very low density, the particle velocity at the surface is nearly in the same direction as that in which the wave is propagating. By contrast, if the surface were a large number of small-diameter tubes packed side by side, the particle velocity would necessarily be only perpendicular to the surface. In general, the direction of the particle velocity at the surface has both a normal (perpendicular) component and a tangential component.

The normal specific acoustic impedance (sometimes called the unit-area acoustic impedance) is defined as the complex ratio of the sound pressure p to the *normal component* of the particle velocity u_n at a plane; in this example, at the surface of the acoustic material. Thus,

$$Z_{sn} = \frac{p}{u_n} \qquad \text{N-sec/m}^3 \text{ (mks rayls)} \qquad (A.1.15)$$

Definition of Complex Admittance   Complex admittance is the reciprocal of complex impedance. In all ways, it is handled by the same set of rules as given by Eqs. (A.1.1) to (A.1.11). Thus, the complex admittance corresponding to the complex impedance of Eq. (A.1.10) is

$$\bar{Y} \equiv \frac{\bar{B}}{\bar{A}} = \frac{|B|e^{j\theta_2}}{|A|e^{j\theta_1}} = |Y|e^{j\phi} \qquad (A.1.16)$$

where $|Y| = 1/|Z|$
$\qquad \phi = -\theta$

The choice between impedance and admittance is sometimes made according to whether $|A|$ or $|B|$ is held constant during a measurement. Thus, if $|B|$ is held constant, $|Z|$ is directly proportional to $|A|$ and is used. If $|A|$ is held constant, $|Y|$ is directly proportional to $|B|$ and is used.

Chapter Two

Levels, Decibels, and Spectra*

LEO L. BERANEK

Sound that propagates through an elastic medium can be described
by a sine wave or by a superposition of waves of different frequencies,
phases, and amplitudes, depending on their complexity. In any region
of space through which a sound travels, variations in sound pressure
are found, motions of the particles may be observed, and, in general, a
flow of energy through the region occurs. Furthermore, because of the
variety of sounds that can be detected by human beings and instruments,
means must be available for describing a wide range of sound pressures,
particle velocities, frequencies, and intensities. Frequencies may range
over a factor of 10^6, sound intensities over 10^{20}, and sound pressures
over 10^{10}. To preserve constant-percentage accuracies in measuring or
describing these quantities, and to avoid large exponents in the numbers
involved, logarithmic scales are used. The logarithm most commonly
associated with acoustics is the decibel.

2.1 Power, Intensity, and Energy Density

2.1.1 Power   A sound source radiates power. A portion of this
power will flow through each small region of the medium that surrounds

* This chapter is derived in part from Chaps. 2 and 3 of *Noise Reduction*. Chapter 2
was prepared by W. J. Galloway and the author and Chap. 3 by Edward M. Kerwin, Jr.

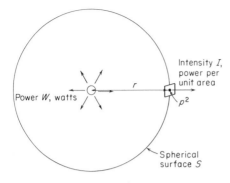

Fig. 2.1 A simple nondirectional source of sound located at the center of a spherical surface produces a sound power of W watts. The intensity of the sound at all points on the surface is $I = W/S$ watts/m^2, where the area of the spherical surface of radius r is $S = 4\pi r^2$.

the source. Let us designate the acoustic power radiated by a source as W watts. If there are no losses in the medium, *all this radiated power must pass through any surface that encloses the source.* The *larger* the enclosing surface, the *less power per unit area* that will pass through any element of the surface.

A simple nondirectional source of sound located at the center of a spherical surface is shown in Fig. 2.1. The total sound power W is equal to the product of the intensity I times the area of the surface.

$$W = IS \qquad \text{watts *} \qquad (2.1)$$

If the sound source is directional, the intensity will vary over the surface and the total power radiated W must be found by integration over the surface S

$$W = \int_S I_S \, dS \qquad \text{watts} \qquad (2.2)$$

where I_S = intensity through each incremented area dS of the surface, watts/m^2

S = area of the enclosing surface, m^2

2.1.2   Intensity   Sound intensity is difficult to measure. Instead, we generally measure the rms sound pressure at a sufficient number of points on the spherical surface, convert to sound intensity at each point of measurement on the spherical surface, and sum the intensity-surface-element products to get total power.

For any *free, progressive wave* there is a unique and simple relation between the mean-square sound pressure and the intensity. This relation [see Eq. (1.20)], at a particular point and in the direction of the wave, is

---

* Obviously, *any consistent set of units* may be used in the equations of this book wherever the constants involved are dimensionless. Alternate consistent sets of units are discussed in Appendix B and are summarized in Table 1.1. For example, in the foot-slug-second system, W is in ft-lb/sec, S in ft^2, and I, therefore, in lb/ft-sec.

$$I = \frac{p_{\text{rms}}^2}{\rho c} \qquad \text{watts/m}^2 \tag{2.3}$$

where I = intensity, watts/m²

p_{rms}^2 = mean-square sound pressure, (N/m²)², measured at that particular point where I is desired in the free progressive wave (mean-square sound pressure is the time average of the square of the instantaneous sound pressure in the acoustic wave)

ρc = characteristic resistance of the medium, mks rayls. For air, at $T = 20°C$ and $p_s = 0.751$ m of Hg, $\rho c = 406$ mks rayls. See Eq. (1.13)

2.1.3 Sound-energy Density

In standing-wave situations, such as inside tubes and rooms, the quantity desired is not intensity, but rather sound-energy density, namely, the energy stored in a small volume of the air in the room owing to the presence of the standing-wave field. The relation between the space-average mean-square sound pressure and the space-average *sound-energy density* is

$$D = \frac{p_{\text{av}}^2}{\rho c^2} = \frac{p_{\text{av}}^2}{\gamma p_s} \qquad \text{watt-sec/m}^3 \tag{2.4}$$

where p_{av}^2 = average of the mean-square sound pressure in space, determined from data obtained by moving a microphone along the tube or around the room, or from samples at various points, N²/m⁴

p_s = atmospheric pressure, N/m². Under normal atmospheric conditions, at sea level, $p_s = 1.013 \times 10^5$ N/m²

γ = ratio of specific heats, C_p/C_v, equal for air to 1.4 (dimensionless)

The quantity D is used in room-acoustic calculations, and elsewhere (see Chap. 9).

2.2   Levels

The fundamental purpose of a level scale is to relate a power, intensity, or energy density to a logarithm of the ratio of that quantity to a reference quantity. The argument of the logarithm is *dimensionless* and the scale is said to give the *level* of the sound in decibels (dB), *above* (or below) the *reference level* that is determined by the *reference quantity*.

2.2.1 Sound-power Level

Sound-power level is defined as

$$L_W = 10 \log_{10} \frac{W}{W_0} \qquad \text{dB } re \ W_0 \tag{2.5}$$

and conversely

$$W = W_0 \text{ antilog}_{10} \frac{L_W}{10} = W_0 \times 10^{L_W/10} \qquad (2.6)$$

where W = sound power, watts

W_0 = reference sound power, watts

The word *level* is used to indicate a logarithm. As seen in Table 2.1, a *ratio* of 10 in the power W corresponds to a *level difference* of 10 dB, regardless of the reference power W_0. Similarly, a ratio of 100 in power corresponds to a level difference of 20 dB. Power ratios less than 1 are allowable; they simply lead to negative levels as shown in Table 2.1 in column 3. For example, a power ratio of 0.1 corresponds to a level difference of -10 dB.

Some sound-power ratios and the corresponding sound-power-level differences are given in Table 2.2. We note from the last line that the sound-power level for the *product of two ratios* is equal to the *sum of the levels* for the two ratios. For example, determine L_W for the quantity 3×9. We get from Table 2.2: $L_W = 4.8 + 9.5 = 14.3$ dB, which is the sound-power level for the ratio 27. Similarly, L_W for a ratio of 8,000 equals the sum of the levels for 8 and 1,000; i.e., $L_W = 9 + 30 = 39$ dB.

TABLE 2.1   Sound Powers in Watts and Sound-power Levels in Decibels

Radiated sound power W, watts		L_W = sound-power level, dB	
Usual notation (1)	Equivalent exponential notation (2)	Relative to 1 watt (3)	Relative to 10^{-12} watt (standard) (4)
100,000	10^5	50	170
10,000	10^4	40	160
1,000	10^3	30	150
100	10^2	20	140
10	10^1	10	130
1	1	0	120
0.1	10^{-1}	-10	110
0.01	10^{-2}	-20	100
0.001	10^{-3}	-30	90
0.000,1	10^{-4}	-40	80
0.000,01	10^{-5}	-50	70
0.000,001	10^{-6}	-60	60
0.000,000,1	10^{-7}	-70	50
0.000,000,01	10^{-8}	-80	40
0.000,000,001	10^{-9}	-90	30

The level differences for fractional power ratios that are shown in Table 2.2 can be obtained, alternatively, from those for power ratios greater than 1. For example, for a power ratio of 0.5 we need only note that 0.5 is the reciprocal of 2, and we may thus determine the level as the negative of the level for a power ratio of 2, i.e., −3 dB.

TABLE 2.2   Some Sound-power Ratios and the Corresponding Power-level Differences

Sound-power ratio (dimensionless), W/W_0 R	Sound-power-level difference,* dB, $10 \log W/W_0$ L_W
1,000	30
100	20
10	10
9	9.5
8	9.0
7	8.5
6	7.8
5	7.0
4	6.0
3	4.8
2	3.0
1	0.0
0.9	−0.5
0.8	−1.0
0.7	−1.5
0.6	−2.2
0.5	−3.0
0.4	−4.0
0.3	−5.2
0.2	−7.0
0.1	−10
0.01	−20
0.001	−30
$R_1 \times R_2$	$L_{W_1} + L_{W_2}$

* To the nearest 0.1 dB.

The Reference Quantity.   Because a level is related to a ratio, a reference quantity must necessarily be involved whenever we describe sound power by a given level.

In acoustics, it has been decided internationally that for sound-power levels the reference power shall be

$$W_0 = 10^{-12} \text{ watt} \tag{2.7}$$

(Note: Earlier texts used 10^{-13} watt as the reference power, which yielded values for L_W that were 10 dB larger than those with 10^{-12} watt as reference.)

Column 4 of Table 2.1 gives sound-power levels relative to this standard reference level.

Example 2.1   Determine the sound-power level of a source radiating 16.3 watts of sound power, using a reference power of 10^{-12} watt.
SOLUTION:

$$L_W = 10 \log (16.3 \times 10^{12}) = 132.1 \text{ dB } re\ 10^{-12} \text{ watt}$$

It is imperative that *the magnitude of a sound-power level never be stated without a definite statement of the associated reference power.* This precaution avoids confusion with levels quoted in the earlier literature, and with references that may be used in other fields.

Example 2.2   Determine the sound-power level of a source radiating 0.063 watt = 63 milliwatts.
SOLUTION:

$$L_W = 10 \log 0.063 + 10 \log_{10} 10^{12}$$

$$= -10 \log \frac{1}{0.063} + 120$$

$$= -10 \log 15.87 + 120$$

$$= -12 + 120$$

$$= 108 \text{ dB } re\ 10^{-12} \text{ watt}$$

Sound-power levels should not be confused with intensity levels or sound-pressure levels, which are also expressed in decibels (see below). Remember, sound-power level is related logarithmically to the total acoustic power radiated by a source. Sound-pressure level specifies the acoustic "disturbance" produced at a point. Sound-pressure level depends on the distance from the source, losses in the intervening air, room effects (if indoors), etc. Perhaps it might be helpful to imagine that sound-power level is related to the total rate of heat production of a furnace while sound-pressure level is related to the temperature produced at a given point in the house.

2.2.2 Sound-intensity Level   Sound-intensity level, in decibels, is defined by

$$\text{Intensity level} = L_I = 10 \log \frac{I}{I_{\text{ref}}} \qquad \text{dB } re\ I_{\text{ref}} \qquad (2.8)$$

where I = sound intensity (power passing in a specified direction through a unit area) whose level is being specified, watts/m²
I_{ref} = reference intensity standardized as 10^{-12} watt/m² (equivalent to 10^{-16} watt/cm²)

Example 2.3 Determine the sound-intensity level at a distance of 10 m from a uniformly radiating source in free space, assuming that the source radiates a power of 0.1 watt.
SOLUTION: First calculate the intensity at 10 m:

$$I = \frac{W}{S} = \frac{0.1}{4\pi r^2} = \frac{0.1}{400\pi} = 7.95 \times 10^{-5} \text{ watt/m}^2$$

Then calculate the intensity level

$$L_I = 10 \log 7.95 - 10 \log 10^5 + 10 \log 10^{12}$$

$$= 9 - 50 + 120$$

$$= 79 \text{ dB } re \ 10^{-12} \text{ watt/m}^2$$

2.2.3 Sound-pressure Level   Most sound-measuring instruments respond to sound pressure, and the word *decibel* is commonly associated with sound-pressure level. More specifically, the decibel is a unit of sound-pressure (squared) level, as well as a unit of power level. In either case, the decibel is associated with some power ratio, because the square of sound pressure is proportional (although not equal) to a sound power. Sound-pressure (squared) level is usually shortened to

$$\text{Sound-pressure level} = L_p = 10 \log \frac{p^2}{p_{ref}^2}$$

$$= 20 \log \frac{p}{p_{ref}} \qquad \text{dB} \qquad (2.9)$$

where p_{ref} = reference rms sound pressure, usually 2×10^{-5} N/m² for airborne sound and sometimes 0.1 N/m² (1 dyn/cm²) or 1μN/m² for sound in other media
p = rms sound pressure in N/m² (note: L_p re 2×10^{-5} N/m² is 94 dB greater than L_p re 1 N/m²; if p is expressed in dynes per square centimeter, L_p re 2×10^{-4} dyn/cm² is 74 dB greater than L_p re 1 dyn/cm²)

Example 2.4 Calculate the sound-pressure level re 2×10^{-5} N/m² for a sound with a rms sound pressure of 2.82 N/m².
SOLUTION: Substitution in Eq. (2.9) gives

$$L_p = 20 \log 2.82 - 20 \log 2 + 20 \log 10^5$$

$$= 9 + 94 = 103 \text{ dB } re \ 2 \times 10^{-5} \text{ N/m}^2$$

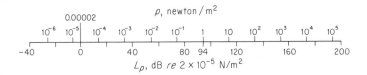

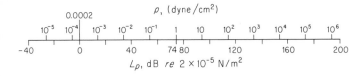

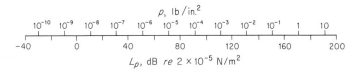

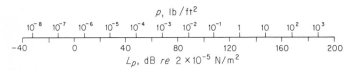

Fig. 2.2   Charts relating L_p (dB $re\ 2 \times 10^{-5}$ N/m²) to p in N/m², dyn/cm², lb/in.², and lb/ft². For example, 1.0 N/m² equals 1.435×10^{-4} lb/in.² equals 94 dB $re\ 2 \times 10^{-5}$ N/m².

To show the relations for given sound-pressure levels ($re\ 2 \times 10^{-5}$ N/m²) for pressures in the mks, cgs, and English systems of units, four nomograms are presented in Fig. 2.2.

2.2.4 Fundamental Rules in Regard to Level   Levels in decibels are basically defined for use with ratios of two powerlike quantities, such as power W, intensity I, mean-square pressure p^2, mean-square voltage e^2. In each case, the level is with reference to a corresponding reference quantity.

$$L_W = 10 \log_{10} \frac{W}{W_{\text{ref}}} \qquad \text{dB } re\ W_{\text{ref}} \qquad (2.10a)$$

$$L_I = 10 \log_{10} \frac{I}{I_{\text{ref}}} \qquad \text{dB } re\ I_{\text{ref}} \qquad (2.10b)$$

$$L_p = 10 \log_{10} \frac{p^2}{p_{\text{ref}}^2} = 20 \log_{10} \frac{p}{p_{\text{ref}}} \qquad \text{dB } re\ p_{\text{ref}} \qquad (2.10c)$$

$$L_e = 10 \log_{10} \frac{e^2}{e_{\text{ref}}^2} = 20 \log_{10} \frac{e}{e_{\text{ref}}} \qquad \text{dB } re\ e_{\text{ref}} \qquad (2.10d)$$

or conversely

$$W = W_{\text{ref}} \text{ antilog}_{10} \frac{L_W}{10} \quad \text{watt} \qquad (2.11a)$$

$$I = I_{\text{ref}} \text{ antilog}_{10} \frac{L_I}{10} \quad \text{watt/m}^2 \qquad (2.11b)$$

$$p = p_{\text{ref}} \text{ antilog}_{10} \frac{L_p}{20} \quad \text{N/m}^2 \qquad (2.11c)$$

$$e = e_{\text{ref}} \text{ antilog}_{10} \frac{L_e}{20} \quad \text{volt} \qquad (2.11d)$$

2.2.5 Other Reference Quantities Used in Noise and Vibration   The
American National Standards Institute has issued a standard [ANSI
Standard S1.8, 1969] on "Preferred Reference Quantities for Acoustical
Levels." The reference quantities are stated in units of the Système
International (SI). The preferred quantities are given in Table 2.3.
Internationally (ISO Draft Recommendation, 508 E), the only major

**TABLE 2.3 American National Standard (ANSI S1.8, 1969) for Acoustical Levels and
Preferred Reference Quantities for Acoustical Levels**

The Reference Quantities Are in Units of the Système International (SI) or the
British System

Name	Definition	Preferred reference quantities *	
		SI units	British units
Sound-pressure level (gases)	$L_p = 20 \log_{10} (p/p_0)$	$p_0 = 20~\mu\text{N/m}^2 = 2 \times 10^{-5}~\text{N/m}^2$	2.90×10^{-9} lb/in.^{2}
Sound-pressure level (liquids)	$L_p = 20 \log_{10} (p/p_0)$	$p_0 = 1~\mu\text{N/m}^2 = 10^{-6}~\text{N/m}^2$	1.45×10^{-10} lb/in.^{2}
Power level †	$L_W = 10 \log_{10} (W/W_0)$	$W_0 = 1~\text{pW} = 10^{-12}~\text{watt}$	8.85×10^{-12} in.-lb/sec
Intensity level	$L_I = 10 \log_{10} (I/I_0)$	$I_0 = 1~\text{pW/m}^2 = 10^{-12}~\text{watt/m}^2$	5.71×10^{-15} lb/in.-sec
Energy-density level †	$L_D = 10 \log_{10}(D/D_0)$	$D_0 = 1~\text{pJ/m}^3 = 10^{-12}~\text{J/m}^3$	1.45×10^{-16} in.-lb/in.^{3}
Vibratory-acceleration level	$L_a = 20 \log_{10} (a/a_0)$	$a_0 = 10~\mu\text{m/sec}^2 = 10^{-5}~\text{m/sec}^2$	39.4×10^{-5} in./sec^2
Vibratory-velocity level	$L_v = 20 \log_{10} (v/v_0)$	$v_0 = 10~\text{nm/sec} = 10^{-8}~\text{m/sec}$	39.4×10^{-8} in./sec
Vibratory-displacement level	$L_d = 20 \log_{10} (d/d_0)$	$d_0 = 10~\text{pm} = 10^{-11}~\text{m}$	39.4×10^{-11} in.
Vibratory-force level	$L_F = 20 \log_{10}(F/F_0)$	$F_0 = 1~\mu\text{N} = 10^{-6}~\text{N}$	0.225×10^{-6} lb
Energy level	$L_E = 10 \log_{10} (E/E_0)$	$E_0 = 1~\text{pJ} = 10^{-12}~\text{J}$	8.85×10^{-12} in.-lb

* By international agreement (ISO/R 1000-1969E), decimal multiples and submultiples of the SI units are
formed by the prefixes: 10^{-1} = deci (d), 10^{-2} = centi (c), 10^{-3} = milli (m), 10^{-6} = micro (μ), 10^{-9} = nano (n),
and 10^{-12} = pico (p). Also, J = joules = watt-sec and N = newtons. Some writers call 1 N/m^2 the pascal (Pa).
Note that 1 lb = 4.448 newtons. In the final printing of S1.8, 1969, the reference quantity d_0 does not appear;
it is included here as a recommendation only.

† In this text, to avoid confusion between power and pressure, we have chosen to use W instead of P for
power; and to avoid confusion between energy density and voltage, we have chosen D instead of E for energy
density. The symbol lb means pounds force.

difference is that the reference pressure is 2×10^{-5} N/m² for all media. As stated in the footnote to the table, in several instances the standard terminology differs from the terminology of this text. The ANSI reference quantities are adopted in this text.

2.2.6 Relations among Sound-power Levels, Intensity Levels, and Sound-pressure Levels   As a practical matter, the reference quantities for sound power, intensity, and sound pressure (in air) have been chosen so that their corresponding levels are interrelated in a convenient way under certain circumstances.

The threshold of hearing at 1,000 Hz for a young listener with acute hearing, measured under laboratory conditions, was determined some years ago as a sound pressure of 2×10^{-5} N/m². This value was then selected as the reference pressure for sound-pressure level.

Intensity at a point is related to sound pressure at that point in a free field by Eq. (2.3). A combination of Eqs. (2.3), (2.8), and (2.9) yields the sound-intensity level

$$L_I = 10 \log \frac{I}{I_{\text{ref}}} = 10 \log \frac{p^2}{\rho c I_{\text{ref}}}$$

$$= 10 \log \frac{p^2}{p_{\text{ref}}^2} + 10 \log \frac{p_{\text{ref}}^2}{\rho c I_{\text{ref}}}$$

$$L_I = L_p - 10 \log K \qquad \text{dB } re \ 10^{-12} \text{ watt/m}^2 \qquad (2.12)$$

where $K = \text{constant} = I_{\text{ref}} \rho c / p_{\text{ref}}^2$, which is dependent upon ambient pressure and temperature. The quantity $10 \log K$ may be found from Fig. 2.3.
$$= \rho c / 400$$
The quantity $10 \log K$ will equal zero, that is, $K = 1$, when ρc equals

$$\frac{p_{\text{ref}}^2}{I_{\text{ref}}} = \frac{4 \times 10^{-10}}{10^{-12}} = 400 \text{ mks rayls} \qquad (2.13)$$

We may also rearrange Eq. (2.12) to give the sound-pressure level

$$L_p = L_I + 10 \log K \qquad \text{dB } re \ 2 \times 10^{-5} \text{ N/m}^2 \qquad (2.14)$$

In Table 2.4, we show a range of ambient pressures and temperatures for which $\rho c = 400$ mks rayls. We see that for average atmospheric pressure, namely, 1.013×10^5 N/m², the temperature must equal 38.9°C (102°F) for $\rho c = 400$ mks rayls. However, if T is 22°C and $p_s =$

TABLE 2.4 Ambient Pressures and Temperatures for Which ρc (Air) = 400 mks rayls

Ambient pressure			Ambient temperature T	
p_s, N/m²	m of Hg, 0°C	in. of Hg, 0°C	°C	°F
0.7×10^5	0.525	20.68	−124.3	−192
0.8	0.600	23.63	−78.7	−110
0.9	0.675	26.58	−27.0	−17
1.0	0.750	29.54	+30.7	+87
1.013×10^5	0.760	29.9	38.9	102
1.1	0.825	32.5	94.5	202
1.2	0.900	35.4	164.4	328
1.3	0.975	38.4	240.4	465
1.4	1.050	41.3	322.4	613

1.013×10^5 N/m², $\rho c \approx 412$. This yields a value of 10 log $(\rho c/400) =$ 10 log $1.03 = 0.13$ dB, an amount that is usually not significant in acoustics.

Thus, for most noise measurements, we neglect 10 log K and in a free progressive wave let

$$L_p \approx L_I \tag{2.15}$$

Otherwise, the value of 10 log K is determined from Fig. 2.3 and used in Eq. (2.12) or Eq. (2.14).

Under the condition that the *intensity is uniform over an area S*, the sound power and the intensity are related by Eq. (2.1). Hence, the sound-power level is related to the intensity level as follows

$$10 \log \frac{W}{10^{-12}} = 10 \log \frac{I}{10^{-12}} + 10 \log \frac{S}{S_0}$$

$$L_W = L_I + 10 \log S \qquad \text{dB } re \ 10^{-12} \qquad \text{watt} \tag{2.16}$$

where S = area of surface, m²

$S_0 = 1$ m²

Obviously, only if the area S equals 1.0 m² will $L_W = L_I$. Also, observe that the relation of Eq. (2.16) is not dependent on temperature or pressure.

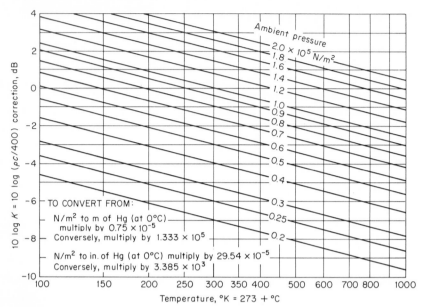

Fig. 2.3 Chart determining the value of $10 \log K = 10 \log (\rho c/400)$ as a function of ambient temperature and ambient pressure. Values for which $\rho c = 400$ are also given in Table 2.4.

2.3 Sound Spectra

Sounds outside the laboratory are rarely pure tones. They may be a combination of tones whose frequencies are harmonically related, such as from a musical instrument, or they may be noise.

The composition of a noise or a musical tone is determined by spectral analysis (also called harmonic or frequency analysis) using techniques discussed in detail in Chap. 5. By spectral analysis, the mean-square pressure (or sound-pressure level) of a sound wave is determined in each band of a set of contiguous frequency bands and is plotted as a function of the center frequency of the band. The bands used for the analysis may be equal in width or they may have widths that are proportional to the center frequencies of their respective bands, or they may obey some other relation. The bandwidth is usually chosen to be only as narrow as is necessary to give an analysis satisfactory for the purpose at hand.

In most noise control work, the relative phases among the components of a noise are not important. Thus we usually obtain the amplitude information needed from a graph in one of the forms of Fig. 2.4.

2.3.1 Line-spectrum Sound   The uppermost graph a of Fig. 2.4 shows the rms sound-pressure spectrum for a group of four components harmonically related with a fundamental frequency of 500 Hz. Such a graph is called a line spectrum.

The combination (summation) of a number of inharmonically related waves (see Fig. 2.4b) produces a resultant wave that is not steady. One can predict the amplitude of such a wave on a statistical basis. Of course, the peak amplitude can never exceed the sum of the amplitudes of all the components. But as the number of components is increased, the probability of achieving the maximum peak amplitude becomes very small. For example, with 10 inharmonically related components of equal amplitude, the maximum possible peak amplitude is 10 times that of one component. However, only 0.02% of the time will the amplitude of the combination exceed 8 times that of one component.

As we showed in Eq. (1.30), the rms value of a combination of waves of different frequencies is computed very simply by taking the square root of the sums of the squares of the rms pressures of the individual components.

2.3.2 Continuous-spectrum Sound   A very common type of sound is one that is built up from a continuum of components. Such a noise is

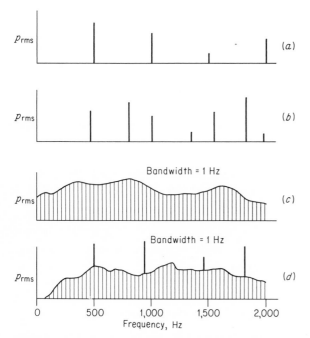

Fig. 2.4   (a) Line spectrum harmonically related; (b) line spectrum inharmonically related; (c) continuous spectrum; (d) combination line and continuous spectrum (complex spectrum).

produced by the exhaust of a jet engine or by the water at the base of Niagara Falls or by the hiss of air. To produce such a noise from pure tones would require an infinite number of waves, each with an infinitesimal amplitude.

An example of a continuous spectrum is shown in graph *c* of Fig. 2.4. When combined with a line spectrum, a complex spectrum of the type shown in *d* is produced.

How do we measure and specify a continuous-spectrum sound? Most simply we could measure the overall sound-pressure level. The answer that we would get would depend, however, on the frequency range of the meter, and no two types of meter would yield the same result. One way to describe a continuous-spectrum noise is by plotting (as a function of frequency) the rms sound pressure existing in increments of frequency along the frequency scale, *each increment 1 Hz in width.* Therefore, graphs *c* and *d* of Fig. 2.4 indicate, as a function of frequency, the rms sound pressure in 1-Hz-wide frequency bands.

There is one difficulty. Generally instruments for measuring effective sound pressure do not have bandwidths as narrow as 1 Hz. Hence, measurements are made with a wider bandwidth and converted to the 1-Hz-bandwidth method of presenting the data.

To develop the philosophy of converting the level for a wider bandwidth to that for a narrower one, let us imagine that we have a machine that produces at a point in space an intensity of 10^{-6} watt/m^2 in a 1-Hz-wide band between 999 and 1,000 Hz (see Fig. 2.5a). Now imagine that we start a second machine that produces the same intensity between 1,000 and 1,001 Hz (see Fig. 2.5b). When the noise from the two machines is combined, we get the spectrum of Fig. 2.5c. Because the two machines produce twice as much sound power as one, the intensity must double. Similarly in Fig. 2.5d 10 such machines would produce 10 times the intensity.

In other words, if the continuous spectrum is "flat" (has the same intensity in every 1-Hz band), the total intensity is given by the product

$$I_{\text{total}} = I_1 \, \Delta f \tag{2.17}$$

If the spectrum is not flat, like that in Fig. 2.5e, we simply substitute for the actual I_1 the average value of I_1 in the band in which we are interested.

Example 2.5   What is the total intensity in the 700- to 800-Hz band of Fig. 2.5e?

SOLUTION: The average intensity for a 1-Hz band within the 700- to 800-Hz band is seen to be 5.5×10^{-6} watt/m^2. Hence, the total intensity in the 700- to 800-Hz band is, from Eq. (2.17),

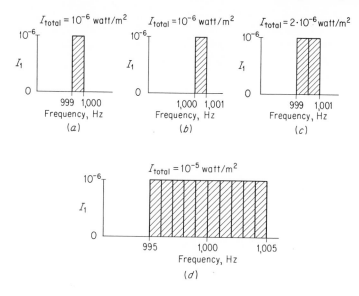

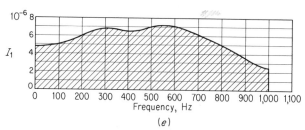

Fig. 2.5 Graphs showing the addition of intensity when the bandwidth of noise is increased.

$$I_{\text{total}} = (5.5 \times 10^{-6}) \times 10^2 = 5.5 \times 10^{-4} \text{ watt/m}^2$$

Conversely, if we have measured an intensity I_T with an analyzer having a bandwidth Δf, the average spectrum intensity I_1 (1-Hz band) in that band is obtained from the formula

$$I_1 = \frac{I_T}{\Delta f} \qquad \text{watt/m}^2 \text{ in 1-Hz band} \tag{2.18}$$

One remaining question needs answering. Because we can measure rms sound pressure easily, but not sound intensity, how do Eqs. (2.17) and (2.18) read when expressed in effective sound pressure? Substitution of Eq. (2.3) in Eq. (2.17) yields

$$\left(\frac{p_{\text{rms}}^2}{\rho c}\right)_{\text{total}} = \left(\frac{p_{\text{rms}}^2}{\rho c}\right)_1 \Delta f \tag{2.19}$$

Cancellation of ρc and extraction of the square root lead to

$$(p_{rms})_{total} = (p_{rms})_1 \sqrt{\Delta f} \qquad (2.20)$$

where $(p_{rms})_1$ is the rms sound pressure in a 1-Hz band.

In words, doubling the bandwidth of a flat continuous-spectrum noise increases the mean-square pressure by a factor of 2 and the rms pressure by a factor of 1.41.

2.4 Multiple Sources

Very often in noise control problems we are concerned with contributions from more than one source of noise. Naturally we wish to be able to use our levels to take account of multiple contributions. It is expected that the reader is familiar with the material in Sec. 1.5.

Example 2.6 Two sound sources having different frequencies produce sound-pressure levels of (a) 88 and (b) 85 dB *re* 2×10^{-5} N/m² at an observer's position. Find the total sound-pressure level *re* 2×10^{-5} N/m².

SOLUTION: Perform the operations:

1. Mean-square sound-pressure *ratio* of (a) is [see Eq. (2.11c)]

$$\frac{p^2}{p_{ref}^2} = \text{antilog } \frac{88}{10} = 6.31 \times 10^8$$

2. Mean-square sound-pressure *ratio* of (b) is

$$\frac{p^2}{p_{ref}^2} = \text{antilog } \frac{85}{10} = 3.16 \times 10^8$$

3. Total mean-square sound-pressure *ratio* is the sum

$$9.46 \times 10^8$$

4. Total sound-pressure *level* is

$$10 \log_{10} 9.46 \times 10^8 = 89.8 \text{ dB}$$

Note that we did not actually need to find the pressures, but only their ratios to the reference pressure.

We may also make use of the information presented in Table 2.2 to combine contributions from two sources. In doing this, we may prepare Fig. 2.6 by interpreting column 1 in Table 2.2 as mean-square-pressure ratio and column 2 as sound-pressure-level difference, respectively. We assume again that the sound waves from two sources have different frequencies. The mean-square sound-pressure ratio when source 2 is combined with source 1 is

$$\left(\frac{p_T}{p_{ref}}\right)^2 = \left(\frac{p_1}{p_{ref}}\right)^2 + \left(\frac{p_2}{p_{ref}}\right)^2 \qquad (2.21)$$

For example, if the second source makes a mean-square-pressure contribution equal to that from the first source, then by Eq. (2.21) the combined mean-square pressure is twice that due to the first source alone. Referring to Fig. 2.6 (or Table 2.2) we find that for a powerlike ratio * of 2 we expect a level difference of 3.0 dB.

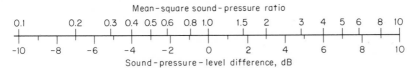

Mean-square sound-pressure ratio

0.1   0.2   0.3 0.4 0.5 0.6   0.8 1.0   1.5   2   3   4   5   6   8   10

-10   -8   -6   -4   -2   0   2   4   6   8   10
Sound-pressure-level difference, dB

Fig. 2.6   Relation between mean-square sound-pressure ratio p_T^2/p_1^2 and sound-pressure-level difference in decibels $L_{p_1} - L_{p_2}$.   This figure is based on Table 2.2.

Likewise, if the second source is contributing twice as much (in mean-square pressure) as the first source, the combination will have three times the mean-square pressure of the first source.   For this case, Fig. 2.6 indicates an L_p increase of 4.8 dB over the level of the first source.

It follows from our earlier discussion that Fig. 2.6 may also be used where more than two sources are combined, provided again that no two have the same frequency.   For example, 10 sources of equal mean-square pressure will give a resultant sound-pressure level 10 dB higher than that for any one of the sources individually.

If any two of the sources have the same frequency, their combined pressure must be determined by Eq. (1.31) † before combining with the others.   For example, two sources of the same frequency, of the same phase, and of equal strength give us a mean-square-pressure ratio of $2^2 = 4$.   The resultant sound-pressure-level increase is 6.0 dB instead of 3.0 dB.

Very often in noise control work we wish to combine the contributions from two or more sound sources each characterized by the sound-pressure level it produces at the observation point.   This combination can, of course, be accomplished as above.

The conversion process can be done once and for all, and a nomogram prepared.   Figure 2.7 can be used to combine any number of levels taken two at a time.   As an example, let us determine the total sound-pressure level for two sources that produce 70 and 75 dB *re* 2×10^{-5} N/m² respectively, at an observer's position.   We choose the larger number as L_1, obtaining a level difference of $L_1 - L_2 = 5$ dB.   Figure 2.7 tells us that the level of the combination is 1.2 dB above L_1 or 76.2 dB *re* 2×10^{-5} N/m².   If more than two sound-pressure levels are to be combined,

---

* In this case we are interpreting "ratio" as the ratio of the combined mean-square pressure from the two sources to the mean-square pressure from one source alone.   This interpretation is a valid one for Table 2.2 and Fig. 2.6.

† $p = \sqrt{p_1^2 + p_2^2 + 2p_1p_2 \cos(\theta_1 - \theta_2)}$.

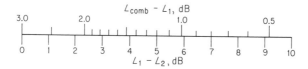

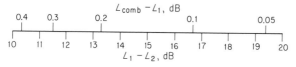

$$L_1 > L_2, \quad L_{\text{comb}} = (L_{\text{comb}} - L_1) + L_1$$

Fig. 2.7   Chart for combining two acoustic levels L_1 and L_2. Levels may be power levels, sound-pressure levels, or intensity levels.
Example: $L_1 = 88$ dB, $L_2 = 85$ dB, $L_1 - L_2 = 3$ dB
Solution: $L_{\text{comb}} = 88 + 1.8 = 89.8$ dB

we combine the first two, then combine the third with our first result, and so forth.

Continuous-spectrum Sounds.   Many sounds that we encounter in noise control are continuous-spectrum sounds.   As we showed above, it is customary, when dealing with them, to consider the mean-square sound pressure in a given frequency band.   Such a frequency band may be divided up into a number of increments, and a mean-square pressure may be assigned to each increment.   Because no two of these increments have the same frequency, the incremental contributions are combined like pure tones of different frequencies.   Therefore, we determine the total mean-square pressure simply by adding the appropriate values of mean-square pressure for the individual frequency increments.

Thus, the mean-square pressure for a frequency band composed of a number of subbands is

$$p_{\text{band}}^2 = p_a^2 + p_b^2 + p_c^2 + p_d^2 + \cdots \tag{2.22}$$

where p_a^2, p_b^2, etc., are the mean-square pressures for the subbands.

A special case of interest occurs when each subband has a width of 1 Hz.   If the mean-square pressure for the subbands is p_1^2 on the average, then the mean-square pressure for the parent band is [see Eq. (2.17)]

$$p_{\text{band}}^2 = p_1^2 \, \Delta f \tag{2.23}$$

where Δf is the width in hertz of the parent band.

Expressed in decibels the sound-pressure level *re* 2×10^{-5} N/m² is

$$L_{p_{\text{band}}} = 10 \log \frac{p_{\text{band}}^2}{p_{\text{ref}}^2} = 10 \log \frac{p_1^2}{p_{\text{ref}}^2} + 10 \log \frac{\Delta f}{(\Delta f)_0} \tag{2.24}$$

or $\qquad L_{p_{\text{band}}} = L_{p_1}(f) + 10 \log \Delta f \qquad$ dB re 2×10^{-5} N/m^2 $\qquad$ (2.25)

where $L_{p_1}(f)$ = the *spectrum level* (dB re 2×10^{-5} N/m^2) at frequency f
$\quad (\Delta f)_0 = 1$ Hz

In other words, *the spectrum level is that average sound-pressure level (for a component band 1 Hz wide) which when added to 10 log $\Delta f/1.0$ yields the sound-pressure level of the parent band.* The spectrum level $L_{p_1}(f)$ is seldom measured directly with an analyzer of unit bandwidth, but rather is determined from the $L_{p\text{band}}$ by

$$L_{p_1}(f) = L_{p\text{band}} - 10 \log \Delta f \qquad \text{dB } re \ 2 \times 10^{-5} \text{ N/m}^2 \qquad (2.26)$$

Determination of Overall Levels from Band Levels. Frequently we measure the sound-pressure levels in a series of contiguous bands and then desire the level as though it had been measured in a single band covering the same frequency range. The level in the all-inclusive band is called the *overall level.*

As an example of the combination of levels in different frequency bands to produce an overall level, let us start with the band levels shown across the top of Fig. 2.8. The frequency limits of the particular bands are not important to the method of calculation. It is sufficient to know that the bands are contiguous and cover the frequency range of interest.

Figure 2.8 indicates schematically the step-by-step computation of the overall sound-pressure level accomplished with the help of Fig. 2.7.

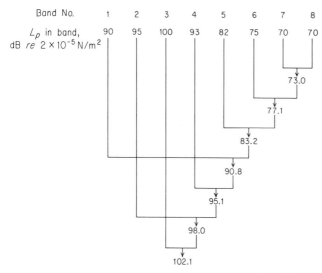

Overall L_p (8 frequency bands) $\approx$ 102 (dB re 2×10^{-5} N/m^2)

Fig. 2.8 Determination of an overall sound-pressure level from levels in frequency bands (see also Fig. 2.9).

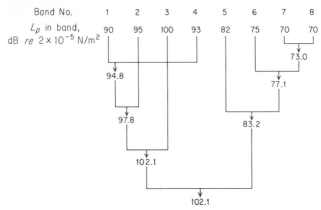

Overall L_p (8 frequency bands) ≈ 102 (dB *re* 2 × 10⁻⁵ N/m²)

Fig. 2.9 Alternate determination of an overall sound-pressure level from levels in frequency bands (see Fig. 2.8).

The order in which we combine the various band levels is unimportant. In Fig. 2.8 the levels were taken in increasing order of magnitude. We see that the seventh and eighth bands combine to give a level of 73. That combination and the sixth band give a level of 77.1, etc.

We may make an interesting observation simply by changing the order in which we combine the levels presented in Fig. 2.8. In Fig. 2.9 we show the same sound-pressure-level spectrum with the levels in bands 1 to 4 combined separately from the levels in bands 5 to 8. The result is seen to be determined entirely by the levels in the first four bands.* This example points up the fact that characterizing a noise spectrum simply by the overall level may be completely inadequate for some noise control purposes, because it ignores a large portion of the frequency spectrum. In a given noise control problem, the sound-pressure levels in bands 5 to 8 might be quite important, say in creating annoyance or hindering communication by voice.

* In almost all noise control problems it makes no sense to deal with small fractions of decibels. Rarely does one require a precision as great as 0.1 dB in his calculations, and in many cases, the results of an analysis—for example, an estimated sound-pressure-level spectrum—are best stated only to the nearest decibel.

Chapter Three

Sound and Vibration Transducers

EDWARD A. STARR

The physical parameters of interest for noise and vibration are related to the properties of wave motion in fluids (sound) or in solid structures (vibration). Quantifiable properties of sound waves at a point include incremental (sound) pressure, particle velocity, particle displacement, intensity, incremental density, and incremental temperature. All these properties can be measured in the laboratory. However, in gases, sound pressure is by far the easiest quantity to measure, particularly under field conditions. For sound in air, we will restrict our attention to devices that measure sound pressure.

Vibration of a solid structure at a point can also be described by many parameters, including displacement, velocity, and acceleration, each in particular directions. These quantities can all be determined successfully, but transducers for the measurement of acceleration are the easiest to construct.

Displacement, velocity, and acceleration are related by a simple derivative; thus once one is known the others can be found.

3.1   Transducer Sensitivity

The transduction mechanisms of microphones and accelerometers must be very sensitive since the force involved is usually very small. For example, a sound pressure $L_p = 100$ dB *re* 2×10^{-5} N/m² (or 0.0002 μbar) will exert about 0.013 N (0.003 lb) of force on a 1-in.-diameter diaphragm. Similarly, the inertial force from a typical seismic mass of an accelerometer excited with 1 *g* rms of vibration is about 0.0044 N (0.001 lb).

Sensitivity is generally one of the first properties of interest in a transducer. Microphone pressure sensitivities are often expressed as dBv/μbar, where dBv is the output-voltage level in decibels *re* 1 volt [see Eq. (2.10*d*)] for an incident-wave sound pressure of 1.0 microbar (1.0 microbar equals 0.1 N/m²). A sound pressure of 1.0 microbar equals a sound-pressure level of 74 dB *re* 2×10^{-5} N/m². Microphone sensitivity is also expressed [1] as dBv/(N/m²). This yields a sensitivity that is 20 dB higher than sensitivity expressed relative to 1 microbar (0.1 N/m²).

Typical microphone sensitivities are between −50 and −125 dBv/μbar. For L_p equal to 74 dB, a microphone with a sensitivity of −50 dBv/μbar will yield an output of −50 dBv (or about 3 millivolts) while one with a sensitivity of −125 dBv/μbar will yield −125 dBv or about 0.5 microvolt.

Accelerometer sensitivity is normally expressed in millivolts per *g*, where a *g* is the unit of gravity equal to 9.81 m/sec² (32.2 ft/sec²). Typical accelerometer sensitivities are from 1 millivolt per *g* to 1 volt per *g*. When expressed in decibels the reference quantities of 10^{-8} m/sec for velocity and 10^{-5} m/sec² for acceleration are standardized (see Table 2.3), although other references are sometimes used.

In general, for both sound and vibration, large transducers have high sensitivities but are useful only at low frequencies, while small transducers have low sensitivities but are useful at both low and high frequencies.

3.2   Microphones

Microphones may be of the condenser, electret, piezoelectric, or dynamic types. Several typical commercially available microphones are shown in Fig. 3.1.

3.2.1   Condenser Microphones   In a condenser microphone sound pressure deflects the diaphragm and changes the capacitance between the diaphragm and a flat electrode parallel to the diaphragm. The change of capacitance is converted to an electrical signal by either (1) maintaining a constant charge on the capacitance so that the voltage

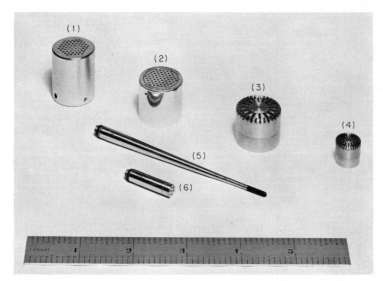

Fig. 3.1 Photograph of typical microphones: (1) General Radio 1560-9605 (piezoelectric), (2) Western Electric 640-AA (condenser), (3) B&K 4131 (condenser), (4) B&K 4133 (condenser), (5) Atlantic Research LC-5 (piezoelectric), (6) BBN 372 (piezoelectric with preamplifier). The scale is in inches.

varies with capacitance, or (2) using the microphone capacitance to establish the frequency of an oscillator so that the frequency is modulated by the capacitance, or (3) using the capacitance as one arm of an ac bridge which is unbalanced by the changing capacitance. The predominant technique is to maintain a constant charge by a polarization voltage supplied through a very high resistance. This technique was initially described by Wente[2] in 1917.

Characteristics of some commercially available condenser microphones are found in Table 3.1.

Typical advantages of condenser microphones are adequate acoustic sensitivity, a well-behaved frequency response, low self-noise, and low sensitivity to mechanical vibration of the unit as a whole. Disadvantages include the small microphone capacitance, a fragile diaphragm, and susceptibility to humidity.

An equivalent electrical circuit for a condenser microphone is shown in Fig. 3.2. Assume that the preamplifier input resistance R and polarization-voltage resistance R_p are near infinite and that the polarizing voltage is constant; thus the charge is constant. The change in voltage across terminals 3–4 because of a deflection of the diaphragm is[3]

$$\Delta E = - kQ_0\Delta x \, \frac{C}{C + C_s \left(1 + \dfrac{\Delta x}{x_0}\right)} \qquad (3.1)$$

TABLE 3.1   Manufacturers' Specifications of Commercially Available Microphones, Circa 1970

Manufacturer	Model	Type	Size Diameter, in.	Length, in.	Sensi- tivity, dBv/ μbar	Maxi- mum, L_p*
Atlantic Research	LC-5	Piezoelectric	0.1	4.0	−124	>180
Brüel and Kjaer	4131	Condenser	0.936	0.75	−46	146
	4133	Condenser	0.520	0.50	−60	160
	4135	Condenser	0.275	0.41	−74	174
	4138	Condenser	0.140	0.268	−86	184
	4117	Piezoelectric	0.936	0.79	−70	140
Bolt Beranek and Newman	372	Piezo with FET †	0.250	0.890	−105	186
Columbia Research	P-200	Piezoelectric	0.83	0.57	−122	200
Laboratory	P-308	Piezoelectric §	0.74	1.0	−104	190
	MCP-200	Piezo with FET †	0.94	2.0	−72	140
Endevco Corp.	2510	Piezoelectric §	0.75	1.5	−112	190
General Radio	1560-P5	Piezoelectric	0.936	1.25	−60	150
Gulton Industries	NQA-2120	Piezoelectric §	0.76	1.1	−99	190
	MVA-2400	Piezoelectric	0.36	1.0	−109	190
Hewlett-Packard	15119A	Condenser	0.50	5.1 ‡	−56	150
	15109B	Condenser	0.97	5.2 ‡	−46	140
Kaman Nuclear	KM-1800	Eddy current	1.06	1.75	−106	180
Kistler	717	Piezo with FET †	0.44	2	−102	180
Sennheiser	MKH105	Condenser	0.78	4.7 ‡	−54	114
Shure	98A108	Piezoelectric	1.25	2.75	−60	134
	578	Dynamic	0.75	7	−61	>150
	SM 60	Dynamic	1.25	6.2	−82	>150
Thermo Electron	T520	Electret with FET †	0.15	0.52	−70	135
Whittaker	404	Condenser	0.936	0.86	−50	140
	504	Condenser	0.50	0.59	−65	160
	814	Condenser	0.125	0.23	−85	175

* dB *re* 0.0002 μbar.
† FET = field-effect-transistor preamplifier.
‡ With preamplifier.
§ Vibration canceling.

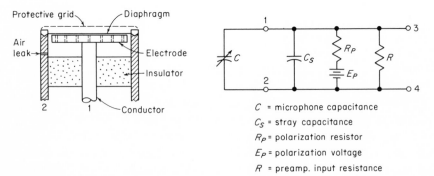

C = microphone capacitance
C_S = stray capacitance
R_P = polarization resistor
E_P = polarization voltage
R = preamp. input resistance

Fig. 3.2   Sketch and equivalent circuit for a condenser microphone.

Operating temperature range, °F	Deviation over temperature	Vibration sensitivity, dB L_p* per g	Air leak	Output impedance	Frequency response, Hz ¶
−40 to +220	0.05 dB/°F	120	No	350 pf	Beyond 100,000
−60 to +140	0.02 dB/°F	90	Yes	70 pf	20 to 18,000; ±2 dB, FF perpend.
−60 to +140	0.02 dB/°F	90	Yes	20 pf	20 to 40,000; ±2 dB, FF perpend.
−60 to +140	0.02 dB/°F	90	Yes	6.4 pf	30 to 100,000; ±2 dB, FF perpend.
−60 to +212	......	90	Yes	3.5 pf	30 to 140,000; ±2 dB, pressure
14 to +122	0.01 dB/°F	100	Yes	4,000 pf	3 to 10,000; ±3 dB, FF perpend.
−60 to +250	±1 dB	115	No	1,500 ohms	2 to 30,000; ±3 dB, FF grazing
−20 to +230	±1 dB	136	No	500 pf	5 to 18,000; ±5%
−65 to +250	+1, −2.5 dB	96	No	3,600 pf	5 to 10,000; FF grazing
−30 to +185	±1 dB	90	No	150 ohms	5 to 10,000
−65 to +500	±1.5 dB	108	No	5,100 pf	2 to 10,000; ±3 dB, pressure
−65 to +140	0.006 dB/°F	100	Yes	400 pf	25 to 12,000; ±3 dB, FF random
−65 to +250	±1.5 dB	90	No	3,500 pf	2 to 10,000; FF grazing
−65 to +250	±2 dB	105	No	1,000 pf	2 to 10,000; FF grazing
+14 to +122	0.02 dB/°F	......	Yes	100 ohms	20 to 25,000; ±1 dB, FF perpend.
+14 to +122	0.02 dB/°F	......	Yes	100 ohms	20 to 16,000; ±1 dB, FF perpend.
−30 to +350	0.003 dB/°F	110	Yes	75 ohms	0.1 to 10,000; ±3 dB, pressure
−65 to +200	±2.5%	115	No	100 ohms	1 to 50,000; ±3 dB
−60 to +140	0.09 dB/°F	......	......	50 pf	20 to 10,000; ±2 dB
−60 to +140	0.09 dB/°F	......	......	20 pf	20 to 10,000; ±2 dB
−60 to +140	0.02 dB/°F	......	......	3 pf	50 to 120,000; ±3 dB
+14 to +158	±1 dB	90	Yes	15 ohms	20 to 10,000; ±2 dB
0 to +135	......	......	Yes	460 pf	20 to 13,000; FF perpend.
0 to +135	±1.5 dB	......	Yes	150 ohms	50 to 15,000; FF perpend.
0 to +135	±1.5 dB	......	Yes	150 ohms	45 to 15,000; FF perpend.
+10 to +180	......	90	Yes	1,500 ohms	50 to 15,000; ±3 dB, FF

¶ FF indicates free-field response (see Sec. 3.4.1 for discussion of free-field and pressure responses); "grazing" indicates sound incident parallel to pressure-sensitive surface; "perpend." indicates sound perpendicular to the surface. Where there is no indication, the manufacturer did not specify the geometry of measurement.

where k = constant (determined by calibration against a known standard), (meters-farads)$^{-1}$

Q_0 = initial charge supplied to the capacitance C, coulombs

Δx = displacement of the diaphragm from its rest position x_0, meters

C = microphone capacitance, farads

C_S = stray capacitance, farads

To maintain sensitivity and linearity of voltage produced, it is necessary that C_s be small compared to C. One solution is that a preamplifier be located essentially at the microphone electrode. Even 3 ft of 20 pf/ft cable would yield a capacitance equal to that of most large con-

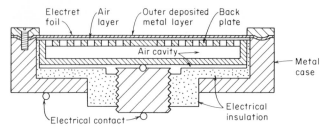

Fig. 3.3   Construction of electret microphone. *(After Sessler and West.)*

denser microphones.* Another solution is to use a guard shield.† A guard shield is particularly important on small-diameter microphones where the microphone capacitance may be only a few picofarads.

Typical values for a condenser microphone are $E_p = 200$ volts and $\Delta E \approx 1$ rms volt for $\Delta x_0 \approx 2.5 \times 10^{-5}$ cm and $x_0 \approx 2.5 \times 10^{-3}$ cm.

The low-frequency response of a condenser microphone to sound pressure is usually limited by an air leakage path to the back of the diaphragm. This leakage path is necessary so that changes in atmospheric pressure will not damage the thin diaphragm. If a microphone is to be installed where it will be subjected to a change of ambient pressure, the experimenter must ensure that the leakage hole is also vented to the same pressure.

3.2.2 Electret Microphones
Self-polarizing condenser microphones have been developed by Sessler and West.[4,5] Instead of an air gap between the two condenser plates, a prepolarized solid dielectric (electret foil) is used. This eliminates the need for a dc bias voltage and provides a higher capacitance for the microphone.

The construction of an electret microphone is very simple, as illustrated in Fig. 3.3. An electret foil less than 0.002 cm thick is stretched above a perforated metal backplate. The foil has a thin metalized layer deposited on its outer surface to provide the other plate of the capacitor. The backplate is perforated to permit the electret diaphragm to move.

Incident sound pressure produces a force on the foil diaphragm, causing it to move. The movement of the foil relative to the perforated backplate alters the charge distribution between foil and backplate, thus producing a voltage.

The operation of the electret microphone depends upon the prepolarized dielectric. Particular concerns in manufacture are retention

---

* NOTE: A picofarad (or pf) is 10^{-12} farad.

† A guard shield is a double-shielded cable where the inside shield is at approximately the same potential as the inner conductor and is held there by negative feedback from the preamplifier. If there is little voltage difference between the conductor and the shield, there will be very little effective capacitance.

of the polarization over long periods of time and change in sensitivity with temperature. Experiments with fluorocarbons[5] indicate an expected polarization life of a few years at room temperature. If the microphone is temperature cycled, the life will be extended greatly, but at some loss in sensitivity. Indications are that with further development the sensitivity at room temperature can be made adequately stable.

The temperature coefficient of sensitivity is larger for the electret microphone than for air-gap condenser microphones; contributors to this are the change of dielectric constant of the electret material with temperature, and the dimensional stability of the foil.

The microphone reported by Sessler and West had a sensitivity of the order of -60 dBv/μbar for a microphone diameter of approximately 4 cm. The capacitance was 500 pf and the resonant frequency 14,000 Hz.

Significant advantages of the electret microphone are (1) elimination of dc bias voltage, (2) simple, rugged construction, and (3) large capacitance.

3.2.3 Piezoelectric Microphones

Piezoelectric microphones use the force produced by pressure to strain a piezoelectric material which generates an electrical charge. Piezoelectric materials may be natural crystals (quartz, lithium sulfate, tourmaline), or may be created by adding impurities to a natural crystalline structure (barium titanates and lead zirconate titanates).[6,7]

A typical construction of a piezoelectric microphone is illustrated in Fig. 3.4a. A diaphragm is used as a force collector. The piezoelectric material is placed behind the diaphragm so that force exerted on the diaphragm strains the crystal. Since the diaphragm is backed by the crystal, the transducer is more rugged than a condenser microphone. Hydrophones for underwater work use piezoelectric materials, but without a diaphragm.

Advantages of a piezoelectric microphone include high capacitance,

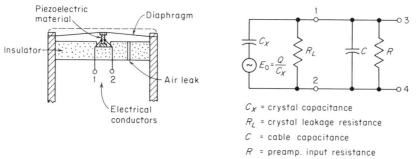

C_X = crystal capacitance
R_L = crystal leakage resistance
C = cable capacitance
R = preamp. input resistance

Fig. 3.4 Sketch and equivalent circuit for a piezoelectric microphone.

ruggedness, wide dynamic range, and the lack of a polarizing voltage. Disadvantages include lower acoustic sensitivity and higher vibration sensitivity. The latter can be reduced by the use of a vibration-canceling crystal within the microphone. Characteristics of some commercially available piezoelectric microphones are also presented in Table 3.1.

An equivalent electrical circuit for the piezoelectric microphone is shown in Fig. 3.4*b*, where R is the input resistance of the preamplifier and R_L is the leakage resistance of the crystal. The voltage generated is

$$E = \frac{Q}{C + C_x} \tag{3.2}$$

where Q = charge generated by the crystal when strained, coulombs
C_x = crystal capacitance, farads
C = capacitance of the cable, farads

The loss of sensitivity from cable capacitance has led to the use of charge amplifiers, as discussed in Sec. 3.5.2.

Low-frequency response is limited electrically by the parallel resistance-capacitance combination of the equivalent circuit. Because C_x for a piezoelectric microphone is considerably larger (of the order of 1,000 pf) than that for condenser microphones, the input impedance of the preamplifier need not be quite as high. It is possible to locate the preamplifier several feet from the microphone, although it is good practice to minimize this distance to reduce the effect of cable noise. In some devices a very small field-effect-transistor preamplifier is incorporated within the microphone case. The microphone output impedance is then low, and a long cable may be used without difficulty.

Many piezoelectric microphones, particularly the less sensitive ones, do not require an air leak behind the diaphragm because of the rugged construction.

3.2.4 Dynamic Microphones   The dynamic (moving-coil) microphone uses the velocity imparted to the diaphragm by the sound pressure to move a coil through a magnetic field, hence inducing a current in the coil. The construction is indicated in Fig. 3.5.

An advantage of the dynamic microphone is low electrical impedance, which permits long cables and gives low internal electrical noise. The configuration also lends itself to balanced-line transformer coupling, which is an effective way to eliminate electrical ground loops.

Disadvantages of the dynamic microphone include its susceptibility to external magnetic fields, a high vibration sensitivity because of the mass of the diaphragm and coil, and a poor low-frequency response.

In the design of the dynamic microphone, three resonant frequencies are considered.[8] The principal resonance of the diaphragm is set at mid-frequencies. Heavy acoustic damping behind the diaphragm causes the microphone to respond as a mechanical resistive element over most of its frequency range, so that the diaphragm velocity is proportional to the sound pressure. The electromotive force induced in the coil is proportional to coil velocity and hence to the sound pressure. Two supplementary resonance circuits act to extend the frequency range at low (M_T,C_V,R_T) and high (M_R,C_G,R_R) frequencies.

The diaphragm coil can be allowed only a limited excursion. Because the displacement of the coil increases as the frequency of the sound pressure is lowered, the maximum sound pressure that can be measured will decrease monotonically with decreasing frequency.

The stability of a dynamic microphone is influenced by both the density and viscosity of the gas, since these affect the acoustic elements used to extend the frequency response. The stability is also influenced by the temperature coefficients of the mechanical elements. The

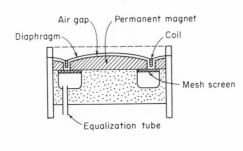

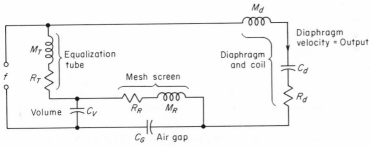

f = force from acoustic pressure on diaphragm
M_d = mass of diaphragm and coil
C_d = compliance of diaphragm and coil
R_d = mechanical resistance of diaphragm and coil
C_G = acoustic compliance of air gap behind diaphragm
R_R = acoustic resistance of mesh screens
M_R = mass of air in mesh screens
C_V = acoustic compliance of large volume V
M_T = mass of air in equalization tube
R_T = acoustic resistance of equalization tube

Fig. 3.5 Sketch and mechanical equivalent circuit of a dynamic microphone.

TABLE 3.2   Manufacturers' Specifications of Typical Piezoelectric Accelerometers, Circa 1970

Manufacturer	Model	Sensitivity, mv/g	Weight, grams	Resonance frequency, kHz	Frequency response, Hz
Brüel and Kjaer	4332	55	30	30	to 10,000 ± 5%
	4333	20	13	45	to 15,000 ± 10%
	4339	10	16	45	to 15,000 ± 10%
	4344	2	2	120	to 40,000 ± 10%
Columbia Research	302-5	200	38	15	to 3,000 ± 5%
Laboratory	504-3	10	10	60	to 12,000 ± 5%
	606	5	1.5	30	to 6,000 ± 5%
	902	10	32	30	to 6,000 ± 5%
Endevco Corp.	2215	16	32	32	to 6,000 ± 5%
	2222A	3	0.5	29	to 8,000 ± 5%
	2233	45	32	32	to 6,000 ± 5%
	2271A	5	27	27	to 5,500 ± 5%
General Radio	1560-P53	70	31	27	to 10,000 ± 10%
	1560-P54	700	90	5	to 1,600 ± 10%
Gulton Industries	AQA-4210	13	10	50	to 8,000 ± 5%
	AQB-4981	125	90	20	to 3,500 ± 5%
	AQB-6054	1	25	28	to 6,000 ± 5%
	AVA-6103	3	1	60	to 8,000 ± 5%
Kistler Instrument	805	5	6	60	to 12,000 ± 5%
Company	808A	11	20	40	to 7,000 ± 5%
	808K1	11	20	40	to 10,000 ± 5%
Shure	61CP	75	45	3	to 1,000
	62CP	580	90	4.5	to 1,200
Wilcoxon Research	90	3	0.25	120	to 40,000 ± 10%
Company	408	50	28	30	to 10,000 ± 10%
	1000	3,000	540	......	to 250 ± 10%

sensitivity and frequency response of a dynamic microphone should be checked at frequent intervals when it is used in a gas other than air, or when it is operated at other than normal laboratory temperatures and barometric pressures.

3.3   Accelerometers

3.3.1   Piezoelectric
In a piezoelectric accelerometer the force exerted by a seismic mass is used to strain the piezoelectric element. In the simplest form, a mass is mounted on one side of a piezoelectric material, and the other side is cemented to the accelerometer base. The

Transverse axis sensitivity, TSR %	Operating temperature range, °F	Sensitivity deviation over range	Maximum acceleration, g's	Impedance, pf	Acoustic sensitivity (equivalent output for 150 dB L_p), g's
4	−100 to +500	0.06%/°F	7,000	1,000	0.03
4	−100 to +500	0.06%/°F	10,000	1,000	0.03
3	−100 to +500	0.06%/°F	10,000	1,000	0.02
8	−100 to +500	0.06%/°F	14,000	900	0.1
5	−100 to +300	±10%	1,000	500	0.1
5	−100 to +350	±10%	2,000	850	0.1
5	−100 to +350	±10%	1,000	300	0.1
5	−100 to +350	±10%	20,000	10,000	0.05
5	−65 to +350	±10%	2,000	10,000	0.02
5	−100 to +350	±10%	1,000	420	......
3	−65 to +350	±10%	2,000	1,000	0.02
2	−300 to +450	±10%	1,000	2,000	......
5	−65 to +290	±20%	1,000	350	0.02
5	0 to +250	±15%	100	700	......
3	−100 to +350	±5%	10,000	750	0.3
3	−300 to +1100	±10%	1,000	40	......
1	−65 to +500	±7%	3,000	1,200	0.05
5	−40 to +250	±10%	10,000	300	0.1
5	−320 to +400	0.01%/°F	100,000	50	0.001
5	−320 to +500	0.01%/°F	10,000	90	0.005
1	−100 to +250	0.01%/°F	1,000	90	0.005
7	−40 to +180	......	100	9,700	2.5
5	−40 to +200	±5%	200	770	1
10	−30 to +150	±10%	2,000	150	......
2	−65 to +350	±10%	5,000	500	0.02
3	−50 to +175	......	15	2,500	0.002

accelerometer base is then rigidly affixed to the sample whose vibration is under study. Motion of the accelerometer base generates an inertial force in the seismic mass, which then strains the crystal. In turn, the crystal generates an output voltage.

The piezoelectric materials used are the same as those employed in piezoelectric microphones. The equivalent circuits of Fig. 3.4 are applicable to the accelerometer. Characteristics of some commercially available accelerometers are shown in Table 3.2.

Usually the resonance of a piezoelectric accelerometer is underdamped and exhibits a peak in excess of 20 dB. Care must be exercised in data taking that the excitation of this peak does not dominate the data.

Many piezoelectric accelerometers are now being manufactured with a field-effect-transistor preamplifier included within the unit. Commercially available units such as these are tabulated in Table 3.3. The advantage of this approach is the low transducer output impedance, which eliminates much of the difficulty with long cables.

3.3.2 Electrodynamic   A transducer similar in principle to the dynamic microphone provides an output signal proportional to velocity. A seismic mass, actually a permanent magnet, is mounted so as to form a spring-mass system with a damped resonance at a frequency below the operating range of frequencies. A stationary coil is mounted near the seismic mass to detect its relative motion.

The low output impedance of an electrodynamic velocity-measuring device is one of its prime advantages. An electrodynamic velocity transducer is susceptible to external magnetic fields, is larger than comparable piezoelectric transducers, and is not as rugged. Typically the resonance frequency of the seismic mass-spring system is of the order of a few hertz and measurements generally can be made up to only a few hundred hertz. This is often inadequate for vibration measurements.

3.4   Some Characteristics of Transducers

3.4.1 Frequency Response of Microphones   Even with an ideal transduction mechanism, the response of a microphone at high frequencies is dependent upon the sound field and the geometry of the microphone. When the size of a microphone exceeds about one-twentieth of a wavelength, its presence disturbs the sound field. The amount of disturbance is dependent upon the wavelength λ of the sound wave compared to the dimensions of the microphone. For a cylindrical microphone with the diaphragm at one end, the disturbance is a maximum when the wavelength is a little less than twice the radius r.

The response of the microphone transduction mechanism itself can be measured with uniform sound pressure (in both phase and amplitude) over the whole diaphragm. This is called the *pressure response* and excludes the geometrical effects in the sound field. Measurements of the pressure response are made in small cavities or with electrostatic actuation of the diaphragm.

Sound diffraction is also dependent upon the angle of incidence of sound onto the microphone when r/λ is greater than 0.05. When r/λ is less than 0.05 a microphone is nearly omnidirectional. Whenever a microphone is used to measure sound above the frequency at which $r/\lambda = 0.05$, the measured results must be corrected for the effect of the presence of the microphone.

An example of the pressure response of a 0.936-in. (2.377-cm) diam-

TABLE 3.3 Manufacturers' Specifications for Typical Piezoelectric Accelerometers with Internal Preamplifier, Circa 1970

Manufacturer	Model	Weight, grams	Sensitivity, mv/g	Maximum acceleration, g's	Resonance frequency, kHz	Output impedance, ohms	Operating temperature range, °F	Deviation over temperature	Transverse sensitivity, TSR %	Frequency response, Hz
Columbia Research Laboratory	1104	60	220	20	40	100	−65 to +185	±5%	5	2 to 8,000 ± 5%
	1106-1	20	0.2	12,500	40	150	−65 to +200	±10%	5	5 to 6,000 ± 5%
	1107-1	50	2,000	2	25	150	−40 to +200	±10%	5	5 to 5,000 ± 5%
	1606	15	10	250	30	100	−65 to +250	±10%	5	2 to 5,000 ± 5%
Gulton Industries	AT-1189	30	30	200		100	−100 to +250	±5%	3	5 to 10,000 ± 5%
Kistler	818	20	10	500	30	100	−65 to +250	0.03%/°F	5	2 to 5,000 ± 5%
Bolt Beranek and Newman	501	2	8	250	90	1,500	−65 to +250	±10%	5	2 to 20,000 ± 5%

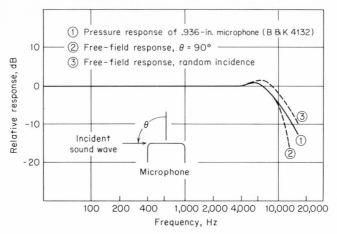

Fig. 3.6 Examples of frequency response for a 0.936-in.-diameter microphone with protective grid.[9] Curve 2 is for grazing incidence.

eter microphone is illustrated by curve 1 in Fig. 3.6.[9] In this design the diaphragm resonance is damped. In a plane-wave free field, the performance of the microphone, which includes the diffraction of sound, is found by applying *free-field corrections* to the pressure response. Since the microphone is not exactly a right cylinder, these corrections are found by measurement.

Figure 3.7 illustrates plane-wave free-field and diffuse-field corrections for a 0.936-in.-diameter microphone. These corrections are typical for microphones of this diameter with a protective grid in front of the microphone. A slightly different set of corrections is required if the protective grid is not used, or for any other item that influences the microphone geometry. For each size and type of microphone the manufacturer's data should be consulted.

The expected free-field response of the 0.936-in. microphone at 90° incidence is found by applying the free-field correction for 90° in Fig. 3.7 to the pressure response of the microphone. This is shown as the line labeled 90° in Fig. 3.6. If the microphone is to be used at other angles of incidence, the appropriate plane-wave free-field corrections are used.

Often the angle of incidence of the sound wave cannot be identified, particularly in reverberant fields. The *diffuse-field* (random-incidence) correction has been developed for measurements in reverberant sound fields where it is assumed that all angles of sound incidence are equally likely to exist. The free-field corrections are then integrated across all angles and averaged. The diffuse-field correction for the 0.936-in.-diameter microphone is shown as the dashed curve in Fig. 3.7.

The response of the microphone illustrated in Fig. 3.6 has been engineered to yield a flat free-field plane-wave response at grazing (90°) incidence. Microphones are available that have been designed for optimum perpendicular (0°), grazing (90°), and random-incidence responses.

American National Standards for sound-level meters (ANSI S1.4, 1961) specify that the microphone used with the sound-level meter should be calibrated for random incidence, which is usually similar to grazing (90°) incidence. European standards for sound-level meters (IEC Publications 129 and 179) indicate that the optimum angle is to be specified by the manufacturer, and perpendicular incidence (0°) is generally used in Europe. Care must be taken that a specific microphone is used at the appropriate angle, since there may be over 15 dB difference at frequencies above 10,000 Hz between various types of microphones.

When measurements are not made in a free-field environment, the free-field response corrections are not applicable. Consider two cases where the pressure on a surface is desired, and assume that the microphone is ideally flush-mounted on that surface. In the first case the sound wave is incident normal to the surface. Here the microphone pressure response is used because the sound will arrive at all parts of the diaphragm in phase.

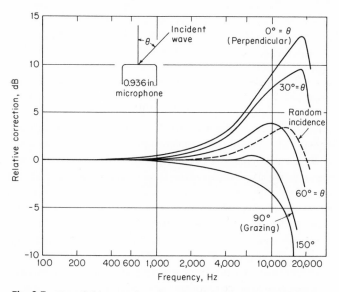

Fig. 3.7   Free-field corrections for a 0.936-in.-diameter microphone (B&K 4132) with protective grid.[9] These corrections (in dB) are to be added to the pressure response (curve 1) of Fig. 3.6.

In the second case, a plane sound wave travels parallel to the surface that holds the microphone and the finite size of the diaphragm spatially averages the sound wave. Because the microphone is mounted in a plane surface, we do not expect the same results as for free-field grazing incidence of Fig. 3.6 where the microphone geometry disturbs the sound wave. If the sound wave is truly a grazing plane wave and the microphone diaphragm is uniformly sensitive over its surface,* the spatial averaging can be described by the magnitude of the familiar $(\sin x)/x$ relationship where $x = \pi d/\lambda$, and d is the microphone diameter. In this case, the $(\sin x)/x$ function has nodes at $d = n\lambda$ where n is any integer and peaks at $d = n\lambda/2$. The magnitudes of successive peaks decrease by $1/n$.

The preceding discussions have focused on the influence of geometry on the frequency response of a microphone. The pressure response of the microphone is the frequency response of the basic transduction mechanism. The pressure response is engineered to yield a near-flat free-field response at a chosen angle of incidence (usually perpendicular or grazing), or for random incidence.

All microphones have resonances, and in the best design these are either critically damped or exist beyond the frequency range of interest for the microphone.

3.4.2 Accelerometer Transverse Sensitivity   Pressure is a scalar quantity, but vibration is not. Although measured at a point, the direction of acceleration is important. Accelerometers are designed to measure the acceleration perpendicular to the plane of the base, and to reject acceleration along the other two axes. The amount of attenuation of the other two axes is termed the *transverse sensitivity ratio* (TSR) by Rockwell and Ramboy [10] and is expressed as

$$\text{TSR} = \frac{\text{mv}/g \text{ for transverse acceleration}}{\text{mv}/g \text{ for axial acceleration}} \tag{3.3}$$

This measurement is usually made for several axes in the plane of the base, and the largest TSR obtained for motion along one of these axes is selected as the TSR rating for the accelerometer. Transverse sensitivity ratios in percent for typical accelerometers are given in Tables 3.2 and 3.3.

Transverse sensitivity can result from a small misalignment of the crystal with the axis during manufacture, and from shear force exerted on the crystal by the seismic mass. The shear force is reduced by the use of low-shear-sensitivity piezoelectric material. The shear force might also be moment-coupled into axial forces if care is not exercised in the mechanical design of the transducer.

---

* Actual thin-stretched diaphragms are most sensitive at the center and least sensitive at the edges, with the equivalent uniform-sensitivity region equaling about 0.6 diameter.

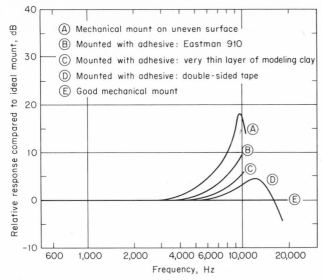

Fig. 3.8 Accelerometer and mount responses for various mounting techniques for 30-gram accelerometer excited at 1 g.[10]

Rockwell and Ramboy discuss methods of measurement of the transverse sensitivity of accelerometers, and they have conducted these measurements up to frequencies of 2,000 Hz. Typical TSRs are of the order of 1 to 5% and are reasonably flat with only minor fluctuations up to frequencies of 2,000 Hz.

3.4.3 Accelerometer Mounting Techniques

It is difficult to mount an accelerometer so that it faithfully measures the vibration of the test sample at high frequencies. Mounting techniques fall into two general categories: *mechanical* (a stud threaded into the test sample) and *adhesive* (cement, tape, or some other technique holding the accelerometer to the sample).

In the adhesive mounting, the elastic properties of the adhesive and the mass of the accelerometer determine the first resonance of the mounting system. In general, it is desirable to have as thin a layer of adhesive as possible. Also, properties of the adhesive (such as clay or tape) may become nonlinear at higher acceleration levels and may vary with temperature. Mounting resonances for several adhesives with a 30-gram accelerometer are illustrated in Fig. 3.8.[11]

In the mechanical mounting, the accelerometer is held onto the test sample with a short stud threaded into the sample. If the accelerometer base is in ideal contact with the surface of the test article, the mount contributes no resonance to the response. Curve E of Fig. 3.8 illustrates an example where no resonance is present out to 20,000 Hz. This is not

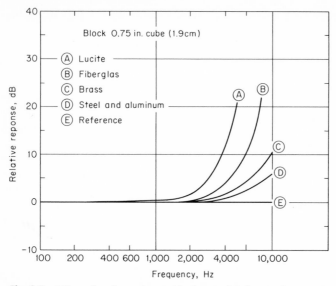

Fig. 3.9   Effect of various adapter block materials for accelerometer mounts with 28-gram accelerometer. (*After Mangolds.*[12])

easily accomplished; the mounting surface and the accelerometer must be both flat and smooth, and the mounting stud must be absolutely perpendicular to the surface.   An example of a mounting on a slightly uneven surface is shown as curve *A* in Fig. 3.8.

Where smooth, flat surfaces do not exist on the test article, adapter blocks are sometimes used to mount the accelerometer.   These blocks add an additional mass and spring between the accelerometer and the measurement.   Mangolds [12] has investigated this, as well as other mounting problems, and Fig. 3.9 summarizes these results.   Cubes, 0.75 in. (1.9 cm) on a side, were made from several materials and tested with a 28-gram accelerometer.   The resonance contributed by the block is dependent upon the Young's modulus of the material, and the resonant frequency is shown to decrease with the modulus.   As illustrated in Fig. 3.9, unless great care is used, adapter blocks can add more difficulties than are solved.

3.4.4   Mass Loading by the Accelerometer

An accelerometer used well below its resonance approximates a mass loading of the test article at its attachment point.   The velocity u_0 of the attachment point is reduced to u_1 by the addition of the mass:

$$\frac{u_1}{u_0} = \frac{Z}{Z + j\omega M} \tag{3.4}$$

where Z = mechanical impedance of the test article at the attachment
point, N-sec/m

M = mass of the accelerometer, kg

ω = radial frequency, radians/sec

This assumes that the test article can, at the point of measurement, be represented by a constant-force generator in series with a lumped impedance Z. Of course, test articles at frequencies of interest often have distributed rather than lumped impedances, and approximations are necessary when Eq. (3.4) is used.

For *plates*, the average impedance of many modes approaches a pure resistance. This assumes that the vibration is wideband and excites many modes of the plate. The average impedance under these conditions is

$$Z_p \approx R_p = \frac{4}{\sqrt{3}} \, \rho_p c_L h^2 \qquad \text{N-sec/m} \qquad (3.5)$$

where ρ_p = density of the plate material, kg/m³

c_L = longitudinal wave velocity in the plate, m/sec

h = thickness of the plate, m

For a plate, the ratio u_1/u_0 for many modes together is down 3 dB when Z_p is equal to ωM in Eq. (3.4).

$$f_{3dB} = \frac{2}{\sqrt{3}} \, \frac{\rho_p c_L h^2}{\pi M} \qquad \text{Hz} \qquad (3.6)$$

Equation (3.6) is plotted in Fig. 3.10 for typical $\frac{1}{8}$- and $\frac{1}{16}$-in. plates. Note that a 10-gram accelerometer effectively mass loads the response of a $\frac{1}{16}$-in. aluminum plate at frequencies above 1,200 Hz; it is necessary to use an accelerometer under 2 grams to obtain a good measure of the plate vibration out to 6,000 Hz. In using Fig. 3.10, the mass is the sum of the mass of the accelerometer, the mounting block or stud (if used), the electrical connector, and that part of the cable which is effectively supported by the accelerometer. Of course, treating all these as a simple mass is an approximation.

The loading indicated by Eq. (3.6) is based upon the simple average of the modal impedance (see Chap. 11), but individual modes which are very responsive will be loaded more than modes with low response. Hence the actual mass loading indicated by Fig. 3.10 is understated somewhat.

If the test article behaves like a large metal block in the frequency range of interest, Schloss [13] has shown that the local stiffness of the test article limits the upper frequency by the following

$$f_{3dB} = \frac{1}{2\pi} \sqrt{\frac{DE}{2M}} \qquad \text{Hz} \qquad (3.7)$$

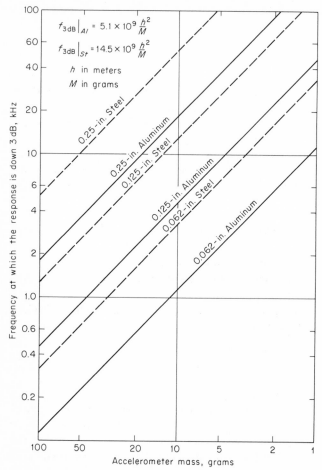

Fig. 3.10 Effect on response of a simple panel by an accelerometer at a point where the panel impedance is a resistance equal to $4\rho_p c_l h^2/\sqrt{3}$. See Eqs. (3.5) and (3.6).

where E = Young's modulus of the test article, N/m²
  D = diameter of the accelerometer base, m
  M = accelerometer mass (assuming the test article mass is large compared to the accelerometer), kg
This formula indicates that the ratio of contact area to mass of an accelerometer should be high.

3.4.5 Vibration Response of Microphones Microphones respond to vibration as well as to sound. The vibration response of a microphone is usually given as an equivalent sound-pressure-level output for 1 g rms excitation, i.e., L_p in dB (re 0.0002 μbar) per g rms. Usually the axis perpendicular to the diaphragm is most sensitive to vibration excitation. The other axes are usually 10 to 20 dB lower in vibration response.

Sensitivity to vibration is caused by inertial forces acting upon the transduction element. The mass of the elements of the transduction mechanism of a microphone must be minimized to reduce vibration response. Condenser microphones have minimum mass consisting simply of the thin diaphragm itself, and they also have the minimum vibration sensitivity. Typical vibration sensitivities for condenser microphones are of the order of 85 to 90 dB L_p per g rms.

In piezoelectric microphones, the effective mass of the crystal must be added to the mass of the diaphragm. The effective crystal mass is usually many times that of the diaphragm itself, resulting in an increase in vibration sensitivity relative to that of condenser microphones. Typical vibration sensitivities for piezoelectric microphones are of the order of 100 to 120 dB L_p per g rms.

The inherent vibration sensitivity of some piezoelectric microphones is canceled by a second crystal. This cancellation is effective up to frequencies where the phase difference between the vibrations of the two crystals is appreciable. The phase difference is dependent upon the internal mechanical design of the unit. For large microphones the upper frequency limit of cancellation is a few kilohertz. Typical vibration sensitivities of piezoelectric microphones using cancellation are of the order of 95 to 105 dB L_p per g rms.

The dynamic microphone is very sensitive to vibration owing to the large mass of its diaphragm and coil, and it should not be used in applications where vibration is expected to be a problem.

The theoretical lower limit of vibration sensitivity of a microphone is found by assuming a diaphragm with zero mass vibrating at 1 g rms and computing the force on the diaphragm produced by the diaphragm motion against still air. A typical 1-in.-diameter condenser microphone has an equivalent acoustic diaphragm diameter of about 0.5 in. (1.25 cm) since all of the diaphragm does not deflect equally. This equivalent diaphragm vibrating at 1 g rms will generate a sound pressure at its surface of approximately 67 dB.[14] Thus the lower limit to vibration sensitivity for the 1-in.-diameter microphone is 67 dB L_p per g rms.

3.4.6 Acoustic Sensitivity of Accelerometers
Accelerometers respond to sound as well as vibration. An acoustic field acting upon the case of an accelerometer can induce vibration of the case, which can couple mechanically to the crystal or produce sound inside the case that will act on the crystal as a pressure. No standard units for expressing the acoustic sensitivity of accelerometers have evolved.[15] Those ratings typically used are equivalent acceleration output in g's rms when the ac-

celerometer is exposed to either a sound-pressure level of 150 dB (6,300 microbars) or 74 dB (1 microbar).

The acoustic sensitivity of an accelerometer is often important when measuring the response of a structure to an acoustic field. If the structure is very responsive to the acoustic field, the measurement of the structure's vibration can usually be made successfully; but if the structure is not responsive, the measurement may simply be that of the acoustic response of the accelerometer to the sound field. Typical acoustic sensitivities of piezoelectric accelerometers depend very much on the mechanical design, and can range from about 0.01 to 1 g rms for an exposure to 150 dB sound-pressure level.

3.5 Signal Conditioning

Following the transducer, the electrical signal is amplified. Amplification may be by voltage gain, charge gain, or impedance reduction. The transducer and preamplifier must be electrically connected so as to achieve a minimum of cable noise or ground loops.

3.5.1 Condenser Microphone Circuits
A condenser microphone requires a bias voltage, unless used with a carrier system. The typical circuit employs a cathode follower and a high impedance voltage source. The low capacitance of a condenser microphone requires a low input capacitance and a high input resistance in the preamplifier to achieve low-frequency response. For example, a microphone with a capacitance of 60 pf requires a preamplifier with a shunt resistance of 300 megohms to achieve a frequency response that is down only 3 dB at 10 Hz. The frequency at which the response is down by 3 dB is given by the simple RC high-pass filter.

The thermal electronic noise can be estimated from the total resistance at the input and is proportional to the square root of the bandwidth and the square root of the resistance. The thermal noise generated by a 1-megohm resistor at room temperature in a 20,000-Hz bandwidth is 18 microvolts.

The capacitance of the microphone filters some of the noise generated by the resistance, which provides a RC low-pass filter for the resistor noise. For example, consider a condenser microphone system with a total resistance R of 300 megohms and a capacitance of 60 pf. The noise generated by 300 megohms in a 10,000-Hz bandwidth is about 220 microvolts. The low-pass filter reduces this noise to about 8 microvolts.

3.5.2 Preamplifiers for Piezoelectric and Electret Transducers
Unlike condenser microphones, piezoelectric and electret microphones

require no polarization voltage and their capacitance is much higher. The higher capacitance allows a much greater physical separation between the microphone and preamplifier.

The frequency at which the response of a piezoelectric transducer is down 3 dB is simply the *RC* high-pass filter, and the electrical noise reduction from the transducer capacitance is also applicable, just as for condenser microphones.

If long cables are to be used with piezoelectric transducers, the cable capacitance reduces the voltage produced by the transducer at the preamplifier by the ratio of the transducer capacitance to the combined capacitance of the cable and transducer. The use of a charge amplifier eliminates this loss of voltage owing to the cable. The output voltage of a charge amplifier is proportional to the change of charge at its input. Since charge generation in the transducer is independent of capacitance, the shunt capacitance of the cable does not reduce sensitivity. The charge sensitivity of a transducer is simply the product of the voltage sensitivity and the capacitance.

Signal-to-noise ratio is not usually improved by the use of a charge amplifier instead of a voltage amplifier, even with long cables. Cable noise is typically generated as charge rather than voltage, and is additive when a charge amplifier is used. With voltage amplification, both the signal and the cable-generated charge are attenuated by the cable capacity. The fundamental advantage of the charge amplification is not an improvement in signal-to-noise, but that the transducer sensitivity calibration is nearly independent of cable length.

3.5.3 Dynamic (Moving-coil) Transducer Circuits   No special problems are presented in preamplification of outputs from dynamic transducers. The transduction element is itself low impedance, and transformer coupling is often used to provide a balanced drive to long cables. Ground loops are isolated and common mode rejection is offered by balanced lines.

Self-generated noise is low from a moving-coil transducer owing to the low electrical resistance, but the output voltage is also low, and the inherent noise filter provided by the capacitance of condenser and piezoelectric transducers is not present. Hence the wideband self-generated noise is of the same order of magnitude relative to the signal level as for other transduction principles. Susceptibility to external magnetic and electric fields is very large.

3.5.4 Cable Noise   A frequent measurement difficulty, particularly with piezoelectric transducers, is the electrical charge generated within the cable that is used between the transducer and preamplifier when the cable is subjected to vibration. The effect is discussed by Perls,[16] and

is related to the relative motion between various internal parts of the cable.

Kamperman [17] found that typical cables, when subjected to a 1-in. peak-to-peak sinusoidal vibration, responded in the range from 1 millivolt to 1 volt per 10 ft of cable. Cable specifically advertised for its low noise properties responded to the range from 10 microvolts to 30 millivolts per 10 ft. The results were reasonably independent of the frequency of excitation from 1 to 10,000 Hz. In addition, for some cables the cable noise increased after use.

Since cable noise is generated by relative motion of parts of the cable, the cable in a measurement setup should be restrained from motion as much as is practical. This is particularly critical for vibration measurements. In no case should a cable be allowed to slap against an object or to lie loose on a vibrating surface during measurements.

The cable noise problem is significantly reduced by providing a preamplifier before the cable. The low output impedance of the preamplifier provides a path for the charge without the generation of significant voltages.

3.5.5 Ground Loops   Transducer signal contamination often results from multiple grounding. A typical "ground loop" is illustrated in Fig. 3.11. The transducer case is connected to the low side of the cable, which in turn is connected through the cable resistance R_C to the ground plane at the amplifier. If the transducer is connected to a nearby ground through a resistance R_I, a ground loop results. Noise or hum is introduced into the loops whenever a voltage e_{CM} exists between the

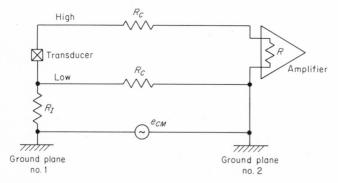

e_{CM} = voltage between ground planes
R = amplifier input resistance
R_C = cable resistance
R_I = isolation resistance between ground and transducer

Fig. 3.11 Equivalent circuit for a ground loop.

transducer ground and the amplifier ground.   This voltage is applied across R_C and is amplified along with the transducer signal.

The potential difference between the two ground planes can have levels ranging from millivolts to several volts, depending upon the particular situation.   If the transducer is tied to ground plane 1 through R_I, and the cable is short, R_I and R_C might have about equal values.   In this case, half of the potential difference e_{CM} would appear in series with the transducer.   This would often be an unacceptable noise level.

Ground loop effects can be reduced by several methods.   The most common is to isolate the transducer return lead from ground.   If the transducer return lead is electrically tied to the transducer case, then the transducer case itself must be isolated from ground.   In other words, R_I is made very large and the ground loop voltage is attenuated by R_C/R_I.

A second method for reducing ground loop noise is to use battery-operated equipment.   Effectively, this means that a large resistance is inserted between the amplifier low lead and ground plane 2 of Fig. 3.11, which is the equivalent of isolating the transducer return conductor. Of course, no part of the electronic equipment tied to the channel can be grounded to more than one point or a ground loop will result.

A third solution is to use a differential amplifier.   A differential amplifier has both input terminals above ground.   It responds only to the difference of the voltage at these input terminals, and rejects any voltage common to both terminals.   A simple example of a differential amplifier is a transformer, often used with dynamic microphones.   At high frequencies, the stray capacitance in the circuit effectively shunts the isolating resistances, which may result in high-frequency ground loop noise.   Fortunately, most common mode voltages between ground planes exist at power generation frequencies.

In summary, the techniques for reducing ground loop effects are all directed toward isolating the circuit from all but one of the ground planes.   If this cannot be achieved, a differential amplifier should be used.

3.6   Microphone Calibration

Microphone calibrations may be made either by absolute or by comparison techniques (see Table 3.4).[18]   For an absolute technique, the level of the sound used for calibration is determined by an independent measurement using a primary standard.   For a comparison technique the level of the calibration sound is measured by a reference transducer, called a secondary standard.   Calibrations are of several types and may be performed in anechoic rooms, reverberation rooms, tubes, and by pistonphones.

TABLE 3.4   Summary of Microphone Calibration Techniques

Technique of calibration	Type of technique	Usual frequency range,* Hz	Type of response	Typical calibration levels, dB re 2×10^{-5} N/m²
Anechoic chamber............	Comparison	400 to 15,000	Free field	75–95
Plane-wave tube	Comparison	10 to $2\dfrac{D}{\lambda}$	Plane-wave field	90–140
Reverberation room.........	Comparison	$1.6\dfrac{c}{V^{1/3}}$ to 15,000	Diffuse field	75–110
Reciprocity.....................	Absolute	10 to 50,000	Pressure	80–90
Fixed-displacement pistonphone.................	Absolute	1 to 250	Pressure	110–170
Other pistonphones.........	Comparison	10 to 10,000	Pressure	110–170

* D = diameter of tube; V = volume of room; c = speed of sound; λ = wavelength.

3.6.1 Anechoic Rooms

An anechoic room used in conjunction with a sound source provides a *plane-wave* free-field condition for the measurement of the sensitivity of microphones. An anechoic room is lined on all six surfaces with deep sound-absorptive materials arranged to reduce the reflected energy in sound waves to very low values. An anechoic environment is essential for a calibration that requires determination of the directional properties, that is to say, the influence of microphone geometry on the sensitivity as discussed in Sec. 3.4.1.

A reference microphone (secondary standard) is used to measure the rms sound pressure produced by a stable source at a specific location in the anechoic chamber. The microphone under test is then substituted at exactly the same position that the reference microphone had occupied, and its electrical output is compared with that of the standard. At higher frequencies where the wavelength of sound is small, the positioning must be done very accurately.

Anechoic chamber measurements are limited at low frequencies by the ability of the chamber to absorb low-frequency sound. For room-sized chambers lined with wedges about 2 ft in length, this lower limit is typically a few hundred hertz. Very large chambers, with wedges of 4 or 5 ft in length, are used at frequencies as low as 60 Hz. The limitations at high frequencies are often due to the properties of the sound source used in the chamber, and to the difficulty of positioning the microphones accurately in a sound field with very short wavelengths.

The *diffuse field* (random incidence) of a microphone may be performed in an anechoic chamber by the inverse process of that used for determining the total sound power radiated by a sound source located in anechoic space. That process is described in Sec. 6.4.

3.6.2 Plane-wave Tube   A plane-wave tube is driven by a loudspeaker at one end and terminated with sound-absorptive material at the other end to eliminate reflections. The sound wave propagated down the tube is plane and may be used for comparison calibrations of microphones at low frequencies where anechoic chambers are not usable. If the tube end is closed by a rigid plate instead of an absorptive material, it can be operated at resonance and high sound pressures can be produced.

As in an anechoic chamber, a reference microphone is used to measure the sound pressure, and the test microphone is substituted at the same position and its output compared to that of the secondary standard.

A plane-wave tube is limited to use at frequencies below those where transverse resonances cause pressure variations across the diameter of the tube. A probe microphone can be used to determine the uniformity of pressure over a plane.

3.6.3 Reverberation Room   Details on the size and shape of the reverberation room, on rotating diffusers, and on where to locate the microphone are given in Chap. 6.

3.6.4 Reciprocity Techniques   The reciprocity theorem for electromechanical transducers [19,20] provides a technique for the absolute calibration of two microphones simultaneously. This technique uses one microphone alternately as a receiver and source, and a second as a receiver. From the voltages and currents, the acoustic sensitivities are established. A reciprocity measurement is usually performed in a small cavity. The technique is also valid in an anechoic space, but the measurement is more difficult to accomplish because of the low sound levels produced by most microphones when used as sources in free space.

By the reciprocity technique the absolute pressure sensitivity of two microphones is obtained with a sound source, a rigid-walled coupling cavity, and the two microphones. One of the microphones must be a reciprocal transducer, i.e., a transducer that operates as either a microphone or a sound source.

In general, condenser and piezoelectric transducers are reversible and linear and satisfy the reciprocal relations. The National Bureau of Standards in Washington performs reciprocity calibrations of 1-in.-diameter condenser microphones (for example, WE-640AA and B & K 4131) for industry. Laboratories active in sound measurement should have on hand a reference microphone (secondary standard) calibrated by the NBS.

3.6.5 Pistonphones   The pistonphone is a straightforward method of generating sound in a closed cavity. A vibrating piston operates into a closed cavity with the sensitive surface of the microphone as one wall of the cavity. The fundamental equation, assuming a rigid-walled cavity and an adiabatic process, is given in Chap. 8.

Two accurate pistonphones have been developed using the fixed-displacement principle; one is a small, battery-operated unit for field use [21] and the other a large laboratory unit for use at very high sound pressures.[22]   Other piston drives, such as hydraulic, electromagnetic, or piezoelectric, are more difficult to calibrate because the displacement is not fixed and must be determined by independent measurement for absolute calibrations; these, however, can be used readily for comparison calibrations.

REFERENCES

1. ANSI S1.2, 1967, "Specifications for Laboratory Standard Microphones," American National Standards Institute, Inc., 1430 Broadway, New York, N.Y. 10018.
2. E. C. Wente, "A Condenser Transmitter as a Uniformly Sensitive Instrument for the Absolute Measurement of Sound Intensity," *Phys. Rev.*, vol. 10, pp. 39–69, 1917.
3. R. W. Leonard, "Generation and Measurements of Sound in Gases," in S. Flügge (ed.), *Encyclopedia of Physics,* Vol. XI/2, Springer-Verlag, Berlin, 1962, and L. L. Beranek, *Acoustics,* Chap. 6, McGraw-Hill, 1954.
4. G. M. Sessler, "Electrostatic Microphones with Electret Foil," *J. Acoust. Soc. Am.*, vol. 35, pp. 1354–1357, 1963.
5. G. M. Sessler and J. F. West, "Foil-Electret Microphones," *J. Acoust. Soc. Am.*, vol. 40, pp. 1433–1440, 1966.
6. D. A. Berlincourt, D. R. Curran, and H. Jaffe, "Piezoelectric and Piezomagnetic Materials and Their Functions in Transducers," in W. P. Mason (ed.), *Physical Acoustics,* Vol. I, Pt. A, Chap. 3, Academic, 1964.
7. T. F. Hueter and R. H. Bolt, *Sonics,* Wiley, 1955.
8. L. L. Beranek, *Acoustics,* Sec. 6.4, McGraw-Hill, 1954.
9. Brüel and Kjaer, "One-inch Condenser Microphones," Naerum, Denmark, pp. 16–17, January, 1963.
10. D. W. Rockwell and J. D. Ramboy, "Measurement of Accelerometer Transverse Sensitivity," *Shock Vibration Bull.* 35, pp. 73–98, Pt. IV, February, 1966.
11. E. A. Starr, "Vibration Measurements," *BBN Instr. Mem.* 12, 1961, Bolt Beranek and Newman Inc., Cambridge, Mass.
12. B. Mangolds, "Effect of Mounting Variables on Accelerometer Performance," *Shock, Vibration, Assoc. Environ., Bull.* 33, Pt. III, March, 1964.
13. F. Schloss, "Inherent Limitations of Accelerometers for High Frequency Vibration Measurements," *J. Acoust. Soc. Am.*, vol. 33, no. 4, 1961.
14. L. L. Beranek, *Acoustics,* Sec. 5.3, McGraw-Hill, 1954.
15. ISA-RP 37.2, "Guide for Specifications and Tests for Piezoelectric Acceleration Transducers for Aerospace Testing," January, 1964, available through Instrument Society of America.
16. T. A. Perls, "Electrical Noise from Instrument Cables Subjected to Shock and Vibration," *J. Appl. Phys.*, vol. 23, June, 1952.

17. G. W. Kamperman, "Measurement of High Intensity Noise," *Noise Control,* vol. 4, no. 5, pp. 22–27, September, 1958.
18. ANSI S1.10, 1966, "American Standard Method for the Calibration of Microphones," American National Standards Institute, Inc., 1430 Broadway, New York, N.Y. 10018.
19. F. Schottky, *Physik,* vol. 36, p. 689, 1926.
20. L. L. Beranek, *Acoustic Measurements,* Chap. 4, pp. 113–148, Wiley, 1949.
21. Brüel and Kjaer, *Instruction Manual on Pistonphone, Model 4220,* Naerum, Denmark.
22. E. A. Starr, "High Pressure–Low Frequency Microphone Calibration," presented at the 69th Meeting of the Acoustical Society of America, Washington, D.C., June, 1965. *J. Acoust. Soc. Am.,* vol. 37, p. 1209(A), 1965.

Chapter Four

Field Measurements: Equipment and Techniques

ROBERT D. BRUCE

The acoustical engineer should do more than simply measure the noise or vibration levels associated with a noise control problem. Prior to embarking on a measurement program he should establish clearly why and for what conditions the data are needed. Much also can be learned from visual, tactile, and aural observations. Identifying the sources of the noise and vibration, the paths by which they are propagated and feasible measures for their control often depends heavily on common sense and on experience.

This chapter discusses the basic components of a field measurement system; it encourages systematic programs of data acquisition and it presents specific examples of measurement techniques. From this information, the noise control engineer should be able to select and utilize field equipment suited to his problem.

74

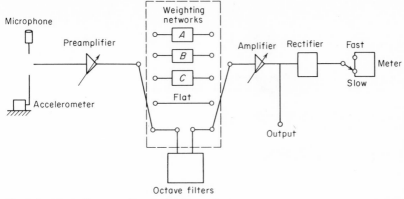

Fig. 4.1 Block diagram of typical noise analyzer.

4.1 Basic Measurement System

A basic system for measuring sound-pressure levels or vibration-acceleration levels consists of an appropriate transducer and calibrator, an octave or one-third-octave noise analyzer, and possibly a tape recorder. To be useful in a field environment, this equipment must not only maintain stable operation; it must also be portable, battery-operated, and reasonably lightweight.

An international standard has been developed for the precision sound-level meter (SLM),[1] and a new American National Standard is now in draft which covers (1) precision, (2) general-purpose, (3) survey, and (4) special-purpose sound-level meters.[2]

A block diagram of a typical noise measuring instrument is presented in Fig. 4.1. The electrical signal from the microphone or accelerometer is fed to the preamplifier. The gain of this amplifier can be varied in calibrated 10-dB steps. The signal then passes to the next amplifier or through one of the spectrum shaping networks (the A, B, or C weighting networks) or through an octave-band filter. Once filtered, the signal is fed to a second amplifier, which usually also has variable gain (10-dB steps), then to two outputs, one for connection to an oscilloscope, tape recorder, headphones, etc., and the other to a rectifier that converts the filtered signal to a direct current, proportional to the quasi-rms value of the noise, that is indicated on the meter. The needle of the indicating meter usually has two speeds of indication: "fast" and "slow." On the "fast" setting, the needle gives true indication of the level within 200 to 250 milliseconds after a 1,000-Hz tone has been fed into the preamplifier. The overshoot before that time is not greater than 1 dB. The "slow" setting averages the level for a greater time period. In Fig. 4.2

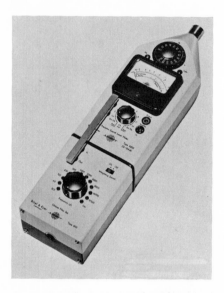

Fig. 4.2 *Upper:* Photograph of sound-level meter with octave-band filter set. (*Brüel & Kjaer.*) *Lower:* Photograph of octave-band noise analyzer. (*General Radio Co.*)

are photographs of the octave-band noise analyzers of two manufacturers.

The frequency response of the A, B, and C weighting networks are presented in Table 4.1. Originally these networks were designed to approximate the loudness level sensitivity of the human ear when listening to pure tones. Today the B weighting network is rarely used. The

**TABLE 4.1   A, B, and C Electrical Weighting Networks
for the Sound-level Meter**
These numbers assume a flat, diffuse-field (random-
incidence) response for the sound-level meter
and microphone (see Ref., 2)

Frequency, Hz	A-weighting relative response, dB	B-weighting relative response, dB	C-weighting relative response, dB
10	−70.4	−38.2	−14.3
12.5	−63.4	−33.2	−11.2
16	−56.7	−28.5	−8.5
20	−50.5	−24.2	−6.2
25	−44.7	−20.4	−4.4
31.5	−39.4	−17.1	−3.0
40	−34.6	−14.2	−2.0
50	−30.2	−11.6	−1.3
63	−26.2	−9.3	−0.8
80	−22.5	−7.4	−0.5
100	−19.1	−5.6	−0.3
125	−16.1	−4.2	−0.2
160	−13.4	−3.0	−0.1
200	−10.9	−2.0	0
250	−8.6	−1.3	0
315	−6.6	−0.8	0
400	−4.8	−0.5	0
500	−3.2	−0.3	0
630	−1.9	−0.1	0
800	−0.8	0	0
1,000	0	0	0
1,250	+0.6	0	0
1,600	+1.0	0	−0.1
2,000	+1.2	−0.1	−0.2
2,500	+1.3	−0.2	−0.3
3,150	+1.2	−0.4	−0.5
4,000	+1.0	−0.7	−0.8
5,000	+0.5	−1.2	−1.3
6,300	−0.1	−1.9	−2.0
8,000	−1.1	−2.9	−3.0
10,000	−2.5	−4.3	−4.4
12,500	−4.3	−6.1	−6.2
16,000	−6.6	−8.4	−8.5
20,000	−9.3	−11.1	−11.2

C network is generally used to limit the low- and high-frequency response of the instrument so that the instrument will not respond easily to signals outside the audible frequency range. As discussed in Chap. 17, the A-weighted reading is used to estimate the probability of hearing damage in industry. In addition, A-weighted values can be correlated with the annoyance caused by traffic and aircraft noise. The flat response of the sound-level meter (see Fig. 4.1) is often used as a calibrated input to a frequency-band filter set or a tape recorder.

The engineer uses filter sets of various bandwidths to segment noise into its component parts. Figure 4.3 presents the frequency response of an octave filter set that meets the international and American standards.

In Table 4.2 the standardized center frequencies and bandwidths of two types of filter sets are given. Typical octave-band filter sets provide filters with the following geometric-mean frequencies: 31.5, 63, 125, 250, 500, 1,000, 2,000, 4,000, 8,000, 16,000 Hz. One-third-octave filter sets are useful in the field as a diagnostic device, especially in situations involving pure tones. The combination of a low-electrical-noise analyzer with one-third-octave and one-tenth-octave filter sets with center frequencies extending down to 2.5 Hz yields an instrument especially useful in complex noise and vibration problems where on-the-spot diagnosis is important. Figure 4.4 is a photograph of such an instrument. Narrower-band analysis, without the availability of a graphic-level recorder, is generally very tedious in the field.

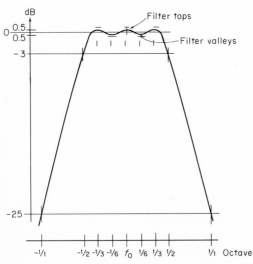

Fig. 4.3 Frequency response of one manufacturer's octave-band filter.

TABLE 4.2 Center and Approximate Cutoff Frequencies for Standard Set of Contiguous-octave and One-third-octave Bands Covering the Audio Frequency Range [3]

	Frequency, Hz					
	Octave			One-third octave		
Band	Lower band limit	Center	Upper band limit	Lower band limit	Center	Upper band limit
12	11	16	22	14.1	16	17.8
13				17.8	20	22.4
14				22.4	25	28.2
15	22	31.5	44	28.2	31.5	35.5
16				35.5	40	44.7
17				44.7	50	56.2
18	44	63	88	56.2	63	70.8
19				70.8	80	89.1
20				89.1	100	112
21	88	125	177	112	125	141
22				141	160	178
23				178	200	224
24	177	250	355	224	250	282
25				282	.315	355
26				355	400	447
27	355	500	710	447	500	562
28				562	630	708
29				708	800	891
30	710	1,000	1,420	891	1,000	1,122
31				1,122	1,250	1,413
32				1,413	1,600	1,778
33	1,420	2,000	2,840	1,778	2,000	2,239
34				2,239	2,500	2,818
35				2,818	3,150	3,548
36	2,840	4,000	5,680	3,548	4,000	4,467
37				4,467	5,000	5,623
38				5,623	6,300	7,079
39	5,680	8,000	11,360	7,079	8,000	8,913
40				8,913	10,000	11,220
41				11,220	12,500	14,130
42	11,360	16,000	22,720	14,130	16,000	17,780
43				17,780	20,000	22,390

The sound-level meter is used frequently as an input to a tape recorder. A tape recorder used for sound measurements must have wide dynamic range (the number of decibels between the noise floor in each band and the overload ceiling), accurate speed control, flat frequency response within the frequency range of interest, and low distortion, flut-

Fig. 4.4 Photograph of narrow band sound and vibration analyzer. (*General Radio Co.*)

ter, and skew.   For use in the field, the tape recorder must also be rugged, lightweight, and battery-powered.

Portable sound-pressure-level calibrators for microphones are of either the loudspeaker or the pistonphone type.   The loudspeaker type consists of a battery-operated oscillator and a small loudspeaker.   Known sound-pressure levels at several frequencies (e.g., 125, 250, 500, 1,000, and 2,000 Hz) are produced at the diaphragm of the microphone, thereby permitting overall calibration of the sound-level meter and tape recorder.   A pistonphone calibrator (see Sec. 8.2) consists of an air cavity, one side of which couples tightly to the microphone and the other side of which is formed by two pistons driven by a cam mounted on the shaft of an electric motor.   The alternating movement of the pistons, which varies the volume of the cavity, is accurately known and, thus, produces an accurately known alternation in the instantaneous pressure at the microphone diaphragm.   A pistonphone usually provides calibration at only one frequency (for example, 250 Hz).

Portable vibration calibrators for vibration-measuring accelerometers usually provide calibration at an acceleration of 1 g at 100 Hz.   Photographs of a pistonphone calibrator, a loudspeaker calibrator, and an accelerometer calibrator are presented in Fig. 4.5.

4.2   Checklists for Field Measurements

Prior to a field trip, the acoustical engineer should carefully consider the purpose of his trip and then choose his measurement system accordingly. He must know the environment in which he is to make measurements and the general properties of the noise or vibration source.   He should prepare checklists and data sheets for equipment and environmental parameters as well as for the noise and vibration data.

4.2.1 Initial Considerations   To understand the complete nature of the problem, the engineer should first determine such information as the following:

1. Why are the measurements to be made?   Hearing-damage considerations, community annoyance, etc.

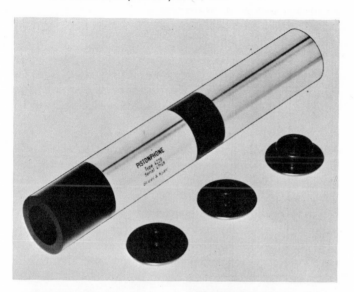

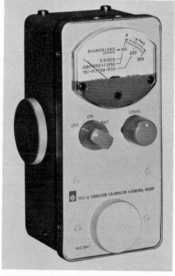

Fig. 4.5 *Upper:* Pistonphone. (*Brüel & Kjaer.*) *Lower left:* Microphone calibrator. (*General Radio Co.*) *Lower right:* Vibration calibrator. (*General Radio Co.*)

2. Where are they to be made?

3. Are there any abnormal environmental conditions?

4. What are the sources of noise or vibration?

5. What are the operating characteristics of the source?

6. When are the noise sources operating?

7. What is the physical size of the source?

8. What are the directional characteristics of the source?

9. What is the character of the noise?

10. What is the temporal pattern of the noise?

The more that is learned beforehand, the more likely the engineer is to make an optimum selection of equipment for the field trip and the better informed he will be concerning the type of measurements to be made.

4.2.2 Equipment Checklist   Once he understands the problem, the engineer can decide what data to take. If the noise is broadband steady noise, most likely a frequency analysis using an SLM with an octave filter set will be sufficient to describe the problem. However, if pure tones are also present in the noise or vibration, it may be necessary to use a narrowband analyzer or to tape-record the signal for later analysis and recording on a strip-chart graphic-level recorder in the laboratory. If the noise or vibration is intermittent, such as with an airplane flyover or a subway passing, it may be necessary to tape-record the signal for future analysis in the laboratory.

After choosing the equipment, the engineer must make certain that he understands the principle of operation of the component instruments. He must know when the equipment is presenting erroneous information and when it is presenting reliable data. The careful operator reads the instruction book and other pertinent literature. He *must* know the answers to:

1. What is the principle of operation of the instrument?

2. Is there danger of damage to the instrument from signal overload, from too much heat, from twisting something too far or too hard?

3. For the complete instrumentation system, what are the lowest and highest sound-pressure levels or acceleration levels measurable in each frequency band and what range of frequencies can be read correctly?

4. How does using an extension cable with the system alter the readings?

5. What is the inherent accuracy of the equipment?

6. How is the accuracy of the calibration of the instrument maintained and checked? (Calibration procedures should be followed meticulously.)

7. Does the meter read the rms, the average, or the peak levels of a complex sound wave? (See Chap. 5.)

8. Will connecting headphones, an analyzer, or a recorder to the output of the instrument load down its output?

9. What is the best way to set gains (on the SLM and the recorder) to get maximum signal-to-noise ratio with assurance that the peak levels of the noise or vibration signal are not being clipped?

10. What are the characteristics of the filters in the analyzer?

11. Does the instrument read differently when it is handheld, when it is placed on a stand, or when the microphone is mounted separately and operated with an extension cable? Which reading is most nearly correct? Are the requirements of the job such that the inaccuracies can be tolerated?

12. What is the electrical noise floor in each frequency band of the equipment?

4.2.3 Operational Checklist

Although the experienced acoustical engineer may not use a written checklist as an aid in obtaining correct data, he automatically follows some routine at the jobsite before making measurements. His routine may include the following items:

1. Perform a visual inspection of the instrument to ascertain that it has not been damaged in transit.

2. Recheck the batteries (they have, of course, been checked before going into the field) to ensure that there is sufficient power for making accurate measurements.

3. Place the acoustic calibrator over the microphone and adjust the gain of the SLM to give the calibrated signal. When placing a close-fitting calibrator such as the Brüel and Kjaer pistonphone over the condenser microphone, be careful not to exert so much pressure that the diaphragm of the microphone will be stretched. The pistonphone should be removed slowly and carefully after calibration. After placing the calibrator over the microphone, compare this gain setting with the gain setting obtained in the laboratory to check whether or not the microphone has been damaged in transit. If the gain settings do not agree within ±0.5 dB use another microphone.

4. Measure the electrical noise floor of the instrumentation (transducer, SLM, filter, or tape recorder) to ascertain a lower limit on the level of the signals that can be analyzed accurately. The noise floor of the SLM can be analyzed in octave bands by replacing the transducer with an equivalent electrical impedance. Such a measurement does not yield information concerning the noise of the microphone. The only way to obtain information about a noisy microphone is to incorporate the microphone into the electrical noise floor measurement. When in-

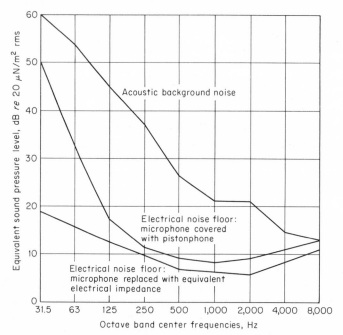

Fig. 4.6   Example of electrical noise floor.

corporated into the system, however, the microphone must be shielded from the acoustic background noise levels present in the space in which one is measuring. To check the electrical noise floor of the complete system at high frequencies, place the calibrator over the microphone. In this position, the acoustic calibrator, especially a pistonphone, provides good isolation at high frequencies. A typical plot of the equivalent sound-pressure level for one system with a pistonphone covering the microphone is shown in Fig. 4.6. The low-frequency response is limited by leakage past the seal around the pistonphone. Also presented in this figure are plots of the acoustic background noise level in the space and the electrical noise level for the sound-level meter with the microphone having been replaced with an equivalent electrical impedance properly shielded.

By monitoring the output of the sound-level meter with a high-quality set of headphones and listening for noise or "popping" sounds in the octave band signals, one can detect erratic behavior of the system.

4.2.4 Environmental Considerations   When the engineer has checked his equipment, found it to be in good operating condition, and knows how to use it properly, he may then consider the problem of obtaining meaningful data. Good data are the result of careful planning.

Of major importance is the influence of external factors upon the measurements, i.e., how the environment affects the readings.

Barometric Pressure. Barometric pressure can affect the calibration of the instrument. A pistonphone that generates a sound-pressure level of 124 dB at an atmospheric pressure of 760 mm Hg would generate a sound-pressure level of 123 dB at 677 mm Hg.

Temperature. Typically the temperature-sensitive elements in the field measuring system are the batteries and the microphone. As the temperature decreases, the life of the batteries decreases. The engineer should know how to interpret the reading of the battery meter; i.e., he should know how much longer reliable measurements can be made with the same batteries. Although a problem at all temperatures, battery life becomes especially important in subzero temperatures.

Some microphones are practically insensitive to temperature; the best grade of condenser microphone typically retains its calibration within 0.5 dB over an operating range of $-40°$ to $150°C$ ($-40°$ to $300°F$). The ceramic piezoelectric microphone is also quite stable with temperature over the range of $-40°$ to $60°C$ ($-40°$ to $140°F$). However, a Rochelle-salt microphone is very sensitive to temperature.[4]

Humidity. Certain types of microphones, in particular condenser microphones, are sensitive to humidity. If moisture has entered and condensed in the space between the cartridge electrode and the diaphragm of a condenser microphone (usually as a result of large variations in temperatures or altitude), a "cracking" or "popping" noise may be observed in the electrical output. The condensation can be dried out by placing the microphone under a light bulb for 5 to 10 min. In Rochelle-salt microphones of early design, the crystal would dissolve if the humidity were too high. Current designs of the Rochelle-salt microphone generally may be exposed to high humidities for periods of several days without permanent damage to the crystal. The modern ceramic piezoelectric microphone may be used in relative humidities of 0 to 100%.

Directivity. (See Chap. 3.) All microphones exhibit directivity effects at high frequencies (the larger the diaphragm, the lower the frequency at which the effects of directivity begin). When performing measurements, the engineer must consider these effects. If measuring a stationary noise source in a free field, he should position the microphone so that optimum free-field response is obtained. For example, the free-field response of a microphone may, because of its design, be flat at 0° sound incidence [sound wave incident perpendicular (normal) to the diaphragm]. If so, the diaphragm of such a microphone should face the source. If, on the other hand, the free-field response of the microphone, again because of its design, is flat at 90° sound incidence, the microphone should be oriented so that sound waves travel across the diaphragm at grazing incidence. If the noise source is moving, the microphone should be oriented so that the directivity effect is constant throughout.

Reflections. Usually objects will be located in the space where measurements are to be made. Should one or more of an object's dimensions be comparable with the wavelength of noise being measured, then the sound field will be affected by the presence of the object. Often, the effect of all nearby objects is required for the purposes of the measurements; at other times, effects of nearby objects may be highly undesirable. One way to minimize interference effects from nearby objects is to place the microphone close to the noise source or to place sound-absorbing material over those objects which are reflecting the noise.

Background Noise. To determine whether the background noise level in a space influences the sound-pressure levels, one should turn off the noise source and measure the background noise level (e.g., distant traffic, air handling equipment, etc.). This measurement should be made with the same analysis system that was used when the noise source was on. If in any frequency band the difference between the background noise level and the source noise level is greater than 10 dB, the background levels will not significantly affect measurement of the source noise. However, if the difference is less than 10 dB, the measured noise levels must be corrected to obtain the level of the source. A list of corrections is given in Table 4.3.[5] For example, if the reading is 100 dB with the source on and 93 dB with it off, the correct reading for the source is 99 dB. If the difference is less than 3 dB or if the ambient sound pressure and the source sound pressure are known to be coherent, one cannot determine with adequate accuracy the noise level attributable to the noise source alone. If, however, the background

TABLE 4.3 Correction for Background Noise to Be Applied Separately for Each Band

dB difference between total noise level and background noise level	dB to be subtracted from total noise level to get the noise level owing to the source
8–10	0.5
6–8	1.0
4.5–6	1.5
4–4.5	2.0
3.5	2.5
3	3.0

noise level can be lowered (e.g., by turning off the mechanical system or by making measurements at night), this latter difficulty may be surmounted.   If the background noise level cannot be lowered, then moving the microphone closer to the source may increase the difference between the background noise levels and the source noise levels.

Microphonics.   When measuring high sound-pressure levels (above about 120 dB octave band), take special precautions that transmission of mechanical vibration from the instrument housing to the microphone or the internal electronic components or both does not generate "microphonics" in the measurement system.   To reduce "microphonic" effects, remove as much instrumentation as possible from exposure to the high noise levels and place the remaining equipment on resilient mountings.

When the microphone is remotely located from the readout device, it is essential to minimize the noise generated by the cable itself.   Relative motion of the wires in the cable with respect to the dielectric results in the generation of noise.   In some cables a conducting coating is used on the dielectric to reduce the noise generated.   At low frequency, the cable noise problem can become acute; therefore, special low-noise cables are often used to connect accelerometers to preamplifiers (see Chap. 3).   Even then, the cable usually increases the electrical noise floor of the system.   Whatever cables are used, they should be thoroughly and regularly checked for noise generation.

Wind Noise.   When air blows by a microphone, turbulence develops on the downstream side and causes fluctuations in the ambient pressure which then generate noise.   The magnitude of the noise increases with an increase in wind speed, as shown in Fig. 4.7.[6]   The top curve of Fig. 4.7 represents the noise level in decibels resulting from wind flowing past a 1-in.-diameter condenser microphone as a function of wind speed in km/hr and mph.   The lower curve represents the wind noise after a spherical windscreen (about $4\frac{1}{2}$ in. in diameter) has been attached to the microphone.   With the windscreen, the airflow also generates noise, but the turbulence is located far enough away from the microphone so that the noise is attenuated by the time it reaches the diaphragm.   A photograph of this type of windscreen is shown in Fig. 4.8.   A frequency spectrum of the wind noise as measured with and without the windscreen is shown in Fig. 4.9 for a wind speed of 30 km/hr (18.6 mph).[6]

Knowing the *turbulent noise reduction* for a given microphone-windscreen combination, one may determine whether or not the measured noise level is caused by the wind or comes from the noise source.   To make this determination, measure the noise level with and without the windscreen.   If no change occurs in the noise level when the windscreen is used, the desired noise level was measured both before and after the windscreen was attached.   If the change in noise level when the wind-

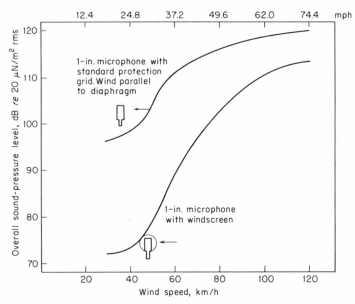

Fig. 4.7 Wind noise as a function of wind speed in the range 20 Hz to 20 kHz. (*Brüel & Kjaer Technical Review, 1966.*)

screen is attached is equal to the experimentally determined turbulence noise reduction for the windscreen, then one concludes that wind noise has dominated in both measurements. If the change in level when the windscreen is used is smaller than the turbulence noise reduction, then the measurement without the windscreen determines wind noise levels and the measurement with the windscreen determines the levels of the noise source.

Windscreens are used primarily for measurements out-of-doors where the wind direction usually is not constant. In ventilating ducts, where the direction of airflow is constant, nose cones are usually attached to the microphone.

RF Pickup. Occasionally, when long runs of cable are used between the microphone power supply and the sound-level meter, the cable serves as an RF antenna so that the sound-level meter may pick up radio stations. This situation can be avoided by using balanced signal lines.

Instrument Stray Pickup. One can usually detect other environmental influences such as electrostatic and magnetic fields by comparing the direct airborne noise with sound that is reproduced in the headphones. Electrostatic and magnetic pickup will usually appear as tones or static not present in the original signal. To minimize magnetic pickup, select the proper orientation of the instrument with respect to the magnetic field.

4.3   Measurement Records

In the matter of keeping records, the novice is not the only one prone to mistakes.   In fact, the more experienced engineer may very easily become careless and record data illegibly or perhaps even lose it.   Although a standard form is slightly more difficult to fill out, it serves the purpose of ensuring that *all* data are recorded and that meaningful permanent records can be kept by a group of people working in one organization.   The following list presents important items that should be included on a complete standard form.

1. Types, models, serial numbers, or other identification characteristics for *all* instrumentation and equipment.

2. Detailed description of the area in which measurements are made.

3. Detailed description of primary noise source including dimensions, type of mounting, location within space, nameplate data, owner's tag number, and other pertinent facts such as speed and power rating at the time of measurement.

4. Description of secondary noise sources including location, type, kinds of operation.

5. Location of engineers, observers (including names), workers, if any, during the measurements.

6. Measurement positions including the orientation of the microphone diaphragm relative to the direction of the source.

7. Barometric pressure, temperature, wind velocity (speed and direction), and humidity, if appropriate.

8. Results of calibration and operational tests.

9. Measured frequency-band levels at each microphone position.

10. Measured frequency-band background noise levels.

11. Date and time.

A form incorporating many of these items is shown in Fig. 4.10.

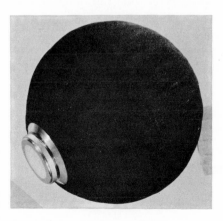

Fig. 4.8 Windscreen. (*Brüel & Kjaer.*) Foamed polyvinyl windscreens of about the same diameter and performance are available from other manufacturers. Diameter: 4.5 in. (11.5 cm).

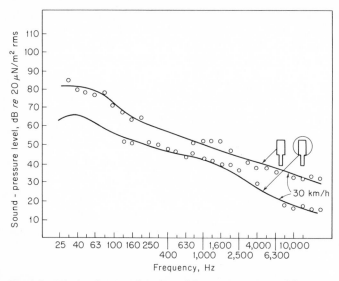

Fig. 4.9 Wind noise as a function of frequency measured in one-third octaves, with the wind direction parallel to the membrane. *Upper curve:* Without windscreen. *Lower curve:* With windscreen. (*Brüel & Kjaer Technical Review, 1966.*)

4.4 Measurement of Steady Noise

One method of reading the sound-level meter is to note and record the central tendency of the meter fluctuations and their range in decibels. If the range of the meter fluctuations on the "slow" setting is less than 6 dB, the noise may be classified as steady noise. If the noise level in a band is about 5 dB greater than the noise levels in the adjacent bands, an audible discrete tone or narrow band of noise probably exists within the measurement band. By feeding the output signal from the sound-level meter to a portable narrowband analyzer, one can determine in the field the frequency of the tone. Such a determination facilitates correlating the discrete tone with its source.

The generation of discrete tones by a noise source often results in standing-wave patterns or large spatial variations in the sound-pressure field. In addition, sources of discrete tones are often directional at the frequency of the discrete tone. To determine the average noise level, move the microphone over a distance of at least one-half the wavelength of the discrete tone and, with the meter setting on "slow," record the central tendency of the meter.

4.4.1 Fluctuating Noise (Less Than 10 dB Range) When measuring noise that fluctuates over a range of less than about 3 dB, it is accepted practice to observe the meter for approximately 10 sec and record

a noise level representing the central tendency (average) of the needle movement. If the fluctuations are greater than 3 dB but less than 10 dB an "average" according to Figs. 5.10 and 5.11 is recommended. For a more accurate statistical analysis,[7] observe the meter deflection for about 5 sec and record both the "average" and the range in decibels of the meter fluctuations. Additional observations of the "average" should then be recorded until the number of observations is somewhat greater than the range in decibels of the meter fluctuations. The arithmetic mean of the "average" readings should be calculated and used as the noise level for that particular position. A more detailed statistical analysis can be performed by playing back a calibrated tape-recorded sample of the noise through appropriate filters either to a statistical-distribution or to a power-spectral-density analyzer (see Chap. 5).

4.4.2 Ambient or Background Noise Measurements   When measuring ambient noise (the total noise associated with a given environment and usually comprising sounds from many sources both near and far) or background noise (the noise without a particular source in operation), the "slow" meter setting of the sound-level meter is generally used in conjunction with the procedures outlined above.

When measuring ambient noise levels for determining hearing-damage hazard, one of the microphone positions should be as close as

SOUND-PRESSURE-LEVEL MEASUREMENTS

DATE _____ SHEET___ OF____ PROJECT# _____	ENGINEER _____

CLIENT		PRIMARY NOISE SOURCE: _____
PROJECT		MANUFACTURER'S NAME AND MACHINE DESIGNATION:___
OBSERVERS		OWNER'S DESIGNATION : _____
INSTRUMENTATION	SLM: TYPE SER.#	OPERATING CONDITIONS:_____
	TRANSDUCER: TYPE SER.#	
	FILTER ANALYZER: TYPE SER.#	SECONDARY NOISE SOURCES:_____
	CALIBRATOR: TYPE SER.#	MANUFACTURERS NAMES AND MACHINE DESIGNATIONS: __

TIME	CALIBRA-TION dB	°F DB	°F WB	% RH	mm Hg	WIND mph	WIND DIRECTION	OWNER'S DESIGNATION:_____
								OPERATING CONDITIONS: _____

TEST NO.	TIME HRS.	POSITION	CONDITIONS	SOUND-PRESSURE LEVEL: dB re 20 $\mu N/m^2$ rms									
				A WEIGHTED LEVEL	OVER-ALL LEVEL	OCTAVE BAND CENTER FREQUENCY: Hz							
						31.5	63	125	250	500	1K	2K	4K	8K

COMMENTS:_____

Fig. 4.10 Suggested form for recording sound-pressure-level measurements.

possible to the normal location of the ears of the listener for whose benefit the measurement is being made, e.g., a factory worker. The location
will vary depending upon the activity of the listener.

The number of measurements required will depend upon the spatial
distribution of the sound field and the purpose of the measurements.
If the ambient noise level within a community is being established, it
might be appropriate to determine the statistical properties of the noise,
both the variations with time and with location. The number of measurement positions would depend upon the size of the community and
the distribution of the noise sources within the community.

4.4.3 Source Noise Measurement   (See Chap. 6.) The manner in
which noise is radiated by a source is directly related to its dimensions.
For example, when the wavelength of the radiated noise is large compared to the dimensions of the source, the source usually radiates noise
uniformly in all directions. When the wavelength of the noise is small
compared to the dimensions of the source, the noise may be radiated in
some directions more than in others. If the size of the source or the frequency range of interest is such that the noise is radiated uniformly, a
measurement a short distance away from each of four sides of the source
is usually sufficient. However, if the frequency range of interest is the
region toward which the source is directional, it may be necessary to
perform measurements at many positions.

In plant measurements, usually the most important source measurement location is the machine operator's position. If the operator stands,
a microphone position of 1.5 m above the floor is preferred. For sitting
positions, 1.1 m above the floor is preferred, while 0.6 m is preferred for
lying down positions. Obviously, other heights may be dictated by
listener location.

For additional measurements,[8] it is recommended that the engineer
mark off an imaginary rectangle within which the source (machine) fits.
Minor projections of the machine, which are not major noise generators,
may cross this rectangle. If only four measurements are to be performed, the microphone should be located midway on each side at a
distance of 1.0 m from the rectangle (provided the largest source dimension is at least 0.25 m) and at a height of 1.5 m (provided the height of
the source is at least 3 m). If the largest source dimension is less than
0.25 m, then the distance from the rectangle to the measurement
position should be four times the largest source dimension. If the
height of the source is less than 3 m, then the measurement height should
be one-half the height of the source. Additional measurements 1.0 m
from the rectangle, at a height of 1.5 m and at a suitable interval from
the adjacent positions (for example, 2.0 m), may be performed as required.

For more distant measurements outdoors and usually for larger sources, it is recommended that the microphone reference line be a circle with a radius of at least two times the largest source dimension. A minimum of four measurements (0°, 90°, 180°, and 270°) should be made. Other measurements at equal intervals (e.g., every 30°) are useful for directional sources. The central tendency of the meter deflection with slow damping on the precision sound-level meter may be recorded as the average level. Where interference patterns or standing waves exist, it is preferable to perform a spatial average, as described earlier.

4.5 Measurement of Nonsteady Noise

A noise with a level that fluctuates over a range greater than 10 dB during observation should be classified as nonsteady, either impulsive or fluctuating. An impulsive noise having a pulse repetition rate of greater than 10 pulses per second may usually be classified as a quasi-steady noise, and the procedures outlined in the section on measurements of steady noise may be used. If the repetition rate is less than 10 pulses per second, however, the noise should be analyzed either by the techniques described below or those described in Chap. 5.

4.5.1 Fluctuating or Intermittent Noises   Cyclically or continuously fluctuating noises or intermittent noises often must be measured with a sound level meter. The source of an intermittent noise is usually "on" (operating) for a period of time longer than the integration time of the ear (about 200 milliseconds). While "on," the noise may be constant in level or may fluctuate in level. If the noise is steady in level during the "on" time, one should record the central tendency of the "slow" meter setting during both the "on" and "off" times. For intermittent noises that fluctuate continuously during the "on" time, one should record the maximum and minimum levels rather than attempt to record a central tendency. If the noise is cyclic in level, fluctuating between two or more well-defined levels, one may either record the central tendency at each level or tape-record the signal for analysis in the laboratory. If the noise fluctuates continuously in level, varying between an undefined maximum and minimum (such as heavy traffic on nearby thoroughfares), a statistical-distribution analyzer should be used.

4.5.2 Impulsive Noise   In the measurement of impulses the parameters of interest are the unfiltered peak-pressure level, the pulse rise time, the duration of the pressure wave, and the frequency spectrum of the pulse. The primary limitation to impulse measurements is the ability of the electronic instrumentation to respond accurately to the

Fig. 4.11 Impact noise analyzer. (*General Radio Co.*)

pressure pulse.   Instrumentation for impulse measurements [9] should have, among other features, a good phase response, a flat frequency response (100 Hz to 70 kHz *), and a rise time of less than 10 microseconds.

If the engineer feeds the output of the sound-level meter to a calibrated oscilloscope, he may use Polaroid camera photographs to measure the peak-pressure level and the duration.   If only the peak-pressure level is required, a true-peak-reading meter can be used in place of the oscilloscope.   To meet tentative regulations of the U.S. Bureau of Labor Standards, the indicating meter would have to have a rise time not greater than 50 microseconds for a response within 1 dB of the peak value for a rectangular-step applied voltage.   Figure 4.11 is a photograph of an impact noise analyzer meeting these requirements.   The use of a peak-reading meter for evaluation of hearing-damage risk is discussed in Chap. 17.

4.6   Measurement of Reverberation Time

A thorough discussion of reverberation time is presented in Chaps. 8 and 9, which deal with sound in small and large enclosures.   For the present, the *reverberation time* of a space may be defined as the time required for the sound pressure within the space to decay to $\frac{1}{1000}$ of its original value; thus, it is the time required for the sound-pressure level to decay 60 dB.   In practice, the reverberation time may be either the 60-dB extent of the actual sound-decay curve, or, if the decay is multisloped, it may be the 60-dB extension of a portion of the actual decay curve.   Currently two basic measurement methods are used to determine reverberation time: (1) the interrupted noise (source) method, using either a graphic-level recorder (GLR) or a decay-rate meter, and

* For hearing-conservation considerations, an upper frequency limit of 20 kHz is considered sufficient.

(2) the Schroeder-Kuttruff integrated impulse method (see Chap. 9). The basic components of each method include a sound source, a microphone, a filter, and a readout device. Usually, a magnetic tape recorder is also required. The GLR method, in connection with an impulsive source, is usually used in making determinations for purposes of noise control. In Fig. 4.12 a schematic for the GLR system is presented. Impulsive sources include 32-cal starting pistols, 10-gauge-shell yachting cannons, and toy rubber balloons. The bursting of a toy rubber balloon, measuring about 14 by 20 in. when fully inflated, generates adequately high peak octave-band sound-pressure levels in the frequency range of 100 to 10,000 Hz.[10] The balloons have the advantage of being easy to transport and store as well as being easy to burst. The explosion due to the firing of the cannon has the tendency to frighten observers and to unsettle dust from room surfaces and ventilation ducts.

As shown in Fig. 4.12, the electrical signal from the microphone representing the impulsive sound is fed through the sound-level meter to a high-quality tape recorder. Later, in the laboratory, the tape-recorded sample is played back; the electrical signal is filtered and fed to the graphic-level recorder. Experience indicates that a graphic-level-recorder writing speed of 300 mm/sec and a paper speed of 30 mm/sec, with a logarithmic potentiometer having a range of 50 dB, proves adequate for most reverberation time measurements. However, for maximum accuracy, the paper speed should be adjusted so that the decay curve lies at an angle of approximately 45° with respect to the base line.

The first step in evaluating decay curves obtained from the graphic-level recorder is the drawing of an idealized slope through the actual decay curve. *The slope chosen should disregard the effects of background noise at the bottom of the slope and the transient effects of "rounding off" at the*

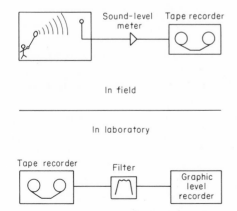

In field

In laboratory

Fig. 4.12 Measurement of reverberation time using tape recorder, filter, and graphic-level recorder.

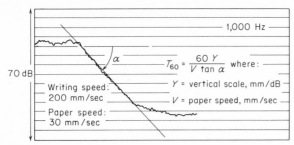

Fig. 4.13 Method for calculating the reverberation time from a sound-decay curve by graphic measurement.

top of the slope. The reverberation time is calculated from the sound-decay curve according to the formula

$$T_{60} \text{ (reverberation time)} = \frac{60\ Y}{V \tan \alpha} \tag{4.1}$$

where Y = logarithmic vertical scale, mm/dB
V = speed of paper, mm/sec
α = angle the decay curve makes with the time axis
This method for determining the reverberation time is shown in Fig. 4.13. Protractors are available which permit the reverberation time results to be read directly from the protractor when it covers the slope.

4.7 Measurement of Noise Reduction

Field transmission loss of a partition is defined as 10 $\log_{10}$ of the ratio of the sound power incident directly upon the partition on the source side to the sound power radiated directly from the quiet side of the partition into the receiving space or, for a two-room situation

$$FR = \bar{L}_1 - \bar{L}_2 + 10\ \log\left(\frac{S_{12}}{A_2}\right) \tag{4.2}$$

where FR = field transmission loss of test partition at a given frequency, dB
$\bar{L}_1$ = average sound-pressure level in the source room at the given frequency, dB *re* 2×10^{-5} N/m^2
$\bar{L}_2$ = average sound-pressure level in receiving room at the given frequency, dB *re* 2×10^{-5} N/m^2
S_{12} = area of test partition, m (or ft)
A_2 = $S\bar{a}_{\text{Sab}}$, total sound absorption in receiving room, expressed in the same units as S_{12} [see Eqs. (9.1) and (9.2)]
When field measurements of transmission loss are being made to determine whether or not the installed partition meets specified values

of transmission loss, it is preferable for FR to be measured in accordance with the American Society for Testing and Materials (ASTM) Standard for Field Transmission Loss (E336 − 67T).[11]

Often, however, the acoustical engineer is primarily concerned with determining the isolation between one space and another; in such cases, a measurement of the noise reduction ($L_{NR} = \bar{L}_1 - \bar{L}_2$) is appropriate.

The average sound-pressure level is 10 times the logarithm of the ratio of the space-average mean-square sound pressure to the square of the reference sound pressure. Thus

$$\bar{L} = 10 \log_{10} \frac{p_1^2 + p_2^2 + \cdots + p_n^2}{np_{ref}^2} \qquad \text{dB} \qquad (4.3)$$

where $p_1, p_2, \ldots, p_n$ = rms sound pressures at the n different measurement positions in the room

p_{ref} = reference sound pressure, 2×10^{-5} N/m²

n = number of measurement positions

In measurements of the noise reduction, the sound fields in rooms should be reasonably diffuse. The sizes of the source room and the receiving room determine to a large extent (for a given measurement bandwidth) the lowest frequency at which the sound fields will be adequately diffuse; the larger the room, the lower the limiting frequency. In a field measurement where the room size is predetermined, the frequency below which a diffuse field does not exist may limit the low-frequency range of measurement. The lower limit of frequency for which this will occur, in one-third-octave or octave bands, may be determined from the volume of the smaller of the two test rooms from Eq. (6.9), which says that $V > 4\lambda_g^3$, where λ_g is the geometric-mean wavelength of the lowest one-third-octave band of interest. In addition, the average absorption coefficient (average for all room surfaces) $\bar{\alpha}_{Sab}$ should ideally be about 0.15, but not greater than 0.2 at each test frequency for either room. If either of these two restrictions is not satisfied, the reader should refer to Appendix B of ASTM E336 − 67T [11] for measurement procedures.

The measurement positions should be located no less than one-third wavelength (at the mean frequency of the test frequency band) from room boundaries or any other large reflecting surface. Also, they should be separated by at least one-half wavelength from each other. No measurement positions in the source room should be in the direct field of the sound source. Thus, for many field measurements of noise reduction, the data may be restricted to the frequency bands with center frequencies of 125 Hz or more.

If large diffusing (rotating) vanes and continuous-spectrum, one-third-octave bands of noise are used, a minimum of five microphone positions should be employed. If very narrow bands (or pure tones) and no diffusing vanes are used, the measurements should be made at enough positions that the result falls within the desired accuracy (see Fig. 6.5).

In summary, to determine accurately the noise reduction or the transmission loss of the common wall between two spaces, we

■ Check the lower-frequency limit as a function of the size of the space; that is, $\lambda_g^3 < 0.25 \, V$, where V is the volume of the smaller room
■ Check the average sound-absorption coefficient $\bar{\alpha}_{Sab}$ to be sure it is less than 0.20, and ideally about 0.15
■ Measure the sound-pressure level at a sufficient number of locations (as required by Fig. 6.5) in both rooms
■ Calculate $\bar{L}$ for each room
■ Calculate $L_{NR} = \bar{L}_1 - \bar{L}_2$, or FR from Eq. (4.2)

4.8 Accessories

The advantage of having various accessories available depends upon the purpose of the measurements, the complexity of the measurements, and the environment in which the measurements are to be made. Some instruments that have been useful on field trips are the following: barometer, camera, manometer, measurement rule, sling psychrometer, stethoscope, stopwatch, stroboscope, tachometer, thermometer, volt-ohm meter and velometer (wind meter). Other valuable accessories include: balloons, spare batteries, spare cables and accessories, clipboard, crayon (for marking positions), Duxseal, hearing protectors, headphones, screwdrivers, pliers, soldering iron and solder, flashlight, rubber sheet to protect instruments against rain, handwarmers, and overshoes.

REFERENCES

1. *Intern. Electrotech. Comm. Publ.* 179, 1965, International Electrotechnical Commission, 1430 Broadway, New York, N.Y. 10018.
2. "Proposed USA Standard for Sound Level Meters," a revision of S1.4, 1961, American National Standards Institute, Inc., 1430 Broadway, New York, N.Y. 10018.
3. ANSI S1.6, 1967, "Preferred Frequencies and Band Numbers for Acoustical Measurements," American National Standards Institute, Inc., 1430 Broadway, New York, N.Y. 10018.
4. Beranek, L. L., *Acoustics*, McGraw-Hill, 1954.
5. Beranek, L. L., *Noise Reduction*, McGraw-Hill, 1960.
6. F. Sköde, "Windscreening of Outdoor Microphones," *Brüel and Kjaer Tech. Rev.* 1, 1966.
7. S3-W-50, "Measurement of Community Noise," third draft Nov. 11, 1969, American National Standards Institute, Inc., 1430 Broadway, New York, N.Y. 10018.

8. S1/168, "Methods for the Measurement of Sound Pressure Levels," fourth draft, Sept. 10, 1969, American National Standards Institute, Inc., 1430 Broadway, New York, N.Y. 10018.

9. R. Ross, A. Coles, Georges R. Garinther, David C. Hodge, and Christopher G. Rice, "Hazardous Exposure to Impulse Noise," *J. Acoust. Soc. Am.*, vol. 43, no. 2, p. 341, 1968.

10. B. G. Watters, "The Sound of a Bursting Red Balloon," *Sound*, vol. 2, no. 2, pp. 8–14, 1963.

11. ASTM E336–67T, 1967, "Tentative Recommended Practice for Measurement of Airborne Sound Insulation in Buildings," American Society for Testing and Materials, 1916 Race Street, Philadelphia, Pa. 19103.

Chapter Five

Data Analysis *

DOUGLAS W. STEELE

In a broad sense, data analysis is the extraction of useful information from data. Data analysis may also be thought of as the estimation of certain properties or measures of a *signal*. These measures may be derived from the total signal or from filtered versions of the signal. The latter yield measures that are a function of frequency.

In sound and vibration, one useful measure of a steady-state signal is the mean-square value. The logarithm of this value is generally taken to yield a *level*. Another measure is a determination of whether the signal is essentially periodic (repeated) or random. For periodic signals the repetition rate, or its inverse the period, is a measure. For random signals, probabilistic parameters are measures. For signals that are intermittent, the duration of the on and off portions are useful measures. For short transient signals the peak value is also a useful measure. In certain situations (like auditoriums) a 'sound level does not commence or cease abruptly, but rises and decays exponentially, so that useful measures are the time rates of rise and decay.

In this chapter we shall think of the signal as a voltage, such as would

* The author wishes to express his thanks to Dr. Denis Noiseux for the many hours of discussions about the material in this chapter. Portions of the chapter have been adapted from material in an unpublished manuscript by Dr. Noiseux.

be found at the output of a transducer. This voltage is assumed to be linearly related to some physical phenomenon such as pressure, displacement, velocity, or acceleration. Frequently, the output of a transducer is recorded on a magnetic tape recorder for later analysis, although real-time analyses are becoming more common. There is no loss of generality in restricting this chapter to voltage signals, because even a sequence of numbers may be transformed into a sequence of voltage impulses with proportional magnitudes.

A measure basic to nearly all signal analysis in sound and vibration is *spectral analysis*. Briefly, spectral analysis is the determination of the distribution of the energy or power in a signal as a function of frequency. This chapter is structured around this topic, although many other aspects of data analysis are introduced and treated briefly. The cursory coverage of these topics is an attempt to acquaint the reader with pertinent portions of the larger subjects of signal processing, linear-system theory, and probability and statistics. All topics covered may be useful to a person who wishes to determine the energy or power spectrum of a signal.

This chapter is divided into three main sections: spectral analysis; practical problems of filtering, measuring, and displaying the signals and spectra; and some of the topics that provide a rigorous mathematical foundation to the concepts of spectral analysis. The latter include the Fourier transform, Parseval's theorem (conservation of energy and power), aliasing and quantization in sampled-data analysis, and a way of considering the uncertainty in the levels of an estimate of a power spectrum.

5.1   Spectral Analysis: The General Problem

A typical data analysis system for obtaining the power spectral density of a signal is indicated in Fig. 5.1. The system comprises the linear functions of filtering and averaging and the nonlinear functions of detection, such as squaring, and logarithmic conversion. The input to the system is a voltage that is the analog of some physical quantity and the output is a display, such as the position of the pointer on a meter, a line on a strip chart, a numerical printout, or a spot on a cathode-ray tube.

5.1.1   Types of Signal

A. Stationary (continuous) signals
   1. Periodic
   2. Random
B. Nonstationary (transient) signals
   1. Impulsive
   2. Slowly varying

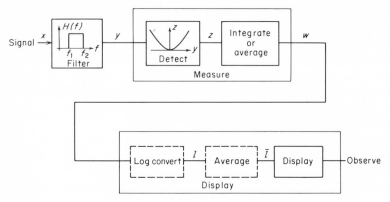

Fig. 5.1   Block diagram of a spectral analysis system showing the major functions of filtering, measuring, and displaying.

Stationary Signals. A signal that is (wide-sense) stationary has a power spectrum that does not change with time. As a practical matter, it is sufficient that it not change during the period of observation. Any signal that is periodic, such as a sine wave or a pulse train, fits this category, as does a large class of random signals such as the noise from Niagara Falls, the vibration of a ship's hull because of the flow of water along the hull, and the noise from an air-conditioning system. These signals are not strictly so but they usually are sufficiently stationary during a time interval of interest that the techniques for analysis of stationary signals may be applied to them.

Nonstationary Signals. The other major category contains nonstationary signals, also called transient signals. A subcategory of nonstationary signals includes impulsive signals which exist only for a relatively short time period. An obvious example is the sound of a gunshot which, compared to the duration of most other sounds, is very short. Another subcategory applies to slowly varying signals, that is to say, signals that have a spectrum that changes slowly. An example is the sound heard on the ground from an airplane flying overhead. Of course, the power spectrum might be determined for successive segments of the flyover signal, each of which contains a signal that is sufficiently constant to be considered a stationary signal. The power spectra of these successive samples show that the power spectrum of the flyover noise signal is nonstationary.

At the extremes, the differences among stationary, slowly varying, and impulsive signals are clear, but there are many cases for which the distinction is not clear. For example, the forces that act to create a mountain may be considered slowly varying, or even stationary on a time scale of a person's lifetime, but may well be considered impulsive on a geological time scale.

The signal processing techniques for stationary signals differ from those for transient signals. For stationary signals, the *power* spectrum is usually of interest, while for transients, the *energy* spectrum is desired. Some techniques of analysis are commonly used for periodic signals which are not suitable for nonperiodic signals. Because energy is power integrated over time, each (necessarily finite) sample of a signal has an energy spectrum. The average power spectrum may be estimated by dividing the energy spectrum by the duration of the signal sample. Conversely, the energy spectrum of an impulsive transient may be estimated by repeating the transient signal periodically (at a period larger than its duration) and using those techniques for analysis applicable to a periodic signal.

At first glance it may not be obvious that a random process should have a particular power spectrum. Later in this chapter it will be shown mathematically that for a stationary random process the power spectrum obtained by averaging the individual power spectra of a sufficiently large number of samples of the process is a good estimate of the true power spectrum of that process. The accuracy with which any of the short-time power spectra predicts the time power spectrum depends on the length of the sample and the width of the frequency band in which the analyses are made.

5.1.2 Mean-square Values of Bandpass-filtered Signals   The block diagram of Fig. 5.1 is redrawn in Fig. 5.2 with the various functions specialized for measuring the portion of the total signal power (mean-square value) that lies in the frequency band extending from f_1 to f_2 Hz. This system is further explored in the next section and a mathematical analysis is found in the third section.

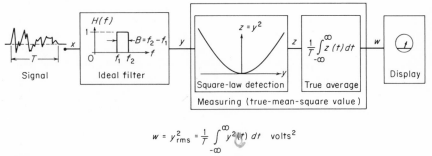

$$w = y_{rms}^2 = \frac{1}{T} \int_{-\infty}^{\infty} y^2(t) \, dt \quad \text{volts}^2$$

Fig. 5.2   Block diagram of the analysis system of Fig. 5.1 particularized for measuring that portion of the power (true-mean-square value) of a signal contained between the frequencies f_1 and f_2. The bandwidth of the ideal filter is $B = f_2 - f_1$.

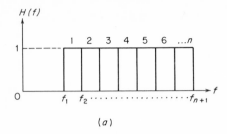

(a)

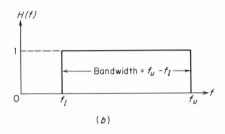

(b)

Fig. 5.3 Filter bands for the analysis of signals: (*a*) *n* narrow contiguous bands; (*b*) one band of the same width as the *n* bands in (*a*).

This signal, which is assumed to be zero except over a time interval of T sec, may be considered as a T-sec-long sample of a continuous (stationary) signal. The measuring system determines a value W, which is the mean-square value of the input signal in the band $f_2 - f_1$ Hz. W is often given units of power, which in the SI (mks) system is the watt. Frequently, the power is not needed and thus W is the mean-square value of the signal; that is, $W = y_{\text{rms}}^2$. In this case, the words *power spectrum* are still used (as though the voltage were across a resistor of 1 ohm), but the unit is volts squared.

Power in Filtered Bands. The measurement process of Fig. 5.2 is usually repeated for n filters with contiguous frequency bands, such as shown in Fig. 5.3*a*. The result of the analysis is n values of $W = y_{\text{rms}}^2$, each value being the average power or voltage squared in one of the bands. It is stated here, and is mathematically demonstrated later, that if the power values for each band measurement are added together, the resulting total value is the same as the value that would have been measured directly with the filter shown in Fig. 5.3*b*, which covers the same frequency range. That is to say, if W_{ai} is the power value for the ith filter of the set in Fig. 5.3*a* and if W_b is the power value using the filter of Fig. 5.3*b*, then

$$W_b = \sum_{i=1}^{n} W_{ai} \tag{5.1}$$

If, in the system of Fig. 5.2, $f_1 = 0$ and $f_2 = \infty$, the measured value of power would be the total power in the signal sample.

Power Spectral Density, psd.   The set of values of W measured with the filters of Fig. 5.3a, when summed, uniquely determines the value of W of the signal with the filter in Fig. 5.3b.   Under one special circumstance, the reverse determination may be made.   If the filter bandwidths in Fig. 5.3a are all equal and if it is valid to assume that the power values measured with these filters are *equal*, then

$$W_{ai} = \frac{W_b}{n} \qquad \text{for } i = 1, 2, \ldots, n \tag{5.2}$$

where $\qquad n = \dfrac{\text{bandwidth of filter in } (b)}{\text{bandwidth of a filter in set } (a)} = \dfrac{B_b}{B_a} \tag{5.3}$

For the equal-power assumption, with the aid of Eq. (5.3) we can rewrite Eq. (5.2) in the form

$$W_{ai} = \frac{W_b}{B_b/B_a}$$

or

$$\frac{W_{ai}}{B_a} = \frac{W_b}{B_b} = \text{constant} \tag{5.4}$$

If Eq. (5.4) is true for all filter bandwidths that are less than B_b, we may call the constant in Eq. (5.4) the *power spectral density* (psd).   The unit is volts^2/Hz.   For cases where W_i/B_i are not constant for all frequencies, the psd is defined as

$$\text{psd} = \lim_{B_i \to 0} \frac{W_i}{B_i} = \lim_{B_i \to 0} \frac{y_{\text{rms}}^2}{B_i} \tag{5.5}$$

Thus *in the limit as the bandwidth approaches zero, the psd is the power ($\sim$voltage^2) in the bandwidth divided by the bandwidth in hertz.*

A plot of psd values for noise signals from a stationary random process generally is a smooth, continuous function of frequency.   The psd function of a periodic function, such as a harmonic series, is not.   For a harmonic series, the signal power occurs at discrete places along the frequency scale, and the spread in power around each frequency is generally much smaller than the bandwidth of the filters in the measuring system.   In this case for any reasonable filter bandwidth, each filter will measure the total power of any frequency component that falls within its bandwidth.   Here it is not valid to estimate a psd function by dividing the measured power in the filter band by the filter bandwidth.

Energy in Bands and Energy Spectral Density. If the signal in Fig. 5.2 is a transient rather than a sample of a continuous signal, the total energy in the transient rather than the average power is a more appropriate measure of the signal. Everything said above about power and power spectral density holds if the $1/T$ factor in the third box of Fig. 5.2 is removed. Since the unit of power is the watt, the unit of energy is watt-sec (or joule) and the unit of the energy spectral density is watt-sec/Hz.

Many of the practical problems in power-spectrum measurement arise because of approximations that are used in actual analysis. Some of these approximations (which will be discussed further) are

1. Real filters do not pass only frequencies within the passband; they also pass other frequencies.

2. Approximations to the true root-mean-square value are used because the equipment employed in measurement may be simpler.

3. The signal duration over which the power spectrum is estimated is finite.

4. The filter bandwidth chosen for the analysis is too small or too large. There is a limit to the amount of information that can be extracted from a finite sample of a signal. For a fixed-duration sample of a signal, the accuracy of the psd and the bandwidth of the filter used are directly related.

5.2 Filtering

Let us now identify the important features of each block of the analysis system of Fig. 5.1. Whether any particular analysis system ought to be analog or digital is dealt with only briefly.

Filtering is the heart of spectral analysis. The filter used is generally a bandpass type which passes only those components of a signal within a specific frequency range called the passband.

5.2.1 Ideal Filters
The magnitude of the amplitude transfer function $|H(f)|$ of an ideal filter is shown in Fig. 5.3b. $|H(f)|$ is zero outside the passband and is unity within the passband. The passband shown extends from f_l to f_u. The frequencies f_u and f_l are called the upper and lower band edges, or cutoff frequencies, respectively. The bandwidth B is the difference in frequency between the upper and lower band-edge frequencies.

$$B = f_u - f_l \qquad (5.6)$$

The location of the passband is specified by its center frequency f_c which is taken as the geometric mean of f_l and f_u.

$$f_c = \sqrt{f_l f_u} \qquad (5.7)$$

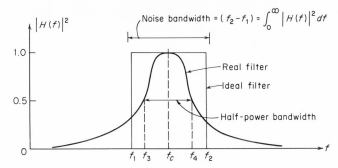

Fig. 5.4   Power transfer function $|H(f)|^2$ of a real filter and an ideal filter having the same noise bandwidth and center frequency. The half-power bandwidth is also called the -3 dB bandwidth.

For bandwidths small compared to either band-edge frequency, f_c is also approximately equal to the arithmetic mean.

5.2.2   Real Filters   The transfer function $|H(f)|$ for real filters does not go immediately to zero outside of the passband nor is it unity at all frequencies within the passband. For purposes of power-spectrum analysis the filter is still, however, characterized by its transfer function $|H(f)|$. A few more definitions are necessary to summarize important characteristics of real filters.

5.2.3   Bandwidth   While the definition of bandwidth for an ideal filter is unambiguous, there are many definitions of bandwidth for real filters. Two of the most important are the -3 dB (half-power) bandwidth and the noise bandwidth. Figure 5.4 shows an example of the power transfer function $|H(f)|^2$ of a real filter.

The *-3 dB bandwidth* $B_{-3\text{dB}}$ is the frequency difference between the points where $|H(f)|^2$ has a value of half the maximum value.

$$B_{-3\text{dB}} = f_4 - f_3 \qquad (5.8)$$

The *noise bandwidth* B_n is defined as the bandwidth of the ideal filter that would pass the same signal power as the real filter when each is driven by a stationary random-noise signal having a power-spectral-density (psd) function that is a constant. In Fig. 5.4 the ideal filter having the same noise bandwidth has the same area under the curve as does the real filter. Thus

$$B_n = f_2 - f_1 = \int_0^\infty |H(f)|^2 \, df \qquad (5.9)$$

Three more descriptors are needed to specify a real-filter transfer function: passband ripple, rolloff slope, and skirt attenuation. In Fig.

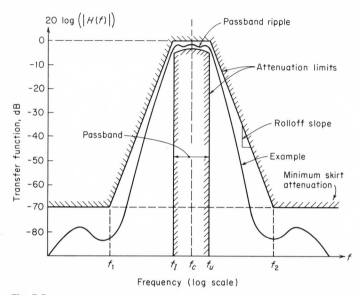

Fig. 5.5   Example of the transfer function $20 \log (|H(f)|)$ of a real filter plotted in decibels vs. frequency on a log scale.   Also plotted is a typical set of limits for an acceptable filter.

5.5 the transfer function of a real filter is plotted in decibels vs. frequency on a log scale.   *Passband ripple* is the difference in decibels between the maximum and minimum values of the transmission function within the passband.   *Rolloff slope* is the minimum average slope of the transfer function vs. frequency between the band-edge frequency and the frequency at which the ultimate skirt attenuation is reached.   *Skirt attenuation* is the minimum value of attenuation outside both the passband and rolloff regions (outside the region f_1 to f_2).

The *skirt attenuation* is of great importance in analysis, because it determines the magnitude of the weakest signal that can be measured in the passband.   For example, if in a measurement one wishes the power passed outside the passband to be 15 dB less than that passed through the passband, the skirt attenuation for a very-sharp-rolloff filter must be at least equal to that given in Table 5.1.   A stationary random noise with a constant psd function extending from 0 to 10,000 Hz is assumed. We observe from Table 5.1 that the skirt attenuation for this case must be 42 dB for a 20-Hz-bandwidth filter.

The required skirt attenuations for a sloping-spectrum noise are much greater than those of Table 5.1.

For example, if a noise slopes off at −6 dB per octave between 20 and 10,000 Hz the skirt attenuation for a 20-Hz-bandwidth filter must be 49 dB if its center frequency is at 1,000 Hz and 63 dB if at 5,000 Hz.   If the noise slope is -12 dB per octave, the required skirt attenuation is 78 dB if its center frequency is at 1,000 Hz and 116 dB if at 5,000 Hz.

5.2.4 Choice of Filters   There are two common classes of narrow-band filters used for spectral analysis: (1) the constant-bandwidth type where the noise bandwidth B_n is the same for all center frequencies f_c, and (2) the constant-percentage-bandwidth type where the noise band-width B_n is a constant percentage of the center frequency f_c. Commonly available filters of the latter type have widths of octaves, one-third octaves, one-tenth octaves, and a few percent.   The choice of which type

TABLE 5.1

Filter noise bandwidth, Hz	Minimum skirt attenuation, dB
1	55
5	48
20	42
100	35
200	32
500	28

will depend upon the type of spectrum expected and the purpose for which the measurement is made.   For example, if the psd is a monotonic function sloping upward or downward only slightly with frequency (when plotted in decibels vs. $\log f$), wideband filters such as the octave-band type are adequate for measurement.   Even if the spectrum is some-what irregular, a wideband filter set is adequate for measurement of noise to study the annoying effects of the noise.   On the other hand, if it is desired to locate the most intense noise components in a frequency spectrum, and these are pure tones, narrowband filters are required.

For octave-, one-half-octave-, and one-third-octave-band filters, ANSI S1.11 1966 provides much useful information.

5.2.5 Implementation of Filter Systems   The most common im-plementations of filter systems are

1. *Swept filter:* One filter is used and its center frequency is slowly swept over the desired frequency range.   If the filter is of the constant-percentage-bandwidth type, the bandwidth will increase in direct pro-portion to frequency.

2. *Serially stepped filter:* Similar to the above except that the center fre-quency is stepped through a series of discrete values generally such that the frequency range is covered completely with no overlaps.

3. *Parallel bank of filters:* The signal is fed simultaneously to the in-

puts of a bank of contiguous filters. The filtered output may then be processed simultaneously. This system is the same as that of (2), except that the analysis can be faster in proportion to the number of filters.

4. *Sped-up serial analysis:* It is possible to record (on tape, digital, or other memory) a signal and then to read it back n times faster. On each playback the signal is analyzed through one of the filter bands.

5. *Digital filtering and the FFT:* It is possible to digitize a signal and then to perform the filtering numerically (digitally). All comments about analog filters hold in the digital case. The fast Fourier transform (FFT) may be thought of as a method for digitally simulating a bank of filters.

5.3 Measuring

In the block diagram of Fig. 5.1, the function called *measure* consists of detection followed by averaging. The function of this block is to convert an input signal to a function that is a desired measure of the magnitude of the signal, i.e., mean square, average, and so forth.

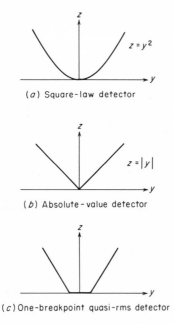

$z = y^2$

(*a*) Square-law detector

$z = |y|$

(*b*) Absolute-value detector

(*c*) One-breakpoint quasi-rms detector

Fig. 5.6 Typical detection functions used in spectral analysis.

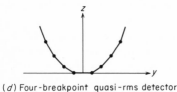

(*d*) Four-breakpoint quasi-rms detector

5.3.1 Detection   (See Fig. 5.2.) As mentioned earlier there is a sound mathematical foundation for using a *square-law*-type detector (see Fig. 5.6a) to detect a signal. Another type of detector is the *absolute-value* detector, which is an approximation to the square-law detector (Fig. 5.6b). In between these two types are two variations of *quasi-rms* detector (Fig. 5.6c and d). The latter two types are generally straight-line approximations to the square-law detector and usually have from one to four breakpoints.

5.3.2 Averaging and Integration   There are two basic types of averaging, the true average given by

$$\overline{z_t(T)} = \frac{1}{T} \int_0^T z(t) \, dt \qquad (5.10)$$

and the "running" average given by

$$\overline{z_r(T)} = \frac{1}{RC} \int_{-\infty}^T z(t) e^{-(T-t)/RC} \, dt \qquad (5.11)$$

If $RC = T$, then

$$\overline{z_r(T)} = \frac{1}{T} \int_{-\infty}^T z(t) e^{-(T-t)/T} \, dt \qquad (5.12)$$

Equations (5.10) and (5.12) may be thought of as the average of the input signal $z(t)$ weighted with a function $h(t)$. This process is illustrated in Fig. 5.7. Note that if $z(t)$ is a constant A, then both averages correctly give A as the average. Where a signal is stationary (periodic or random) both methods give correct averages if T is long enough. Where the signal is an impulsive transient or the power spectrum of the signal is changing with time, the results can differ.

The running average is also called the RC average because the usual realizations involve a resistance R and capacitance C.

If integration rather than averaging is required, either method will give the correct answer if the transient is zero outside the interval $(0,T)$ and if RC is chosen to be much greater than the duration of the transient [so that the weighting function is approximately unity during the interval $(0,T)$].

5.3.3 Differences in Measured Values with Different Detectors   Detectors other than square-law detectors are often used for economic reasons (see Fig. 5.6b to d). The measured values obtained will in general not be the same as with square-law detectors, even though they are adjusted to give the same value for one type of signal, usually a sine wave.

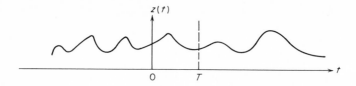

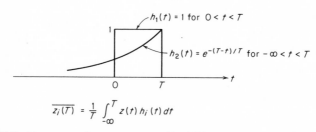

$$\overline{z_i(T)} = \frac{1}{T} \int_{-\infty}^{T} z(t)\, h_i(t)\, dt$$

Fig. 5.7 Two types of average. Use of h_1 yields a true average value of the signal $z(t)$ while use of h_2 will result in a running average. For $z(t)$ equal to a constant, the results are the same.

Let us again assume that all averaging times are long enough to obtain the correct average reading. Consider two measuring systems: (1) a root-mean-square (rms) system and (2) a mean-absolute-value (mav) system. As the names imply, the rms value is the square root of the mean of the signal squared and the mav is the mean of the absolute value of a signal. So far, we have dealt only with the mean-square value, which is proportional to power, rather than the rms value, which is proportional to the amplitude of a signal.

If we calculate the rms value and mav of a sine wave, a periodic pulse train, a square wave, and a Gaussian random signal as shown in Fig. 5.8, we see that the ratio y_{rms}/y_{mav} depends on the properties of the signal (actually on the probability distribution). Table 5.2 presents these ratios. Although the rms value of a signal is generally desired, most simple meters measure the mav. Since the sine wave is the most commonly measured signal, these simple meters are nearly always calibrated to read alike when measuring a sine wave; i.e., they indicate its rms value. This may be thought of as multiplying the mav of Fig. 5.8 by 1.11. The second column of Table 5.2 shows the ratio $y_{rms}/1.11\, y_{mav}$, for which the ratio for a sine wave is unity. To estimate the rms value of a Gaussian signal, however, the measured value on the meter must be multiplied by 1.13 which is 1.05 dB. One solution to the problem of estimating the rms value of a sine wave and a Gaussian random noise has been the use of quasi-rms detectors such as are shown in Fig. 5.6c and d. Their important characteristic is that the y value of the break-points is adjusted to a fixed fraction of the output value. By proper choice of the fraction and of the slope after the breakpoint, the output can be made to approximate the true rms value for a large class of signals which have a ratio of peak value to rms value which is not too large ($\leqslant 5$).

5.3.4 Measurement Uncertainty Caused by a Finite-duration Signal
Sample   Until now, we have assumed when measuring a signal that the averaging time T is long enough to allow the measured estimate of the rms value of the signal to converge to the actual rms value. In this section we shall estimate the measurement uncertainty as a function of the averaging time T for a sine wave and for a sample of random noise.

Uncertainty of the rms Value of a Sine Wave. Consider the problem of measuring the rms value of a sine wave:

$$y(t) = A \sin \omega t \tag{5.13}$$

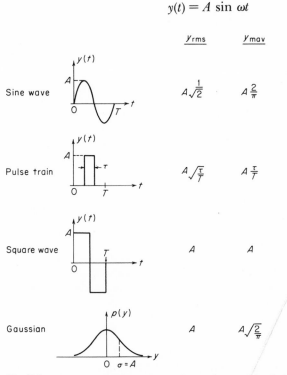

Fig. 5.8   Root-mean-square (rms) value and mean absolute value (mav) of four signals. One period of each of the three periodic signals is shown. The Gaussian signal is specified by its probability-density function, with a standard (rms) deviation $\sigma = A$.

TABLE 5.2   Ratios of rms Value to mav for the Four Signals Shown in Fig. 5.8

Signal	$\dfrac{y_{rms}}{y_{mav}}$	$\dfrac{y_{rms}}{1.11 y_{mav}}$	$20 \log \dfrac{y_{rms}}{1.11 y_{mav}}$, dB
Sine wave	$\dfrac{\pi}{2\sqrt{2}} = 1.11$	1.00	0
Pulse train (for $\tau/T = 0.1$)	$\sqrt{\dfrac{T}{\tau}} = 1.73$	1.56	3.85
Square wave	1.00	0.90	-0.91
Gaussian signal	$\sqrt{\dfrac{\pi}{2}} = 1.25$	1.13	1.05

Substituting the signal into Eq. (5.10) with $z = y^2$ we obtain the mean-square value y_{rms}^2

$$y_{rms}^2 = \frac{1}{T} \int_0^T A^2 \sin^2 \omega t \, dt \tag{5.14}$$

$$y_{rms}^2 = \frac{A^2}{2} - \frac{A^2}{4\omega T} \sin (2\omega T) \tag{5.15}$$

Note that as T becomes very large the last term vanishes, leaving the mean value of y_{rms}^2 as

$$\overline{y_{rms}^2} = \frac{A^2}{2} \tag{5.16}$$

which we recognize as the average power of a sine wave. Now consider the last term in Eq. (5.15). It is an oscillating term with an amplitude that decreases as T increases. The largest absolute value of this term is

$$\left| \frac{A^2}{4\omega T} \sin (2\omega T) \right|_{max} = \frac{A^2}{4\omega T} \tag{5.17}$$

Thus for a particular averaging time T, the largest error possible is $A^2/4\omega T$. We shall define the maximum uncertainty of this measurement as the ratio of this (maximum) error to the mean value of y_{rms}^2 (power). For notational convenience let $\overline{y_{rms}^2} = \overline{W}$.

$$\text{Max uncertainty} = \frac{\text{max error}}{\text{mean value}} = \frac{|W - \overline{W}|_{max}}{\overline{W}} \tag{5.18}$$

For the sine wave measurement the max uncertainty is

$$\text{Max uncertainty} = \frac{1}{2\omega T} \tag{5.19}$$

Consider now, instead of a single measurement of power W, a large group (ensemble) of measurements, each with an averaging time approximately equal to T but randomly deviating from it enough to cover at least one-half period of the sine wave. It is possible to compute the rms measurement error where the averaging is performed over the ensemble of measurements rather than over time. Using the brackets $\langle\ \rangle$ to denote this ensemble average, the rms error is

$$\text{rms error} = \sqrt{\langle(W - \overline{W})^2\rangle} \qquad (5.20)$$

Using this definition, we can define the rms uncertainty as

$$\text{rms uncertainty} = \frac{\text{rms error}}{\text{mean value}} \qquad (5.21)$$

Borrowing from the notation used in probability theory we will use the following notation

$$\sigma_W = \text{rms error of } W \qquad (5.22)$$

$$m_W = \text{mean value of } W \qquad (5.23)$$

For the previously considered ensemble of measurements of the sine wave,

$$\frac{\sigma_W}{m_W} = \frac{1}{\sqrt{2}\,(2\omega T)} \qquad (5.24)$$

which is a little smaller than the max uncertainty of Eq. (5.19).

Uncertainty of the rms Value of a Finite Sample of Random Noise. The following results have been derived for a Gaussian noise (a noise whose amplitudes obey a "normal" probability distribution) but are applicable to most common noises. Let W_n ($= y_{\text{rms}}^2$) be the average power of a sample of random noise T sec long. If another sample is taken from the same random process, the value of W_n for that sample will not generally be the same as for the first sample due to the nature of random processes. If, as we did in the sine wave example, we take a large enough ensemble of T-sec-long samples from the random process and average the values of W_n, we will get the correct average value of the power (mean-square value) of the noise. We can also calculate the rms error σ_W of the ensemble of W_n values. For Gaussian noise that has been filtered through an ideal filter of bandwidth B_n, the rms uncertainty of the mean-square value for a T-sec-long sample is given as

$$\frac{\sigma_W}{m_W} = \frac{\sigma_{y_{\text{rms}}^2}}{m_{y_{\text{rms}}^2}} = \frac{1}{\sqrt{B_n T}} \qquad (5.25)$$

The term $B_n T$ is called the time-bandwidth product. Generally the larger the time-bandwidth product the smaller the error.

If the "bandwidth" of a sine wave is considered to be the frequency of the sine wave, the uncertainty of the mean-square value of a sample of random noise is much larger than for a sample of a sine wave.

One caution is in order. In deriving Eq. (5.25) the random signal was a "white" (constant psd), Gaussian, random noise filtered through an ideal bandpass filter of bandwidth B_n. If the noise psd already contains peaks, and it is then filtered by a bandpass filter with a bandwidth greater than the effective bandwidth B_e of the psd function, B_e should be used in Eq. (5.25) rather than the filter bandwidth B_n.

When true averaging (of the square) is used, the value of T is unambiguous. However, when RC averaging is used, the value of T in Eq. (5.25) should be taken as RC. Thus,

$$T_{\text{equivalent}} = RC \tag{5.26}$$

Degrees of Freedom. A more satisfactory means of expressing uncertainty than that of Eq. (5.25) would be obtained if we could find a more convenient and representative distribution for the values of y_{rms}^2 measured from a set of samples of stationary random noise. The chi-square * distribution is commonly chosen to represent the distribution of the value of y_{rms}^2. To make use of this distribution we first identify the number of degrees of freedom n as

$$n = 2B_n T \tag{5.27}$$

and obtain the same result as for Eq. (5.25) for large values of n.

$$\frac{\sigma_{y_{\text{rms}}^2}}{m_{y_{\text{rms}}^2}} = \frac{1}{\sqrt{n/2}} \tag{5.28}$$

In a practical measurement situation we would like the value of y_{rms}^2 derived from a sample of a signal to be a good approximation to the true value of y_{rms}^2. We call this true rms value $\overline{y_{\text{rms}}^2}$. A measure of the closeness of this approximation is the function $y_{\text{rms}}^2 / \overline{y_{\text{rms}}^2}$.

As an example consider the measurement of y_{rms}^2 for a large group of signal samples each 5 sec long which have been filtered in a 1-Hz-bandwidth filter. For this measurement, the time-bandwidth product is 5 and from Eq. (5.27) the number of degrees of freedom n is 10. From a table of the chi-square distribution we find that the value of $y_{\text{rms}}^2 / \overline{y_{\text{rms}}^2}$

* The chi-square distribution is the distribution of the function

$$y_{\text{rms}}^2 = \sum_{n=1}^{n} x_i^2$$

where the x_i are Gaussian random variables and where the average value of x_i^2 is 1. For a set of x_i which are statistically independent (which we assume is true in psd estimation) the mean value $y_{\text{rms}}^2 = n$.

will exceed 1.83 with a 5% probability and will fall below 0.39 with a 5% probability. This means that there is a 90% probability that that value will lie between 0.39 and 1.83. We will say that the 90% *confidence limits* for the value are 0.39 and 1.83. Conversely, we can say that the 90% confidence limits of $\overline{y_{rms}^2}/y_{rms}^2$ are $1/1.83 = 0.55$ and $1/0.39 = 2.6$. We can express these confidence limits in decibels as

$$\text{Confidence limits} = 10 \log_{10} \frac{\overline{y_{rms}^2}}{y_{rms}^2} \qquad (5.29)$$

Substituting in the above example we get limits of $10 \log 0.55 = -2.6$ dB and $10 \log 2.6 = 4.1$ dB. These values can also be seen in Fig. 5.9, which is a plot of the 90% confidence limits in decibels vs. the number of degrees of freedom for the chi-square distribution.

TABLE 5.3

Confidence limits, %	G
68	6.1
80	7.8
90	10.0
96	12.6
98	14.3

Blackman[2] has given the following approximations to these confidence limits as

$$10 \log_{10} \frac{\overline{y_{rms}^2}}{y_{rms}^2} \approx G \left(\frac{2}{3n - 1} \pm \frac{1}{\sqrt{n - 1}} \right) \qquad \text{dB} \qquad (5.30)$$

for $n \geq 3$. G depends on the confidence limits as shown in Table 5.3.

For $n \geq 100$ notice that the last term in Eq. (5.30) dominates. Direct substitution will show that, for large n, this term is identical to Eq. (5.28).

In planning a measurement, the required accuracy and one's confidence in that accuracy determine the required number of degrees of freedom n which in turn determines the required time-bandwidth product. If the sample duration T is already fixed, the required n determines the minimum filter bandwidth with which this sample should be analyzed. If the filter bandwidth is already set, the required n determines the required sample length T. For example if an accuracy of ± 1 dB is required, Eq. (5.30) or Fig. 5.9 gives a value of n of approximately 100. If the analysis filter bandwidth is 1 Hz, then the required duration T is $n/2B_n = 50$ sec.

If only a 25-sec sample were available, then 2 Hz is the minimum

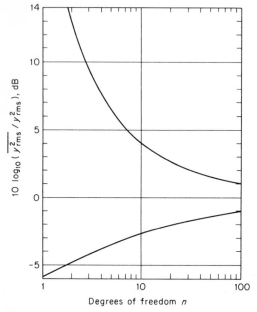

Fig. 5.9 90% confidence limits from the chi-square distribution.

bandwidth which will yield the required accuracy (±1 dB). If a 1-Hz filter were used for the 25-sec sample, n would be $2 \times 1 \times 25 = 50$ and Fig. 5.11 would give 90% confidence limits of about ±1.4 dB.

Uncertainty of a Swept-filter Analysis. A bandpass filter which has its center frequency continuously varied from one frequency to another is called a swept filter. When analyzing random noise of a fairly uniform spectrum the effective duration T of the sample is the time that any one frequency component spends within the passband B_n of the filter. This maximum sweep rate is

$$\left.\frac{\Delta f}{\Delta t}\right|_{max} = \frac{B_n}{T} \quad \text{Hz/sec} \tag{5.31}$$

Accuracy requirements determine the number of degrees of freedom n, which in turn determines the time-bandwidth product $(n = 2B_n T)$. Substituting we get

$$\left.\frac{\Delta f}{\Delta t}\right|_{max} = \frac{2B_n^2}{n} \quad \text{Hz/sec} \tag{5.32}$$

The total analysis time to cover a frequency range $f_2 - f_1$ will be

$$t_{analysis} = \frac{f_2 - f_1}{\Delta f/\Delta t \big|_{max}} = \frac{(f_2 - f_1)n}{2B_n^2} \quad \text{sec} \tag{5.33}$$

If a 2-Hz-bandwidth analysis of a stationary Gaussian random noise were required over the frequency range from 0 to 1,000 Hz, with 98% confidence limits of ± 0.5 dB, we go to Fig. 5.9 and find that $n \geqslant 100$. This allows us to use the last term of Eq. (5.30) with $G = 14.3$, and we get

$$\pm \frac{14.3}{\sqrt{n-1}} = \pm 0.5 \text{ dB}$$

or

$$n \approx \left(\frac{14.3}{0.5}\right)^2 \approx 820$$

Substituting into Eq. (5.33) we get

$$t_{\text{analysis}} = \frac{1,000 \times 820}{2(2)^2} \approx 10^5 \text{ sec } (28 \text{ hr})$$

which is a *very* long time. The analysis of any one bandwidth interval (2 Hz) would take $2 \times 820/8 = 205$ sec (3.5 min). This is the analysis time that would be required with a parallel filter bank.

5.4 Displaying

The last of the major functions shown in Fig. 5.1 is the display of the measured results. As shown in that figure, "display" is separated into three parts: (1) log conversion, (2) averaging of the log value, and (3) linear displaying.

Log conversion and averaging are not always performed but are often used because data that cover wide ranges are more clearly understood if presented logarithmically (in decibels).

Averaging is always present to some degree, either explicitly or as a lag in the response of the display system. We shall combine both forms into the averaging block.

The result of the display process is generally an x-y plot of, say, the psd function vs. frequency or a sequence of numerical values (printout) that represent the function. The x-y display might be seen on the face of an oscilloscope or be plotted on an x-y or strip-chart recorder. Numerical values may be taken from a meter recording or a numerical printer, or may be the binary representation from the output of an analog-to-digital converter.

Averaging of the Log. After log conversion there is generally some amount of averaging. This raises the question as to the meaning of the average value of the log of a function. Direct substitution into the defining equations will show that the average of the log of a set of positive values is the log of the geometric mean of the values. It is well known that the geometric mean is always less the arithmetic mean. Cox [1] has derived a lower bound on the log of the ratio of the geometric mean to the arithmetic (true rms) mean for the case where the data are confined to a range of values between an upper limit U and a lower limit L. Figure 5.10 gives this bound on $L_g - L_{true}$ in decibels vs. 10 log U/L in decibels for the case where the data are mean-square values.

If it can be assumed that a fluctuating level y_i^2 is bounded between an upper and lower value U and L respectively and is distributed about a mean value $(U + L)/2$, the true mean-square level L_{true} in decibels above the upper limit level $L_U = 10$ log U is given by

$$L_{true} - L_U = -10 \log \frac{10^{L_U/10} + 10^{L_L/10}}{2} \qquad (5.34)$$

The upper curve of Fig. 5.11 gives $L_{true} - L_U$. Notice that if $L_U - L_L$ is very large (>20 dB), $L_{true} - L_U$ takes on the minimum value of -3 dB.

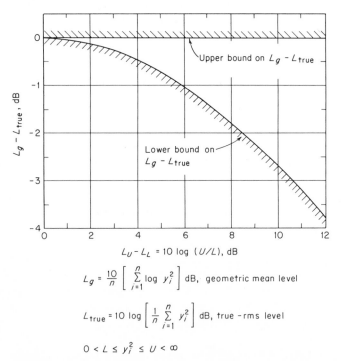

Fig. 5.10 Upper and lower bounds for the difference between the geometric mean level L_g and the true rms level L_{true}, for a set of points bounded between an upper and lower bound U and L, respectively. Derived from Cox.[1]

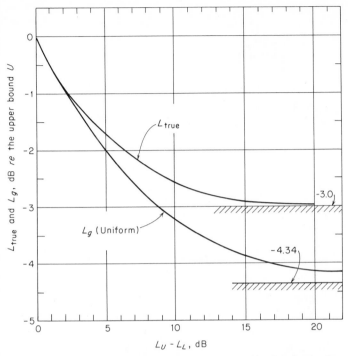

Fig. 5.11   Curve for estimating the true rms level of a fluctuating signal $y^2(t)$ which has fluctuations that are confined between an upper and lower value U and L with mean value $y^2(t) = (U + L)/2$. L_g and L_{true} are defined in Fig. 5.10. The term "uniform" means that the function $y^2(t)$ is uniformly distributed over the range of amplitude from L to U. The ordinate is either $L_{true} - L_U$ (associated with the upper curve) or $L_g - L_U$ (associated with the lower curve).

If $L_U - L_L$ is less than 1 dB, $L_{true} - L_U$ is approximately half of $L_U - L_L$.

If the averaging is done after converting to decibels, $L_g - L_U$ for the case where the square values are *uniformly* distributed between U and L is shown by the lower curve in Fig. 5.11.

As an example consider estimating the maximum true rms level of the signal displayed in Fig. 5.12. The fluctuation $L_U - L_L$ at the peak of the trend is about 7 dB. Figure 5.11 shows that for this value L_{true} is 2.2 dB below the upper bound level L_U. Because L_U is about 61.8 dB, this means that L_{true} is about 59.6 dB.

5.5   Theoretical Background Material

The discussions above have avoided lengthy mathematics in order to assist the reader who is without appreciable background in signal processing or electrical signal theory to understand the basic concepts of

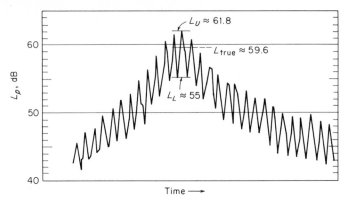

Fig. 5.12   Display of a signal level vs. time showing level fluctuations of 7 dB. The true mean-square level at the peak of the trend lies about 2.2 dB below the highest level.

data analysis.   This section provides some of the necessary foundation material.

5.5.1   Fourier Transform   The heart of spectral analysis is the Fourier transform.   It exists for all transient signals of finite energy which include all signals of finite duration and finite amplitude.   The definition of the transform and its inverse (called a Fourier-transform pair) are:

$$X(\omega) = \int_{-\infty}^{\infty} x(t)e^{-j\omega t} \, dt \tag{5.35}$$

$$x(t) = \int_{-\infty}^{\infty} X(\omega)e^{j\omega t} \, \frac{d\omega}{2\pi} \tag{5.36}$$

The functions $X(\omega)$ and $x(t)$ are equivalent ways of representing the same signal; the former is a complex function in the frequency domain and the latter a real function in the time domain.

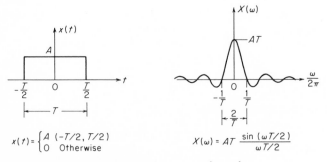

Fig. 5.13   Fourier transform of a rectangular pulse.

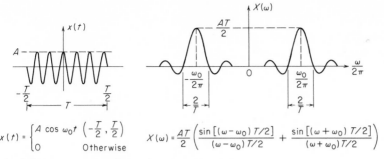

$$x(t) = \begin{cases} A \cos \omega_0 t & \left(-\frac{T}{2}, \frac{T}{2}\right) \\ 0 & \text{Otherwise} \end{cases}$$

$$X(\omega) = \frac{AT}{2}\left(\frac{\sin\left[(\omega - \omega_0)\, T/2\right]}{(\omega - \omega_0)\, T/2} + \frac{\sin\left[(\omega + \omega_0)\, T/2\right]}{(\omega + \omega_0)\, T/2}\right)$$

Fig. 5.14   Fourier transform of a tone burst.

For convenience, we shall use the shorthand notation

$$X(\omega) = F[x(t)] \tag{5.37}$$

$$x(t) = F^{-1}[X(\omega)] \tag{5.38}$$

where F and F^{-1} stand for the direct and inverse Fourier-transform (FT) operations.

If $x(t)$ has dimensions of volts, then $X(\omega)$, which is called the voltage-spectral-density function, has units of volt-sec or, more usefully, volts/Hz.

The reader should be aware that there are other valid definitions of the FT pair; for example, other conventions place the $(2\pi)^{-1}$ factor in the other equation or change the sign of j. Thus $X(\omega)$ would have dimensions scaled differently.

$X(\omega)$ is, in general, complex; that is to say

$$X(\omega) = |X(\omega)|e^{j\theta(\omega)} \tag{5.39}$$

For real $x(t)$ the magnitude is an even function of frequency and the phase angle is an odd function of frequency. This means that $X(\omega)$ and $X(-\omega)$ are complex conjugates *

$$X(-\omega) = X^*(\omega) \tag{5.40}$$

Figures 5.13 and 5.14 show examples of FT pairs where the time signals $x(t)$ are a square pulse and a tone burst, respectively. Note from Fig. 5.13 that the value of $X(\omega = 0)$ is equal to the area under $x(t)$. This can be seen from Eq. (5.35) by setting $\omega = 0$

$$X(\omega = 0) = \int_{-\infty}^{\infty} x(t)\, dt \tag{5.41}$$

Similarly from Eq. (5.36)

$$x(t = 0) = \int_{-\infty}^{\infty} X(\omega)\, \frac{d\omega}{2\pi} \tag{5.42}$$

* If a complex function is represented by $|A|e^{j\phi}$, its complex conjugate has the same magnitude but opposite sign of phase angle, $|A|e^{-j\phi}$.

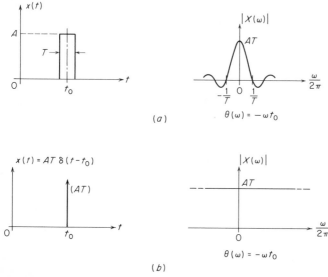

Fig. 5.15 Time signals and their Fourier transforms for (a) a finite-duration pulse of area AT and (b) an impulse of the same area.

The value of $x(t = 0)$ depends not on the amplitude of $X(\omega)$ but only on the integral of $X(\omega)$, that is to say, the area under the function when plotted against frequency in hertz.

5.5.2 Impulses   Practically, an impulse is a pulse with a duration short enough that, during the pulse, all other signals of interest are essentially constant. The important measure of an impulse is its area, not its amplitude.

Consider Eq. (5.41) and Fig. 5.15a. If we hold the area of the rectangular pulse constant and let T approach zero so that the pulse becomes very tall and very narrow, $X(\omega = 0)$ remains constant at a value of AT while the first zero crossings at $\omega/2\pi = \pm T^{-1}$ move outward toward $\pm\infty$, as shown in Fig. 5.15b. In the limit, $|X(\omega)|$ is a constant of amplitude AT. Under this condition the rectangular pulse becomes an impulse located at $t = t_0$.

The notation for an impulse of unit area is $\delta(t - t_0)$, sometimes called the Kronecker or dirac delta function. The $t - t_0$ means that it is located at time t_0 on the time scale. Substituting $x(t) = \delta(t - t_0)$ into Eq. (5.35) we get, because its area is unity and it exists only at $t = t_0$, the complex result

$$X(\omega) = F[\delta(t - t_0)] = e^{-j\omega t_0} \tag{5.43}$$

Obviously Eq. (5.43) has unit magnitude independent of frequency.

The term $e^{-j\omega t_0}$ represents a phase shift $\theta = -\omega t_0$ which varies linearly with frequency corresponding to the location of the impulse at $t = t_0$ along the time scale.

Similarly we can define a unit frequency impulse

$$X(\omega) = \delta(\omega - \omega_0) \tag{5.44}$$

such that it has unit area when integrated vs. $\omega/2\pi$

$$\int_{-\infty}^{\infty} X(\omega) \frac{d\omega}{2\pi} = 1 \tag{5.45}$$

From Eq. (5.36) we get the inverse FT of $\delta(\omega - \omega_0)$ or

$$F^{-1}[\delta(\omega - \omega_0)] = e^{j\omega_0 t} \tag{5.46}$$

Obviously Eq. (5.46) has unit magnitude independent of time and a phase shift that varies linearly with time.

Rectangular Pulse.   Consider the rectangular pulse of Fig. 5.16a. Remember we said in Eq. (5.42) that $x(t = 0)$ is equal to the area under $X(\omega)$ vs. $\omega/2\pi$.   If we let T increase toward infinity while holding the amplitude of $x(t)$ constant, $X(\omega)$ will get very large (approaching ∞), but the area will remain constant.   In the limit, as $T \to \infty$, we get a frequency impulse with a magnitude equal to the area A.   This is illustrated in Fig. 5.16b.

Tone Burst.   Now consider the tone burst of Fig. 5.17a as T increases toward infinity.   The FT, $X(\omega)$, approaches two impulses, one each at $\pm\omega_0/2\pi$, each of area $A/2$.   This is illustrated in Fig. 5.17b where

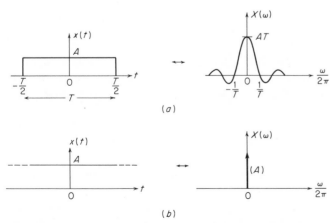

Fig. 5.16   Time signals and their Fourier transforms for (a) a finite-duration pulse and (b) an infinite-duration pulse (a constant).

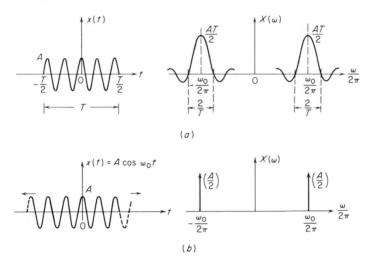

Fig. 5.17   Time signals and their Fourier transforms for (a) a tone burst and (b) a tone burst of infinite duration (a continuous sine wave).

$$x(t) = A \cos \omega_0 t \tag{5.47}$$

and

$$X(\omega) = \frac{A}{2} \left[\delta(\omega - \omega_0) + \delta(\omega + \omega_0) \right] \tag{5.48}$$

Noting that $F^{-1}[\delta(\omega - \omega_0)] = e^{j\omega_0 t}$, we get from Eq. (5.48) that

$$x(t) = \frac{A}{2} \left(e^{j\omega_0 t} + e^{-j\omega_0 t} \right) \tag{5.49}$$

which we recognize from trigonometry as being equal to $x(t)$ in Eq. (5.47).

5.5.3 Parseval's Theorem   Parseval's theorem is a statement of conservation of energy in the two representations of a signal $x(t)$ and $X(\omega)$. It can be derived from the definition of the FT pair in Eqs. (5.35) and (5.36). We find that

$$J_x \equiv \int_{-\infty}^{\infty} x^2(t) \, dt = \int_{-\infty}^{\infty} |X(\omega)|^2 \, \frac{d\omega}{2\pi} \tag{5.50}$$

The identity of Eq. (5.50) is valid as long as the integrals are *finite*, which means as long as the signal has finite energy. We recognize that the left-hand side of Eq. (5.50) is the energy in the signal $x(t)$. $|X(\omega)|^2$ is a function of frequency which, when integrated over $\omega/2\pi$, also yields the same energy. This gives a rationale for calling $|X(\omega)|^2$ the *energy spectral density function* of $x(t)$.

If $x(t)$ is in volts, $|X(\omega)|^2$ is in units of volts²-sec or volts²/Hz. A word of caution on units. Equation (5.50) shows that total energy is obtained when $|X(\omega)|^2$ is integrated vs. $\omega/2\pi$ which is in units of hertz. Other definitions of FT pairs could lead to the units of $|X(\omega)|^2$ being volts²/(radians/sec), a quantity smaller by the factor 2π.

5.5.4 Power Spectral Density (Two-sided)

If the signal $x(t)$ is a sample of duration T sec taken from a stationary process, it makes sense to calculate the *average power* in the sample as the total energy divided by the sample length

$$W_x = x^2_{\text{rms}} = \frac{\text{energy}}{T} = \frac{J_x}{T} \tag{5.51}$$

Dividing Eq. (5.50) by T we get

$$\frac{J_x}{T} = x^2_{\text{rms}} = \int_{-\infty}^{\infty} \frac{|X(\omega)|^2}{T}\frac{d\omega}{2\pi} \tag{5.52}$$

Let us define

$$S_x(\omega) \equiv \frac{|X(\omega)|^2}{T} \tag{5.53}$$

$S_x(\omega)$ is called the *two-sided power spectral density (psd) function* because it exists for both positive and negative frequencies.

5.5.5 Filtering as an Estimate of the psd Function (One-sided)

In Sec. 5.2 and in earlier chapters, we discussed the filtering of a signal through a filter and then measuring the mean-square value (average power) of the filtered signal. We called this mean-square value the *band level* when expressed in decibels. We also heuristically applied the concept that there was a psd function, and we tried to estimate it by dividing the mean-square value of the filtered signal by the bandwidth of the filter (this would give the right units of power per hertz).

An important difference between the discussions of Secs. 5.2 and 5.5.4 becomes obvious. The psd function estimated in the former section exists only for positive values of frequency. This filtered estimate of the psd function will be called the *one-sided psd function* to distinguish it from the two-sided psd function of Sec. 5.5.4.

The practical difference between the two-sided and the one-sided psd is that for real signals the magnitude of the former is one-half that of the latter. It will not be proved here but it can be demonstrated that if a signal $x(t)$ with FT $X(\omega)$ is passed through a linear system which has a

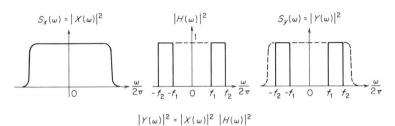

$$|Y(\omega)|^2 = |X(\omega)|^2\,|H(\omega)|^2$$

Fig. 5.18   Filtering of a time signal $x(t)$ by a linear filter.

transfer function $H(\omega)$, the output $y(t)$ is related to the input $x(t)$ through the transforms by

$$|Y(\omega)|^2 = |X(\omega)|^2|H(\omega)|^2 \tag{5.54}$$

$$S_y(\omega) = S_x(\omega)|H(\omega)|^2 \tag{5.55}$$

where $|H(\omega)|^2 =$ square of the transfer function
$S_x(\omega) = |X(\omega)|^2 =$ two-sided input power spectral density
$S_y(\omega) = |Y(\omega)|^2 =$ two-sided output power spectral density

In Fig. 5.18 this process is illustrated.   Remember that for a real signal $x(t)$, the transform $|X(\omega)|$ is symmetrical about $\omega = 0$.

From the fact that real linear systems also have a symmetrical magnitude transfer function, we can sensibly relate the one-sided psd, which we will call $W_1(\omega)$, to the two-sided psd $S(\omega)$

$$|W_1(\omega)| = 2|S(\omega)| \tag{5.56}$$

where $W_1(\omega) =$ one-sided spectral density.

Let us prove Eq. (5.56).   If the linear system in Fig. 5.18 is an ideal filter as shown in the plot of $|H(\omega)|^2$, then the *total power* W_y in the filtered signal $y(t)$ is

$$W_y = \int_{-\infty}^{\infty} |H(\omega)|^2 S_x(\omega)\,\frac{d\omega}{2\pi} \tag{5.57}$$

$$W_y = \int_{-f_2}^{-f_1} S_y(\omega)\,\frac{d\omega}{2\pi} + \int_{f_1}^{f_2} S_y(\omega)\,\frac{d\omega}{2\pi} \tag{5.58}$$

or

$$W_y = 2\int_{f_1}^{f_2} S_y(\omega)\,\frac{d\omega}{2\pi} = \int_{f_1}^{f_2} W_{1y}(\omega)\,\frac{d\omega}{2\pi} \tag{5.59}$$

5.5.6   Sampled Data   Data that are a sequence of numerical values representing the instantaneous values of a signal at a discrete set of times are called sampled data.   There are two important considerations that must be kept in mind if sampled data are to be a good representation of the original data.   By "good representation" we mean that the original signal can be recovered from the set of samples with a sufficiently small error.   One source of error is a phenomenon called *aliasing* which is the generation of spurious signals caused by a sampling rate that is too small compared to the bandwidth of the signal being sampled.   The other source of error is the *quantization* of the signal values into integer values.

In introducing the subject of aliasing it will be useful to cover first the subject of convolution.

Convolution.   It can be shown by direct substitution into the defining equations for Fourier transforms that, for reasonably well-behaved functions, if

$$x_3(t) = x_1(t) \, x_2(t) \tag{5.60}$$

then

$$F[x_3(t)] = X_3(\omega) = \int_{-\infty}^{\infty} X_1(v)X_2(\omega - v) \frac{dv}{2\pi} \tag{5.61}$$

where $X_1(v)$ and $X_2(v)$ are the Fourier transforms of $x_1(t)$ and $x_2(t)$ respectively.

The function on the right-hand side of Eq. (5.61) is called a convolution integral.   In words, *the Fourier transform of the product of two time-domain functions is the convolution of the Fourier transforms of the respective*

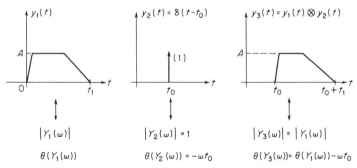

Fig. 5.19   Convolution of a signal $y_1(t)$ with a unit impulse located at $t = t_0$.   Note that the effect of the convolution is to shift $y_1(t)$ by an amount t_0.   The Fourier transform of $y_3(t)$ has the same magnitude as for $y_1(t)$ but has an added phase term $= -\omega t_0$.   The graphs show the time signals. Beneath each graph is the magnitude and phase of its complex Fourier transform.

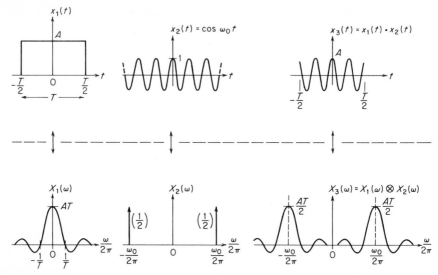

Fig. 5.20 Illustration that the Fourier transform of the product of two time signals is the convolution of the Fourier transforms of the signals. The Fourier transform of a tone burst was shown in Fig. 5.14. Here a tone burst is formed from the product of a continuous sine wave $x_2(t)$ and a rectangular pulse $x_1(t)$. The Fourier transform in Fig. 5.14, $X(\omega)$, is identical to $X_3(\omega)$ above.

functions. Remember that, in general, Fourier transforms are complex quantities.

Functionally, Eq. (5.61) may be written as

$$X_3(\omega) = X_1(\omega) \otimes X_2(\omega) \tag{5.62}$$

where the symbol $\otimes$ denotes convolution.

The time signal $x_3(t)$ may be calculated by taking the inverse FT of $X_3(\omega)$.

A similar relation holds for the inverse Fourier transform of the product of two Fourier transforms. If

$$Y_3(\omega) = Y_1(\omega)Y_2(\omega) \tag{5.63}$$

then

$$F^{-1}[Y_3(\omega)] = y_3(t) = \int_{-\infty}^{\infty} y_1(\tau)y_2(t - \tau) \, d\tau \tag{5.64}$$

In shorthand notation we can summarize Eqs. (5.60) to (5.64) as

$$F[x_1(t)x_2(t)] = X_1(\omega) \otimes X_2(\omega) \tag{5.65}$$

$$F^{-1}[Y_1(\omega)Y_2(\omega)] = y_1(t) \otimes y_2(t) \tag{5.66}$$

Using Eq. (5.64), we can calculate the convolution of the function

$y_1(t)$ with a unit impulse $y_2(t) = \delta(t - t_0)$ as shown in Fig. 5.19. Direct substitution into the defining Eq. (5.66) shows that the results of the convolution are as shown for $y_3(t)$ in Fig. 5.19. The effect of the convolution is to shift $y_1(t)$ by an amount t_0.

As an example, recall from Eq. (5.65) that when two time signals are multiplied, and the FT of their product is taken, this product is the same as the convolution of their respective FTs. Recall also the examples shown in Figs. 5.13 and 5.14. We may consider the tone burst of Fig. 5.14 as a continuous sine wave multiplied by the square pulse of Fig. 5.13. The spectrum of the tone burst must, therefore, be the convolution of the FT of the sine wave with the FT of the square pulse. This is illustrated in Fig. 5.20. Observe that the convolution of the square-pulse spectrum $X_1(\omega)$ with the two unit-frequency impulses of $X_2(\omega)$ shifts the square-pulse spectrum to the frequencies of the two impulses.

Sampling of a Signal. The process of sampling a signal at discrete times may be considered as the multiplying of the signal $x_1(t)$ (see Fig. 5.21) by a sampling signal $x_2(t)$ which is a periodic train of equal-area pulses. The resulting sampled signal $x_3(t)$ is a train of pulses, the area of each of which is proportional to the value of the signal $x_1(t)$ at the sample instant. We shall now show that this result follows directly from the definition of convolution given by Eqs. (5.60) and (5.61).

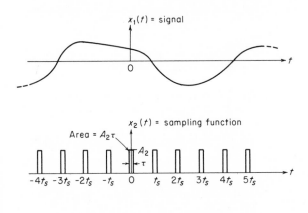

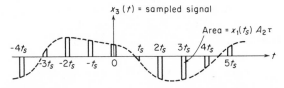

Fig. 5.21 Sampling of a signal $x_1(t)$ by multiplying it by a periodic train of pulses.

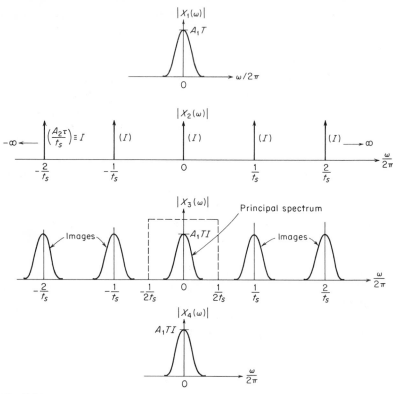

Fig. 5.22 $X_1, X_2,$ and X_3 are Fourier transforms of the signals $x_1, x_2,$ and x_3 of Fig. 5.21. X_4 is the signal X_3 filtered by an ideal filter from $-1/2t_s$ to $1/2t_s$.

The FTs $X_1(\omega)$, $X_2(\omega)$, and $X_3(\omega)$ of the respective time functions in Fig. 5.21 are shown in Fig. 5.22. Note that $X_3(\omega)$ is the convolution of $X_1(\omega)$ and $X_2(\omega)$ [see Eq. (5.65)].

We said above that a set of samples adequately represents a signal only if the signal can be recovered accurately from the samples. To recover the signal, the sampled signal $x_3(t)$ in Fig. 5.21 is passed through an ideal filter which has a transfer function equal to unity in the frequency range from $\omega/2\pi = -(2t_s)^{-1}$ to $(2t_s)^{-1}$. The spectrum $X_4(\omega)$ in Fig. 5.22 is the result of this filtering. It is clearly equal to the spectrum $X_1(\omega)$, also of Fig. 5.22. The recovered signal $x_4(t)$ is therefore equal to the original signal $x_1(t)$ and thus a complete recovery has been effected. This was true only because the spectrum $X_1(\omega)$ was zero outside the frequency range $(-\frac{1}{2}t_s, \frac{1}{2}t_s)$ and an ideal filter was used to reject completely all the "images" which occur at integer multiples of the sampling frequency $1/t_s$ Hz.

Aliasing.   Refer to the spectrum $X_3(\omega)$ in Fig. 5.22 of the sampled signal $x_3(t)$ in Fig. 5.21.   If the principal spectrum (within the dashed lines) were to have components at frequencies higher than $1/2t_s$, then the image spectra would have components which would extend into and overlap the principal frequency spectrum.   These components could not be separated from the principal spectrum by filtering.   These spurious spectral components are called aliases and the phenomenon is called aliasing.   An example of aliasing is shown in Fig. 5.23.   The original spectrum $X_1(\omega)$ is a continuous frequency function similar in principle to the one of Figs. 5.21 and 5.22.   The sampled spectrum $X_2(\omega)$

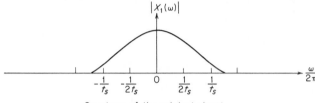

Spectrum of the original signal

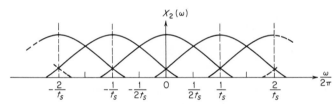

Spectrum of the sampled signal

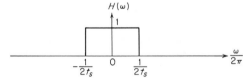

Transfer function of an ideal filter used to recover the signal

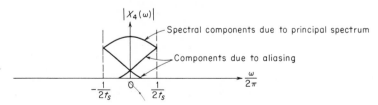

Fig. 5.23   Example of aliasing.   The original spectrum $X_1(\omega)$ is broader than half the sampling frequency $1/t_s$.   Thus the spectrum of the sampled signal, $X_2(\omega)$, is composed of overlapping spectra.   When filtered, the original spectrum within the filter band is not recovered, but contains also components due to aliasing.

consists of the principal spectrum plus the overlapping images.   When $X_3(t)$ is filtered the confused spectrum $X_4(\omega)$ results.

Aliasing cannot occur if the sampling rate is higher than twice the highest-frequency spectral component in the signal.   This just-adequate sampling rate is called the Nyquist rate.

Periodic Signals.   A periodic signal of period T is one which repeats every T sec; i.e.,

$$y(t + nT) = y(t) \tag{5.67}$$

Another way to consider the signal is as the convolution of one period of the signal with an impulse train of period T sec.   If $y(t)$ is the periodic signal, let

$$x(t) = y(t) \qquad \text{on } (C,C + T)$$
$$= 0 \qquad \text{outside } (C,C + T)$$

where C is an arbitrary constant.   Thus,

$$y(t) = x(t) \otimes \sum_{n=-\infty}^{\infty} \delta(t - nt)$$

Taking the FT of both sides using Eq. (5.35) we get

$$Y(\omega) = X(\omega) \left(\sum_{n=-\infty}^{\infty} e^{-j\omega nT} \right) \tag{5.68}$$

It can be shown that the last term in Eq. (5.68) is a periodic train of frequency impulses.   We can rewrite Eq. (5.68) as

$$Y(\omega) = X(\omega) \frac{1}{T} \sum_{n=-\infty}^{\infty} \delta \left(\omega - n \frac{2\pi}{T} \right) \tag{5.69}$$

The last term of Eq. (5.69) is periodic in frequency with period $1/T$ Hz.

The familiar Fourier-series pair for describing periodic signals is defined as follows

$$y(t) = \sum_{n=-\infty}^{\infty} a(n\omega_0) e^{jn\omega_0 t} \tag{5.70}$$

$$a(n\omega_0) = \frac{1}{T} \int_0^T y(t) e^{-jn\omega_0 t} \, dt \tag{5.71}$$

where T is the period and $\omega_0 = 2\pi/T$ is the fundamental frequency.   The Fourier transforms of both sides of Eq. (5.70) yield

$$Y(\omega) = \sum_{n=-\infty}^{\infty} a(n\omega_0) \, \delta(\omega - n\omega_0) \tag{5.72}$$

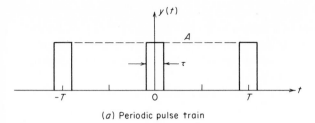

(*a*) Periodic pulse train

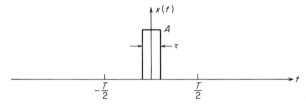

(*b*) One period of the periodic pulse train in (*a*)

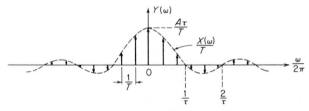

(*c*) Fourier transform of the periodic pulse train in (*a*)

Fig. 5.24   A periodic pulse train and its Fourier transform showing that the magnitude of the frequency impulses is given by the FT of the signal over one period divided by the period.

which upon substituting $a(n\omega_0) = X(\omega)/T$ yield the same result as Eq. (5.69).

As an example of Eq. (5.69), consider the periodic pulse train in Fig. 5.24*a*. The FT of the signal is shown in Fig. 5.24*c* and consists of a series of frequency impulses every $1/T$ Hz. From previous examples we notice that the magnitude of the impulses is the FT of the signal over one period divided by the period T, i.e., $X(\omega)/T$.

The Discrete Fourier Transform.   When a signal is both periodic and sampled at discrete time intervals, the Fourier-transform representation of it is called the discrete Fourier transform, DFT. Recall that a signal which is periodic with period T has a FT which is nonzero only at frequencies that are multiples of $1/T$ Hz. Recall also that a signal which has been sampled every t_s sec also has an FT which repeats itself periodically along the frequency scale every $1/t_s$ Hz. A sampled, periodic function will have both properties.

Consider a signal $x(t)$ which is sampled every t_s sec until N samples are accumulated. Now consider a new signal $y(t)$ which is a train of impulses at the sample time whose sizes equal the values of $x(t)$.

$$y(t) = \sum_{n=0}^{N-1} x(nt_s)\ \delta(t - nt_s) \qquad (5.73)$$

The FT of $y(t)$ is

$$Y(\omega) = \sum_{n=0}^{N-1} x(nt_s)e^{-j\omega nt_s} \qquad (5.74)$$

The DFT of a sequence of samples $y(n\tau)$ which could have arisen from sampling a signal every τ sec is defined as the sequence

$$Y(k\Omega) = \sum_{n=0}^{N-1} y(n\tau)e^{-j\Omega\tau nk} \qquad k = 0, 1, \ldots, N-1 \qquad (5.75)$$

where $\Omega = 2\pi/N\tau$. The inverse DFT is defined as

$$y(l\tau) = \frac{1}{N}\sum_{k=0}^{N-1} Y(k\Omega)e^{j\Omega\tau lk} \qquad l = 0,1, \ldots, N-1 \qquad (5.76)$$

There are many similarities to the regular FT of sampled, periodic signals. The signal $y(l\tau)$ returned as the inverse DFT is periodic in l with period N. The DFT is also periodic in k with period N. This periodicity in k corresponds to sampling in a regular FT while the discrete-frequency values separated by Ω correspond to the discrete values in the FT of a periodic signal.

The fast Fourier transform is merely an efficient method of calculating a DFT and produces exactly the same values. Many problems which were not practical with the straightforward evaluation of the DFT become practical due to the increased computation speed of the FFT.

The DFT and many other types of digital (numerical) filtering have been covered very well in the literature and will not be covered here. The purpose of these topics has been to provide an introduction and quick reference to some of the more important topics in signal processing.

Amplitude-quantization Considerations.   When a time-sampled signal is converted to an integer, there is generally an error in the representation. This error may be thought of as digital noise. In many ways it may be handled in the same manner as noise in analog systems. The dynamic range for an analog system is the ratio of the largest to the smallest value of a signal which can be measured or analyzed adequately. Various definitions of dynamic range exist, but in general, all are ratios of a largest useful signal to a value of the system noise, which obscures small signals. In a numerical representation of a value, one definition of dynamic range might be the ratio of the largest value expressible to the maximum error possible. For example, with a two-decimal-digit representation, the maximum number expressible is 99 and the maximum error in expressing a continuous variable is 0.5 if rounding is used. With the above definition, the dynamic range is 99/0.5, which is approximately 46 dB.

We must also recognize that in quantizing a series of signals, the relative accuracy of quantization of the instantaneous values of the signal decreases as the values decrease. For example, an instantaneous value equal to 99 will be in error by only ±0.5%, or ±0.04 dB. On the other hand, a signal with an instantaneous value of 5 will be in error by 5 ± 0.5, or ±10%, or ±0.8 dB. Thus, although the dynamic range of the system above is 46 dB, a signal of peak amplitude 5 uses only the lower 26 dB of the available dynamic range.

REFERENCES

1. H. Cox, "Linear versus Logarithmic Averaging," *J. Acoust. Soc. Am.,* vol. 39, pp. 688–690, April, 1966.
2. R. B. Blackman, *Linear Data-smoothing and Prediction in Theory and Practice,* p. 150, Addison-Wesley, 1965.

SUGGESTIONS FOR FURTHER READING

B. Gold and C. M. Rader, *Digital Processing of Signals,* McGraw-Hill, 1969.
S. H. Crandall and W. D. Mark, *Random Vibration in Mechanical Systems,* Academic, 1963.
S. J. Mason and H. J. Zimmermann, *Electronic Circuits, Signals, and Systems,* Wiley, 1960.
S. H. Crandall (ed.), *Random Vibration,* Technology Press and Wiley, 1958.
W. B. Davenport and W. L. Root, *Introduction to Random Signals and Noise,* McGraw-Hill, 1958.
J. D. Robson, *Random Vibration,* Elsevier, 1964.

Chapter Six

The Measurement of Power Levels and Directivity Patterns of Noise Sources

LEO L. BERANEK

Sound-pressure level in decibels is the physical quantity that is usually measured to describe sound waves quantitatively. To describe the strength of a noise source, sound-pressure level alone is not a satisfactory quantity because it varies with the distance between the source and the observer and with the environment for both.

Two quantities are needed to describe completely the strength of a source, *sound-power level* and *directivity*. Sound-power level measures the total sound power radiated by the source in all directions and is usually stated as a function of frequency; for example, it is usually measured with a set of contiguous constant-percentage filter bands. Directivity is a measure of the difference in radiation with direction, and is usually stated as a function of angular position around the acoustical center of the source, also as a function of frequency. Obviously a

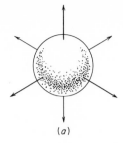

(a)

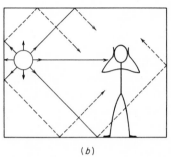

(b)

Fig. 6.1 A sound source that radiates uniformly in all directions: (a) Sound source in free space; (b) the same source in an enclosure showing reflections from interior surfaces. The solid lines show the direct sound; dashed lines, the reflected (reverberant) sound.

complete description of the noise characteristics of a complicated source, such as a jet engine, requires considerable data.

Some sources of sound radiate nearly uniformly in all directions. These are called nondirective sources (see Fig. 6.1). Generally, such sources are small in size compared to the wavelength of the sound they are radiating. Most practical sources are somewhat directive (see Fig. 6.2); in other words, they radiate more sound in some directions than in others.

The concepts of sound-power level and directivity index make it possible to calculate the sound-pressure levels created by the source in its acoustical environment. Moreover, the sound-power level alone is useful in comparing the relative strengths of different noise sources. Unfortunately, any method for rating a noise source is not entirely unique because both the radiated sound power and the directivity are influenced by nearby reflecting surfaces, such as floors and walls. Also, a source may set a nearby surface into vibration if the source is rigidly joined to it, causing more sound power to be radiated than if the source were vibration-isolated.

6.1   The Radiation Field of a Sound Source

6.1.1   Near Field, Far Field, and Reverberant Field
The character of the radiation field of a typical noise source may vary with distance from the source. In the vicinity of the source, the particle velocity is

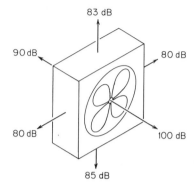

83 dB

90 dB

80 dB

80 dB

100 dB

85 dB

Fig. 6.2 A sound source with directive radiation into free space. This behavior is typical of equipment noise.

not necessarily in the direction of travel of the wave, and an appreciable tangential velocity component may exist at any point.   We speak of this as the *near field*.   It is frequently characterized by appreciable variations of the sound-pressure position along a given radius even when the source is in free space (anechoic chamber).   Moreover, in the near field the acoustic intensity is not simply related to the mean-square sound pressure.

The extent of the near field of a noise source depends on the frequency, on a characteristic source dimension, and on the phases of the radiating parts of the surface.   The characteristic dimension may vary with frequency and angular orientation.   It is difficult, therefore, to establish general limits for the near field of an arbitrary source with accuracy.   It is often necessary to explore the sound field experimentally.

In the *far field* the sound-pressure level decreases 6 dB for each doubling of distance * if the sound source is in free space (anechoic chamber) or if the absorption in an enclosure is great enough that the *reverberant field* has not yet been reached (see Fig. 6.3).   In this *free-field part of the far field,* the particle velocity is primarily in the direction of propagation of the sound wave and the sound intensity is related to mean-square sound pressure by Eq. (1.20).

If the source is radiating within an enclosure, one also finds considerable fluctuations of sound pressure with position in the reverberant part of the far field, i.e., in the region where the waves reflected from the boundaries of the enclosure are superimposed upon the incident field (see Fig. 6.3).   Many reflected wavetrains are generally present and the sound pressure, averaged over several feet, reaches a level that is essentially independent of distance from the source.   The reverberant

---

* These distances are measured from the acoustic center which is located somewhere within the region occupied by the source.

field is called a *diffuse field* if a great many reflected wavetrains cross from all possible directions and the sound-energy density is very nearly uniform throughout the field.

6.1.2 Anechoic Chamber and Reverberation Chamber   Two special types of acoustic environment in which sound sources are measured are of interest.   In an *anechoic chamber,* where all boundaries are highly absorbent, the free-field region extends very nearly to the boundaries. In a *reverberation chamber* all boundaries are hard and the reverberant field extends over nearly the entire room volume with the exception of a small region around the source.

6.2   Sound-power Level

The sound powers radiated in air by sources of practical interest range from less than a microwatt to megawatts.[1]   Because of this great range, sound power is expressed on a decibel scale using an internationally selected power as the reference for the logarithm (see Chap. 2).   By definition the sound-power level is

$$L_W = 10 \log_{10} \frac{W}{W_0} \quad \text{dB } re \ W_0 \text{ watt} \tag{6.1}$$

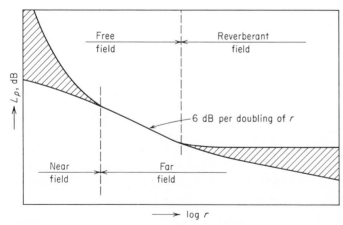

Fig. 6.3   The variation of sound-pressure level in an enclosure along a radius r from a typical noise source.   Shown are the free field, reverberant field, near field, and far field.   The shaded areas indicate regions where the sound-pressure level L_p fluctuates most with distance.   In between the near and reverberant fields the L_p decreases at the rate of 6 dB for each doubling of distance from the acoustic center of the source.

where W = acoustic power, watts
W_0 = acoustic reference power, internationally agreed on as 10^{-12} watt *

Consequently

$$L_W = 10 \log W + 120 \qquad \text{dB } re\ 10^{-12} \text{ watt} \qquad (6.2)$$

Example 6.1   A sound source radiates an acoustic power of 3 watts. Find the sound-power level relative to the ISO reference power.
SOLUTION:

$$L_W = 10 \log 3 + 120$$
$$= 4.8 + 120 \approx 125 \text{ dB } re\ 10^{-12} \text{ watt}$$

When stating a sound-power level, it is important to state the reference level also to avoid confusion.

6.3   Sound Intensity and Sound Power [2,3]

The sound power radiated by a source spreads over a greater area as the wave travels outward. At any point removed from the source the sound field has a strength, called the *sound intensity,* which, for a non-directive source, is

$$I \text{ (at radius } r) = \frac{W}{A_s} = \frac{W}{4\pi r^2} \qquad \text{watts/m}^2 \qquad (6.3)$$

where W = sound power, watts
$A_s = 4\pi r^2$ = area of a sphere, m²
r = radius of sphere, m

Conversely,

$$W = (\text{area of sphere}) \times (\text{intensity at } r) = 4\pi r^2 I \qquad \text{watts} \qquad (6.4)$$

If the source is directive, the sound intensity will not be the same at all points on a spherical surface of radius r, and Eqs. (6.3) and (6.4) are not valid. Consequently, the acoustic power radiated by a directive source must be determined from the sum of the products of area segments on any closed (usually spherical) surface that contains the source and the intensities at those areas, respectively. Mathematically

$$W = \sum_i I_i S_i \qquad \text{watts} \qquad (6.5)$$

* Prior to international agreement, 10^{-13} watt was used in the United States because the sound-pressure level and the sound-power level are nearly alike if the sound passes uniformly through 1 ft² of area. Obviously, sound-power levels expressed $re\ 10^{-13}$ watt are 10 dB greater than those expressed $re\ 10^{-12}$ watt. Great care must be taken to ascertain which reference power was used when working with published data or serious errors will result.

where I_i = sound intensity averaged over the ith segment of area, watts/m^2

S_i = ith segment of area, m^2

$\sum_i S_i$ = total area of the closed surface, m^2

As seen from Eq. (1.20), for the far field of a source radiating into free space

$$I(\text{at } r) = \frac{p^2_{\text{rms}} \,(\text{at } r)}{\rho c} \qquad (6.6)$$

where ρc = characteristic resistance of air (see appendix to Chap. 1), N-sec/m^3, equal to about 406 mks rayls at normal room conditions

p_{rms} = root-mean-square sound pressure, N/m^2

Combining Eqs. (6.5) and (6.6) yields

$$W = \frac{1}{\rho c} \sum_i p_i^2 S_i \qquad \text{watts} \qquad (6.7)$$

where p_i = average rms sound pressure over the area segment S_i, N/m^2.

Using Eqs. (6.1) and (2.9) we may express Eq. (6.7) logarithmically

$$L_W = 10 \log \sum_i S_i \left[\text{antilog} \frac{L_{pi}}{10} \right] - 10 \log \frac{\rho c}{400} \qquad \text{dB} \qquad (6.8)$$

where L_W = see Eq. (6.1), dB re 10^{-12} watt

L_{pi} = sound-pressure level over ith area segment, dB re 2×10^{-5} N/m^2

S_i = area of ith segment, m^2

antilog $(L_{pi}/10)$ = $(p_i/p_{\text{ref}})^2$

p_{ref} = 2×10^{-5} N/m^2

10 log $(\rho c/400)$ = $10 \log K$ = values for this correction may be found from Fig. 2.3. At normal temperatures and atmospheric pressures this term is negligible

The computation of sound power from sound-pressure measurements can also be made in a diffuse reverberant sound field. In a diffuse sound field the sound-energy density is constant and the net intensity is near zero. But the sound-energy density is directly related to the mean-square sound pressure and, therefore, to the sound power of the source. Hence it is again possible to compute the acoustic power from sound-pressure-level measurements.

Sufficiently close engineering approximations to free-field conditions can be achieved in properly designed anechoic rooms, or outdoors. Approximately diffuse sound fields can be obtained in rooms that are fairly reverberant and, preferably, irregularly shaped. Where these environments are not available or where it is not possible to move the noise source under test, other techniques valid for field situations are described below for the determination of sound power.

All procedures described in this chapter apply to the determination of sound power in octave or, where greater frequency resolution is necessary, one-third-octave bands. The techniques are independent of bandwidth. For impulsive (transient) sounds, such as are produced by drop hammers or gunshots, the techniques of Chaps. 4 and 5 must be used.

Most practical noise sources, when mounted and supported in the same way relative to surfaces, radiate very nearly the same power in a free field as in a reverberant field, if the reverberation room is large enough.

6.4   Sound-power Measurement: Comparison Method

Reference Source.   In the last decade, the reference-source method of determining the sound power radiated by a noise source has become standard in some industries.[4-9]   By this method, a reference source (see Fig. 6.4) is used as a comparison with the source under test. The reference source was designed by ASHRAE committee members who arranged to make it commercially available at nominal cost.[8]   Its sound-power levels in one-third-octave and octave bands as determined from free-field measurements are given in Table 6.1. The diffuse-field power levels are average values determined in a number of reverberation chambers.

Fig. 6.4   Reference sound source. Manufacturer: ILG Industries, Inc., 2826 North Pulaski Road, Chicago, Ill. 60641.

TABLE 6.1   Calibration of ILG Reference Sound Source

One-third-octave-band center frequency, Hz	Sound-power level, dB re 10^{-12} watt		Octave-band center frequency, Hz	Sound-power level, dB re 10^{-12} watt	
	Free field	Diffuse field		Free field	Diffuse field
50	78	72 ± 3			
63	77	73 ± 2	63	82	76 ± 2
80	76	72 ± 1			
100	76	71			
125	76	72	125	81	77
160	76	73			
200	76	74			
250	76	74	250	81	78
315	76	75			
400	76	75			
500	76	75	500	81	79
630	76	74			
800	76	74			
1,000	76	75	1,000	81	79
1,250	76	76			
1,600	76	75			
2,000	76	75	2,000	81	80
2,500	76	74			
3,150	75	74			
4,000	75	73	4,000	79	78
5,000	74	73			
6,300	74	73 ± 1			
8,000	73	75 ± 2	8,000	78	76 ± 2
10,000	71	71 ± 3			

Experimental Setup. Generally, measurements of sound power produced by a device or machine are performed in a laboratory-grade reverberant room.

The determination of sound power in such a room is based on the premise that the measurements are performed entirely in the diffuse (reverberant) sound field (see Fig. 6.3). No microphone should be placed less than one major source dimension, and at least a distance of $V^{1/3}/6$, away from the source. If the source is highly directive, microphone positions that fall within that directive region should be avoided. In the reverberant field the average sound-pressure level is essentially uniform, although there are fluctuations from one point to another, and it is simply related to the sound power radiated by the source. It is clear that information about the directivity of the source cannot be obtained in a diffuse sound field.[10,11]

A test room of adequate volume and suitable shape is required whose boundaries, over the frequency range of interest, can be considered reasonably hard. The volume of this room should be large enough so that the number of normal modes of vibration (see Chap. 9) is large enough to permit a satisfactory state of sound diffusion. It has been generally accepted that the volume of the room should be chosen from the formula

$$V > 4\lambda_g^3 \qquad (6.9)$$

where V = total volume of reverberation room, m³

λ_g = wavelength of sound, in meters, evaluated at the geometric center of the lowest one-third-octave frequency band in which measurements are to be made

Waterhouse and Lubman[12] have drawn two conclusions in regard to determining the space-average mean-square sound pressure in a reverberant sound field: (1) Discrete averaging (averaging the output of a set of fixed microphones) is generally to be preferred to continuous averaging (one microphone swept about the room) as the former is simpler to perform experimentally and yields more accurate results; (2) the individual microphones of a discrete set should be at least a half wavelength apart in all directions within the volume at the lowest frequency of interest; i.e., the spacing should be at least $\lambda_g/2$.

Stationary Diffuse Sound Field of Narrow Bandwidth. Assuming the conditions of Eq. (6.9) and (1) and (2) above, Fig. 6.5 gives the number of measurement positions M that are needed to determine the space-average mean-square sound-pressure level for a stationary diffuse sound field of narrow bandwidth (or of discrete frequencies) to within $\pm 2, \pm 1$, and ± 0.5 dB of the true-average value in the reverberant field. The approximate number of measurement positions located $\lambda_g/2$ apart that can be found within a volume V is

$$M \text{ (possible)} \approx \frac{12V}{\lambda_g^3} \qquad (6.10)$$

Equation (6.10) is a good approximation for determining M (possible) only if $V/\lambda_g^3 \geq 16$. Below this limit, the number of possible microphone positions is a little higher.

Nonstationary Diffuse Sound Field of Various Bandwidths. The number of microphones required to obtain the average mean-square sound-pressure level to within an accuracy of ± 1 dB in a stationary diffuse sound field is too large to be practical (for example, 50 microphones spaced $\lambda_g/2$ apart). Fortunately, the number of measurement positions can be reduced greatly by creating a nonstationary, i.e., a modulated, diffuse sound field.

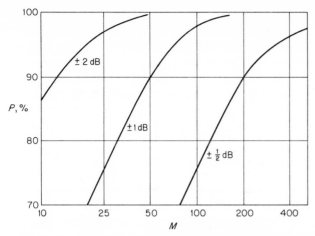

Fig. 6.5   Percent probability P of the mean-square sound pressure, averaged over space in a reverberant room, lying within ± 2, ± 1, or ± 0.5 dB range of the true-average, mean-square sound pressure plotted vs. the number of microphones M located at points at least $\lambda/2$ apart.   A very narrow band or pure tone is assumed.   Approximately one-tenth as many microphones may be used if rotating diffusers are employed and if the source radiates random noise within a one-third-octave bandwidth above the lowest frequency of interest.

A modulated diffuse sound field is obtained by rotating one or more large surfaces, each with dimensions greater than the longest wavelength, within the reverberation room.   Such surfaces are called *rotating diffusers*.   They serve the same function as would time variations in the dimensions of the room, with volume held constant.   To be more explicit, the normal frequencies of the resonances (see Chap. 8) of the reverberant room are modulated.   Modulation of the normal frequencies means that the mean-square sound pressure at a microphone position fluctuates enough to reduce the deviation of that measurement from the true, space-average, mean-square sound pressure.   Thus fewer measurement positions are required.

A further advantage of a modulated diffuse sound field is that the average impedance of the sound field presented to a sound source varies less from one position to another in the room than if the field were not modulated.

Two shapes of rotating diffusers have been described recently in the literature.[13,14]   These diffusers are sections of cones and are covered with small-scale irregularities.   They rotate in the room at speeds of 15 to 30 revolutions per minute, with the upper value being better for averaging, although noisier and creating some fanning of the air.

Lubman's experimental data[15] indicate that with a rotating diffuser the variation of mean-square sound pressures may be as much as a factor of 10 lower than those for a stationary diffuse sound field. Thus, for the same accuracy, only about one-tenth the number of microphone positions is needed.

A further reduction in number of microphone positions results when the source produces a broadband noise, free of discrete components, and when a wide filter band is used in measurement. If we assume that Fig. 6.5 applies to a filter-bandwidth-reverberation-time product $B_f T_{60}$ of 1.0 or less, then increasing this product to 10 by increasing B_f decreases the variation in mean-square pressure by a factor of about 1.5. Further increasing $B_f T_{60}$ to 100 decreases the variation by a factor of about 10. The value of T_{60} should ideally be about 1 sec so that the room resonances will overlap sufficiently to provide an adequate ratio of modal bandwidth to modal frequency spacing. It is generally accepted that the average sound-absorption coefficient $\bar{\alpha}_{\mathrm{Sab}}$ for the test room should ideally be about 0.15, which corresponds to $T_{60} \approx 1$ sec for a room with a volume of about 200 m³ (7,000 ft³).

The above analysis indicates that with rotating diffusers, measurements may be made down to frequencies that satisfy Eq. (6.9) using as few as about six microphones, spaced half wavelengths apart at the lowest frequency. With this number of microphones, it is feasible to electronically switch from one microphone output to another to obtain the average mean-square sound pressure.

Other Considerations. Ideally, a reverberation room should be nonrectangular. If near-rectangular, the room should be designed so that the ratio of any two dimensions is not equal or close to an integer. The proportions $1 : 2^{1/3} : 4^{1/3}$ are frequently used. Customary test-room volumes for measurements down to 90 Hz are in the range of 180 to 280 m³ (6,500 to 10,000 ft³).

If the noise source under test is normally associated with a hard floor, wall, edge, or corner, it should be placed in a corresponding position in the reverberation room.[11] It is not advantageous, generally, to place the source near the geometric center of the room or of a wall of the room since this positioning limits the available test radius (see below) and an appreciable portion of the resonance modes in the room may not be excited.

In selecting the microphone positions one should also bear in mind that near the room boundaries the sound field departs from the ideal state of diffusion.[16] Assuming a source radiating continuous-spectrum noise and an analyzing filter one-third octave wide, it is good engineering practice to place the microphone at least $\lambda_g/4$ ft away from all room surfaces, which was the assumption in regard to Fig. 6.5 and Eq. (6.10).

Comparison Procedure. The procedure for determining the sound-power level of a source by the comparison method with a reference source, assuming that the measuring microphones are in the reverberant field, is as follows: (1) With the source under test at a suitable location in the room, determine, in each frequency band, the space-average mean-square sound-pressure level $\bar{p}^2$ in the reverberant field using an adequate number of fixed microphones as detailed above.   Convert $\bar{p}^2$ into a sound-pressure level $\bar{L}_p$ for each frequency band.   (2) Replace the source under test with the reference sound source and repeat the measurement to obtain $(p')^2$ for each frequency band.   Convert into $\overline{L_p'}$.

The sound-power level of the source under test L_W for a given frequency band is given by

$$L_W = L_W' + (\bar{L}_p - \overline{L_p'}) \qquad \text{dB } re \ 10^{-12} \text{ watt} \qquad (6.11)$$

where L_W' = sound-power level for the reference source, dB (see Table 6.1)

$\bar{L}_p$ = space-average sound-pressure level in the reverberant field for the source under test (in the same frequency band), dB. The procedure for averaging sound-pressure levels to obtain the average mean-square sound-pressure level is illustrated in Sec. 6.6

$\overline{L_p'}$ = same for the reference source, dB

Example 6.2   Assume that in the 1,000-Hz one-third-octave band the average sound-pressure level produced by the source under test is 100 dB and that produced by the ILG reference source is 70 dB.   What is the sound-power level of the source under test?

SOLUTION: The sound-power level of the ILG source in the 1,000-Hz band is 75 dB re 10^{-12} watt (see Table 6.1).   Thus,

$$L_W = 75 + (100 - 70) = 105 \text{ dB } re \ 10^{-12} \text{ watt}$$

6.5 Sound-power Measurement: Absolute Method

By this technique, a reference sound source is not used. Instead, the sound-absorbing properties of the room are determined by measuring, for each frequency band, the reverberation time for the room (see Sec. 4.6).

The space-average, mean-square sound pressure, for each frequency band, of the source under test is determined and converted to the sound-pressure level $\bar{L}_p$.   The sound-power level L_W, for the same band, is found from

$$L_W = \bar{L}_p + 10 \log V - 10 \log T_{60} + 10 \log \left(1 + \frac{S\lambda}{8V}\right) - 13.5 \text{ dB} \quad (6.12)$$

where V = total volume of test room (with volume of source subtracted), m^3

T_{60} = reverberation time of test room with the source in place, sec

λ = wavelength of the sound at the geometric frequency of the test band, m

S = area of all boundary surfaces in the room, m^2

$\bar{L}_p$ = see following Eq. (6.11). Note that $\bar{L}_p$ is determined with no microphone closer to a wall or the floor than $\lambda/4$, where λ is the wavelength at the test frequency

Example 6.3 Assume a room with a volume of 163 m^3 and surface area $S =$ 200 m^2 and a reverberation time at 1,000 Hz of 2 sec. The space-average sound-pressure level in the stationary diffuse field with a given machine operating is 100 dB. Find the sound-power level for this machine. Assume a discrete-frequency spectrum.

SOLUTION:

$$L_W = 100 + 10 \log 163 - 10 \log 2 + 10 \log \left(1 + \frac{200 \times 0.344}{8 \times 163}\right) - 13.5$$

$$= 100 + 22.1 - 3 + 10 \log (1.053) - 13.5$$

$$\approx 106 \text{ dB}$$

6.6 Sound-power Measurements in Anechoic Hemispherical Space

The sound power radiated by a device or machine may be measured directly in free (spherical) space and in hemispherical space. Hemispherical space is often easy to provide as it may be the open air above a paved area, distant from reflecting surfaces such as buildings. Or it may be a semianechoic room, in which only the floor is rigid. This environment is suited to machinery that is mounted on a concrete floor in a room that has sound-absorptive treatment on ceiling and walls.

The determination of the sound power radiated in hemispherical space is based on the premise that the reverberant field is negligible at the positions of measurement and that the total radiated sound power is obtained by a space integration of the sound intensity over a hypothetical test hemisphere centered on the source of noise. The surface of the test hemisphere must, in addition, be in the far field of the source. Then, the space average of the mean-square sound pressure over the hemisphere is determined. This may be done by choosing an array of

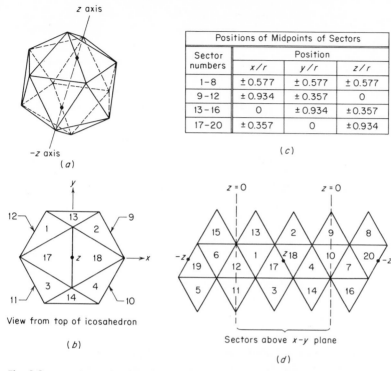

Positions of Midpoints of Sectors			
Sector numbers	Position		
	x/r	y/r	z/r
1–8	±0.577	±0.577	±0.577
9–12	±0.934	±0.357	0
13–16	0	±0.934	±0.357
17–20	±0.357	0	±0.934

(c)

View from top of icosahedron

(b)

Sectors above x-y plane

(d)

Fig. 6.6 Drawings of an icosahedron, a near-spherical surface with 20 equal areas of identical shape. (a) Isometric view; (b) view along z axis looking toward x-y plane; (c) coordinates of midpoints of sectors; (d) planar development of icosahedron with numbering of all sectors. To avoid disecting four of the area segments by the x-y plane, the icosahedron may be tilted 20.9°. The coordinates for the tilted hemispherical icosahedron are given in Table 6.2.

microphone locations over the hemisphere. How many points are needed will depend on the accuracy required and the directivity of the source. The more directive the source, the greater the number of points required.

As one example, a half of an icosahedron may be used (see Fig. 6.6).[1] Assume that it is situated such that its x-y plane rests on the floor of the test space. Measurements are made at the center of eight equilateral triangles, Nos. 1–4, 13, 14, 17, and 18, and at the center of four half-equilateral triangles, Nos. 9–12. Symmetry in the radiation pattern could be used to reduce the number of measurement positions, say, by aligning the axis of symmetry with the z axis, but there is danger in too few measurement positions. Reflections of the sound from the hard floor may cause a far-field interference pattern. If a fair percentage of the measurement points fall in the regions of reduced sound pressure in this pattern, the average mean-square sound pressure may be in error. Thus a large number of points, not at only a few heights above the floor, enhances the probability of accuracy of the results.

An alternate to the orientation of the icosahedron of Fig. 6.6 would be to tilt the array by 20.9° so that when it is used for hemispherical measurements no area sector is bisected by the x-y plane. The coordinates of the midpoints of the sectors for the tilted icosahedron are given in Table 6.2.

To obtain the average mean-square sound-pressure level over the test hemisphere $\bar{L}_H$, the following procedure is used:

1. The sound-pressure level at each microphone position is converted into mean-square sound-pressure ratio, using Eq. (2.10c), and is multiplied by the area S_i associated with that position.

$$S_i \frac{p_i^2}{p_{\text{ref}}^2} = S_i \text{ antilog } \frac{L_{pi}}{10} \qquad (6.13)$$

where S_i = area of the ith test segment, m^2

$(p_i/p_{\text{ref}})^2$ = mean-square sound-pressure ratio for that segment

L_{pi} = sound-pressure level at the center of the test segment, dB

2. The S_i $(p_i/p_{\text{ref}})^2$ products are added, and the result is divided by the total area of the test hemisphere, that is, $2\pi r^2$, to get $(\bar{p}_H/p_{\text{ref}})^2$.

3. The summed value $(\bar{p}_H/p_{\text{ref}})^2$ is converted into the space-average

TABLE 6.2   Coordinates of a 10-point Hemispherical Icosahedral Array *

No.	$\dfrac{x}{r}$	$\dfrac{y}{r}$	$\dfrac{z}{r}$
1	−0.872	0.357	0.333
2	−0.333	0.577	0.745
3	0.127	0.934	0.333
4	0.745	0.577	0.333
5	0.667	0	0.745
6	0.745	−0.577	0.333
7	0.127	−0.934	0.333
8	−0.333	−0.577	0.745
9	−0.872	−0.357	0.333
10	0	0	1

* This array is derived from the 20-point array of Fig. 6.6. by tilting it by 20.9°. Thus, two of the four area sectors that are bisected by the x-y plane are elevated to full area, and the other two are dropped altogether.

TABLE 6.3   Measured Sound-pressure Level in 500-Hz Band

Point	z/r position	L_{pi}dB	$(p_i/p_{ref})^2$, dimensionless
1	0.576	60	1×10^6
2	0.576	60	1×10^6
3	0.576	65	3.16×10^6
4	0.576	55	0.32×10^6
9	0	61(58)	(0.63×10^6)
10	0	60(57)	(0.50×10^6)
11	0	61(58)	(0.63×10^6)
12	0	62(59)	(0.79×10^6)
13	0.357	55	0.32×10^6
14	0.357	55 .	0.32×10^6
17	0.934	64	2.51×10^6
18	0.934	61	1.26×10^6
		Sum $=$	12.44×10^6

sound-pressure level by

$$\bar{L}_{pH} = 10 \log \left(\frac{\bar{p}_H}{p_{ref}}\right)^2 \quad \text{dB} \qquad (6.14)$$

Example 6.4   Sound-pressure levels are measured at 12 points on the 12-sector hemispherical array of Fig. 6.6 around a small source located at $x = y = z = 0$. Column 3 of Table 6.3 gives the levels measured in the 500-Hz band in decibels *re* 2×10^{-5} N/m². Find $\bar{L}_{pH}$, the level in dB of the space-average, mean-square sound pressure of the noise produced by the source.

SOLUTION: In this array of points, eight have equal areas, say S_1, and four have half areas, $S_1/2$. These four points are Nos. 9–12, in boldface type. It is not necessary to know the actual area of each sector; we may call it 1 m². Sectors 9–12 then have an area of $\frac{1}{2}$ m². The total area of the hemisphere is 10 m². However, for the half sectors we can reduce the sound-pressure level L_{pi} by 3 dB [the same as reducing $(p_i/p_{ref})^2$ by a factor of 2] and use the same area for all 12 sectors, i.e., a total area of 12 m². (Note: If the coordinate of the half icosahedron had been inclined as given in Table 6.2, two of the four points would have been elevated to full weighting and two would have disappeared.) The values of $(p_i/p_{ref})^2$ so obtained are given in column 4 of Table 6.3. Multiply each item in column 4 by the area (here assumed unity). Addition of these numbers (i.e., column 4 because the areas are unity) gives

$$\sum_{1-12} \left(\frac{p_i}{p_{ref}}\right)^2 \times 1 = 12.44 \times 10^6 \text{ m}^2$$

Division by the total area of 12 m² to obtain the space-average mean-square sound pressure yields

$$\left(\frac{\bar{p}_H}{p_{\text{ref}}}\right)^2 = \sum \left(\frac{p_i}{p_{\text{ref}}}\right)^2 \times \frac{1}{12} = \frac{12.44}{12} \times 10^6 = 1.037 \times 10^6$$

Application of Eq. (2.9) gives

$$\bar{L}_{pH} = 10 \log \left(\frac{\bar{p}_H}{p_{\text{ref}}}\right)^2$$

$$= 10 \log 1.037 \times 10^6 \approx 60 \text{ dB}$$

Note that $\bar{L}_{pH}$ is the space-average sound-pressure level in decibels over the surface of the hemisphere.

Alternatively, because we are using the numbers in parentheses in column 3 of Table 6.3, we can combine the sound-pressure levels directly, taking two levels at a time, with the help of Fig. 2.7. The procedure yields

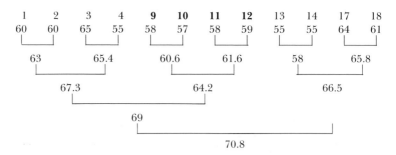

Because the 12 areas of Table 6.2 were assumed to be equal to 1 m² each, we must subtract 10 log 12 = 10.8 dB from the 70.8 dB obtained. Thus, as before,

$$70.8 - 10.8 = \bar{L}_{pH} = 60 \text{ dB}$$

For quick but less accurate results, simply average the levels in dB (using the numbers in parentheses) and add 1 dB. This procedure works if the levels are fairly near each other and if no level is lower than the highest by more than 10 dB. In the example of column 3 in Table 6.3, the result is 58.5 + 1 = 59.5 ≈ 60 dB. The largest error by this quick procedure would occur if half the numbers were 10 dB greater than the other half. Assume 6 at 60 dB and 6 at 50 dB. By our approximate formula, we get 55 + 1 = 56 dB, compared to the correct value of 57.5 dB, an error of 1.5 dB.

The test hemisphere, which determines the location of the microphone positions, has its center on the reflecting plane beneath the acoustic center of the sound source. In order that the measurements be carried out in the far field, the hemisphere radius should be equal to at least two major source dimensions or four times the average source height above the reflecting plane, whichever is the larger.   It is good engineering practice for the radius of the test hemisphere always to be greater than about 0.6 m (2 ft).   No microphone position should be closer to the room boundaries than $\lambda/4$, where λ is the wavelength of sound at the center frequency of the lowest frequency band of interest. Outdoors, atmospheric effects are likely to influence the measurements if the radius of the test hemisphere is much greater than about 15 m (50 ft), even in favorable weather.

After the space-average hemispherical sound-pressure level has been determined from the measured data, the sound-power level in each frequency band is computed

$$L_W = \bar{L}_{pH} + 20 \log r + 10 \log 2\pi \qquad \text{DB} \qquad (6.15) *$$

where L_W = sound-power level, dB re 10^{-12} watt
    $\bar{L}_{pH}$ = average mean-square sound-pressure level, over test hemi-sphere, dB re 2×10^{-5} N/m²
    r = radius of text hemisphere, m
$10 \log 2\pi = 8$ dB
L_W should be computed for all frequency bands of interest.
The radiated sound power W may be found from Eq. (2.11a)

$$W = 10^{-12} \text{ antilog } \frac{L_W}{10} \text{ watt} \qquad (6.16)$$

The value of L_W should be rounded off to conform to the measure-ment conditions (wind, etc.) and the limits of the instrumentation, generally to the nearest 0.5 dB.   Accordingly, W would be rounded off to within 12%.

6.7   Sound-power Measurements in Anechoic Full Space

Sometimes the noise source under test is not associated with a hard surface and perhaps is small enough to be placed near the center of an anechoic chamber.   In this true free-field case, it is necessary to deter-mine the space-average mean-square sound-pressure level over a test

---

* In Eqs. (6.15), (6.17), (6.27), and (6.28), a term on the right-hand side which is depend-ent on atmospheric pressure and temperature is not included.   This term equals 10 log $(\rho c/400)$ and should be added algebraically to Eqs. (6.27) and (6.28) and subtracted alge-braically from Eqs. (6.15) and (6.17).   It may be determined directly from Fig. 2.3.   As an example, $\rho c = 413$ mks rayls at a barometric pressure of 30 in. (0.762 m) Hg and $T = 70°F$ (21.1°C).   Thus $10 \log_{10} (4.13/4) \approx 0.1$ dB, which is negligible in practical problems.   This neglected term may become significant when the ambient pressure or temperature or both differ appreciably from normal.   For example, let $p_s = 20$ in. (0.508 m) Hg and $T = 140°F$ (60°C), yielding $\rho c = 259$.   Thus $10 \log (259/400) = -1.9$ dB.

sphere ($\bar{L}_s$). The general principles stated above for the hemispherical case apply, except that the test area is doubled.

After the free-field measurements have been performed in spherical space, the sound-power level in each frequency band is computed as follows

$$L_W \approx \bar{L}_{pS} + 20 \log r + 10 \log 4\pi \qquad \text{DB} \qquad (6.17)*$$

where L_W = sound-power level, dB re 10^{-12} watt
$\bar{L}_{pS}$ = mean-square sound-pressure level for the noise source, dB re 2×10^{-5} N/m², over a test sphere (see Fig. 6.6)
r = radius of test sphere, m
$10 \log_{10} 4\pi = 11$ dB
L_W should be computed for all frequency bands of interest. The radiated acoustic power W is found with Eq. (6.16).

6.8   Sound-power Measurements under Field Conditions

In those cases where the noise source cannot be moved from its location into a special test environment, the measurements must be carried out in a situation where the walls and ceiling are not highly absorbing and the room is not highly reverberant. A hard floor is almost always present.

No specific assumptions are made concerning the room except that it should be large enough so that the microphone can be placed in the far field of the source without being too close to the room boundaries. The noise source under test should be mounted as it is normally used. This mounting condition will typically include the hard floor. Unusually long or narrow rooms are best avoided.

In order to compute the sound-power level, the total room absorption of the test room must be known. To this purpose, the test room is "calibrated." A reference sound source (Fig. 6.4) is used.

The radius r of the test hemisphere that will be used around the machine is chosen. For this radius the mean-square sound-pressure levels in the various frequency bands for the reference source (as though it were in a free field) are calculated by subtracting $10 \log 2\pi r^2$, where r is in meters, from the sound-power levels in the frequency bands (taken from Table 6.1). *Let us designate these levels* $\bar{L}_{pR_1}$.

The reference source is now placed in the field situation on the floor at what one judges is an acoustically equivalent location in the room. The mean-square sound-pressure band levels over a hemisphere of the

<hr>

* See footnote after Eq. (6.15).

same radius are determined, using the techniques of Example 6.4. *Let us designate these levels* $\bar{L}_{pR_2}$.

The expression for determining a, the total sound absorption of the test room, is computed from

$$\frac{4}{a} = \frac{1}{S_H}\left[-1 + \text{antilog}\left(\frac{\bar{L}_{pR_2} - \bar{L}_{pR_1}}{10}\right)\right] \quad (\text{m}^2)^{-1} \qquad (6.18)$$

where $a = S\bar{\alpha}_{\text{Sab}}$ = total absorption in the room, m² [see Eqs. (9.1) and (9.2)]

$S_H = 2\pi r^2$, the area of the test hemisphere, m²

The noise source under test is now operated in the test room and the mean-square sound-pressure level $\bar{L}_{pS}$ over a test hemisphere of the same radius or above is determined, again proceeding as in Example 6.4. The band power levels of the source under test are computed from

$$L_W \approx \bar{L}_{pS} - 10\log\left(\frac{1}{S_H} + \frac{4}{a}\right) \qquad \text{dB } re \ 10^{-12} \text{ watt} \qquad (6.19)$$

where $\bar{L}_{pS}$ = mean-square sound-pressure level of source under test in a frequency band, dB re 2×10^{-5} N/m², determined in semireverberant test room over test hemisphere of surface S_H

$\dfrac{4}{a}$ = as computed from Eq. (6.18)

Obviously, if the source under test could be moved and the reference source substituted for it, and $\bar{L}_{pR}$ determined, the values of $\bar{L}_{pS}$ and $\bar{L}_{pR}$ could be substituted for $\bar{L}_p$ and $\bar{L}_p'$, respectively, in Eq. (6.11) to obtain L_W.

6.9 Sound-power Measurements in a Duct

The sound-power level in a duct can be computed easily from sound-pressure-level measurements, provided one has essentially a plane progressive wave, using the equation

$$L_W \approx L_{pD} + 10\log S_D \qquad \text{dB} \qquad (6.20)$$

where L_W = level of the total sound power traveling down the duct, dB re 10^{-12} watt

L_{pD} = sound-pressure level measured just off the center line of the duct, dB re 2×10^{-5} N/m²

S_D = cross-sectional area of the duct, m²

The above relation assumes not only a nonreflecting termination for the end of the duct opposite the source, but also a uniform sound

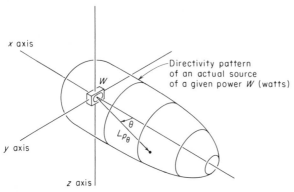

x axis

W

Directivity pattern
of an actual source
of a given power *W* (watts)

θ

L_{p_θ}

y axis

z axis

Fig. 6.7   Directivity pattern of a noise source radiating acoustic power *W* into free space.   A particular sound-pressure level L_{p_θ} is shown as the length of a vector terminating on the surface of the directivity pattern at angle θ.   The sound-pressure levels were measured at various angles θ and at a fixed distance *r* from the actual source in free space.

intensity across the duct.   At frequencies above the first cross resonance of the duct the latter assumption is no longer satisfied.   Equation (6.20) can still be used provided L_{pD} is replaced by a suitable space average $\bar{L}_{pD}$ over the cross-sectional area S_D obtained from sound-pressure-level measurements at several points across the duct.   The number of measurement positions across the cross section used to determine $\bar{L}_{pD}$ will depend on the accuracy desired and the frequency.

In practical situations, reflections occur at the open end of the duct, and the effect of branches and bends must be considered.[16]   It is also necessary to surround the microphone by a suitable windscreen (see Chap. 4) to reduce the aerodynamic noise that might interfere with the measurements.

6.10   Directivity Pattern [1,2]

Most sources of sound of practical interest are somewhat directive.   If one measures the sound-pressure level in a given frequency band a fixed distance away from the source, different levels will generally be found for different directions.   A plot of these levels in polar fashion at the angles for which they were obtained is called the *directivity pattern* of the source.   A directivity pattern forms a three-dimensional surface, a hypothetical example of which is sketched in Fig. 6.7.   The particular pattern shown exhibits rotational symmetry about the direction of maximum radiation, which is typical of many sources of noise of practical interest. It is also in the nature of things that at low frequencies many sources of noise are nondirective, or nearly so. As the frequency increases, directivity also increases. The directivity pattern is usually determined in the far (anechoic) field (see Fig. 6.3) and in the absence of obstacles and reflecting surfaces other than those associated with the source itself.

6.10.1 Directivity Factor Q

A numerical measure of the directivity of a sound source is the directivity factor Q, a dimensionless quantity. To understand the meaning of the directivity factor we must first compare Figs. 6.7 and 6.8. We see in Fig. 6.8 the directivity pattern of a nondirective source. It is a sphere with a radius equal in length to L_{pS}, the sound-pressure level in decibels measured at distance r from a source radiating a total sound power W watts. In other words, the sources of Figs. 6.7 and 6.8 both radiate the same total acoustic power W, but because the source of Fig. 6.7 is directive, it radiates more sound than that of Fig. 6.8 in some directions and less in others.

In order to derive a directivity factor Q, we must assume that the directivity pattern does not change shape, regardless of the radius r at which it is measured. For example, if $L_{p\theta}$ at a particular angle is 3 dB greater than at a second angle, the 3 dB difference should be the same whether r is 1, 2, 10, or 100 m. This is only true in the far field of a source located in anechoic space.

The directivity factor Q_θ is defined as the ratio of (1) the mean-square sound pressure $(N/m^2)^2$ measured at angle θ and distance r from an actual source radiating W watts to (2) the mean-square sound pressure measured at the same distance from a nondirective source radiating the same acoustic power W. Alternatively, Q_θ is defined as the ratio of the intensity $(watts/m^2)$ measured at angle θ and distance r from an actual

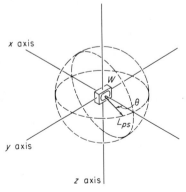

x axis

y axis

z axis

Fig. 6.8 Spherical directivity pattern of a nondirectional source radiating acoustic power W into free space. At all angles θ and distance r the sound-pressure levels equal L_{pS}, where $L_{pS} = 10 \log_{10}(10^{12}W/4\pi r^2)$.

source to the intensity measured at the same distance from a nondirective source, both sources radiating the same power W. Thus

$$Q_\theta = \frac{p_\theta^2}{p_s^2} = \frac{I_\theta}{I_s} = \frac{\text{antilog } (L_{p\theta}/10)}{\text{antilog } (L_{pS}/10)} \quad \text{(dimensionless)} \quad (6.21)$$

or

$$Q_\theta = \text{antilog } \frac{L_{p\theta} - L_{pS}}{10} \quad (6.22)$$

where $L_{p\theta}$ = sound-pressure level, measured at distance r and angle θ from a source radiating power W into anechoic space (see Fig. 6.7)

L_{pS} = sound-pressure level, measured at distance r from a nondirective source of power W radiating into anechoic space (see Fig. 6.8)

We emphasize that Q_θ is for the angle θ at which $L_{p\theta}$ was measured and that $L_{p\theta}$ and L_{pS} are for the same distance r.

6.10.2 Directivity Index The directivity index (DI) is simply defined as

$$\text{DI}_\theta = 10 \log Q_\theta \quad \text{dB} \quad (6.23)$$

or

$$\text{DI}_\theta = L_{p\theta} - L_{pS} \quad \text{dB} \quad (6.24)$$

Obviously, a nondirective source radiating into spherical space has $Q_\theta = 1$ and $\text{DI}_\theta = 0$ at all angles θ. Conversely

$$Q_\theta = \text{antilog } \frac{\text{DI}_\theta}{10} \quad (6.25)$$

6.10.3 Relations among $L_{p\theta}$, Directivity Factor, and Directivity Index The sound-pressure level for the nondirectional source of Fig. 6.8 is

$$L_{pS} = 10 \log \frac{p^2(\text{at distance } r)}{4 \times 10^{-10}} \quad \text{dB} \quad (6.26)$$

From Eq. (6.6), assuming $\rho c = 400$,* the area of a sphere $= 4\pi r^2$ and $I = W/\text{area}$, we get

$$L_{pS} = 10 \log \frac{W \times 10^{12}}{4\pi r^2} \quad \text{dB} \quad (6.27)*$$

* See footnote on p. 155.

From Eqs. (6.22) and (6.27) we find that

$$L_{p_\theta} \approx 10 \log \frac{WQ_\theta 10^{12}}{4\pi r^2} \quad \text{dB} \qquad (6.28)*$$

Remember that W is in watts and r is in meters.   In logarithmic form [see Eqs. (6.1) and (6.23)]

$$L_{p_\theta} \approx L_W + \text{DI}_\theta - 20 \log r - 11 \quad \text{dB} \qquad (6.29)$$

6.10.4 Determination of Directivity Index in Spherical Space   The directivity index DI_θ of a sound source in free space at angle θ and for a given frequency band is computed from

$$\text{DI}_\theta = L_{p_\theta} - \bar{L}_{pS} \quad \text{dB} \qquad (6.30)$$

where L_{p_θ} = sound-pressure level measured at distance r and angle θ from the source, dB

$\bar{L}_{pS}$ = space-average mean-square sound-pressure level determined over a test sphere of area $4\pi r^2$ surrounding the source, using the technique of Example 6.4, dB

6.10.5 Determination of Directivity Index in Hemispherical Space
The directivity index DI_θ of a sound source on a rigid plane at angle θ and for a given frequency band is computed from

$$\text{DI}_\theta = L_{p_\theta} - \bar{L}_{pH} + 3 \quad \text{dB} \qquad (6.31)$$

where L_{p_θ} = sound-pressure level measured at distance r and angle θ from the source, dB

$\bar{L}_{pH}$ = space-average mean-square sound-pressure level determined over a test hemisphere of area $2\pi r^2$ surrounding the source, using the technique of Example 6.4, dB

The 3 dB in this equation is added to the $\bar{L}_{pH}$ because the measurement was made over a hemisphere instead of a full sphere, as defined in Eq. (6.24).   The reason for this is that the intensity at radius r is twice as large if a source radiates into a hemisphere as compared to a sphere.   In other words, if a source were to radiate uniformly into hemispherical space, $\text{DI}_\theta = \text{DI} = 3$ dB.

6.10.6 Determination of Directivity Index in Quarter-spherical Space
Some pieces of equipment are normally associated with more than one reflecting surface.   An air conditioner standing on the floor against a wall is a case in point.   The power level of noise sources of this type should be measured with those surfaces in place.   This is done best in a

* See footnote on p. 155.

test room with anechoic walls but with one hard wall forming an "edge" with the hard floor. The general considerations of the preceding paragraphs apply here as well. One determines the space-average mean-square sound-pressure level for the quadrant $\bar{L}_{pQ}$ and also determines L_{p_θ} as before. The directivity index is given by

$$\mathrm{DI}_\theta = L_{p_\theta} - \bar{L}_{pQ} + 6 \quad \mathrm{dB} \tag{6.32}$$

6.11 Cautions Regarding Use of Sound-power Levels

The concept of sound-power level is very useful for describing sources that are to be located in a reverberant room, outdoors, or in a duct. If the source is to be located outdoors, the directivity index is also needed.

There are situations, such as the noise at the ear of a person operating a machine while standing in the near field, where sound-power level is of no value. There, sound-pressure levels should be measured at all positions where the ears of the operator (or operators) might be located.

If a source is strongly directional, both the sound-power level and the directivity index are needed, if the operator is to stand in the direction of high directivity. Here again, it may be simpler to measure only the sound-pressure level at the operator's ear.

There are also cases in which it is necessary to obtain at least an approximate sound-power level for relatively large equipment in a room with a high background noise level caused by other equipment. If the machine is not highly directive, this can be done if many measurements are taken on an envelope relatively close to the machine.

REFERENCES

1. A. P. G. Peterson and E. E. Gross, *Handbook of Noise Measurement*, General Radio Company, West Concord, Mass. (Revised editions are issued frequently.)
2. L. L. Beranek, *Acoustics*, McGraw-Hill, 1954.
3. ANSI S1.2, 1962, "Physical Measurement of Sound," American National Standards Institute, Inc., 1430 Broadway, New York, N.Y. 10018.
4. ISO/R495-1966, "General Requirements for the Preparation of Test Codes for Measuring the Noise Emitted by Machines," available through American National Standards Institute, Inc., 1430 Broadway, New York, N.Y. 10018.
5. ADC Test Code 1062 R1. Also, "Testing Air Diffusion Equipment under the ADC Equipment Test Codes" and "Equipment Test Codes," Air Diffusion Council, 435 North Michigan Ave., Chicago, Ill. 60611.
6. ASHRAE 36-62, "Measurement of Sound Power Radiated from Heating, Refrigerating and Air Conditioning Equipment," American Society of Heating, Refrigerating and Air-Conditioning Engineers, Inc., United Engineering Center, 345 East 47th St., New York, N.Y. 10017.
7. ARI 443-66, "Sound Rating of Room Fan-coil Air Conditioners"; ARI 270-67, "Sound

Rating of Outdoor Unitary Equipment"; ARI 446-68, "Sound Rating of Room Air-induction Units," Air Conditioning and Refrigeration Institute, Washington, D.C.

8. P. K. Baade, "Standardization of Machinery Sound Measurement," *ASME Publ.* 69-WA/FE-30, 1969, The American Society of Mechanical Engineers, United Engineering Center, 345 East 47th Street, New York, N.Y. 10017.

9. British Standard 4196:1967, "Guide to the Selection of Methods of Measuring Noise Emitted by Machinery," British Standards Institution, British Standards House, 2 Park St., London, W. 1, England.

10. P. M. Morse and K. U. Ingard, *Theoretical Acoustics*, McGraw-Hill, 1968.

11. R. V. Waterhouse, "Output of a Sound Source in a Reverberation Chamber and Other Reflecting Environments," *J. Acoust. Soc. Am.*, vol. 30, p. 4, 1958.

12. R. Waterhouse and D. Lubman, "Discrete versus Continuous Space Averaging in a Reverberant Sound Field," *J. Acoust. Soc. Am.*, vol. 48, pt. 1, pp. 1–5, 1970.

13. E. D. Lawler, "Rotating Diffusers for Reverberation Rooms," *S/V Sound and Vibration*, vol. 2, no. 9, pp. 16–20, 1968.

14. C. Ebbing, "Experimental Evaluation of Moving Sound Diffusers for Reverberation Rooms," *J. Acoust. Soc. Am.*, vol. 45, p. 343 (A), 1969.

15. D. Lubman, "Fluctuations of Sound with Position in a Reverberant Room," *J. Acoust. Soc. Am.*, vol. 44, pp. 1491–1502, 1968.

16. R. V. Waterhouse, "Interference Patterns in Reverberant Sound Fields," *J. Acoust. Soc. Am.*, vol. 27, p. 247, 1955.

Chapter Seven

Sound Propagation Outdoors

ULRICH KURZE and LEO L. BERANEK

7.1 Introduction

Sound waves travel from source to receiver outdoors through an atmosphere that is in constant motion. Turbulence, temperature and wind gradients, viscous and molecular absorption, and reflection from the earth's surface all affect the amplitude and create fluctuations in the sound received. The longer the transmission path through the atmosphere, the less certain the average amplitude and the greater the fluctuations in the received sound.

Often, in a noise control problem, the receiver is near the ground. The noise source is also frequently near the ground. An exception is aircraft noise, one of today's important sources of noise "pollution," which usually travels a considerable distance through the atmosphere, from the aircraft in flight. The three primary propagation paths for sound outdoors are shown in Fig. 7.1.

The discussion in this chapter is limited to propagation of sound in the audible frequency range over the air-to-ground and ground-to-ground transmission paths. Acoustic sky-wave transmission is not discussed, although enormous distances are occasionally bridged under favorable wind and temperature conditions.

164

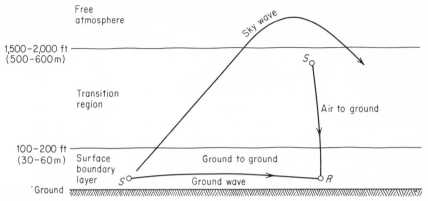

Fig. 7.1   The three principal types of sound propagation in the atmosphere to a receiver located near the ground.

Sound transmission by any of the paths is strongly affected by the weather. Only since World War II have adequate micrometeorological instrumentation techniques been available to assist in predicting transmission characteristics. The material reported here is a combination of theory and empiricism. In this context, prediction of sound propagation is a combination of formulas and of engineering judgment.

7.2 Wave Divergence

The sound-pressure level generated by a sound source at a receiver located in the far field of the source (see Fig. 6.3) decreases as some function of the distance from the source owing to wave divergence. The sound-pressure level also is affected by environmental conditions.

Let us assume that the sound source is directive and is located very near a hard, flat area on which there may be trees, shrubbery, or some types of barrier (wall). If this source has a sound-power level L_W, it will produce [see Eqs. (6.29) and (6.30)] a sound-pressure level of

$$L_{p\theta} = L_W + \mathrm{DI}_\theta - 20 \log r - A_e - 11 \qquad \mathrm{DB} \qquad (7.1)$$

where $L_{p\theta}$ = sound-pressure level at a receiver located in the direction θ, a distance r (in m) from the source, dB *re* 2×10^{-5} N/m^2

L_W = sound-power level of the source, dB *re* 10^{-12} watt

DI_θ = directivity index of source in the direction θ, dB (note: even if the source is nondirective, DI = 3 dB for hemispherical radiation)

r = distance of receiver from source, m

A_e = excess attenuation caused by environmental conditions, dB (see one or more of the factors listed in Sec. 7.3)

11 dB $\approx 10 \log 4\pi$

For hemispherical wave divergence, in a homogeneous, loss-free ($A_e = 0$) atmosphere, and over loss-free ground, free of barriers, the sound-pressure level drops off 6 dB for each doubling of distance or 20 dB for each factor of 10 in distance. Remember, for a nondirective source, DI = 3 dB because of the presence of the plane, which, by reflection, doubles the power radiated into half space.

If the sound-power level of the source is not known but the sound-pressure level in a particular direction (angle) from a point source is known, it is possible to calculate the sound-pressure level at a new position (in the same direction, i.e., the same angle, from the source) by

$$L_{p_2} = L_{p_1} - 20 \log \frac{r_2}{r_1} - A_e \qquad \text{dB} \qquad (7.2)$$

where L_{p_1} = sound-pressure level at an angle θ and distance r_1 from the source, dB re 2×10^{-5} N/m^2

L_{p_2} = sound-pressure level at an angle θ and distance r_2 from the source, dB re 2×10^{-5} N/m^2

r_1, r_2 = distances on the same line from the source that must lie in the far field, m

A_e = excess attenuation for the distance $r_2 - r_1$, dB

Point Source. For the special case of a point source * radiating sound into a homogeneous, loss-free atmosphere above a flat rigid surface, Eq. (7.1) can be simplified to

$$L_p = L_W - 20 \log r - 8 \qquad \text{dB} \qquad (7.3)$$

Discrete Sources on a Line. A common line source is freely flowing automobile traffic on a straight flat highway. Let us assume *n point sources on such a highway a distance b apart* each radiating different (incoherent) sounds, but all of the same strength (see Fig. 7.2). The sound-pressure level L_p at an observer for loss-free air and ground is determined from †

$$L_p = L_{W_1} + 10 \log \left(\frac{\alpha_n - \alpha_1}{r_0 b} \right) + \Delta L - 8 \qquad \text{dB} \qquad (7.4)$$

where L_{W_1} = sound-power level of each of the incoherent sources, assumed equal in power, dB re 10^{-12} watt

$\alpha_n - \alpha_1$ = aspect angle of the n sources, radians (Note: α_n and α_1 may take on both positive and negative values, but $\alpha_n > \alpha_1$.)

r_0 = perpendicular distance from the line to the observer, m

b = distance between two adjacent sources, m

n = number of sources

---

* A point source is one that is small compared to the wavelength of the emanated sound, and at the surface of it all points have the same sign of radial velocity. It is, therefore, nondirectional.

† Consideration of excess attenuation is found in Secs. 7.4 and 7.6.

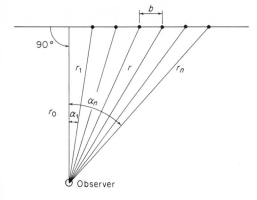

Fig. 7.2 Set of n incoherent point sound sources (indicated by the dots) separated by distance b, arranged on a line and located at a perpendicular distance r_0 relative to an observer. This situation is equivalent to a line of n automobiles passing by on a highway. The angles α_1 and α_n may take on both positive and negative values, but $\alpha_n > \alpha_1$.

and

$$\Delta L = 10 \log \left\{ \frac{b}{r_0} \frac{\cos^2 \alpha_1}{\alpha_n - \alpha_1} \times \right.$$

$$\left. \sum_{m=1}^{n} \frac{1}{1 + (m-1)\dfrac{b}{r_0} \cos \alpha_1 [(m-1)\dfrac{b}{r_0} \cos \alpha_1 + \sin \alpha_1]} \right\} \quad (7.5)$$

The correction ΔL is less than 1 dB if both *

$$n \text{ (the number of sources)} \geq 3 \quad (7.6)$$

and

$$\frac{r_0}{b \cos \alpha_1} \geq \frac{1}{\pi} \quad (7.7)$$

Equation (7.6) represents no real limitation because calculations for one or two sources can be done straightforwardly [see Eq. (2.20)] rather than with Eq. (7.5).

Equation (7.7) gives the condition, derived by Rathe,[1] for which a number of point sources on a line is equivalent to a continuous line source. If the condition is met, the observer's distance from the closest point source is larger than one-third the distance between two of the point sources. If the condition in Eq. (7.7) is not met, the observer's place is so close to the nearest point source that the level can be calculated by considering exclusively the radiation from this nearest point source with an error of less than 1 dB.

For an *infinite number n of point sources spaced at distance b apart,*[1] $\alpha_n - \alpha_1 = \pi$ radians and we get from Eq. (7.4)

---

* For randomly distributed sources on a line (freely flowing road traffic), Eq. (7.4) gives the energy mean level with $\Delta L = 0$.[41]

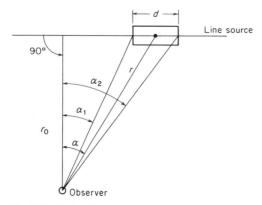

Fig. 7.3   Finite line source of length d, subtended by the angles α_1 and α_2 located on a line at a perpendicular distance r_0 relative to the observer. The source sound-power level L_{WL} is specified per unit length. This is equivalent to a long line of automobiles (rush hour) or a train at a distance from an observer.

$$L_p = L_{W_1} - 10 \log r_0 b - 3 \qquad \text{for } r_0 \geq \frac{b}{\pi} \qquad \text{dB} \qquad (7.8)$$

or

$$L_p = L_{W_1} - 20 \log r_0 - 8 \qquad \text{for } r_0 < \frac{b}{\pi} \qquad \text{dB} \qquad (7.9)$$

For an *infinite number n of point sources within a finite length d* (which form an incoherent line source), Eq. (7.4) is generally valid with $\Delta L = 0$ if the definition of L_{W_1} is altered (see Fig. 7.3). We have [1]

$$L_p = L_{WL} + 10 \log \left(\frac{\alpha_2 - \alpha_1}{r_0 d} \right) - 8 \qquad \text{dB} \qquad (7.10)$$

where L_{WL} = sound-power level for unit length (1 m) of the line source, dB

α_1 = angle of the vector to the start of the line source (Note: α_1 and α_2 may take on both positive and negative values.)

α_2 = angle of the vector to the end of the line source; α_2 must be greater than α_1

Close to the line source, where $\alpha_2 - \alpha_1 \to \pi$, Eq. (7.8) can be used instead of Eq. (7.10) with L_{W_1} replaced by L_{WL} and with b replaced by d.

Far from the line source, where $\alpha_2 - \alpha_1 \approx d/r_0 \cos \alpha \approx d/r$, one obtains (with the above definition of L_{WL})

$$L_p = L_{WL} - 20 \log r - 8 \qquad \text{dB} \qquad (7.11)$$

Note that the distance r_0 from the observer to the line is replaced by the distance r to the center of the line source.

Such simple expressions cannot be derived for the general case of m parallel lines, nor for a finite plane source because of the more complicated directivity index. The sound-pressure level can be expressed in simplified form only on the axis of symmetry and within a rather large error equal to a maximum of 3 dB. One configuration of a plane source is covered by Rathe.[1]

Utilizing the two basic building blocks, namely, the point source and the series of point sources arranged in a line, one should be equipped to analyze most practical problems relating to the transmission of sound out-of-doors. At this point in the chapter, the reader should be able to estimate the noise levels as a function of distance from a source under ideal conditions of sound propagation.

7.3 Excess Attenuation A_e

The excess attenuation A_e owing to environmental and other conditions is the attenuation beyond that caused by wave divergence. It includes one or more of the following:

A_{e_1} = effect of the difference in value of ρc from 400 mks rayls when the ambient temperature and barometric pressure differs appreciably from the values given in Table 2.4. The value of A_{e_1} may be determined from Fig. 2.3 for ambient temperature and pressure conditions that differ appreciably from the combinations in Table 2.4. Units are dB

A_{e_2} = attenuation by absorption in the air, dB

A_{e_3} = attenuation by rain, sleet, snow, or fog, dB

A_{e_4} = attenuation by barriers, dB

A_{e_5} = attenuation by grass, shrubbery, and trees, dB

A_{e_6} = attenuation and fluctuation owing to wind and temperature gradients, to atmospheric turbulence, and to the characteristics of the ground, dB

These attenuation terms are to some extent interrelated in that the effect of one may be dependent upon the presence of one or more of the others. In the case of air-to-ground sound propagation, the presence of barriers, trees, and plantings on the ground and the effects of ground-created wind and temperature gradients are usually not important except for aircraft flying low just after takeoff or prior to landing.

A basic assumption throughout this chapter is that the sound power from the source is not so great that the air is "overdriven"; thus, it is not necessary to consider the problems of nonlinearity in the medium.

7.4 Excess Attenuation by Air Absorption, A_{e_2}

Sound absorption in quiet, isotropic air is caused by two processes:[2] First, energy is extracted from a sound wave by losses arising from heat conduction and viscosity in the air. This so-called classical absorption, which is proportional to frequency squared, is of significance only at very low temperatures. Second, energy is extracted from a sound wave by rotational and vibration relaxation of the oxygen molecules in the air. The vapor content of the air determines the time constant of the vibration relaxation, which is more important than rotational relaxation. In addition, this "molecular absorption" depends in a major way on temperature.

The dependence of the sound absorption on frequency, temperature, and humidity has been determined theoretically,[3] as well as in the laboratory[4] and in the field.[5] Extensive tables and graphs of atmospheric sound attenuation have been published from laboratory tests[6] in the frequency range from 125 to 12,500 Hz, for temperatures between -10 and 30°C (14 to 86°F), and for relative humidities between 10 and 90%.

Atmospheric attenuation at a temperature of 20°C (68°F) may be calculated from[7]

$$A_{e_2} = 7.4 \frac{f^2 r}{\phi} \, 10^{-8} \quad \text{dB} \qquad (7.12)$$

where f = geometric-mean frequency of band, Hz
ϕ = relative humidity, %
r = distance between source and receiver, m
For other temperatures the approximation following may be used

$$A_{e_2}(T, \, \phi = 50\%) = \frac{A_{e_2}(20°\text{C}, \, \phi = 50\%)}{1 + \beta \Delta t f} \quad \text{dB} \qquad (7.13)$$

where ΔT = temperature difference in either °C or °F between the actual temperature and the reference value of 20°C (68°F). This difference must not exceed ±10°C (±18°F) for this equation to be valid
β = constant (4×10^{-6} for ΔT in °C or 2×10^{-6} for ΔT in °F)
Equations (7.12) and (7.13) are useful engineering estimates of A_{e_2} for atmospheric conditions near standard, provided propagation takes place in isotropic quiet air. More accurate information can be found in Ref. 6, which also covers more extreme conditions such as very low temperatures or very low humidities. Also, air-to-ground sound-attenuation charts as standardized by the aircraft industry for sound propagated through normal air are given in Figs. 7.4 and 7.5.[38]

Atmospheric attenuation outdoors will have the same values only if the air is still and homogeneous. Micrometeorological inhomogeneities, such as turbulences and gradients of wind, temperature and humidity, result in considerable deviations of attenuation from those found in the laboratory, particularly for sound propagation over distances of a mile (1.6 km) or so. Over short distances of a few hundred feet, atmospheric absorption generally plays a minor role in outdoor noise control problems. More information on anomalies in attenuation is given in Sec. 7.8.

The influence of atmospheric attenuation on the sound-pressure level of a noise depends mainly upon the strength of the high-frequency components in the noise spectrum. As an example, we show in Fig. 7.6 the distance between the observer and the noise source at which A-weighted noise level is reduced by 3 dBA due to atmospheric attenuation. The results for three types of noise spectra are plotted in Fig. 7.6 as a function of the relative humidity, based on calculations made from Eq. (7.12). For "white" noise, which has constant sound-pressure-squared (p^2) per unit bandwidth, the high-frequency components are strongest and atmospheric absorption causes a 3-dBA reduc-

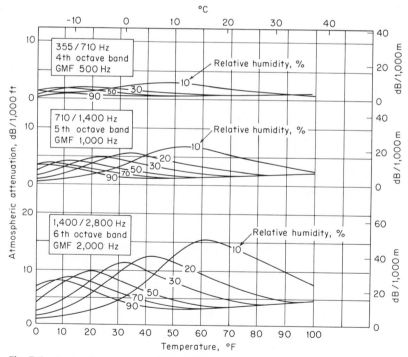

Fig. 7.4 Atmospheric attenuation for aircraft-to-ground propagation in dB/1,000 ft (or dB/1,000 m) for octave bands with center (geometric-mean) frequencies at 500, 1,000, and 2,000 Hz.

tion at a distance of only a few hundred feet (100 m). More typical of traffic noise sources is "pink" noise, with constant p^2 per constant-percentage bandwidth, and "reddish" noise with a hyperbolically decreasing p^2 per constant-percentage bandwidth (see formulas associated with curves A and B of Fig. 7.6). Noise of the latter two types experiences 3 dBA excess attenuation in the atmosphere at distances between 1,000 and 10,000 ft (300 to 3,000 m).

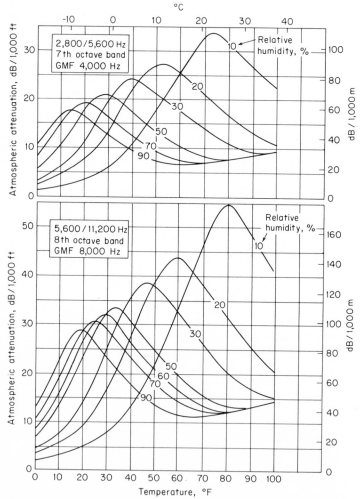

Fig. 7.5 Atmospheric attenuation for aircraft-to-ground propagation in db/1,000 ft (or dB/1,000 m) for octave bands with center (geometric-mean) frequencies at 4,000 and 8,000 Hz.

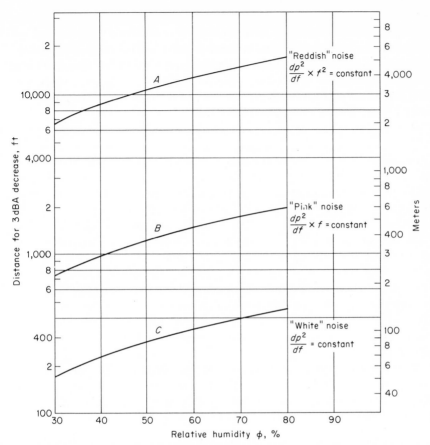

Fig. 7.6   Means for determining the distance a continuous-spectrum random noise must travel to be attenuated 3 dBA as measured on the A scale of an American National Standard sound-level meter.   The abscissa is relative humidity in percent. Curve *A* is for a noise with a spectrum that when measured with a set of constant-percentage bands (e.g., octave or one-third-octave bands, flat-response instrument) yields a plot that decreases 3 dB for each doubling of band mid-frequency.   Curve *B* is for a noise that, when measured the same way, has a spectrum that remains constant with frequency.   Curve *C* is for a noise that, when measured as for *A*, has a spectrum that rises 3 dB for each doubling of band mid-frequency.

7.5   Excess Attenuation by Fog, Rain, or Snow, A_{e_3}

It is commonly said that on days of fog or light precipitation sound carries exceptionally well.   While this observation is true, it is not attributable to any remarkable acoustic property of fog or rain, but to secondary effects.   During light precipitation, the gradients of temperature and wind (measured vertically above the ground) tend to be small so that sound "carries" farther outdoors than on a sunny day with the attendant micrometeorological inhomogeneities resulting from the sun's heating. Another factor that contributes to this observation is background noise level. When there is fog, the noise of traffic, birds, aircraft, children, and other outdoor activities diminishes appreciably. Also, with snow on the ground, background levels diminish.

There would appear to be conflicting evidence concerning the effect of fog on sound propagation in air. Some laboratory experiments with (extremely heavy) artificial fog show appreciable excess attenuation beyond that of molecular absorption in dry air.[9-11] Field measurements show that excess attenuation of sound obtained downwind in fog of visibility varying between about 1,000 and 3,000 yards is not statistically different from that obtained downwind in the absence of fog.[12] The field measurements are consistent with theoretical predictions of the excess sound attenuation in natural fog that has typical diameters of water droplets of about 25 microns (2.5×10^{-5} m) with a total liquid water content of 10^{-4} kg/m^3. It appears that molecular absorption [11,13] is entirely negligible in the frequency range of interest even for dense fog and that viscous and heat loss absorption [14] for fog becomes comparable to absorption in clear atmosphere only below a frequency of about 300 Hz.

If we neglect excess attenuations of less than 0.5 dB per 1,000 m, and if we desire conservative estimates of excess attenuation in noise control problems, it is recommended that no excess attenuation be assigned to fog. That is to say

$$A_{e_3} \approx 0 \tag{7.14}$$

According to observations described in Ref. 12, the same assumption applies to light falling precipitation.

7.6　Attenuation by Barriers, A_{e_4}

7.6.1　Walls　Nonporous walls of sufficient mass (minimum of 20 kg/m^2 or 4 lb/ft^2), if interposed between source and receiver, can result in appreciable noise reduction, because sound can reach the receiver only by diffraction around the boundaries of the obstacle.

The analysis in this section is based on an analytical approximation of experimental data [18,19] which is consistent with asymptotic results of optical-diffraction theory.[16,17,20,21] In the theory of Fresnel diffraction, only that region of an incident wavefield that is close to the top edge of a barrier contributes appreciably to the wavefield that is diffracted over the barrier.[15] For an observer in the shadow zone of the barrier at some distance away, the diffracted sound field appears to be radiated from a line source along the top edge of the barrier. The strength of this virtual line source is proportional to the strength of the incident wave at the edge of the barrier. The sound pressure in the incident wave, of course, decreases in direct proportion to the distance of the barrier from a point source or in proportion to the square root of the distance between an infinite parallel line source and the barrier. If the distance between the observer and the edge is much smaller than that between the edge and a point source, all parts of the section of the virtual line source nearest the observer radiate with almost equal strength, which results in cylindrical wave propagation from the edge to the observer. On the other hand, if the point source is much closer to the edge than is the observer, the strongest section of the virtual line source is very short (as viewed by the observer), which results in spherical wave propagation from the wall to the observer.

The simplified model just described requires corrections in the near field of the line source, i.e., close to the top edge of the barrier, and at low frequencies. It requires further corrections, mainly at the high frequencies, for micrometeorological inhomogeneities which distort the presumed phase correlation of the incident wave.

The diffracted wave emanating from the top edge of the barrier is not restricted to the shadow zone and, therefore, also affects a small transition region close to the shadow zone by interfering with the direct wave.

Point Source. The following formula for the excess attenuation of a rigid straight barrier for sound incident from a point source includes the approximate result from optical-diffraction theory, a correction for the near field at the edge, and an approximation for the transition region between the shadow zone and the bright zone (see Fig. 7.7).

$$(A_{e_4})_{\text{point}} = \begin{cases} 20 \log \dfrac{\sqrt{2\pi N}}{\tanh \sqrt{2\pi N}} + 5 \text{ dB} & \text{for } N \geq -0.2 \\ 0 & \text{otherwise} \end{cases} \qquad (7.15)$$

N is the Fresnel number (dimensionless):

$$N = \pm \frac{2}{\lambda}(A + B - d) \qquad (7.16)$$

where λ = wavelength of sound, m
  d = straight-line distance between source and receiver, m
  $A + B$ = shortest path length of wave travel over the wall between source and receiver (see Fig. 7.7a), m
  + sign = receiver in the shadow zone
  − sign = receiver in the bright zone
Equation (7.15) is plotted in Fig. 7.8.

In the *bright zone*, which is defined in Fig. 7.8 for $N < -0.2$, the diffracted wave can be neglected and the excess attenuation taken as zero.

In the *transition zone*, where $z_R + |x_R/x_S|z_S < 0$ (see Fig. 7.7b), the edge of the barrier approaches the line of sight from source to receiver and the excess attenuation is between 0 and 5 dB (see Fig. 7.8). In Eq. (7.15), it is appropriate to replace N by its absolute value and the hyperbolic tangent (tanh $\sqrt{2\pi N}$) by the trigonometric tangent (tan $\sqrt{2\pi |N|}$) to obtain the excess attenuation.

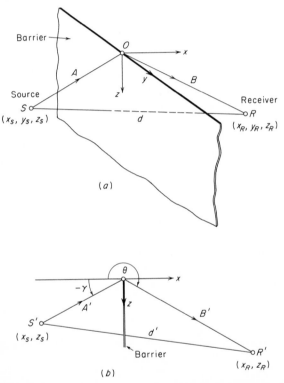

Fig. 7.7 Geometry of sound-propagation path over a barrier wall: (a) perspective view and (b) projection plane perpendicular to the barrier. The source and receiver are S and R, respectively. The coordinate system is shown in (a) and the source coordinates are x_s, y_s, z_s and the receiver coordinates are x_r, y_r, z_r. The shortest path that the sound can travel over the barrier between source and receiver is distance $A + B$. The angle $(\gamma + \theta)$ gives, on the projection plane, the direction of the refracted wave relative to that of the incident wave. Ground reflections are neglected, which is permissible if the source and receiver are very near to a completely reflective ground. For other heights of source and receiver over partially absorptive ground, sound can travel to the receiver along three additional paths, with one or two reflections where phase shifts may occur, causing interference effects.

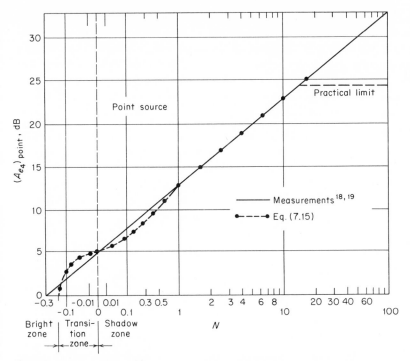

Fig. 7.8 Excess attenuation $(A_{e_4})_{point}$ of the sound from a point source by a rigid barrier as a function of Fresnel number N [see Eqs. (7.15) and (7.16)]. A negative N refers to the case where the receiver is able to see the source; i.e., $(\theta - \gamma) < \pi$.

In the *shadow zone,* the excess attenuation ranges from 5 dB, which is the case for very low frequencies or close to the line of sight, to an upper limit, which is determined by atmospheric irregularities and turbulence. Experimental data show that this upper limit to the excess attenuation is about 24 dB.[1,22]

If *both the source and the receiver are relatively close to the barrier,* so that the shortest path length $A + B$ of the diffracted wave is considerably longer than the straight-line distance d, one has to account for the stronger spherical divergence of the diffracted wave with an additional term $20 \log (A + B)/d$ in Eq. (7.15).

The *shortest path length* of the direct plus diffracted wave can be found from the length $A' + B'$ in the projection plane perpendicular to the barrier (see Fig. 7.7b) and from the path length $y_s - y_r$ parallel to the barrier

$$A + B = \sqrt{(A' + B')^2 + (y_s - y_r)^2} \tag{7.17}$$

If the source or the receiver or both are not very close to the barrier

$(A + B - d \ll d)$, it is sufficient to determine the *maximum Fresnel number*

$$N_{\text{max}} = \pm \frac{2}{\lambda} (A' + B' - d') \qquad (7.18)$$

where λ = wavelength of sound, m

d' = straight-line distance between source and receiver in the projection plane perpendicular to the barrier (see Fig. 7.7b), m

$A' + B'$ = shortest path length of the diffracted wave between source and receiver in the projection plane (see Fig. 7.7b), m

N_{max} determines the maximum excess attenuation when the line connecting the source and receiver is perpendicular to the barrier ($y_s = y_r = 0$). The smaller excess attenuations in other directions, which are defined by

$$\cos \alpha = \frac{d'}{d} \qquad (7.19)$$

are calculated from Eq. (7.15) with the Fresnel number

$$N = N_{\text{max}} \cos \alpha \qquad (7.20)$$

Coherent Line Source. For sound that emanates from a coherent line source parallel to the edge of a barrier, the barrier attenuation also can be calculated from Eq. (7.15) with $N = N_{\text{max}}$. If source and receiver are close to the barrier, an additional term $10 \log [(A' + B')/d']$ has to be considered, which accounts for the stronger cylindrical divergence of the diffracted wave.

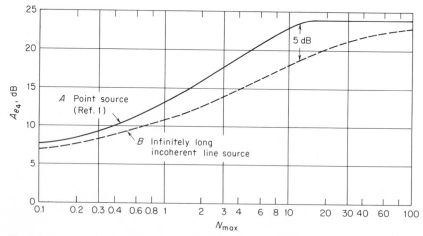

Fig. 7.9 Excess attenuation by a rigid barrier vs. Fresnel number N_{max} for sound waves that arrive from (A) a point source on that line nearest the receiver and (B) an infinitely long, incoherent line source parallel to the edge of a rigid barrier.

Incoherent Line Source. The excess attenuation by a barrier of the sound from an incoherent line source parallel to the edge of the barrier follows from integration over the aspect angle $\Delta\alpha$ (see sketch in Fig. 7.10) that encompasses the two ends of the line source. The result for an infinitely long line source $\Delta\alpha = \pi$ is plotted as B in Fig. 7.9. It shows that the excess attenuation of sound from an incoherent line source by a barrier is always smaller than the excess attenuation for the closest point of the line. The maximum difference is 5 dB, when considering an upper limit of 24 dB for the barrier attenuation and neglecting atmospheric absorption.

The maximum difference between (1) the barrier attenuation of sound from a finite-length line source and (2) the attenuation of sound from the closest point of the same line source is only 1 dB (and, hence, negligible), if the aspect angle $\Delta\alpha$ is *confined* to those numbers that are

TABLE 7.1

i	α_i(radians)	$\Delta\alpha_i$(radians)
1	0	1.05
2	1.05	0.20
3	1.25	0.12
4	1.37	0.067
5	1.437	0.067
6	1.504	0.067

plotted in Fig. 7.10 as a function of the angle α_1. The excess attenuation of sound from a line source [long enough so that it cannot be treated as a point source (see table above)] is calculated from

$$(A_{e4})_{\text{line}} = -10 \log \sum_i \frac{1}{\Delta\alpha_i} \, 10^{-[(A_{e4_i})\text{point}/10]} \qquad (7.21)$$

where i = summation index over parts of the line source
$\Delta\alpha_1 = \alpha_2 - \alpha_1$, aspect angle of the closest part of the line source
$\Delta\alpha_2 = \alpha_3 - \alpha_2$, aspect angle of the next part of the line source
$\Delta\alpha_i$ = aspect angle of the ith part of the line source
$(A_{e4_i})_{\text{point}}$ = excess attenuation of sound from a point source in the direction α_i [see Eq. (7.15)], dB

The above table, which can be applied to parts of a finite-length line source and uniform propagation conditions within $\Delta\alpha_i$, gives upper limits for the aspect angles $\Delta\alpha_i$, according to Fig. 7.10.

Measurements of attenuation of barriers along models of highways have yielded results consistent with those from the theory given here.[23]

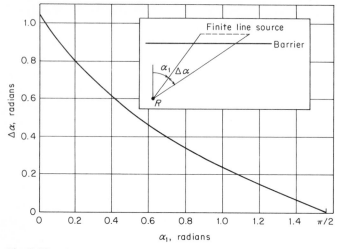

Fig. 7.10   Finite line source and infinite barrier: particular values of the aspect angle $\Delta\alpha$ as a function of α_1, for which the excess attenuation is not greater than 1 dB less for the whole line compared to the excess attenuation for a source located at the point on the line nearest the receiver.

Discussion.   In many practical cases, the barrier attenuation calculated by Eq. (7.15) yields conservative (low) numbers.   Maekawa [18] found experimentally higher values of excess attenuation for *absorptive* barriers, i.e., barriers with a sound-absorbing covering on the source side, consistent with theoretical predictions from a geometrical theory of diffraction. [17]   The theory predicts, for a completely absorptive barrier, an increase of the excess attenuation equal to $20 \log \{[\sin 0.5(\theta + \gamma) - \cos 0.5(\theta = \gamma)]/[\sin 0.5(\theta + \gamma) + \cos 0.5(\theta - \gamma)]\}$ dB.   The angles θ, γ refer to sound waves in the projection plane sketched in Fig. 7.7b.   Higher barrier attenuation may also be expected for buildings and similarly shaped thick barriers, where double diffraction at two edges is found.

Lower excess attenuation than given by Eq. (7.15) has been observed for barriers placed in the middle of areas with absorbing ground. [24,25]   The reason for this phenomenon is that destructive interference of a ground reflection with the direct wave, which might occur without a barrier (most important in the frequency range from 300 to 600 Hz as discussed in Sec. 7.8.2), can be inhibited by a barrier, thus giving rise to a sound level at the observer with the barrier that might be even higher at these frequencies than without.

7.6.2   Buildings   Measurements have been made of the penetration of freely flowing traffic noise into residential areas. [26]   Sound-pressure-level measurements were made along three lines perpendicular to a busy highway and extending into the residential areas a distance of approximately 2,000 ft (600 m). All measurement positions beyond 100 ft (30 m) were blocked visually from the highway by at least one set of buildings. The highway was elevated about 20 ft (6 m) at two of the measurement positions and was at ground level at the third.

It is seen from the data that the presence of intervening buildings introduces an excess attenuation of 15 to 20 dB. Most surprising, this excess attenuation is essentially independent of the number of intervening structures. In other words, the first shielding structure introduces the level reduction, and subsequent structures do not further reduce the sound-pressure level significantly. An attenuation of 10 dB was found by comparison of field data with theoretical results from a theoretical model for urban noise.[27] Of course, in addition to this excess attenuation, the sound-pressure levels measured along a line decreased by about 3 dB per doubling of distance owing to cylindrical spreading of the sound wave [see Eq. (7.8)].

7.6.3 Streets   A set of measurements[28] was made in a city by placing a large loudspeaker in the center of a street, facing down the street. Both sides of the street were lined with buildings. Sound-pressure levels in octave bands were measured along the street at distances of up to 1,000 ft (300 m). The levels fell off at a rate slightly in excess of 6 dB per doubling of distance at frequencies below 2,000 Hz. Above 2,000 Hz, the excess attenuation (beyond the 6-dB rate) was 6 dB per 1,000 ft at 3,000 Hz, 10 dB per 1,000 ft at 4,000 Hz, and 20 dB per 1,000 ft at 6,000 Hz. The levels did not drop off at a faster rate when the measuring position passed a cross street located 620 ft (190 m) from the loudspeaker.

Measurements were also made at positions along the cross street at distances up to 300 ft (91 m) from the intersection. Very near the intersection, just out of line of sight of the loudspeaker, the levels in the bands below 2,000 Hz suddenly dropped 8 to 10 dB. At 300 ft from the intersection, the levels below 2,000 Hz dropped an additional 10 dB. At 4,000 Hz, the initial drop just out of line of sight was 13 dB and the next drop was 12 dB.

It is concluded from the data that the sound attenuation both upwind and downwind along city streets is accounted for primarily by the 6-dB-per-doubling-of-distance law and by the usual attenuation of sound in air at the higher frequencies.

7.6.4 Depressed and Elevated Highways   One form of barrier is a side wall of a highway that is depressed below the level of the surrounding ground.[26] Measurements show that depressing a highway by about 4 m (12 ft) yields an excess attenuation, measured on the A-scale of a standard sound-level meter, of 7 to 10 dBA at all distances from the highway. For an elevated highway, excess attenuations of 2 to 10 dBA are found within 100 m (300 ft) of the highway. Beyond 300 ft the noise radiated by elevated and ground-level highways is the same.

7.7   Ground Effects, Plantings, and Trees, A_{e_5}

Ground attenuation is a function of the structure and the covering of the ground, both of which influence its acoustic properties, and of the heights of the source and receiver above the ground. In general, ground attenuation cannot be stated in terms of excess attenuation per unit of distance. When the source and the receivers are above the ground, the reflected wave from a nonrigid boundary will interfere with the direct wave in the far field at frequencies that are most likely in the frequency range between 300 and 600 Hz.[8,9,29]

To a first approximation, excess attenuation due to ground absorption can be neglected within a distance of 30 to 70 m (100 to 200 ft) of the source. In the range of 70 to 700 m, the excess attenuation can be re-

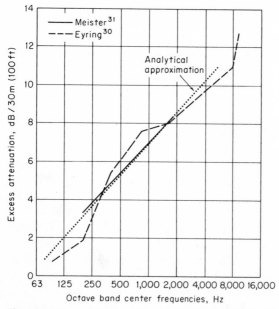

Fig. 7.11   Excess attenuation for sound propagation through shrubbery and over thick grass.

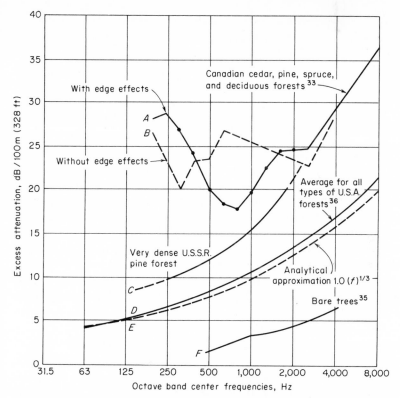

Fig. 7.12 Measured excess attenuation for sound propagation in tree zones (forests). For very thick forests, the upper two curves show excess attenuations of 10 to 25 dB per 100 m (328 ft) at frequencies below 2,000 Hz. For less thick forests, the average curve shows excess attenuations of 5 to 12 dB per 100 m at these frequencies. Deciduous trees without their leaves (bare) yield excess attenuations of 0 to 5 dB/100 m. Curves A and B are Canadian; C is Russian; D is American; E is Eq. (7.23); and F is German.

ferred to a unit distance of, say, 100 m (328 ft) unless the total attenuation exceeds 30 dB.

For a source and receiver height of 2.5 m (8 ft) an excess attenuation of 5 to 10 dB over a distance of 250 m has been observed in the frequency range from 100 to 6,300 Hz,[9] with values as high as 50 dB in the range of 300 to 600 Hz. (See Sec. 7.8.2.)[8,29]

The excess attenuation over thick grass and through shrubbery is higher. The excess attenuation at 1,000 Hz for this situation can be as high as 23 dB per 100 m, and usually increases roughly at the rate of 5 dB per 100 m for each doubling of frequency (Fig. 7.11).[30,31] The excess attenuation is approximately

$$(A_{e_5})_{\text{shrubbery or grass}} = (0.18 \log f - 0.31)r \qquad \text{dB} \qquad (7.22)$$

where f = frequency of sound, Hz

r = path length through shrubbery or over grass, m

Sound propagation through tree zones (forests) has been investigated by several workers.[30,32,33,35] Values for excess attenuation at $f = 1{,}000$ Hz vary from 23 dB per 100 m (Ref. 33) for dense evergreen woods to 3 dB per 100 m (Ref. 35) or less for bare trunks above absorbing ground. An average value for various trees [36] is given by

$$(A_{e_5})_{\text{forest}} = 0.01 \ (f)^{1/3} r \qquad \text{dB} \qquad (7.23)$$

This equation, along with the measured data, is plotted in Fig. 7.12. Equation (7.23) does not indicate the effects of sound entering a forest from outside—the so-called edge effects. One observer [33] measured the edge effect for four types of forest (cedar, pine, spruce, and deciduous) and obtained puzzling results—the excess attenuation with edge effects was higher at low frequencies (below 400 Hz) and was lower at middle frequencies (500–1,500 Hz) (see Fig. 7.12). At frequencies above 1,500 Hz, the excess attenuation is about the same with and without edge effects. No evidence was found by the same observer for a correlation between sound attenuation and visibility as suggested by Eyring [29] based upon measurements in the jungles of Panama.

There is no satisfactory explanation for the large differences among the curves for Canadian, Russian, and American forests. The data were taken by qualified observers in all cases and their reports describe the forests as being somewhat similar.

7.8   Atmospheric Inhomogeneities, A_{e_6}

7.8.1   Wind and Temperature Gradients   Appreciable vertical temperature and wind gradients almost always exist over open level ground; the former because of heat exchange between ground and atmosphere, the latter because of friction between the moving air and ground. Because of these gradients, the speed of sound varies with height above the ground and sound waves are refracted, that is to say, bent upward or downward. Under such conditions, it is possible to have a "shadow zone" into which no direct sound can penetrate or, conversely, propagation over unusually long distances.[8,9,32] Shadow zones are never sharp in the sense of light because acoustical diffraction effects are associated with much longer wavelengths and, in addition, sound energy is scattered into the shadow zone by turbulence.

Wind gradients near the ground are nearly always positive; that is, the windspeed increases with height. As a result, a shadow zone is most

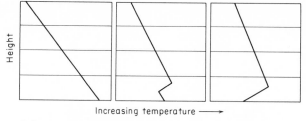

Increasing temperature ⟶

(*a*) Clear afternoon   (*b*) Early morning   (*c*) Clear night

Fig. 7.13 Example of temperature profiles at three times of day: (*a*) normal lapse; (*b*) high inversions with normal lapse above and below; and (*c*) ground inversion with normal lapse above.

commonly encountered upwind from a source, because there the wind gradient bends the sound rays upward. Downwind, the sound rays are bent downward, and no shadow zone is produced. Crosswind, there is a zone of transition. This asymmetric behavior is characteristic of wind-induced sound refraction.

Temperature-induced sound refraction tends to be symmetrical about the source. A shadow may completely encircle a source in the presence of a strong negative temperature gradient (large temperature lapse, as shown in Fig. 7.13*a*) and low windspeed, such as may be expected on a calm, clear afternoon. On the other hand, there will be no shadow at all, within a mile or two of the source, in the presence of a strong positive temperature gradient (large temperature inversion, as shown in Fig. 7.13*c*) and low windspeed, such as may be expected on a clear calm night. A double gradient, such as shown in Fig. 7.13*b*, may cause a channeling of sound, with slowly diminishing level over a long distance.

7.8.2 Excess Attenuation over Ground—No Shadow Zone   Two
anomalies, even when wind and temperature gradients are not factors in sound attenuation, and, as a result, there is no shadow zone, are commonly observed.[8] The first is an excess sound attenuation at comparatively low frequencies, say 300 to 600 Hz, which often may be *greater* by 10 to 50 dB than the value predicted from geometrical spreading and air absorption of sound. The second is at high frequencies, say at 6,000 Hz, where the sound attenuation may be 10 to 50 dB *less* than that predicted.

The low-frequency excess attenuation is probably caused [8,40] by interference between the direct sound (say from a source 3 m above the ground) and the reflected sounds that reach the receiver. The amplitude and phase of the reflected sounds are determined by the acoustic impedance of the ground, which may be regarded as a very deep porous medium. Because the impedance of the ground varies with frequency, the condition for destructive interference is met only in a narrow frequency band. Typical data for ground impedance indicate that the

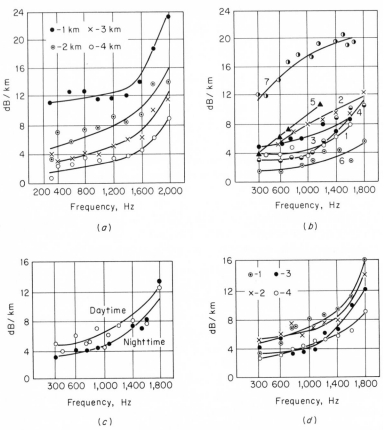

Fig. 7.14 Values of excess attenuation measured near Leningrad over a period of 3 years: (*a*) excess attenuation in dB/km (dB/3,280 ft), in the absence of acoustic shadowing, for four path lengths; (*b*) excess attenuation for (1) straight highway, (2) field covered with sparse shrubbery, (3) open road in forest, (4) open field, (5) field with jagged and broken profile, (6) lake, and (7) very dense pine forest with treetops higher than 6 m (20 ft); (*c*) excess attenuation for daytime and nighttime; and (*d*) excess attenuation for (1) summer, (2) spring, (3) autumn, and (4) winter. It is apparent from (*a*) that one cannot speak of anomalous excess attenuation in decibels per unit distance, because the rate of attenuation is greater for short distance than long probably owing to atmospheric scattering.

anomalous increase of sound attenuation should be most in the range of 300 to 600 Hz.

The high-frequency unusually low attenuation compared to that calculated for the average air humidity over a range of heights above the ground, or for the humidity obtaining at ground level, is often caused by the fact that both temperature and water-vapor content usually vary with position, particularly with height above the ground.[8]   The acoustic transmissivity, therefore, is different for different paths between the source and the receiver. Exceptionally low attenuation results from the occurrence of a path, or "channel," of unusually high transmissivity.

To help quantify the anomalous excess attenuation, a series of 159 measurements were made in the Leningrad region [34] over a 3-year period, during all seasons, at night and during the day, over seven distinct stretches of terrain, at distances ranging from 1.5 to 5 km (1 to 3 miles), and at frequencies up to 2,000 Hz. Atmospheric measurements of temperature, humidity, wind velocity, and temperature and wind gradients were made at heights up to 250 m (800 ft). By calculation, the attenuation owing to geometric spreading and molecular absorption was subtracted out to yield the *anomalous excess attenuation.*

The results are illustrated in Fig. 7.14. In (*a*) the excess attenuation in dB/km (dB/3,280 ft) is shown for four measurement distances and is seen to be a strongly nonlinear function of distance. In (*b*) the anomalous excess attenuation is shown for the seven different propagation paths. In (*c*) it is shown that the anomalous excess attenuation is greater in daytime than at night. From (*d*), we see that the anomalous excess attenuation is greater in summer (1), decreases in spring and fall, (2) and (3), and is least in winter (4). The graphs also show that the anomalous excess attenuation increases sharply above 1,000 Hz. No correlation between anomalous excess attenuation and wind velocity or direction was found.

Figure 7.15 shows an empirical design chart, based on all available data in the literature, appropriate for the determination of A_{e_6} for sound propagation in the *non-shadow zone* (downwind) over open level terrain with low ground cover and for frequencies in the range of about 300 to 5,000 Hz and distances of 2 to 4 km (1.25 to 2.5 miles). Afternoon and average season conditions are assumed. The abscissa is plotted in terms of the product $f_m r$, where r is the distance from the source to the receiver (lower scale in feet; upper scale in meters) and f_m is the center frequency of the band in question in hertz. For shorter or longer distances, and for other conditions, these values of excess attenuation may be modified by estimates determined from Fig. 7.14.

7.8.3 Excess Attenuation over Ground — Shadow Zone [37]

Consider Fig. 7.16 where a source and receiver are shown a distance r apart. The average direction from which the wind is blowing is indicated by a wind vane; that is, the wind blows from the upwind sector to the downwind sector. The angle between the direction of the wind vane and the line connecting the source and receiver is called ϕ. There will generally be a shadow zone (shaded region) produced on the upwind side of the source because the sound waves that travel upwind tend to be bent upward by the wind. Oftentimes the air near the ground is warmer

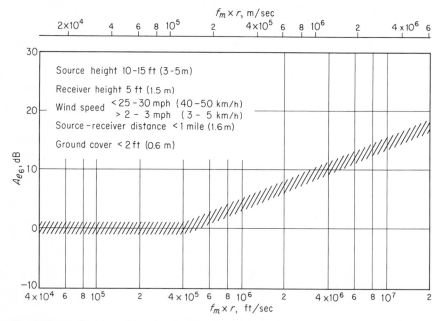

Fig. 7.15 Design chart for estimating the excess attenuation for non-shadow-zone (downwind) propagation over open level terrain (subject to confirmation at the very low and very high audio frequencies).

than that farther up, so that, in addition, there is a tendency for the sound waves to be refracted upward on all sides of the sound source. However, any wind present tends to bend the sound waves downward in the downwind direction. At some critical angle ϕ_c the effects of wind and temperature gradients may cancel each other and the shadow zone vanishes. As a result the plane is divided into an upwind sector $2\phi_c$ and a downwind sector $360° - 2\phi_c$.

Experiments have shown that the excess attenuation is frequently radically different upwind and downwind, with a gradual transition at the boundaries $\phi = \pm\phi_c$. On a sunny day, with moderate winds, the excess attenuation upwind inside the shadow zone, at distances in excess of $2X_0$ from the source, is typically 20 to 50 dB higher than that for the same distance downwind. For very low frequencies, it can be assumed that $A_{e_6} = 0$ to 3 dB upwind, crosswind, and downwind up to distances of about 0.5 mile from the source.

Figure 7.17 shows a design chart, giving values of A_{e_6}, for straight upwind propagation. The abscissa is plotted in terms of source-receiver distance, normalized to the minimum distance to the shadow zone X_0. The distance X_0 can be obtained either by direct measurement in the field or estimated from Table 7.2.

TABLE 7.2   Estimates of X_0 Upwind, 300 to 5,000 Hz, $\phi = 0$
Source height 10–15 ft, receiver height 5 ft

Time		Sky		Temperature profile			Wind, mph	X_0, ft
Day	Night	Clear	Overcast	Lapse	Neutral	Inversion		
	x	x				x	2–4	2,000
x			x		x		10–15	400
x		x		x			10–18	250

It should be noted that Figs. 7.15 and 7.17 do not necessarily apply for source and receiver heights appreciably different from those shown. This is especially true for upwind propagation, since the distance X_0 to the shadow zone increases with both source and receiver height.

7.8.4 Turbulence   Unstable turbulent atmosphere, which is more likely at daytime than at nighttime, mainly causes fluctuations of the received sound propagated along the ground (see Sec. 7.10). The excess attenuation by turbulence is of minor importance. In keeping with conservative estimates, assume [9]

$$(A_{e_6})_{\text{turbulence}} \approx 0 \qquad (7.24)$$

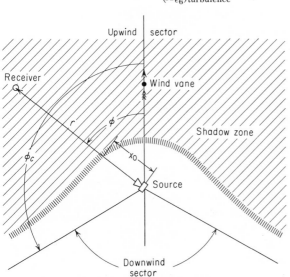

x_0 Distance from source to shadow zone
ϕ Angle between wind and sound
ϕ_c Critical angle

Fig. 7.16   Geometry of sound propagation over open level terrain (plan view). Average daytime conditions are shown. The wind blows from the upwind sector toward the downwind sector.

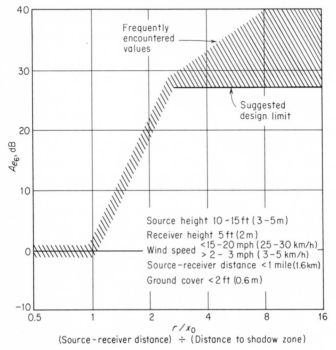

Fig. 7.17   Design chart for estimating the excess attenuation up-wind over open level terrain (subject to error at the very low and very high audio frequencies).

7.9   Propagation of Sound over Hilly Terrain

Experiments where sound is propagated from hilltop to hilltop across a valley indicate that wind direction, temperature, and wind gradients play a much smaller role than if sound propagation had taken place over level terrain.   After accounting for the excess attenuation resulting from molecular absorption (see Sec. 7.4), the experimental data show a residual excess attenuation that is roughly independent of frequency.[32]   Residual attenuations of 10 to 20 dB have been measured in the frequency range of about 400 to 2,000 Hz for distances of 1 to 1.5 miles (1.5 to 2.5 km). This residual attenuation is possibly due to turbulence in the air.   For conservative engineering estimates and until more systematic data are available, it seems best to disregard an excess attenuation A_{e_6} in sound-control problems involving propagation over hilly terrain.

7.10 Fluctuations of the Sound-pressure Level for Waves That Have Propagated through the Atmosphere

Fluctuations of the received sound-pressure level have been investigated [8,32,39] with the following typical results:

(1) For downwind propagation the magnitude of the fluctuations increases with the frequency of the signal and with distance; (2) for upwind propagation the magnitude of the fluctuations is greatest near the shadow boundary; (3) in a stable atmosphere (clear night, weak winds) the range of fluctuations is typically about 5 dB; (4) in an unstable atmosphere (clear sunny day, strong winds) the range of fluctuations is typically 15 to 20 dB; (5) the spectrum of the fluctuations measured over open level ground encompasses components from fractions of a hertz to several hertz; and (6) sound propagation from hilltop to hilltop and from air to ground is frequently characterized by large low-frequency fluctuations in the received sound-pressure level in addition to the faster fluctuations observed over level terrain.

REFERENCES

1. E. J. Rathe, "Note on Two Common Problems of Sound Propagation," *J. Sound Vibration*, vol. 10, pp. 472–479, 1969.
2. M. C. Henderson, "Sound in Air: Absorption and Dispersion," *Sound*, vol. 2, no. 6, pp. 28–36, 1963.
3. H. O. Kneser, "Interpretation of the Anomalous Sound Absorption in Air and Oxygen in Terms of Molecular Collision," *J. Acoust. Soc. Am.*, vol. 5, pp. 122–126, 1933.
4. C. M. Harris, "Absorption of Sound in Air in the Audio-frequency Range," *J. Acoust. Soc. Am.*, vol. 35, pp. 11–17, 1963.
5. *Boeing Co. Transportation Div. Doc.* 6, D6-4084 TN; *Rolls-Royce Ltd., Aero Engine Div., Tech. Rept.*, Mar. 7, 1962; *Douglas Aircraft Co. Rept.* SM-23641, December, 1959.
6. C. M. Harris, "Absorption of Sound in Air versus Humidity and Temperature," *J. Acoust. Soc. Am.*, vol. 40, pp. 148–159, 1966.
7. L. Cremer, *Raumakustik*, vol. II, p. 27, S. Hirzel Verlag, 1961.
8. U. Ingard, "On Sound-transmission Anomalies in the Atmosphere," *J. Acoust. Soc. Am.*, vol. 45, pp. 1038–1039, 1969.
9. U. Ingard, "A Review of the Influence of Meteorological Conditions on Sound Propagation," *J. Acoust. Soc. Am.*, vol. 25, pp. 405–411, 1953.
10. V. O. Knudsen, J. V. Wilson, and N. S. Anderson, "The Attenuation of Audible Sound in Fog and Smoke," *J. Acoust. Soc. Am.*, vol. 20, pp. 849–857, 1948.
11. W. L. Nyborg and D. Mintzer, "Review of Sound Propagation in the Lower Atmosphere," *Wright Air Develop. Center, Ohio, Tech. Rept.* 54-602, May, 1955, ASTIA No. AD-67880.
12. F. M. Wiener, "Sound Propagation over Ocean Waters in Fog," *J. Acoust. Soc. Am.*, vol. 33, pp. 1200–1205, 1961.
13. K. L. Oswatitsch, "Die Dispersion und Absorption des Schalles in Wolken," *Phys. Z.*, vol. 42, p. 365, 1941.

14. P. S. Epstein and R. R. Carhart, "The Absorption of Sound in Suspensions and Emulsions, I: Water Fog in Air," *J. Acoust. Soc. Am.,* vol. 25, pp. 553–565, 1953.
15. L. D. Landau and E. M. Lifschitz, *Klassische Feldtheorie,* p. 166, Akademie-Verlag, 1967.
16. S. W. Redferan, "Some Acoustical Source-Observer Problems," *Phil. Mag.,* ser. 7, vol. 30, pp. 223-236, 1940.
17. J. B. Keller, "Geometrical Theory of Diffraction," *J. Opt. Soc. Am.,* vol. 52, pp. 116–130, 1962.
18. Z. Maekawa, "Noise Reduction by Screens," *Memoirs of Faculty of Eng., Kobe Univ., Japan,* 11, pp. 29–53, 1965.
19. Z. Maekawa, "Noise Reduction by Screens of Finite Size," *Memoirs of Faculty of Eng., Kobe Univ., Japan,* 12, pp. 1–12, 1966.
20. R. O. Fehr, "The Reduction of Industrial Machine Noise," *Proc. 2d Natl. Noise Abatement Symp., Armour Res. Found., Chicago, 1951,* pp. 93–103.
21. U. J. Kurze and G. S. Anderson, "Sound Attenuation by Barriers," *Applied Acoustics,* vol. 4, pp. 56–74, 1971.
22. F. Fleischer, "Zur Anwendung von Schallschirmen," *Lärmbekämpfung,* vol. 14(6), pp. 131–136, 1970.
23. J. M. Rapin, "Études des modes de protection phonique aux abords des voies rapides urbaines," vol. I, A, B, Centre Scientifique et Technique du Bâtiment, Paris, July 1, 1969.
24. U. Ingard, private communication.
25. W. E. Scholes, "Noise Reduction by a Barrier," paper 70/67, presented at British Acoustical Society Meeting, Newcastle on Tyne, April, 1970.
26. Bolt Beranek and Newman Inc., "Noise in Urban and Suburban Areas," January, 1969, FHA and HUD Technical Studies Program. Order from Superintendent of Documents, U.S. Government Printing Office, Washington, D.C. 20402 (price 40 cents).
27. E. A. G. Shaw and N. Olsen, "Theoretical Model for Steady State Urban Noise" (abstract), *J. Acoust. Soc. Am.,* vol. 46, p. 99, 1969.
28. F. M. Wiener, C. I. Malme, and C. M. Gogos, "Sound Propagation in Urban Areas," *J. Acoust. Soc. Am.,* vol. 37, pp. 738–747, 1965.
29. P. H. Parkin and W. E. Scholes, "The Horizontal Propagation of Sound from a Jet Engine Close to the Ground, at Hatfield," *J. Sound Vibration,* vol. 2, pp. 353–374, 1965.
30. C. F. Eyring, "Jungle Acoustics," *J. Acoust. Soc. Am.,* vol. 18, pp. 257–270, 1946.
31. F. J. Meister and W. Ruhrberg, "The Influence of Green Areas on the Propagation of Noise," *Lärmbekämpfung,* vol. 3, no. 1, 1959.
32. F. M. Wiener and D. N. Keast, "An Experimental Study of the Propagation of Sound over Ground," *J. Acoust. Soc. Am.,* vol. 31, p. 724, 1959.
33. T. F. W. Embleton, "Sound Propagation in Homogeneous Deciduous and Evergreen Woods," *J. Acoust. Soc. Am.,* vol. 35, pp. 1119–1125, 1963.
34. I. A. Dneprovskaya, V. K. Iofe, and F. I. Levitas, "On the Attenuation of Sound as It Propagates through the Atmosphere," *Soviet Phys.-Acoust.,* vol. 8, pp. 235–239, 1963.
35. H. Kuttruff, "Über Nachhall in Medien mit unregelmässig verteilten Streuzentren, insbesondere in Hallräumen mit aufgehängten Streuelementen," *Acustica,* vol. 18, p. 131, 1967.
36. R. M. Hoover, "Tree Zones as Barriers for the Control of Noise due to Aircraft Operations," *Bolt Beranek and Newman Inc. Rept.* 844, February, 1961.
37. F. M. Wiener, "Sound Propagation Outdoors," *Noise Control,* vol. 4, pp. 224ff, July, 1958.
38. Aerospace Recommended Practice, ARP 866, "Standard Values of Atmospheric Absorption as a Function of Temperature and Humidity for Use in Evaluating Aircraft

Flyover Noise," Aug. 31, 1964, Society of Automotive Engineers, 485 Lexington Ave., New York, N.Y.

39. P. Baron, "Propagation du son dans l'atmosphère et audibilité des signaux avertisseurs dans le bruit ambiant," Cahiers d'Acoustique, *Ann. Télécommun.,* vol. 9, pp. 258–274, 1954.

40. P. B. Oncley, "Low-frequency Ground Attenuation in Outdoor Noise Measurements" (abstract), *J. Acoust. Soc. Am.,* vol. 47, p. 122, 1970.

41. U. J. Kurze, "Statistics of Road Traffic Noise," to be published.

Chapter Eight

Sound in
Small Spaces

LEO L. BERANEK

8.1 Introduction

Only in an anechoic room may sound waves travel outward in any direction without encountering reflecting surfaces. In practice, we must deal with every shape and size of enclosure containing an infinite variety of sound-diffusing and absorbing objects and surfaces. To prepare for noise control in practical situations, we shall study in this chapter sound fields in several special types of small enclosures.

Some mathematics is required to understand the analysis. To ease the way for the nonmathematically inclined engineer, each step and symbol is defined as fully as space permits. The engineer with a good mathematical background is encouraged to read books by Morse and Ingard [1] and by Rayleigh.[2] Somewhat easier books to follow are by Stephens and Bate,[3] Kinsler and Fry,[4] Beranek,[5] and Malecki.[6]

8.2 Sound Pressure in a Very Small Enclosure

Frequently, in noise control problems, a source of sound is enclosed in a very small box. If the source has a frequency low enough so that the wavelength of the sound is long compared to the largest dimension of the box, the sound pressure produced by the source will be uniform throughout the box. A uniform sound-pressure field will also be found in a small box, one or more sides of which are vibrated at a low frequency. Either of these situations is represented conceptually by Fig. 8.1.

By the Charles-Boyle gas law,

$$\frac{\Delta p}{p_s} = -\gamma \frac{\Delta V}{V} \tag{8.1}$$

where p_s = average or ambient pressure of the gas in the box. For air at sea level, $p_s \approx 1.013 \times 10^5$ N/m²

Δp = incremental variation in pressure in the box, N/m²

V = volume of gas in the box undergoing compression, m³

$\Delta V = S\xi$ = incremental variation in volume owing to movement of the vibrating piston, equal to the area of the piston times its displacement, m³

γ = ratio of specific heats. For air, $\gamma = 1.4$, dimensionless

We may write

$$U(t) = S \frac{\partial \xi(t)}{\partial t} = Su(t) = -\frac{\partial V(t)}{\partial t} \quad \text{m/sec} \tag{8.2}$$

where $U(t) = Su(t)$ = volume velocity of the piston, m³/sec

Substitution in Eq. (8.1) yields

$$\frac{\partial p(t)}{\partial t} = \frac{\gamma p_s}{V} U(t) \quad \text{N/m²-sec} \tag{8.3}$$

Let

$$U(t) = \sqrt{2} \, Q_0 \cos \omega t \tag{8.4a}$$

$$= \text{Re} \, \sqrt{2} \, Q_0 e^{j\omega t} \quad \text{m³/sec} \tag{8.4b}$$

$$p(t) = \text{Re} \, \sqrt{2} \, p e^{j\omega t} \quad \text{N/m²} \tag{8.5}$$

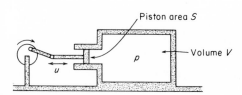

Fig. 8.1 Small air volume V driven by a piston of area S vibrating sinusoidally with an rms velocity u at a frequency f. The largest dimension of the cavity is assumed here to be small compared to a wavelength so that p is uniform throughout the interior. This is the basic principle underlying a pistonphone calibrator for microphones or the buildup of sound pressure in a box whose dimensions are small compared to a wavelength.

Piston area S

Volume V

where $Q_0 =$ rms volume velocity of piston, m³/sec

$p =$ rms sound pressure, N/m²

Substitution in Eq. (8.3) yields the rms sound pressure

$$p = \frac{1}{j\omega}\left(\frac{\gamma p_s}{V}\right) Q_0 \qquad N/m^2 \qquad (8.6)$$

where $V/\gamma p_s =$ acoustic compliance of the gas in the enclosure, m⁵/N

$1/j = -j$ and indicates that the sound pressure reaches its maximum value one-fourth cycle after the velocity reaches
its maximum

It is apparent from Fig. 8.1 that when the piston displacement is at its positive maximum (measured to the right from its neutral value), the sound pressure is at its peak value. That is to say, the displacement and sound pressure are in phase. Obviously, the piston has zero velocity at the instant that the displacement is greatest, which is why pressure and velocity are 90° out of phase. If $U(t)$ is given by Eq. (8.4a), then for air

$$p(t) = \sqrt{2}\, Q_0\, \frac{1.4p_s}{\omega V} \sin (\omega t) \qquad N/m^2 \qquad (8.7)$$

Example 8.1   The piston of Fig. 8.1 has an area of 1 cm² and is driven with a peak-to-peak displacement of 4 millimeters at a frequency of 100 Hz. The volume of the cavity is 0.0125 m³. What is the sound-pressure level in the cavity?

SOLUTION: Remembering from Eq. (8.2) that $Q_0 = \omega S \xi_{rms}$, we get from Eq. (8.6) for the rms pressure (disregarding phase)

$$|p| = \frac{1.4p_s\omega S\xi_{rms}}{\omega V} = \frac{1.4p_s S\xi_{rms}}{V}$$

$$= \frac{1.4 \times 1.013 \times 10^5 \times 10^{-4} \times (2/\sqrt{2}) \times 10^{-3}}{1.25 \times 10^{-2}}$$

$$= 1.61 \text{ N/m}^2 \text{ or } 98 \text{ dB } re\ 2 \times 10^{-5} \text{ N/m}^2$$

8.3   Rise and Decay of Sound in a Long Tube

It will help to visualize sound fields in large rooms by considering the rise and decay of sound in a tube that is long compared to a wavelength. The tube is terminated at both ends by sound-absorbing surfaces T (see Fig. 8.2).

When the source is turned on, the sound field builds up during a transient period of time after which it reaches a steady state. The factor of power removed by each sound-absorbing surface T each time a sound wave is reflected from it is α_0, defined as the power-absorption coefficient for normal (perpendicular) sound incidence.

The sound source S is located at one end of the tube and the microphone M_1 is located at the center of the tube. The source is driven by a random-noise generator that produces, in a frequency band Δf, a large number of components closely spaced in frequency and equal in amplitude.

8.3.1 Growth of the Sound Field

Initially the energy in the tube is zero. At time $t = 0$ the source begins to emit sound energy at a rate of W_A watts. The energy density (see Chap. 2) in the sound field at the center of the tube remains zero until $t = t_1 = l_x/2c$, at which time the wavefront traveling at the speed c reaches the center of the tube. The energy density at M_1 then rises abruptly to a value

$$D_0 = \frac{W_A}{cS_1} \text{ watt-sec/m}^3 \qquad (8.8)$$

where D_0 = energy density in the direct sound wave, watt-sec/m^3

W_A = power radiated by the source, watts

As the wave progresses to the right, the energy density remains constant at the value D_0 until the wavefront is reflected from the right-hand termination and returns to the center at a time $t = 3t_1 = 3l_x/2c$. The

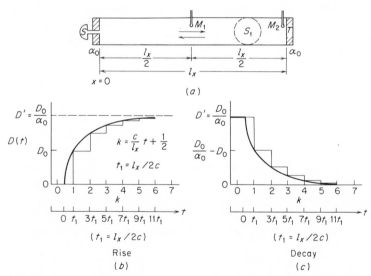

(a)

(b) Rise ($t_1 = l_x/2c$)

(c) Decay ($t_1 = l_x/2c$)

Fig. 8.2 (a) A rigidly closed tube of cross-sectional area S_1 with lateral dimension small compared to a wavelength so that no transverse standing waves occur. The sound source is coupled to the main tube through a tube so small that it does not change the wall impedance at $x = 0$; (b) and (c) rise and decay characteristics of sound in the center of a uniform tube of length l_x terminated at both ends by absorbing materials having an absorption coefficient α_0.

energy density at the midpoint then rises to a value

$$D_0 + D_0(1 - \alpha_0) \tag{8.9}$$

because the wave returning from the right has now been diminished in energy by the factor $(1 - \alpha_0)$. The total energy density at the point $l_x/2$ after the succeeding reflection from the left-hand termination rises to the value

$$D_0[1 + (1 - \alpha_0) + (1 - \alpha_0)^2] \tag{8.10}$$

at a time $t = 5t_1 = 5l_x/2c$.

Thus, after k reflections the total energy density at the midpoint of the tube is given by the series

$$D(n) = D_0 \sum_{k=0}^{n-1} (1 - \alpha_0)^k \tag{8.11}$$

A plot of this step-wise function is shown in Fig. 8.2b. The energy density approaches a limiting (steady-state) value D' as the number of reflections increases. This limiting value is found by summing the series of $n - 1$ terms and taking the limit as n gets to be very large. The summation of Eq. (8.11) is equivalent to

$$D(n) = \frac{D_0}{\alpha_0} [1 - (1 - \alpha_0)^n] \quad \text{watt-sec/m}^3 \tag{8.12}$$

Because the quantity $(1 - \alpha_0)$ has a value less than 1 by definition, the limit as n becomes infinite is

$$D' = \frac{D_0}{\alpha_0} = \frac{W_A}{cS_1\alpha_0} \quad \text{watt-sec/m}^3 \tag{8.13}$$

where D' = steady-state energy density in the tube, watt-sec/m^3
   α_0 = normal sound-absorption coefficient for each end, dimensionless

It is convenient in many calculations to divide the total steady-state energy density, Eq. (8.13), into two components: one associated with the direct field, $D_0 = W_A/cS_1$ [see Eq. (8.8)], and the other associated with the sound field set up by all reflections, namely, the reverberant field D_R. Starting with Eqs. (8.8) and (8.13), we obtain for the steady-state energy density

$$D' = D_0 + D_R = \frac{W_A}{cS_1} + \frac{W_A}{c} \frac{1 - \alpha_0}{S_1\alpha_0} \quad \text{watt-sec/m}^3 \tag{8.14}$$

The reverberant component is frequently written in the form

$$D_R = \frac{2W_A}{cR} \tag{8.15}$$

where

$$R \equiv \frac{2S_1\alpha_0}{1 - \alpha_0} = \frac{S\alpha_0}{1 - \alpha_0} \tag{8.16}$$

where $S = 2S_1$ = total area of absorbing material in the tube, m^2

The quantity R is analogous to the *room constant* in the acoustics of large enclosures (see Chap. 9).

Mean Free Path. We see from Fig. 8.2 that for large values of time the number of reflections $k = n - 1$ is nearly equal to n and to ct/l_x, where statistically the length of the tube now may be interpreted as the *mean free path*, the average distance that the sound wave travels between reflections. Equation (8.12) becomes

$$D(t) = \frac{W_A}{cS_1\alpha_0} [1 - (1 - \alpha_0)^{ct/l_x}] \qquad \text{watt-sec/m}^3 \tag{8.17}$$

This function is plotted as a smooth curve in Fig. 8.2b. Strictly speaking the sound field within any enclosure rises in a series of steps as shown in the figure. However, this discontinuous function can be closely approximated by a smooth curve if the number of reflections is large. In the limit, as t becomes very large, $D(t)$ becomes $D' = D_0/\alpha_0$.

8.3.2 Decay of the Sound Field   In reverberation measurements the sound-energy density is allowed to reach a steady-state value and then the source is suddenly turned off, say at time $t = 0$. Returning again to the tube of Fig. 8.2, the energy density $D' = D_0/\alpha_0$ persists until $t = t_1 = l_x/2c$, at which time the trailing edge of the direct wave reaches $l_x/2$. The energy density suddenly decreases to a value $(D_0/\alpha_0) - D_0$ and remains at this value until the once-reflected wave arrives at a time $t = 3t_1 = 3l_x/2c$. At this instant the energy density drops to a value $D_0[1/\alpha_0 - 1 - (1 - \alpha_0)]$. After sufficient time has elapsed for the edge of the $n - 1$ reflected wave to reach the point $l_x/2$, the energy density has been reduced to a value of

$$D(n) = D_0 \left[\frac{1}{\alpha_0} - \sum_{k=0}^{n-1} (1 - \alpha_0)^k \right] \qquad \text{watt-sec/m}^3 \tag{8.18}$$

which, through the use of Eqs. (8.11) to (8.13), may be expressed in the convenient form

$$D(n) = \frac{D_0}{\alpha_0} (1 - \alpha_0)^n = D'(1 - \alpha_0)^n \tag{8.19}$$

This function is plotted vs. n as a stepwise function in Fig. 8.2c. Clearly, the energy density approaches zero as n approaches infinity. Just as before, when t is large we replace n by ct/l_x and obtain a smooth exponential decay curve

$$D(t) = D'(1 - \alpha_0)^{ct/l_x} = \frac{W_A}{cS_1\alpha_0} e^{-(c/l_x)[-\ln(1-\alpha_0)]t} \qquad \text{watt-sec/m}^3 \quad (8.20)$$

where ln is the logarithm to the base e.

The function $D(t)$ is plotted as a smooth curve in Fig. 8.2c. It will be observed in Fig. 8.2 that approximating the actual step response by a smooth rise or decay curve is grossly in error for the first few reflections. As the number of reflections increases, the fit becomes better, and finally, after many reflections, the fit is almost perfect.

One striking fact emerges from the analysis above, namely, that *the sum of the rise and decay curves of Fig. 8.2 is at every instant a constant* $D' = D_0/\alpha_0 = W_A/cS_1\alpha_0$. This fact can be shown by adding Eqs. (8.11) and (8.18). It is also true for the sum of Eqs. (8.17) and (8.20).

Decay Constant k_m. It is customary in acoustics [2] to define a decay constant for the rms decaying sound-pressure field

$$p(t) = p_{(t=0)}e^{-k_m t} \qquad (8.21)$$

From Eq. (2.4) we find that the average energy density is

$$D = \frac{p^2}{\rho c^2} \qquad \text{watt-sec/m}^3 \qquad (8.22)$$

Thus

$$D(t) = D_{(t=0)}e^{-2k_m t} \qquad (8.23)$$

or, for the tube of Fig. 8.2, from Eq. (8.20)

$$k_m = \frac{c}{2l_x}[-\ln(1 - \alpha_0)] \qquad \text{sec}^{-1} \qquad (8.24)$$

8.3.3   Power Supplied by Constant-volume-velocity Source   In the derivations above we have assumed that the sound source S of Fig. 8.2 radiates a constant power into the tube regardless of the value of the sound-absorption coefficient. It can be shown that if, instead, the source has a constant volume velocity $Q_0(\text{m}^3/\text{sec})$, the sound power W_{A_2} supplied by the source would be

$$W_{A_2} = \frac{Q_0^2 \rho c}{S_1 \alpha_0} \qquad (8.25)$$

If we define W_{A_0} as the sound power that would be radiated by a source of strength Q_0 into a tube with the $x = 0$ end rigid and the other end perfectly absorbing, then

$$W_{A_0} = \frac{Q_0^2 \rho c}{S_1} \qquad (8.26)$$

The ratio W_{A2}/W_{A0} is

$$\frac{W_{A2}}{W_{A0}} = \frac{1}{\alpha_0} \qquad (8.27)$$

We observe that when the source power is constant, $D'/D_0 = 1/\alpha_0$ [see Eq. (8.13)], while when the source volume velocity is constant, $W_{A2}/W_{A0} = 1/\alpha_0$. The latter relation tells us what every xylophone manufacturer knows: a low-loss resonator greatly enhances the power radiated by a vibrating (constant-volume-velocity) bar.

For the special case where α_0 is small (i.e., $\alpha_0 < 0.1$), so that $-\ln(1 - \alpha_0) \approx \alpha_0$, we use Eqs. (8.24), (8.26), and (8.27) (and also $W_{A2} = p_2 Q_0$ is the source power fed into the many resonances, where p_2 is the sound pressure at $x = 0$) to obtain

$$p_2 \approx \frac{Q_0 \rho c^2}{2 k_m V} \qquad \text{N/m}^2 \qquad (8.28)$$

where $k_m =$ damping constant for the resonances in the tube in the frequency band Δf, sec^{-1}

$V = l_x S_1 =$ volume of the tube, m^3

Equation (8.28) says that when the source volume velocity is constant, the rms pressure in the tube is inversely proportional to the damping constant k_m for small values of α_0.

8.4 Standing Waves in a Tube

Let us again consider the tube of Fig. 8.2, but allow its length to take on any value l_x, i.e., not restrict the analysis to a very long tube. Also, let us locate the microphone at M_2, the end opposite the source.

The general single-dimensional wave equation with a source present at some point x along the tube [1] is

$$\frac{\partial^2 p(x, t)}{\partial x^2} - \frac{1}{c^2} \frac{\partial^2 p(x, t)}{\partial t^2} = \frac{\rho}{V} \frac{\partial Q(x, t)}{\partial t} \qquad (8.29)$$

where $V = S_1 l_x =$ volume of tube, m^3

$p(x, t) =$ instantaneous sound pressure at plane x in the tube, N/m^2

In the steady state, we may set

$$Q(x, t) = \sqrt{2}\, Q_0(x) \sin \omega t \qquad (8.30)$$

$$p(x, t) = \sqrt{2}\, p(x) \cos \omega t \qquad (8.31)$$

where $Q_0(x) =$ rms volume velocity of a tiny sound source which may be located at any point x along the length of the tube, m^3/sec

$p(x) =$ rms sound pressure at any point x, N/m^2

Substitution of these quantities into Eq. (8.29) yields

$$\frac{\partial^2 p(x)}{\partial x^2} + \frac{\omega^2}{c^2}\, p(x) = \frac{\rho\omega}{V}\, Q_0(x) \tag{8.32}$$

8.4.1 Sound Pressure in the Tube   The steady-state rms sound pressure at any point x along the tube with the tiny *source located at* $x = 0$ is a sum of terms of the form

$$p(x) = \sum_m p_m(x) = \sum_m |p_m(x)| e^{j\theta_m} \qquad \text{N/m}^2 \tag{8.33}$$

Each m term is the sound pressure vs. x for a particular *normal mode of vibration*. This sound pressure has an rms magnitude equal to $|p_m(x)|$ and a phase angle θ_m. Each normal mode of vibration has its own *normal frequency (natural frequency)* ω_m, and a *damping constant* k_m, both with units of radians/sec.

If the end walls of the tube are not very absorbent, the value of the rms sound pressure associated with one of the terms (a particular value of m) at a point x along the tube is [1]

$$|p_m(x)| = \frac{c^2 Q_0 \rho\omega[\cos\,(m\pi x/l_x)]}{V[4\omega_m^2 k_m^2 + (\omega^2 - \omega_m^2)^2]^{1/2}} \qquad \text{N/m}^2 \tag{8.34}$$

The phase angle of a particular term in m relative to the source velocity at $x = 0$ is

$$\theta_m = \tan^{-1} \frac{\omega_m^2 - \omega^2}{2k_m\omega_m} \qquad \text{radians} \tag{8.35}$$

It is obvious that $|p_m(x)|$ for a particular mode of vibration has its greatest value whenever the frequency of the source ω is nearly equal to ω_m.

In order for Eq. (8.40) to satisfy the wave equation (8.32) the following must hold

$$\omega_m = \frac{cm\pi}{l_x} \qquad m = 0,\, 1,\, 2,\, 3,\, \ldots \text{ radians/sec} \tag{8.36}$$

or

$$f_m = \frac{cm}{2l_x} \qquad m = 0,\, 1,\, 2,\, 3,\, \ldots \text{ Hz} \tag{8.37}$$

8.4.2 Tube Driven at Resonance   If the source is driven by a steady-state generator of frequency f, and if that frequency equals a *normal frequency* f_m, then ω will equal ω_m and Eq. (8.34) becomes

$$|p_m(x)| = \frac{c^2 Q_0 \rho}{2Vk_m} \cos \frac{m\pi x}{l_x} \qquad \text{N/m}^2 \tag{8.38}$$

Actually, the maximum value of $|p_m(x)|$ is not reached exactly when $\omega = \omega_m$, but rather at a frequency slightly higher than ω_m, called the *resonance frequency*. The resonance frequencies of the tube are found by maximizing Eq. (8.34) as a function of frequency.

The exact frequency ω_{0m} at which $|p_m(x)|$ becomes a maximum is

$$\omega_{0m} = \sqrt{\omega_m^2 + k_m^2} \tag{8.39}$$

Comparison of Eq. (8.38) with Eq. (8.28), which was derived in the previous section by a different procedure, reveals the same result for the sound pressure at $x = 0$.

It is also interesting to explore the normal mode of vibration for which $m = 0$. This normal mode covers the case of very low frequencies, well below the first resonance of the tube. In general, this means that $l_x < \lambda/16$.

Substitution of $\omega_m = 0$ into Eq. (8.34) yields

$$|p_0| = \frac{c^2 \rho}{V\omega} Q_0 \qquad \text{N/m}^2 \tag{8.40}$$

From the definition of the speed of sound, $c = \sqrt{\gamma p_s/\rho}$, we get

$$|p_0| = \frac{1}{\omega} \frac{\gamma p_s}{V} Q_0 \qquad \text{N/m}^2 \tag{8.41}$$

Comparison of Eq. (8.41) with Eq. (8.6) reveals that the zero mode of vibration for the tube is simply the case of a very small enclosure driven by a source with a constant volume velocity Q_0.

Example 8.2 Given a tube 0.688 m (2.26 ft) long, speed of sound equal to 344 m/sec, and a source of sound whose frequency f can be varied. (1) Find the first four normal frequencies of the tube; (2) find the first four resonance frequencies of the tube, assuming that the damping constants for the tube are $k_1 = 20\pi$, $k_2 = 40\pi$, $k_3 = 100\pi$, $k_4 = 200\pi$; (3) for $x = l_x$ plot a resonance curve for the second resonance frequency; and (4) plot the sound-pressure distributions in the tube as a function of x for the first and fourth normal modes of vibration.

SOLUTION: (1) The first four *normal frequencies* f_m occur for $m = 1, 2, 3$ and 4 in Eq. (8.37)

$$f_1 = \frac{344}{2 \times 0.688} = 250 \text{ Hz}$$

$$f_2 = 2 \times 250 = 500 \text{ Hz}$$

$$f_3 = 3 \times 250 = 750 \text{ Hz}$$

$$f_4 = 4 \times 250 = 1{,}000 \text{ Hz}$$

(2) The first four *resonance frequencies* f_{0m} are given by Eq. (8.39)

$$f_{0m} = \sqrt{f_m^2 + \left(\frac{k_m}{2\pi}\right)^2}$$

$$f_{01} = \sqrt{250^2 + 10^2} \quad = 250.2 \text{ Hz}$$

$$f_{02} = \sqrt{500^2 + 20^2} \quad = 500.4 \text{ Hz}$$

$$f_{03} = \sqrt{750^2 + 50^2} \quad = 751.7 \text{ Hz}$$

$$f_{04} = \sqrt{1{,}000^2 + 100^2} = 1{,}005 \text{ Hz}$$

(3) The sound-pressure-vs.-frequency curve for the $m = 2$ mode of vibration at $x = l_x$ is computed from the equation below and is plotted in Fig. 8.3. The value of f_2 is 500 Hz. Because we do not know the value of Q_0, we have normalized the calculation relative to $|p_2(\omega_2)|$.

$$\left|\frac{p_2(\omega)}{p_2(\omega_2)}\right|_{x=l_x} = \frac{2\omega k_2}{[4\omega_2^2 k_2^2 + (\omega^2 - \omega_2^2)^2]^{1/2}}$$

$$= \frac{f/f_2}{\left[1 + \dfrac{\omega_2^2}{4k_2^2}\left(1 - \dfrac{f^2}{f_2^2}\right)^2\right]^{1/2}}$$

$$= \frac{f/f_2}{\left[1 + \dfrac{(3{,}140)^2}{(80\pi)^2}\left(1 - \dfrac{f^2}{f_2^2}\right)^2\right]^{1/2}} \qquad \text{(dimensionless)}$$

(4) The curves of pressure distribution as a function of x for $m = 1$ and $m = 4$ are computed from

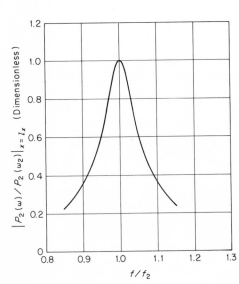

Fig. 8.3 Plot of a resonance curve as computed for Example 8.2 for $x = l_x$ and $m = 2$.

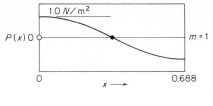

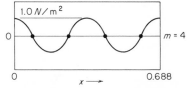

Fig. 8.4   Plot of the cosinusoidal sound-pressure distribution in the tube of Example 8.2 for $m = 1$ and $m = 4$.

$$p_1(x) = [p_1(x = 0)] \cos \frac{\pi x}{0.688}$$

$$p_4(x) = [p_4(x = 0)] \cos \frac{4\pi x}{0.688}$$

Because we do not know the value of p_1 or p_4 at $x = 0$, we shall assume them equal to 1.0 N/m². The actual values would come from Eq. (8.38) provided we knew Q_0. Usually a microphone located at the edge of the tube at $x = 0$ is used to determine $c^2 Q_0 \rho / 2 V k_m$ experimentally. The curves are plotted in Fig. 8.4.

Summary.   A tube closed at both ends will resonate at an infinite number of normal frequencies, harmonically related to each other. For each normal mode of vibration, the magnitude of the sound pressure is distributed in the tube according to a cosine function [Eq. (8.38)] with maximum values at the two ends, $x = 0$ and $x = l$, and at every half-wavelength point in between (Fig. 8.4). The value of m equals the number of *nodal (zero) sound-pressure points* along the length of the tube.

If the frequency f of the source of sound is about equal to one of the normal frequencies f_m, resonance is said to occur, and $p_m(x)$ becomes very large. If the source has many frequencies, the tube will resonate at every frequency where the source and normal frequencies are nearly alike. As we increase frequency, the value of $p_m(x)$ increases along one side of the resonance curve, goes through a maximum at resonance (f_0), and then decreases along the other side. As frequency continues to increase, $p_m(x)$ will soon begin to increase along the left side of the resonance curve for the next higher normal frequency, and so on.

The maximum value to which the pressure rises at resonance is inversely proportional to the damping constant or the total absorption in the tube [see Eq. (8.38)].

8.4.3 Particle Velocity in the Tube

The particle velocity u at any point in the tube is always related to the *sound-pressure gradient* $\partial p/\partial x$ at that point by the basic equation [1]

$$u = -\frac{1}{\rho} \int \frac{\partial p}{\partial x}\, dt \qquad (8.42)$$

If we again assume the steady state, then p is given by Eq. (8.31), and u is given by

$$u(x, t) = \sqrt{2}\ u(x)\cos(\omega t + \theta) \qquad (8.43)$$

where θ is a time phase angle relative to the sound pressure.

In the steady state (and assuming a particular value of m) the insertion of Eqs. (8.31) and (8.43) in Eq. (8.42) yields

$$u_m(x) = \frac{-1}{\omega\rho}\frac{\partial p_m(x)}{\partial x} \qquad \angle_{-90°} \qquad (8.44)$$

The symbol $\angle_{-90°}$ means that the phase angle introduced into $u_m(x)$ by the time integral is $-90°$. The total angle θ [see Eq. (8.43)] is the sum of $-90°$ and the phase angle introduced by the derivative of $p_m(x)$ with respect to x.

To obtain the magnitude of the particle velocity $u(x)$ in the tube of Fig. 8.2, we substitute Eq. (8.34) into Eq. (8.44). Remembering Eq. (8.36), this operation yields

$$u_m(x) = \frac{|p_m(x = 0)|}{\rho c}\left[\frac{\omega_m}{\omega}\right]\sin\frac{m\pi x}{l_x} \qquad \angle\theta \qquad (8.45)$$

where, as we can see from Eq. (8.38),

$$|p_m(x = 0)| = \frac{c^2 Q_0 \rho\omega}{V[4\omega_m^2 k_m^2 + (\omega^2 - \omega_m^2)^2]^{1/2}} \qquad (8.46)$$

Equation (8.45) contains three pieces of information that we should know: (1) in the vicinity of a normal frequency, the particle velocity follows the same type of resonance curve as that shown for the sound pressure in Fig. 8.3. (2) The particle velocity is at a maximum at those points in the tube where the sound pressure is near zero and vice versa. This can be seen by the presence of the $\sin(m\pi x/l_x)$ in Eq. (8.45) as compared to the $\cos(m\pi x/l_x)$ in Eq. (8.34). In particular, at $x = 0$ and $x = l_x$, the particle velocity is near zero because the end walls are nearly rigid. (3) The maximum value of $u_m(x)$, i.e., when $\sin(m\pi x/l_x) = 1$, is related to the maximum value of $p_m(x)$, i.e., when $\cos(m\pi x/l_x) = 1$, by the factor (ω_m/ω) $(1/\rho c)$. In this respect, when one is near resonance, there is similarity with a free progressive wave, wherein $u/p = 1/\rho c$.

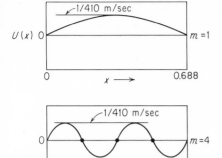

Fig. 8.5 Plot of the particle velocity distribution in the tube of Fig. 8.2 for $m = 1$ and $m = 4$; note that at the walls the particle velocity must be near zero if the walls are near rigid.

Example 8.3 Assume the same tube and speed of sound as for Example 8.2. The normal frequencies, resonance frequencies, and the resonance curve are, of course, the same as for that example. Plot the rms particle velocity distributions in the tube for the first and fourth normal modes of vibration. SOLUTION: The curves of rms particle velocity distribution as a function of x for $m = 1$ and $m = 4$ are computed from

$$u_1(x) = \frac{p_1(x = 0)}{410} \sin \frac{\pi x}{0.688}$$

$$u_4(x) = \frac{p_4(x = 0)}{410} \sin \frac{4\pi x}{0.688}$$

Assuming as we did in the previous examples that $p_1(x = 0) = p_4(x = 0) = 1$ N/m², we obtain the results of Fig. 8.5. The mode number m equals the number of *antinodes (maxima) in particle velocity* along the length of the tube.

8.5 Waves in Small Rectangular Rooms

Having already studied the acoustical conditions in a tube, it is easy to extrapolate to the case of a three-dimensional enclosure. Let us start with a rectangular enclosure with smooth parallel walls. Let us assume that the walls have very small sound-absorption characteristics. Furthermore, let us put our tiny source at a corner of the room, say at $x = y = z = 0$ (see Fig. 8.6).

When the source of sound is started, a sound wave spreads out in all directions because the wave is no longer confined by the small cross section of a tube. It can reflect obliquely off the walls at $x = l_x$, $y = l_y$, and $z = l_z$ and travel around the room in every direction. Now, if each path that the wave takes is traced around the room, there will be certain paths of travel that will repeat on themselves to form normal modes of vibration. When the frequency of the source equals one of the normal frequencies, resonance occurs.

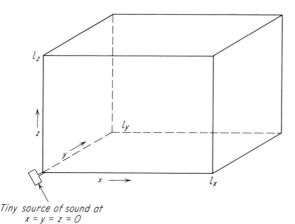

Tiny source of sound at
x = y = z = 0

Fig. 8.6   Rectangular enclosure with dimensions l_x, l_y, and l_z.

There are three types of normal modes of vibration in a rectangular enclosure:

1. *Axial modes* — in which the component waves move parallel to an axis (one-dimensional), the $(n_x, 0, 0)$, $(0, n_y, 0)$, and $(0, 0, n_z)$ modes of vibration.

2. *Tangential modes* — in which the component waves are tangential to one pair of surfaces, but are oblique to the other two pairs (two-dimensional), the $(n_x, n_y, 0)$, $(n_x, 0, n_z)$, and $(0, n_y, n_z)$ modes of vibration.

3. *Oblique modes* — in which the component waves are oblique to all three pairs of the walls (three-dimensional), the (n_x, n_y, n_z) modes of vibration.

8.5.1   Steady-state Sound Pressure in Rectangular Enclosures   For each normal mode of vibration, the rms sound pressure in the rectangular room of Fig. 8.6 at the point x, y, z is given by an equation almost exactly like Eq. (8.34), namely [1]

$$|p_n(x, y, z)| = \frac{c^2 Q_0 \rho \omega \psi_n(x, y, z)}{V[4\omega_n^2 k_n^2 + (\omega^2 - \omega_n^2)^2]^{1/2}} \qquad (8.47)$$

where Q_0 = rms volume velocity of source, m^3/sec

V = volume of room, m^3

ω_n = normal angular frequency for nth mode, radians/sec

k_n = damping constant, units of inverse time, for nth mode

The factor $\psi(x, y, z)$ gives the sound-pressure distribution in the room for each normal mode of vibration.

$$\psi(x, y, z) = \cos \frac{n_x \pi x}{l_x} \cos \frac{n_y \pi y}{l_y} \cos \frac{n_z \pi z}{l_z} \qquad (8.48)$$

The normal frequencies of the room are given by

$$f_n = \frac{\omega_n}{2\pi} = \frac{c}{2} \sqrt{\left(\frac{n_x}{l_x}\right)^2 + \left(\frac{n_y}{l_y}\right)^2 + \left(\frac{n_z}{l_z}\right)^2} \qquad \text{Hz} \qquad (8.49)$$

where n_x, n_y, and n_z can independently take on all integral values of 0, 1, 2, 3, . . . , ∞; c is the speed of sound in meters per second; and l_x, l_y, and l_z are the dimensions of the room in meters.

Example 8.4   Assume that the dimension l_z is less than 0.1 of all wavelengths being considered. This corresponds to n_z being zero at all times. Hence,

$$f_{n_x, n_y, 0} = \frac{c}{2} \sqrt{\left(\frac{n_x}{l_x}\right)^2 + \left(\frac{n_y}{l_y}\right)^2} \qquad \text{Hz} \qquad (8.50)$$

Let $l_x = 1.5$ m and $l_y = 1.0$ m. (1) Find the normal frequencies of the $n_x = 2$, $n_y = 0$; the $n_x = 1$, $n_y = 1$; and the $n_x = 2$, $n_y = 1$ normal modes of vibration. (2) Plot the sound-pressure distribution for these three normal modes of vibration.

SOLUTION: (1) From Eq. (8.50) we have

$$f_{2,0,0} = \frac{344}{2} \sqrt{\left(\frac{2}{1.5}\right)^2} \qquad = 229 \text{ Hz}$$

$$f_{1,1,0} = \frac{344}{2} \sqrt{\left(\frac{1}{1.5}\right)^2 + \left(\frac{1}{1}\right)^2} = 207 \text{ Hz}$$

$$f_{2,1,0} = \frac{344}{2} \sqrt{1.78 + 1} \qquad = 287 \text{ Hz}$$

(2) The pressure distribution is calculated from Eq. (8.48) with $m_z = 0$. The results are shown graphically in Fig. 8.7, assuming $p_n = 1$ N/m² at $x = y = z = 0$.

An interesting thing is seen from Eq. (8.48) and from Fig. 8.7, namely, that for every mode of vibration the sound pressure is at a maximum at the corners of the box. Also, for every mode of vibration for which one of the numbers n_x, n_y, or n_z is *odd*, the sound pressure is zero at the *center* of the room; hence at the geometrical center of the room only one-eighth of the modes of vibration have a finite sound pressure. Extending this further, at the center of any one wall the modes for which two of the n's are odd will have zero pressure, so that only one-fourth of them are detectable. Finally, at the center of one edge of the room, the modes for which one of the n's is odd will have zero pressure, so that only one-half of them are detectable there. The latter case would also occur for a microphone at $x = l_x/2$ in the tube of Fig. 8.2. Note from Eq.

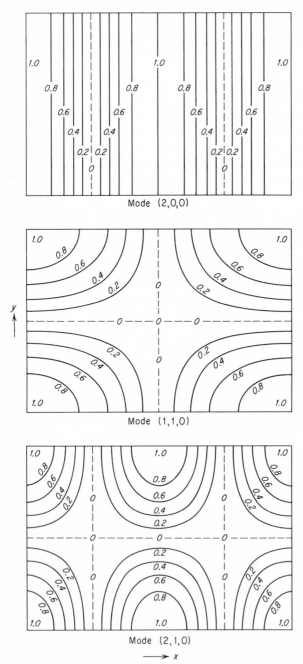

Fig. 8.7 Sound-pressure contour plots on a section through a rectangular room for three different modes of vibration. The numbers on the plots indicate the relative sound pressure.

(8.34) that at the $l_x/2$ plane in the middle of the tube all modes of vibration for which m is odd produce zero sound pressure because $\cos{(m\pi/2)} = 0$. The inverse is also true. No mode of vibration may be excited from a point source placed where Eq. (8.48) has zero value, e.g., along the "0" lines of Fig. 8.7.

Examples of the sound pressure as a function of frequency produced in a small rectangular enclosure measured with the microphone at one corner and either (1) the source at the diagonally opposite corner or (2) the source in the center of the room are given respectively in Figs. 8.8 and 8.9. The curves were obtained by slowly varying the frequency (a pure tone) of the loudspeaker and simultaneously recording the output of the microphone. The eightfold increase in the number of modes of vibration that were excited with the source at the corner over that with the source at the center is apparent. It is apparent also that the addition of sound-absorbing material on the 16- by 30-in. wall decreases the height of resonance peaks, particularly at the higher frequencies where it is most effective.

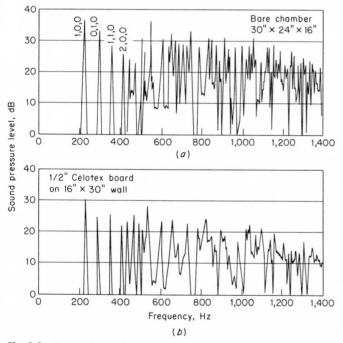

Fig. 8.8 Comparison of two sound-pressure-level curves for a constant-volume-velocity source. The microphone was in one corner, and the source was diagonally opposite. (*a*) Bare chamber; (*b*) with a sound-absorbing sample on a 16- by 30-in. wall. Zero decibels is about 78 dB *re* 2×10^{-5} N/m².

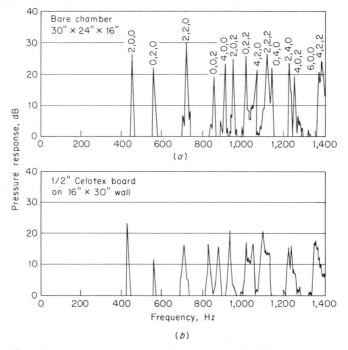

Fig. 8.9 Comparison of two sound-pressure-level curves for a constant-volume-velocity source. The microphone was in one corner, and the source was in the center of the room. (*a*) Bare chamber; (*b*) with a sound-absorbing sample on a 16- by 30-in. wall. Zero decibels is about 78 dB *re* 2×10^{-5} N/m².

8.6 Sound Decay in Three-dimensional Enclosures

When the source of sound in a room is turned off, the sound dies out, or decays, at a rate that depends on the dissipation, or damping in the room. In fact, each mode of vibration behaves independently of the others. The total process of sound decay is the rms summation of the sound pressures associated with all the individual modes of vibration that fall within the frequency band of interest. The discussion that follows applies either to the tube or the box, where neither is large enough for the sound field to be treated statistically.

At the time $t = 0$, when the sound source is turned off, the rms magnitude of the sound pressure associated with a particular mode at microphone position x, y, z is given by Eq. (8.47). A particular mode of vibration has its angular normal frequency ω_n and its damping constant k_n. The sound pressure at $t = 0$ is called $p_{n(t=0)}$.

The sound pressure for each normal mode of vibration, after $t = 0$,

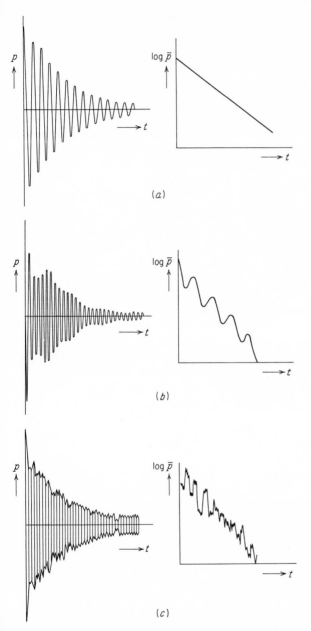

Fig. 8.10   Sound-pressure decay curves: (*a*) for a single mode of vibration; (*b*) for two closely spaced modes of vibration with the same decay constant; (*c*) for a number of closely spaced modes of vibration with the same decay constant.   The graphs on the left show the course of the instantaneous sound pressure, and those on the right show the curve of the envelope of the left graphs plotted in log $\bar{p}$ vs. t coordinate system.

decays at its own normal frequency at an exponential rate of k_n. That is to say,

$$p_n(t) = p_{n(t=0)}e^{-k_n t} \cos \omega_n t \tag{8.51}$$

Let us take the rms time average of $\cos \omega_n t$ and designate the resultant $p_n(t)$ as $\bar{p}_n(t)$. Then Eq. (8.51) says that on a log $\bar{p}_n$ scale vs. time, the *envelope* of the sound pressure decays linearly with time (see Fig. 8.10a).

When many normal modes of vibration (each with its own amplitude, normal frequency, and damping constant) decay simultaneously, the total rms sound pressure is given by

$$\bar{p}(t) = \sqrt{\bar{p}_1^2 + \bar{p}_2^2 + \cdots + \bar{p}_n^2} \tag{8.52}$$

where n represents any combination of the indices n_x, n_y, n_z describing the modes involved in the decay. The logarithms of $\bar{p}(t)$ as a function of time for two combinations of decaying modes of vibration are shown in Fig. 8.10b and c. We see that the decay is irregular because the different modes of vibration have different frequencies and beat with each other during decay.

This brief description of sound decay in small rooms is applicable in concept to rooms of any size and shape. The fact that the sound fields in large spaces are too complicated to calculate does not mean there are not distinct normal modes of vibration, each with its own normal frequency and damping constant. The means for handling large rooms on a statistical energy basis will be treated in the next chapter.

8.7 Complex Modes of Vibration in Cylindrical Ducts

In Secs. 8.3 and 8.4 we discussed the propagation of plane sound waves along the axis of a cylindrical duct. It was assumed that the diameter of the duct was less than about one-eighth wavelength, and thus no normal modes of vibration could occur other than those involving the x dimension. When the diameter of a cylindrical duct approaches a wavelength, additional modes of vibration occur that involve the radial r and the angular θ dimensions. These types of resonance are of particular importance in modern jet or in rocket engines where acoustic excitation of the duct is by rotating vaneaxial fans or by combustion. Also, ducts in engines often have lengths that are approximately the same as their diameters so that the simple axial mode may be of lesser importance.

When the radius of a cylindrical duct is less than one-sixteenth wavelength, plane-wave propagation occurs and the pressure is uniform across the tube for all values of x. If the tube is closed at both ends, the axial pressure distribution is given by a cosine function as shown in

Eq. (8.38) and resonances occur at the frequencies given by Eq. (8.37).

At higher frequencies, transverse modes of vibration occur which can be excited into resonance by the proper kind of source operating at the appropriate resonance frequencies. The modes may be tangential modes, for which the particle velocities are tangential to the wall of the tube, or radial, for which the particle velocities are perpendicular to the tube wall, or combined tangential and radial. Each tangential mode can exist in two forms: the standing form, in which the nodal surfaces are stationary and the spinning form, in which the nodal surfaces rotate. The two forms of the first tangential mode are illustrated in Fig. 8.11,[7] which shows, schematically, pressure and velocity patterns at four instants during a period of oscillation. Spinning modes can be excited, for example, by a rotating vaneaxial fan.

It is simpler mathematically to discuss the normal modes of vibration for a space between two concentric cylinders, as shown in Fig. 8.12.[8] We see examples of a plane wave and of spiral waves. Below a certain "cutoff" frequency, no energy is propagated down the annular duct. Above this cutoff frequency energy propagates better as the wave spirals "tighten." In open-ended ducts, part of the energy is reflected back and part is radiated outward and the radiation efficiency is dependent on the radiation impedance at the end.

Briefly, the theory [8] says that the tangential pressure distribution is

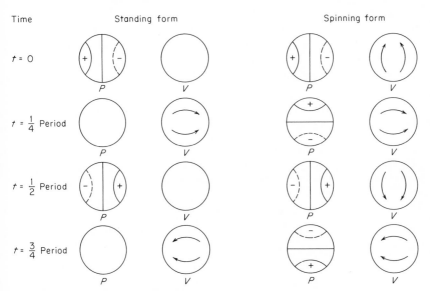

Fig. 8.11 Standing and spinning forms of the first tangential mode of vibration in a cylindrical duct. The pressure and particle velocity are shown at four instants of one period of the vibration. (*After Ref. 7.*)

given by the relation

$$H(\theta) = e^{-jm\theta} \qquad m = 0, 1, 2, 3, \ldots \qquad (8.53)$$

The radial pressure distribution is given by the function

$$G(r) = \frac{J_m}{Y_m}(k^*r) \qquad (8.54)$$

where $k^* = \dfrac{\omega^*}{c} \equiv \sqrt{\dfrac{m^2}{a^2} + \dfrac{n^2\pi^2}{h^2}}$ \qquad (8.55)

$m =$ mode number, which is equal to the number of pressure maxima as viewed from the end of the duct. For example, $m = 6$ in Fig. 8.12b and c.

$n = 0, 1, 2, 3, \ldots$

$J_m, Y_m =$ forms of Bessel functions

$r =$ radial coordinate, m

$a =$ rms radius of the annular duct, m, that is, $a = (0.5 \; a_1^2 + 0.5 \; a_2^2)^{1/2}$

$h =$ distance between the walls, m

The *cutoff frequency* occurs at the frequency

$$\omega_c = \omega^* \qquad (8.56)$$

Above the cutoff frequency, energy propagates down the duct. Below it, the energy in the wave is attenuated very rapidly with distance x.

In Fig. 8.12, the solid lines indicate regions of maximum sound pressure. The number of "starts" on the spiral "screw thread" is the mode number m. For any mode, the pitch of the spiral depends on frequency. If the frequency is reduced toward the cutoff value, the wavelength separating the "plane" waves increases. To accommodate this, the spiral pitch has to increase until in the limit the wavefronts are spinning around at right angles to the axis of the duct. This limiting case is illustrated by the lowest sketch of Fig. 8.12.

Wavelength

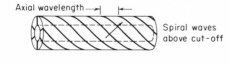

Plane waves

Axial wavelength

Spiral waves above cut-off

Fig. 8.12 Illustration of plane waves and higher-order modes of vibration in an annular duct. The lowest drawing shows the case of "cutoff," which occurs at a lower frequency than for the middle drawing. (*After Ref. 8.*)

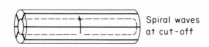

Spiral waves at cut-off

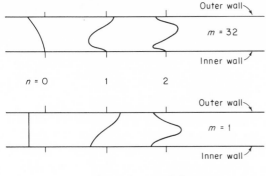

(*a*) Hub-tip ratio 3/4

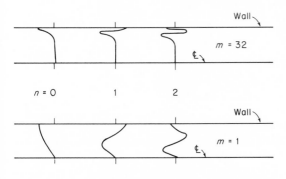

(*b*) Hub-tip ratio 0

Fig. 8.13 Radial profiles of sound pressure in annular-duct modes. (*a*) Ratio of diameters of inner to outer cylinders equals 0.75. (*b*) Single cylindrical tube. The upper edge of each sketch is the outer wall. The lower edge is in (*a*) the inner wall and in (*b*) the center line. (*After Ref. 9.*)

Examples of radial profiles of sound pressure in ducts are shown in Fig. 8.13.[9] The upper drawings (*a*) show a case where the inner cylinder has a radius three-fourths that of the outer cylinder. The lower drawings (*b*) show a case of a plane cylinder (no inner cylinder). In the latter case the pressure variations for higher mode numbers are seen to concentrate near the wall of the duct.

Plots from which cutoff frequencies f_c may be determined for two ratios of inner to outer radii of the concentric cylinders are given in Fig. 8.14. The ordinate is the dimensionless quantity $2\pi f_c a/c$.

A more quantitative discussion of the behavior of higher-order modes in circular ducts would require knowledge of the source, air flow, radiation impedance at the open end, temperature and density inhomogeneities in the duct, and so on.[8]

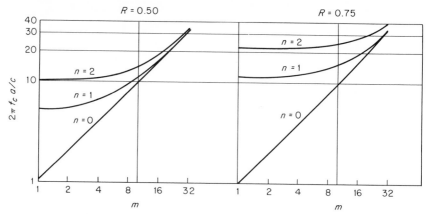

Fig. 8.14   Charts for determining the cutoff frequencies f_c of annular ducts. The left graph is for a ratio of radii R equal to 0.5 and the right for $R = 0.75$. The ordinate is $2\pi f_c a/c$, where a is the rms value of the two radii. The abscissa equals the mode number.

REFERENCES

1. P. M. Morse and K. U. Ingard, *Theoretical Acoustics*, McGraw-Hill, 1968.
2. Lord Rayleigh, *The Theory of Sound*, Vols. I and II, Dover, 1945. (Volume I was first published in 1877, revised in 1894; Vol. II was first published in 1878, revised in 1896. The 1945 edition is a reprint, bound in one book.)
3. R. W. B. Stephens and A. E. Bate, *Acoustics and Vibrational Physics*, Wm. Clowes and Sons, Ltd., 1966.
4. K. E. Kinsler and A. R. Fry, *Fundamentals of Acoustics*, 2d ed., Wiley, 1962.
5. L. L. Beranek, *Acoustics*, McGraw-Hill, 1954.
6. I. Malecki, *Physical Foundations of Technical Acoustics*, Pergamon, 1969.
7. F. H. Reardon, L. Crocco, and D. T. Harrje, "Velocity Effects in Transverse-mode Rocket-combustion Instability," *AIAA J.*, vol. 2, pp. 1631–1641, 1964.
8. C. L. Morfey, "Rotating Pressure Patterns in Ducts: Their Generation and Transmission," *J. Sound Vibration* (British), vol. 1, pp. 60–87, 1964.
9. J. M. Tyler and T. G. Sofrin, "Axial Flow Compressor Noise Studies," Society of Automotive Engineers, Inc., Aeronautical Meeting, 1961, preprint 345D.

Chapter Nine

Sound in Large Rooms

T. F. W. EMBLETON *

A large room is one with dimensions that are many times the wavelength of any sound wave being considered. Hence a source emitting sound, even within a narrow band of frequencies, will excite many normal modes of vibration of the enclosure.

The sound field at any measurement position in a large room may be described in terms of the mean-square sound pressure averaged in space over a volume enclosing this nominal position. Another useful description of the sound field is the intensity I, or power flow per unit area, in the progressive sound waves that travel around the room.

The mean-square sound pressure is influenced at any instant by the detailed content of the signal emitted by the source as well as by the properties of the enclosure. In order to make it a unique function of the enclosure and the location of the source and not of the varying properties of the source it is necessary to average the energy density over all possible phase relations that can exist.

The mean-square sound pressure, averaged in space and over all possible phase relations, is proportional to the *ensemble-average energy den-*

* Division of Physics, National Research Council, Ottawa, Canada.

sity, called, for brevity, the *energy density D* (see also Chap. 2). The mean-square sound pressure can be measured with conventional instruments. In SI units, D is expressed in watt-seconds per cubic meter.

The sound field in a large room will be "known" when certain of its properties can be described in all regions of importance. Near a source the sound field has a relatively high level and in such regions there is an easily identifiable flow of energy away from the source. The existence of various scattering objects in the enclosure results in a breaking up of the outward flow of energy from the source and redistribution into other directions. The location, distribution, and relative effectiveness of sound-absorbing surfaces and objects also influence the direction of net flow of energy. Far from the source, in a room that is fairly reverberant, the sound directly radiated becomes weak in comparison with the reverberant sound, that is to say, that sound which has experienced reflections from the various surfaces and boundaries of the room. In very large enclosures we also have to take into account the absorption of sound in the air through which it travels.

In a large, irregular enclosure it is possible, in principle, to have a diffuse sound field, one that consists of a superposition of sound waves traveling in all directions with equal probability. This characteristic ensures that the average energy density is the same at all points. If this were actually so, there would be no net flow of power in any direction. Hence, a diffuse sound field never actually exists because there is always a net flow of power away from the source to the places where the energy is ultimately absorbed. Nevertheless, the concept of a diffuse sound field is useful in rooms that are not highly absorptive and where, in addition, the measurement position is neither in the vicinity of the source nor near any small area that is highly absorptive.

9.1 Sound-absorption Coefficients

When sound waves fall on a surface or object, their energy is partially reflected and partially absorbed. The sound-absorbing efficiency of the surface involved is given in terms of an absorption coefficient designated by the symbol α. There are several types of such coefficients.

9.1.1 Definitions

Sound-absorption Coefficient for a Given Angle of Incidence, α_θ. The sound-absorption coefficient α_θ is defined as the ratio of the sound energy absorbed by a surface to the sound energy incident upon that surface at a given angle θ. Accordingly, α_θ may take on all numerical values between 0 and 1.

As sound waves travel around inside an enclosure they undergo many reflections at various angles of incidence. For most sound-absorbing materials a different absorption coefficient is associated with each angle of incidence.

Statistical (Energy) Sound-absorption Coefficient, α. With a view to obtaining a single-number index for general use, a *statistical absorption coefficient* α is defined (for an absorbing surface of infinite extent) as the ratio of sound energy absorbed by the surface to the sound energy incident upon the surface when the incident sound field is perfectly diffuse.[1] Statistical sound-absorption coefficients (sometimes called random-incidence sound-absorption coefficients) range from about 0.01 for marble up to about 0.996 for a surface covered with long, glass-fiber wedges such as are used in anechoic chambers. The maximum statistical value for a flat surface of absorbing material can be shown theoretically to be about 0.96.[1]

Laboratory-measured Sound-absorption Coefficient (Sabine Absorption Coefficient) α_{Sab}. Most of the sound-absorption coefficients that are published [2] are obtained by measuring the time rate of decay of the sound-energy density in an approved reverberation room with and without a patch of the sound-absorbing material under test laid on the floor. The correct design and operation of a reverberation room is such as to ensure a near-diffuse reverberant sound field as the test environment.*

An absorption coefficient determined in a reverberation room invariably exceeds the true statistical (energy) absorption coefficient α defined above, sometimes by as much as 50% at frequencies near 500 Hz and 20% at frequencies as high as 4 kHz. For this reason the absorption coefficient measured from decay rates should be denoted by a distinctive symbol.[3] In this text we shall refer to it as the *Sabine absorption coefficient* α_{Sab}.

Noise Reduction Coefficient (NRC). Where a curve or set of values for a commercial material is not given, its performance is sometimes specified only at 500 Hz or more often by a *noise reduction coefficient (NRC)* obtained by averaging (to the nearest multiple of 0.05) the Sabine absorption coefficients at 250, 500, 1,000, and 2,000 Hz.

9.1.2 Why Differences between α and α_{Sab}?

The differences between the values of α_{Sab} and α have been discussed for many years and are evidently caused by several factors.[4] One of the more interesting aspects of the problem is the fact that, for a highly absorbing material, the value of α_{Sab} can exceed unity, sometimes by as much as 20 or 30%.†

* ASTM C423−66. See also Chap. 6.

† Because it is difficult and usually unnecessary to explain how it is possible for a surface to absorb more sound power than its geometrical area would seem to warrant, the Acoustical and Insulating Materials Association [2] has followed a policy of rounding off all sound-absorption coefficients to 0.99 whenever the measured results exceeded this value. Some testing laboratories have different methods for this rounding-off process, for example by

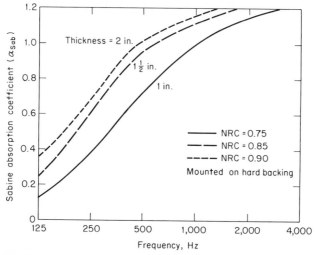

Fig. 9.1 Absorption coefficients α_{Sab} and noise reduction coefficients NRC of porous rigid glass-fiber form board mounted directly on hard backing. The results show that when the thickness of the board is less than one-sixteenth wavelength the absorption halves for each halving of frequency, approximately.

Diffraction of the sound field when incident on an absorbing patch of finite size is an important factor [5] which seems to account for most of the difference $(\alpha_{Sab} - \alpha)$ at low frequencies, but probably for not more than a small part of the difference at high frequencies. The area of absorbing material used to measure its absorption coefficient in a reverberation room is a matter of compromise: if the area is very large, the total absorption in the room becomes too great and the incident sound field is not sufficiently diffuse. If the area is small then the correction of the measured values to account for diffraction becomes large. These considerations argue for large reverberation rooms and large patches of test material—the practice in North America is to use an area of 72 ft² (6.7 m²) in a single patch 8 × 9 ft near the center of the floor, while in Europe somewhat larger areas of 10 or 12 m² are more common.

In laboratory reverberation rooms, in which the Sabine coefficients are measured, stationary irregular reflectors and moving vanes tend to redistribute the energy continuously among the modes of vibration to help ensure diffuseness of the sound field during the decay measurements except in the vicinity of highly absorbing surfaces. The degree of diffusion in a reverberation room containing 72 ft² of sound-absorbing material is much greater than that in most inhabited rooms and offices.

---

using a variable scale of reductions extending down to $\alpha = 0.90$ or 0.65. ASTM C423−66 recommends that no such adjustments be made and requires the measuring laboratory to report the results as measured, or if any rounding off is done it must be explained clearly.

9.1.3 Other Factors Affecting Sound-absorption Coefficients   The

absorption of a given acoustical material depends on the manner in which it is mounted. Furthermore, all materials have absorption coefficients that vary with frequency. Thus to specify the absorbing qualities of a material, a curve of α or α_{Sab} as a function of frequency is necessary for each of several mounting conditions. Values for laboratory-measured Sabine absorption coefficients, usually labeled α but by the convention in this chapter labeled α_{Sab}, of sound-absorbing materials are generally tabulated at 125, 250, 500, 1,000, 2,000, and 4,000 Hz.

9.1.4 Typical Laboratory Coefficients   Typical Sabine absorption

coefficients of porous commercial materials used in noise reduction applications are shown in Figs. 9.1 and 9.2. Further data and discussion of the acoustical performance of homogeneous materials as a function of their structure and mounting are given in Chap. 10.

9.1.5 Average of Absorption Coefficients in a Room at Each Frequency [3]   There is still another type of averaging of sound-absorption

coefficients that is necessary for calculations of sound fields in rooms. In calculating the average sound-pressure level and reverberation time at each frequency for a room, an average sound-absorption coefficient for the room as a whole is needed. The most widely used room-average absorption coefficient for the total enclosure is labeled $\bar{\alpha}_{Sab}$ in this chapter, and is calculated by weighting the Sabine absorption coefficients $(\alpha_{Sab})_i$ of the individual surfaces of the room according to their respective areas and taking the arithmetic average as follows:

$$\bar{\alpha}_{Sab} = \frac{S_1(\alpha_{Sab})_1 + S_2(\alpha_{Sab})_2 + \cdots + S_i(\alpha_{Sab})_i + \cdots + S_q(\alpha_{Sab})_q}{S} \qquad (9.1)$$

and

$$S = S_1 + S_2 + \cdots + S_i + \cdots + S_q \qquad \text{m}^2 \text{ (or ft}^2) \qquad (9.2)$$

where $S_1, S_2, \ldots, S_i, \ldots, S_q =$ areas of the individual sound-absorptive surfaces, m^2 (or ft^2)

$(\alpha_{Sab})_1, (\alpha_{Sab})_2, \ldots, (\alpha_{Sab})_i, \ldots, (\alpha_{Sab})_q =$ respective individual Sabine absorption coefficients (dimensionless)

If $\bar{\alpha}_{Sab}$ is to be useful, it is necessary that no one part of the room be heavily absorbing. A near-diffuse sound field cannot exist in an enclosure if $\bar{\alpha}_{Sab}$ is large, because the traveling waves die out too rapidly.[6]

Absorbing objects such as chairs, seats, tables, desks, and even people must be included when calculating $\bar{\alpha}_{Sab}$, but such objects generally have ill-defined areas. Hence, it is common practice to assign a value of Sabine absorption A_i to each object, where Sabine absorption is equivalent to

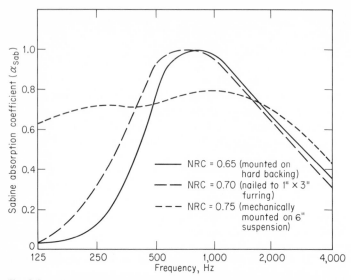

Fig. 9.2   Effect of mounting conditions on absorption coefficient and noise reduction coefficient of 3/4-in. textured acoustic tile.   The decrease in absorption at high frequencies is due to the nonopen nature of the textured surface of this particular type of tile.   Note that when nailed to 1 × 3 in. furring, there is an airspace of about 3/4 in. between the material and the hard ceiling.   With the 6-in. ceiling suspension, the airspace is 6 in. deep.   These data were probably adjusted at 500 and 1,000 Hz to prevent α_{Sab} from exceeding 1.0.

$S_i(\alpha_{Sab})_i$, the product of area times Sabine absorption coefficient.   All values of absorption for people and objects are summed, $A_0 = \Sigma A_i$, and are added into the numerator of Eq. (9.1).   *No modification of the total area S is made.*   In other words, the total area S is taken to be that of the boundaries of the room, including the floor but excluding objects and people.   In the case of a regular array of closely spaced absorbers, e.g., an audience seated in an auditorium, caution is necessary since the absorption per item or person is dependent on the area of floor devoted to each seated person.[7]   Also, when there are small patches of absorbing material on an otherwise sound reflecting surface, the absorption coefficients will be higher than those published (because the α_{Sab}'s were obtained from measurements made on larger patches of material).[8]

9.2   Steady-state Sound-pressure Levels in Direct and Reverberant Fields

The steady-state sound-pressure level in a room that contains a steady source of sound decreases as the total absorption is increased, provided the measurement is not made too near the source.   The ensemble-average energy density is obtained by averaging the mean-square pressure in both space and time. The period of time must be long enough to encompass a statistically valid sample of the signal radiated from the sound source. The space average must encompass at least a half wavelength.

9.2.1   The Direct Field

We shall assume that the sound source is far removed from any walls and emits energy at a constant rate of W watts.

The energy density of the direct field (see Chap. 6) is

$$D_0 = \frac{WQ_\theta}{4\pi r^2 c} \quad \text{watt-sec/m}^3 \tag{9.3}$$

where Q_θ = directivity factor of the source in a particular direction θ (dimensionless)

r = distance from the source, m

As the distance from the source increases, the energy density of the direct sound field becomes equal to, and ultimately less than, that of the reverberant field.

9.2.2   The Reverberant Field

The reverberant sound field differs in that all waves have been reflected at least once, with different parts of the wavefronts having been incident (and reflected) at different angles on different elements of the boundaries of the enclosure.

Let us focus attention on a small area ΔS chosen arbitrarily somewhere in the ideal reverberant sound field, Fig. 9.3. The energy density D_R of the reverberant field is due to waves traveling in all directions with equal probability; hence only half is due to waves traveling through ΔS from left to right. Of these, some contribute more than others to the energy flow through ΔS, depending on the angle of incidence θ. Waves incident normal to the area contribute a rate of energy flow (power) equal to $I \Delta S$, whereas those incident at an angle θ provide a reduced amount equal to $I \Delta S \cos \theta$. Remembering the basic assumption that waves are traveling in all directions with equal probability, one can also show that the average amoung of power flowing through ΔS from left to right is half what it would be if all the energy were arriving at normal incidence. Gathering up these two factors of $\frac{1}{2}$ we can express the power flow in one direction

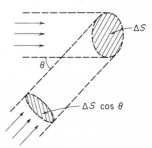

Fig. 9.3 An area ΔS located arbitrarily somewhere in a diffuse sound field. Half of the sound energy passes through ΔS in each direction. Waves incident at an angle contribute less to the energy flow— $I \Delta S \cos \theta$ —than those incident perpendicularly.

through ΔS in terms of the energy density D_R due to waves traveling in all directions as $(D_R c/4)\,\Delta S$.

If the area ΔS lies on one of the walls of the enclosure, and if this particular element of wall has a statistical absorption coefficient α, the power absorbed is

$$\frac{D_R c}{4}\,\Delta S \alpha \qquad \text{watts} \qquad (9.4)$$

The total power absorbed from the reverberant field can be obtained by summing Eq. (9.4) over all the absorbing surfaces, or more conveniently by making use of the arithmetic average absorption coefficient $\bar{\alpha}$ determined by an equation like Eq. (9.1) in which α_i appears in place of $(\alpha_{\text{Sab}})_i$. The power loss from the reverberant field is, therefore, given by

$$\frac{D_R c}{4}\,S\bar{\alpha} \qquad \text{watts} \qquad (9.5)$$

In the steady state the power lost at the walls must be balanced exactly by power supplied to the reverberant field. This power comes initially from the sound source. By definition, that power supplied to the reverberant field is equal to the power remaining after the first reflection. Hence the power balance is

$$W(1 - \bar{\alpha}) = \frac{D_R c}{4}\,S\bar{\alpha} \qquad \text{watts} \qquad (9.6)$$

The steady-state energy density in the reverberant field is given by rearranging Eq. (9.6)

$$D_R = \frac{4W}{cS\bar{\alpha}}\,(1 - \bar{\alpha}) \qquad \text{watt-sec/m}^3 \qquad (9.7)$$

This may also be written in the form

$$D_R = \frac{4W}{cR} \qquad \text{watt-sec/m}^3 \qquad (9.8)$$

where R *is the room constant* [3,6] defined as

$$R \equiv \frac{S\bar{\alpha}}{1 - \bar{\alpha}} \qquad \text{m}^2 \text{ (or ft}^2 \text{ in English units)} \qquad (9.9)$$

The room constant R of a large enclosure having the approximate proportions $1:1.5:2$ is plotted vs. volume of the enclosure in Fig. 9.4 for a range of average statistical (energy) absorption coefficients $\bar{\alpha}$. As we shall see later, we may avoid using R in practical applications by substituting $S\bar{\alpha}_{\text{Sab}}$.

9.2.3 Total Steady-state Sound-pressure Levels

At a point located a distance r from a single source the direct field is given by Eq. (9.3) and the reverberant field by Eq. (9.8). The total energy density is

Room volume V, m^2

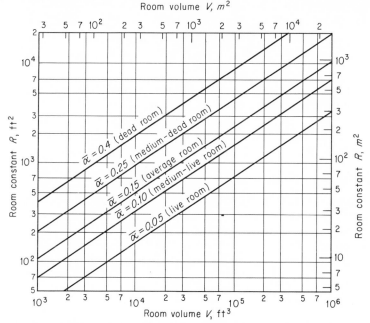

Fig. 9.4 Calculated values of room constant R as a function of room volume for rooms with proportions of about $1:1.5:2$. The left-hand and lower coordinates are in English units. The right-hand and upper coordinates are in SI units. $\bar{\alpha}$ is the average statistical sound-absorption coefficient for the room. These proportions give $S = 6.25\,V^{2/3}$. The ordinate R is also approximately equal to the total Sabine absorption of the room, $S\bar{\alpha}_{Sab}$. This chart must be used with caution, because for large room volumes and at frequencies above 1,000 Hz, sound absorption in the air of the room must be considered (see Sec. 9.5). (*After Beranek.*[6])

given by $D = D_0 + D_R$, or in terms of mean-square sound pressure [see Eq. (1.18)]

$$|p|^2 = W\rho c \left(\frac{Q_\theta}{4\pi r^2} + \frac{4}{R} \right) \qquad \text{N}^2/\text{m}^4 \qquad (9.10)$$

The sound-pressure level L_p *re* 2×10^{-5} N/m^2 is

In mks units *

$$L_p = L_w + 10 \log_{10}\left(\frac{Q_\theta}{4\pi r^2} + \frac{4}{R} \right) \qquad \text{dB} \qquad (9.11)$$

In English units, * 10 dB must be added because r^2 and R are each larger

* Sound power is not related solely to sound pressure and area. The right-hand sides of Eqs. (9.11) and (9.12) should properly be increased by a term $+10 \log (\rho c/400)$ (see Fig. 2.3 for values). In most problems of room acoustics this additional term amounts to less than 0.3 dB and may be neglected.

in magnitude by a factor of 10.8, so that

$$L_p \approx L_w + 10 \log_{10} \left(\frac{Q_\theta}{4\pi r^2} + \frac{4}{R} \right) + 10 \qquad \text{dB} \qquad (9.12)$$

Equations (9.11) and (9.12) describe the sound field at all points in a room regardless of the relative magnitudes of the direct and reverberant sound fields. However these equations contain the room constant R, which depends on the theoretical (and difficult to measure) average energy-absorption coefficient $\bar{\alpha}$, rather than on the corresponding Sabine absorption coefficient $\bar{\alpha}_{Sab}$ that usually is tabulated for acoustical materials. One observes from Eq. (9.9) that R is always greater than $S\bar{\alpha}$ by a factor that varies between 1 and 1.67 as $\bar{\alpha}$ varies from 0 to 0.4. It happens that the Sabine absorption coefficient $\bar{\alpha}_{Sab}$ is greater than $\bar{\alpha}$ by a factor that is often close enough to that just stated that it may be used instead for solution of practical problems where values of α for individual materials are not known.* Thus one substitutes the total sound absorption $S\bar{\alpha}_{Sab}$ for R in each of the Eqs. (9.10) to (9.12).

In mks units, Eq. (9.11) then becomes

$$L_p = L_w + 10 \log_{10} \left(\frac{Q_\theta}{4\pi r^2} + \frac{4}{S\bar{\alpha}_{Sab}} \right) \qquad \text{dB} \qquad (9.13)$$

where $L_w = 10 \log_{10} W + 120$ dB *re* 10^{-12} watt

W = radiated power, watts

S = total area of all surfaces of the room, m^2

$\bar{\alpha}_{Sab}$ = average Sabine absorption coefficient for the room, including objects and people. Also, air absorption must be included at frequencies above 1,000 Hz, whereupon $(\bar{\alpha}_{Sab})_T$ is substituted as discussed in Sec. 9.5, Eq. (9.31).

r^2 = squared distance from source, m^2

Q_θ = directivity factor of the source in direction θ

In English units, where r^2 and S are in ft^2, the equation is the same except that 10 dB is added to the right-hand side.

The relative importance of the direct- and reverberant-field contribu-

---

* The approximation is $S\bar{\alpha}_{Sab} \approx S\bar{\alpha}/(1 - \bar{\alpha}) = R$. The published values of absorption coefficient α_{Sab} are sometimes corrected [2] to make an approximate allowance for the finite size of sample measured in the reverberation room. To be consistent one should also apply a "reverse" correction to allow for diffraction when using a finite area of material in a practical situation. O. L. Angevine, Jr. (private communication) argues that for a small area of material $\bar{\alpha}/(1 - \bar{\alpha})$ should be used. For a large area he believes $\bar{\alpha}$ alone should be used instead of $\bar{\alpha}/(1 - \bar{\alpha})$. For intermediate areas $\bar{\alpha}_{Sab}$ should probably be used instead of $\bar{\alpha}/(1 - \bar{\alpha})$. In practical problems we recommend $\bar{\alpha}_{Sab}$ as the best choice.

tions at a given point is readily evident by comparison of the magnitudes of the quantities $Q_\theta/4\pi r^2$ and $4/S\bar{\alpha}_{\text{Sab}}$. If the former predominates then the sound-pressure level is largely due to direct radiation. If $4/S\bar{\alpha}_{\text{Sab}}$ is large compared with $Q_\theta/4\pi r^2$, the room absorption is the important factor in determining L_p. A practical example is the case of a factory worker. If he is standing next to a noisy machine, he is probably in the near field and sound-absorbing materials on the walls are of no help. However, if he stands far from the machine, sound-absorbing materials will benefit him. In either case, sound-absorbing materials will benefit other workers at a distance from his machine, provided they are not in the near field of their own machines.

A generalized graph showing $L_p - L_w$ vs. $r/\sqrt{Q_\theta}$ is shown in Fig. 9.5. R_T and $S(\bar{\alpha}_{\text{Sab}})_T$ on the graph may be considered interchangeable with R and $S\bar{\alpha}_{\text{Sab}}$ at frequencies below about 1,500 Hz. Above 1,500 Hz and for large rooms R_T and $S(\bar{\alpha}_{\text{Sab}})_T$ differ from R and $S\bar{\alpha}_{\text{Sab}}$ in that they take into account the sound absorption in the air in the manner discussed in Sec. 9.5, Eq. (9.31).

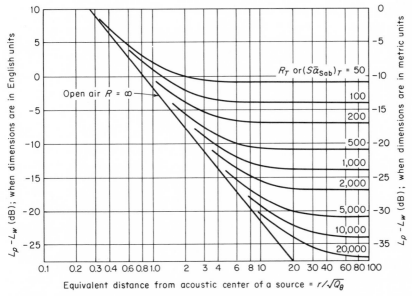

Fig. 9.5 Chart for determining the sound-pressure level in a large irregular enclosure at a distance r from the center of a source of directivity factor Q_θ. The ordinate is $L_p - L_w$ dB (where L_w is referred to 10^{-12} watt) and is calculated from $10 \log_{10} (Q_\theta/4\pi r^2 + 4/R_T)$. When all dimensions are in English units use the ordinate on the left-hand side; for dimensions in SI units the right-hand ordinate should be used. The parameter R_T (or R if air absorption is negligible) is approximately equal to $(S\bar{\alpha}_{\text{Sab}})_T$ (or $S\bar{\alpha}_{\text{Sab}}$, neglecting air absorption).

There are many cases such as offices where nearly all the absorption is on one surface and consequently the sound field is not diffuse. One would expect some deviations in the measured curves from those calculated using the results of Eqs. (9.11) and (9.13), and this is often the case. Figure 9.6 shows an experimental plot of the relative sound-pressure level vs. distance for a large classroom with a 15-ft-high ceiling treated with acoustical tile. The room constant R (or $S\bar{\alpha}_{Sab}$) was estimated to be no larger than about 8,000. In spite of small fluctuations, the measured curve does not appear to level off at 40 ft but continues to decrease by another 3 dB at 100 ft.

9.2.4 Decrease in Sound-energy Density Near a Highly Absorptive Surface
From the discussion that precedes Eq. (9.4) we can understand situations where a diffuse field does not exist. For example, in the extreme case where the statistical absorption coefficient α of a surface were unity, there would be no waves from the direction of this surface to contribute to the energy density of the field at a point of measurement. In this case the energy-density level and related quantities such as sound-pressure level would fall by 3 dB in going from some average point in the room up to this surface.[9] This illustration applies in lesser degree to any surface where there is absorption of sound.

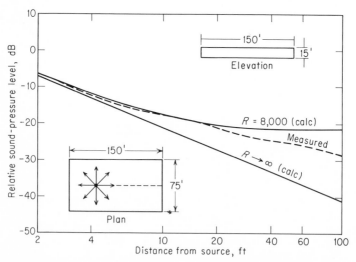

Fig. 9.6 Measured and calculated relative sound-pressure levels in a room approximately 150 by 75 ft with a 15-ft ceiling, plastered walls, wooden floor, acoustic-tile ceiling, and numerous scattering objects in the room. Source and receiver both at a height of 6 ft from floor. The intention here is to illustrate a room having acoustic characteristics that do not satisfy the simple diffuse-room theory.

9.2.5 Reaction of Enclosure on Source [10, 11]

If a given source has a directivity factor Q_θ (see Chap. 6) of unity in free space, the actual Q_θ value within a room may increase or decrease in certain directions because of reflections from nearby surfaces. This effect is most pronounced when the nearby surfaces are highly reflecting. If the room is large, the Q_θ for a *nondirectional* source will still be 1 when the source is located near the center of the room, provided the power output is the same. If this source is located very near the center of a wall, ceiling, or floor, it will have a value of about 2 since the power is now being radiated into a hemisphere. Near an edge of the enclosure (the intersection of two hard walls) it will have a value of 4, and at a corner (three hard walls) it will go as high as 8, i.e., a 9-dB increase, *always provided the power output is the same.**

9.2.6 Extended and Multiple Sources

Practical noise sources are sometimes distributed over regions of considerable size—either as a single large machine or as an array, regular or otherwise, of smaller machines. Near-field effects are avoided only when the distance between the source and observation point is greater than the largest dimension of the source (for a single large machine a good rule of thumb is 4 times this dimension). When studies involving a sound source in an enclosure are made, the variation of sound-pressure level, L_p, with distance should be studied by means of direct measurements. By this means one can determine the extent of the free-field region (the region in which L_p falls by 6 dB per doubling of distance) and the reverberant region (the region of "uniform" L_p †).

Figure 9.7 shows the theoretical prediction of the sound field in a large room 20 ft high for a rectangular array of small machines 8 ft apart in each direction.[12] With changes in dimensions the results are applicable to such diverse situations as nail-making machines in a steel

---

* Some authors talk about *constant-volume-velocity sources* not in terms of changes of directivity but in terms of changes of power output with changes of source location, i.e., a 9-dB power output increase if the source is moved into a trihedral corner where the same volume velocity delivers power to eight times the number of normal modes (compared to the number that can be excited at the center of the room).[10, 11] This follows from the discussion following Example 8.4.

† Remember that spatial uniformity only occurs for broadband sources; pure tones will show space variations of about ±6 dB even in a perfectly diffuse sound field, with a mean spacing of about $\lambda/2$ between points of high pressure. Kuttruff and Thiele (*Acustica*, vol. 4, pp. 614–617, 1954) have shown that, for a *fixed* microphone location, the response of an enclosure to a pure tone that is swept through a range of frequencies also exhibits fluctuations, the mean level difference between successive sound-pressure maxima and minima being 9 or 10 dB and independent of room volume or reverberation time. The *number* of maxima in any frequency range is approximately proportional to the reverberation time.

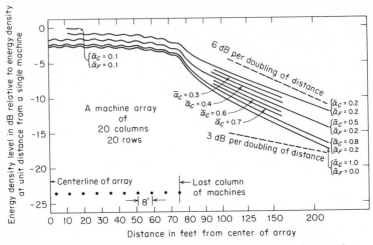

Fig. 9.7 Theoretical relation between energy density and distance for a square 20×20 array of incoherent point sources 8 ft on center in a room 20 ft high. The curves show the effects of varying combinations of absorption coefficients, $\bar{\alpha}_C$ and $\bar{\alpha}_F$, on ceiling and floor respectively. Where no value is specified $\bar{\alpha}_F$ has the value 0.2.

mill, looms in a textile factory, or the keypunch room of a computation center. The location of the point of observation, i.e., the microphone, is measured from the center of the array and moves outward in a straight line between the rows of machines; i.e., it never comes closer than 4 ft to any machine.

The levels are, on the whole, remarkably constant, within less than 1.5 dB, anywhere inside the array, and it is not until the machines in the row on the edge of the array are passed that any significant reduction occurs. In the next few feet a considerable drop in level takes place. Thereafter the curves take on a smaller slope as the relative contributions to the sound field from the different machines become more nearly equal.

The curves of Fig. 9.7 are given for several different absorption coefficients at the floor α_F and at the ceiling α_C and show that a given amount of absorption produces a greater reduction of sound level when one is far from the last row of machines than when one is within the array of machines. It also appears beneficial in this case, other practical considerations aside, to concentrate the total absorption on one surface, say the ceiling, rather than to distribute it between floor and ceiling.

9.3   Decay of the Sound Field

9.3.1   Energy-density Relations   At the instant the source of sound is switched off certain relations exist among the normal modes of the en-

closure and, with no more energy being supplied to the field, these relations are a factor in determining the characteristics of the sound field in the subsequent period during which the field is decaying. Averaging the energy density at each instant—even if it could be done by arresting time in some mysterious way—would not dispose of this problem as it did in the steady-state case. In principle, one must perform the experiment of cutting off the sound field an infinite number of times in order to obtain an average of the decay curves over all possible sets of phase relations that are present in the signal at the instant of cutoff. Fortunately, a simpler technique is available for determining the ensemble-average value.[13-15] By this technique, the sound field in the enclosure is excited impulsively and the measured signal is recorded on magnetic tape. This tape is played backward and on playback the output is first squared and then fed to an integrating circuit. The output from the integrator gives, on a reversed time scale, the ensemble average of all decay curves for noise having the same bandwidth as the impulsive sound source used.

9.3.2 The Direct Field
When the point of measurement is close to the source the direct sound field predominates over the reverberant sound field; consequently when the source ceases to radiate, the main effect is the loss of the direct field. This is not detected immediately but only after an interval of time t_r, which is the length of time taken for the last of the radiated sound to travel from the source to the receiver. As the distance between source and receiver increases, this time interval becomes greater. Also, because the energy density of the direct field becomes smaller, the drop in the total sound field after the time t_r is less. After t_r sec, the total sound field drops immediately to the reverberant field which had coexisted at that point and which has not yet been affected (see Fig. 9.8, steps a and b).

9.3.3 The Reverberant Field
After the direct sound suddenly drops off, the various components of the reverberant field disappear one by one as the "end" of each field passes the point of observation. In general, the larger components of the field disappear first since these have usually traveled a shorter distance and have undergone fewer absorbent reflections. The energy density of the reverberant field at any instant during the decay is determined by the remaining components of the sound field. A "stepped" decay curve of the ensemble-average energy density vs. time is shown as curve c in Fig. 9.8.

In order to retain the simplicity offered by the statistical approach, one may consider an alternative picture. In this picture the reverberant sound field contains component waves traveling in many different directions; these component waves fall continuously on various surfaces, both the bounding surfaces and other objects within the enclosure, and the energy of the sound field is absorbed by a continuous process at a rate which depends on the combined effect of all the individual absorbing surfaces. The Sabine absorption coefficient of an absorbing material is computed from its effect on the rate of decay of the sound-energy-density level in a reverberation room. Of prime interest, therefore, is the sound-energy density remaining in the room at any given instant, and how this quantity varies with time.

First, let us define a *statistical reflection coefficient* r_i as the ratio of sound energy reflected from the ith surface of the room to the sound energy incident on it, when the incident sound field is diffuse. Hence, r_i is the exact counterpart of the statistical absorption coefficient defined earlier; and $r_i + \alpha_i = 1$.

Let us follow the passage of a given bundle of sound energy, say J_0, as it proceeds around the enclosure, being reflected many times from many different surfaces. After q such reflections at various surfaces, the sound energy J that remains of the original bundle of energy J_0 may be written

$$J = J_0(r_1^{S_1/S} r_2^{S_2/S} r_3^{S_3/S} \ldots) \tag{9.14}$$

where S_i/S is the proportion that the area of the surface i bears to the total area of the room, and the product of the reflection coefficients r_i encompasses all surfaces.[16,17]

The energy incident on all the surfaces of the enclosure S in a small interval of time Δt is given by [see discussion preceding Eq. (9.4)]

$$\frac{D_R c}{4} S \, \Delta t \tag{9.15}$$

The total energy in the whole enclosure at any instant equals the average energy density D_R times the volume of the room V. Equating Eq. (9.15)

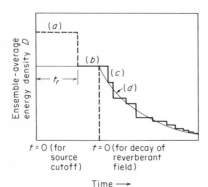

Fig. 9.8 Steady-state values and the decay of the sound field in an enclosure. (a) The steady-state direct field; (b) the steady-state reverberant field; (c) the decay shown as a series of discontinuous steps as component fields disappear; (d) the decay shown as a smooth curve owing to the sound energy being continuously absorbed from the reverberant field.

to $D_R V$ and solving for the time Δt required to give the total energy a chance to undergo one absorbent reflection only

$$\Delta t = \frac{4V}{cS} \tag{9.16}$$

During this time interval Δt the energy is striking the various surfaces *simultaneously*, whereas Eq. (9.14) relates to the energy that strikes the various surfaces *sequentially*. However, we shall overlook this inconsistency and assume that the ratio of the energy density $D_R(\Delta t)$ at the end of the time interval Δt to its value D_0 at the beginning, $t = 0$, is equal to the ratio of energies in Eq. (9.14). Hence

$$D_R(\Delta t) = D_0(r_1^{S_1/S} r_2^{S_2/S} r_3^{S_3/S} \ldots) \tag{9.17}$$

In keeping with the distinction drawn above between simultaneous and sequential events, the value of the reflection coefficient r_i for a surface S_i that consists of several parts having different reflection coefficients should be the area-weighted arithmetic-average value of its parts; it is this value that should be multiplied by the reflection coefficients of other surfaces in Eq. (9.14) et seq. This applies, for example, to a patch of absorbing material occupying only part of an otherwise reflecting surface.

After another such interval of time Δt all the energy will, on the average, have undergone one more reflection and partial absorption at the room surfaces.

$$D_R(2\,\Delta t) = D_0(r_1^{S_1/S} r_2^{S_2/S} r_3^{S_3/S} \ldots)^2 \tag{9.18}$$

After a length of time t the energy of the reverberant field will have been reduced by n encounters with the walls, where $n\,\Delta t = t$ or $n = t/\Delta t$. Hence one finds

$$D_R(t) = D_0(r_1^{S_1/S} r_2^{S_2/S} r_3^{S_3/S} \ldots)^n \tag{9.19}$$

The essence of the diffuse-field approach is that absorption of the reverberant field more closely resembles a continuous process. This may be emphasized by combining Eqs. (9.19) and (9.16), remembering that $n = t/\Delta t$, to yield

$$D_R(t) = D_0(r_1^{S_1/S} r_2^{S_2/S} r_3^{S_3/S} \ldots)^{\frac{cSt}{4V}} \quad \text{watt-sec/m}^3 \tag{9.20}$$

and hence

$$D_R(t) = D_0 e^{-(c/4V)[-\Sigma S_i \ln r_i]t} \tag{9.21}$$

In this equation ln is the logarithm to base e.

Thus the reverberant sound field decays exponentially from the time that sound energy ceases to be supplied to it, as shown by the smooth curve d in Fig. 9.8. The start of the decay begins at that instant when the last of the direct field is being reflected at the walls of the enclosure. Until this time, the reverberant field has its steady-state value (b of Fig. 9.8).

As in the steady-state case, one is usually more interested in the sound-pressure level than in the mean-energy density. Therefore we rewrite Eq. (9.21) as the logarithm

$$(L_p)_t = (L_p)_0 - \frac{4.34c}{4V} [-\Sigma S_i \ln r_i]t \qquad \text{dB} \qquad (9.22)$$

because 10 log $e^x = 4.34x$. Hence the sound-pressure level, or related quantities such as the energy-density level, decays at a rate given by

$$\text{Decay rate} = 1.086 \frac{c}{V} [-\Sigma S_i \ln r_i]$$

$$= 1.086 \frac{c}{V} [-\Sigma S_i \ln (1 - \alpha_i)] \qquad \text{dB/sec} \qquad (9.23)$$

From Eq. (9.23) and the argument that precedes it, it is evident that the influence of any particular surface on the decay rate of the sound field in the enclosure is determined by the product of its surface area S_i and the natural logarithm of the reciprocal of its energy reflection coefficient $\ln (1/r_i)$ or $-\ln r_i$.

Finally, we note that the decay curve is more likely to follow curve d if the sound field is diffuse and if the absorption is well distributed over the various surfaces. It will approximate a stepwise curve c more closely as these two requirements are satisfied to a lesser degree. The earlier "steps" of curve c are associated with components of the field that have encountered the walls of the enclosure once, and later steps in the curve relate to waves reflected two or more times from the walls; in the same way we may think of curve a as representing an earlier member of the series that has been reflected a zero number of times.

9.3.4 The Reverberation Equations

The reverberation time of the enclosure is defined as the time required for the sound-pressure level to fall 60 dB. This time corresponds to about that required for a sound to diminish from a fairly loud level in a room to the threshold of audibility.

Millington-Sette Equation. The Millington-Sette equation [16,17] for reverberation time T is obtained from Eq. (9.22) by replacing $(L_p)_0 - (L_p)_t$ by 60 dB and calculating the required time in seconds, replacing t by T. In terms of the statistical absorption coefficient of each surface, α_i, the equation is

$$T = \frac{60V}{1.086c[-\Sigma S_i \ln (1 - \alpha_i)]} \qquad \text{sec} \qquad (9.24)$$

Caution. It is explicit in the derivation of Eqs. (9.22) and (9.24) that all the sound energy in the enclosure is incident on all the surfaces in proportion to their area during the process of decay. In general this does not happen in a real enclosure, especially when the total absorption is large, though it is more likely to happen when the flow of sound energy is broken up by scattering objects within the enclosure. Thus the "absorbing power" of a surface may be overestimated by the theoretical expression $-\ln(1 - \alpha_i)$. The Millington-Sette model seems to get into difficulty if any surface, no matter how small, has a coefficient approaching unity. Then, the equation says the reverberation time should go to zero, which does not happen. To avoid this trouble one must average such highly absorbing small surfaces into larger surfaces of low absorption to keep all average values of r_i reasonably large, as required for a near-diffuse sound field.

Sabine Equation. It is apparent that by measuring the decay rate of the sound field the "absorbing power" of a surface may be found directly by experiment, even though it may not agree exactly with the statistical energy value $-S_i \ln(1 - \alpha_i)$ given by Eq. (9.23). By analogy with Eq. (9.24) one obtains the Sabine equation for reverberation time

$$T = \frac{60V}{1.086c\Sigma S_i(\alpha_{\text{Sab}})_i} \quad \text{sec}$$

or, by Eqs. (9.1) and (9.2),*

$$T = \frac{60V}{1.086cS\bar{\alpha}_{\text{Sab}}} \quad \text{sec} \qquad (9.25)$$

For most engineering problems, published sound-absorption coefficients (see Refs. 2 and 7) used in the Sabine equation predict reverberation times with adequate accuracy. *We recommend Eq. (9.25), or its practical forms Eqs. (9.32) and (9.33), for use in engineering design.*

In further support of the substitution of α_{Sab} for $-\ln(1 - \alpha)$ we observe that the published Sabine absorption coefficient α_{Sab} for any surface is always calculated from measurements of decay rate in reverberation chambers according to a formula analogous to Eq. (9.25); decay rate and reverberation time are reciprocals. Thus it follows that the use of Eq. (9.25) is correct, unless the "absorbing power" of a surface is partially dependent on the absorption of other surfaces.

Norris-Eyring Equation. In both the Millington-Sette and the Sabine equations for reverberation time the absorption of each surface enters individually into the denominator and it is apparent that these must be summed to obtain the total absorption for the whole room. From Eq. (9.25) this summation is equal to $S\bar{\alpha}_{\text{Sab}}$. By analogy with Eq. (9.23),

---

* See Eqs. (9.32) and (9.33) for the practical forms of Eq. (9.25), including air absorption.

$S\bar{\alpha}_{Sab}$ may be replaced by $-S \ln (1 - \bar{\alpha})$ where $\bar{\alpha}$ is an average statistical (energy) absorption coefficient for the whole enclosure. The well-known *Norris-Eyring* [18] equation for reverberation time is then obtained

$$T = \frac{60V}{-1.086cS \ln (1 - \bar{\alpha})} \quad \text{sec} \quad (9.26)$$

From Eq. (9.20) one may write $\bar{r}_g$ as the area-weighted geometric mean of the energy reflection coefficients r_i for the whole enclosure. To obtain Eq. (9.26) one must define the $\bar{\alpha}$ in it as $1 - \bar{r}_g$. It is then apparent that this $\bar{\alpha}$ is *neither* the arithmetic mean nor the geometric mean of the statistical absorption coefficients α_i. It is only in the special case when all the values of α_i are equal [18] that the $\bar{\alpha}$ in Eq. (9.26) is the arithmetic-mean value.

The Numerical Factor. The factor $60/1.086c$, which occurs in each of the equations for reverberation time, Eqs. (9.24) to (9.26), has the following values at 20°C (68°F):

In mks units: $60/1.086c = 0.161$
  where $V = $ volume of room, m³
  and $S = $ area of surface, m²
In English units: $60/1.086c = 0.049$
  where $V = $ volume of room, ft³
  and $S = $ area of surfaces, ft²

The unit of absorption $A(= S\alpha_{Sab})$ is the *sabin* in English units, or *metric sabin* in mks units.

9.3.5 Discussion
Using a computer model and a ray-tracing technique it has been shown [19] that the effect of an absorbing surface on the decay rate of the sound field in a two-dimensional room is greater (by 20 to 30%) than would be expected if the absorbing power were calculated from its statistical absorption coefficient α rather than its Sabine coefficient α_{Sab}. This technique also shows that the decay rate depends on the shape of the room and the distribution of absorbing surfaces [20] — factors not taken into account in any of the Eqs. (9.22) to (9.26). A recent analysis [21] also indicates that the absorption of power from a decaying sound field depends partly on the difference between the absorption coefficients α_i of individual surfaces and their arithmetic-mean value taken over all surfaces.

Example 9.1 Consider a large irregular room having a volume of 400,000 ft³ and a total surface area of about 45,000 ft², including the area occupied by seats. Let the average Sabine absorption coefficient for the room as a whole be $\bar{\alpha}_{Sab} = 0.16$ at a frequency of 2,000 Hz including the seats but without occupants. Find the reverberation times with and without audience.

SOLUTION: The absorption $S\bar{\alpha}_{Sab}$ is about 7,200 sabins. The reverberation without audience is given by Eq. (9.25)

$$T = \frac{0.049 \times 400,000}{7,200} = 2.7 \text{ sec}$$

If the seating capacity of the enclosure is 2,000 and the area occupied by it is 14,000 ft², and if we assign an incremental absorption of 0.31 per unit area of audience owing to the audience entering the room, the total absorption for full occupancy will be

$$S\bar{\alpha}_{Sab} = (45,000 \times 0.16) + (14,000 \times 0.31)$$
$$= 7,200 + 4,350 = 11,550 \text{ sabins}$$

An additional absorption of 4,350 sabins is contributed by the audience. The reverberation time with audience is

$$T = \frac{0.049 \times 400,000}{11,550} = 1.7 \text{ sec}$$

Thus, there is roughly 1.0 sec difference in reverberation time at 2,000 Hz between the empty room and the fully occupied room.

9.4 Rise of the Sound Field

Some of the ideas discussed in earlier sections of this chapter are perhaps easier to understand when one considers the buildup of the composite sound field from its parts after the source begins to radiate. In the earlier stages of this process the sound field consists of only a few components. As time passes the number of the components increases. In many practical cases the number of components is too small to represent adequately all directions of propagation. This situation is improved to some degree when the enclosure is of irregular shape or contains many sound scatterers. Both a low absorption coefficient and either an irregular shape or many scatterers are required in order to achieve a good approximation to a perfectly diffuse field.

Each time the sound source is switched on, the detailed shape of the curve of sound-pressure level vs. time at any location is slightly different depending on the relative phases with which the various frequencies are being emitted. If one considers instead the ensemble average of the energy density—a property of the enclosure alone which is unaffected by the detailed phases of the source—then a very simple relation, which we discussed in the previous chapter, emerges between the rise of the field in the moments after the source is switched on and the decay of the field just after the source is switched off.[22] The abrupt beginning and abrupt end of the signal both propagate through the enclosure with the same velocity c and are reflected and scattered according to the same laws.   Hence the ensemble-average energy densities during rise and decay of the field at the same instant of time after switching the source on or off are related by

$$D_{\text{rise}} = D_{\text{steady state}} - D_{\text{decay}} \qquad (9.27)$$

The rise of the field for the same hypothetical situation as existed for Fig. 9.8 is shown in Fig. 9.9: it is seen that at any point along the horizontal time axis, the sum of the two curves is a constant equal to the steady-state value. (See also Fig. 8.2.)

9.5   Air Absorption

Sound energy is absorbed by the air itself in a room or auditorium to a degree ranging from an almost negligible amount below 1,000 Hz to fairly sizable amounts at frequencies above.   This fact has not been taken into account in the previous sections of this chapter.   One can see clearly that if the enclosure is small the number of reflections (and partial absorptions) at the walls is large and the distance which waves travel between reflections is comparatively small.   In this case air absorption is generally not as important as boundary absorption.   On the other hand, for the same number of reflections at the walls in a very large enclosure the distance traveled by the sound waves through the air is much greater and the energy loss in the air cannot be ignored.

While waves are propagating through the air a distance x, the *energy density* is decreasing exponentially owing to air absorption, according to the decay formula

$$D(x) = D_0 e^{-mx} \qquad \text{watt-sec/m}^3 \qquad (9.28)$$

where $m =$ energy attenuation constant in units of reciprocal length
$\quad x =$ distance of propagation, in the same units of length as for m
Measured values of $4m$ under some typical atmospheric conditions, with

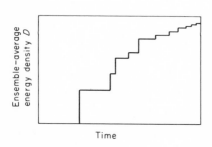

Time

Fig. 9.9   Rise of the sound field in the same hypothetical enclosure as considered in Fig. 9.8. Note that the sum of the rising curve of this figure and the decaying curve of Fig. 9.8 is equal at all times to the steady-state value. This is true for the ensemble-average energy density (but not the sound pressure). [This is also shown to be true for sound decay in a long tube by adding Eqs. (8.11) and (8.18) or Eqs. (8.17) and (8.20).]

frequency as a parameter, are shown in Tables 9.1 and 9.2.[23] It is observed that the value of $4m$ depends strongly upon frequency and relative humidity. To a somewhat lesser degree it is also temperature dependent.

The room constant may be modified to take into account the effect of air absorption as follows [6]

$$R_T = \frac{S\bar{\alpha}_T}{1 - \bar{\alpha}_T} \quad \text{m}^2 \text{ (or ft}^2) \tag{9.29}$$

where

$$\bar{\alpha}_T = \bar{\alpha} + \frac{4mV}{S} \tag{9.30}$$

and R_T is defined as the total room constant. In deriving Eq. (9.29) it is assumed that $4mV/S$ is less than about 0.4.

When the reverberant or total sound field is expressed in terms of the total Sabine absorption of the enclosure $S\bar{\alpha}_{\text{Sab}}$ [as, for example, in Eq. (9.13)], modification of the expression to allow for air absorption is made very simply by adding the term $4mV$ to $S\bar{\alpha}_{\text{Sab}}$. One has

$$(S\bar{\alpha}_{\text{Sab}})_T = S\bar{\alpha}_{\text{Sab}} + 4mV \quad \text{m}^2 \text{ (or ft}^2) \tag{9.31}$$

where $(S\bar{\alpha}_{\text{Sab}})_T$ is the total absorption.

Obviously, when the total boundary absorption $S\bar{\alpha}_{\text{Sab}}$ is large in comparison with $4mV$, that is to say, the absorption per reflection is large compared with air absorption between reflections, air absorption may be ignored. This is nearly always the case at frequencies of 1,000 Hz or less, or in small rooms at all frequencies.

When the reverberation time formulas, Eqs. (9.24) to (9.26), are corrected to account for air absorption, they become, for example from Eq. (9.25),

In mks units: $T = 0.161 \dfrac{V}{S\bar{\alpha}_{\text{Sab}} + 4mV} \quad$ sec $\tag{9.32}$

In English units: $T = 0.049 \dfrac{V}{S\bar{\alpha}_{\text{Sab}} + 4mV} \quad$ sec $\tag{9.33}$

Example 9.2 If in Example 9.1 the room has a relative humidity of 30% at a temperature of 68°F, the value of four times the energy attenuation coefficient, $4m$, has a value of 0.0036 ft^{-1} at 2,000 Hz as determined from Table 9.1. This yields $4mV = 0.0036 \times 400,000 = 1,440$. The reverberation time for the empty room now becomes

$$T = \frac{0.049 \times 400,000}{7,200 + 1,440} = \frac{19,600}{8,640} = 2.3 \text{ sec}$$

Thus, appreciable air absorption will decrease the reverberation time of the

TABLE 9.1   Values of Energy Attenuation Constant Multiplied by 4 for Air, $4m$, ft^{-1}

Relative humidity	Temper- ature, °C (°F)	2,000 Hz	4,000 Hz	6,300 Hz	8,000 Hz
30%	15° (59°)	0.0044	0.0148	0.0322	
	20° (68°)	0.0036	0.0116	0.0256	0.041
	25° (77°)	0.0035	0.0095	0.0209	
	30° (86°)	0.0034	0.0086	0.0172	
50%	15° (59°)	0.0030	0.0087	0.0191	
	20° (68°)	0.0029	0.0074	0.0153	0.026
	25° (77°)	0.0029	0.0072	0.0135	
	30° (86°)	0.0028	0.0071	0.0130	
70%	15° (59°)	0.0027	0.0068	0.0138	
	20° (68°)	0.0026	0.0065	0.0122	0.0184
	25° (77°)	0.0026	0.0064	0.0118	
	30° (86°)	0.0025	0.0063	0.0117	

TABLE 9.2   Values of Energy Attenuation Constant Multiplied by 4 for Air, $4m$, m^{-1}

Relative humidity	Temper- ature, °C (°F)	2,000 Hz	4,000 Hz	6,300 Hz	8,000 Hz
30%	15° (59°)	0.0143	0.0486	0.1056	
	20° (68°)	0.0119	0.0379	0.0840	0.136
	25° (77°)	0.0114	0.0313	0.0685	
	30° (86°)	0.0111	0.0281	0.0564	
50%	15° (59°)	0.0099	0.0286	0.0626	
	20° (68°)	0.0096	0.0244	0.0503	0.086
	25° (77°)	0.0095	0.0235	0.0444	
	30° (86°)	0.0092	0.0233	0.0426	
70%	15° (59°)	0.0088	0.0223	0.0454	
	20° (68°)	0.0085	0.0213	0.0399	0.060
	25° (77°)	0.0084	0.0211	0.0388	
	30° (86°)	0.0082	0.0207	0.0383	

enclosure from the value based upon boundary absorption alone. At higher frequencies the effect of air absorption is much greater and may govern the high-frequency reverberation time in very large rooms. This is because [from Eq. (9.25) with Eq. (9.31) substituted for $S\bar{\alpha}_{\text{Sab}}$] the maximum value of the reverberation time is given by $13.8/mc$ no matter how large or small the hall.

We can also calculate the total average Sabine absorption coefficient from Eq. (9.31). Considering air absorption, the total Sabine absorption for the *occupied* room at 68°F, with a relative humidity of 30% and a frequency of 2,000 Hz, is

$$(S\alpha_{Sab})_T = 11{,}550 + 1{,}440 = 12{,}990 \text{ ft}^2$$

from which $\bar{\alpha}_{Sab_T} = 12{,}990/45{,}000 = 0.29$.

For a nondirectional source, the difference between the power level of the source and the sound-pressure level in the reverberant field ($r > 60$ ft) is found from the left-hand ordinate of Fig. 9.5 to be about 25 dB. Thus, if the source has a power level of 95 dB re 10^{-12} watt at 2,000 Hz, the sound-pressure level in the reverberant part of the enclosure is $95 - 25 = 70$ dB re 2×10^{-5} N/m².

REFERENCES

1. P. M. Morse and K. U. Ingard, *Theoretical Acoustics*, Chap. 9, Paragraph 5, McGraw-Hill, 1968.
2. *Performance Data, Architectural Acoustical Materials*, Acoustical and Insulating Materials Assoc., 205 West Touchy Avenue, Park Ridge, Ill. 60068. (This bulletin is published annually.)
3. R. W. Young, "Sabine Reverberation Equation and Sound Power Calculations," *J. Acoust. Soc. Am.*, vol. 31, pp. 912–921, 1959.
4. R. Huntley, T. D. Northwood, H. J. Sabine, R. W. Young, and others, a discussion by private communication, 1969.
5. T. D. Northwood, "Absorption of Diffuse Sound by a Strip or Rectangular Patch of Absorptive Material," *J. Acoust. Soc. Am.*, vol. 35, pp. 1173–1177, 1963.
6. L. L. Beranek, *Acoustics*, Chap. 10, McGraw-Hill, 1954.
7. L. L. Beranek, "Audience Absorption in Large Halls, II," *J. Acoust. Soc. Am.*, vol. 45, pp. 13–19, 1969.
8. E. D. Daniel, "On the Dependence of Absorption Coefficients upon the Area of the Absorbent Material," *J. Acoust. Soc. Am.*, vol. 35, pp. 571–573, 1963.
9. T. J. Schultz, private communication, 1967.
10. R. V. Waterhouse, "Output of a Sound Source in a Reverberation Chamber and Other Reflecting Environments," *J. Acoust. Soc. Am.*, vol. 30, pp. 4–13, 1958.
11. R. V. Waterhouse and R. K. Cook, "Interference Patterns in Reverberant Sound Fields, II," *J. Acoust. Soc. Am.*, vol. 37, pp. 424–428, 1965.
12. T. F. W. Embleton and I. R. Dagg, "Sound Radiation from a Rectangular Array of Incoherent Sources," *Sound*, vol. 1, pp. 32–36, 1962.
13. M. R. Schroeder, "New Method of Measuring Reverberation Time," *J. Acoust. Soc. Am.*, vol. 37, pp. 409–412, 1965.
14. H. Kuttruff, "On Autocorrelation Measurements in Room Acoustics" (in German), *Acustica*, vol. 16, pp. 166–174, 1965/66.
15. H. Kuttruff and M. J. Jusofie, "Measurement of Reverberation Decay in Many Rooms, Using the Method of the Integrated Impulse Function," *Acustica*, vol. 21, pp. 1–9, 1969.
16. G. Millington, "A Modified Formula for Reverberation," *J. Acoust. Soc. Am.*, vol. 4, pp. 69–82, 1932.
17. W. J. Sette, "A New Reverberation Time Formula," *J. Acoust. Soc. Am.*, vol. 4, pp. 193–210, 1933.
18. C. F. Eyring, "Reverberation Time in Dead Rooms," *J. Acoust. Soc. Am.*, vol. 1, pp. 217–241, 1930.

19. M. R. Schroeder, "Computers in Acoustics: Symbiosis of an Old Science and a New Tool," *J. Acoust. Soc. Am.*, vol. 45, pp. 1077–1088, 1969.

20. B. S. Atal and M. R. Schroeder, "Study of Sound Decay Using Ray-tracing Techniques on a Digital Computer" (abstract), *J. Acoust. Soc. Am.*, vol. 41, p. 1598, 1967, and M. R. Schroeder, "Digital Simulation of Sound Transmission in Reverberant Spaces," *J. Acoust. Soc. Am.*, vol. 47, pp. 424–431, 1970.

21. T. F. W. Embleton, "Absorption Coefficients of Surfaces Calculated from Decaying Sound Fields," *J. Acoust. Soc. Am.*, vol. 47, p. 116(A), 1970.

22. M. R. Schroeder, "Complementarity of Sound Buildup and Decay," *J. Acoust. Soc. Am.*, vol. 40, pp. 549–551, 1966.

23. C. M. Harris, "Absorption of Sound in Air," *J. Acoust. Soc. Am.*, vol. 40, pp. 148–159, 1966.

Chapter Ten

Acoustical Properties of Porous Materials

DAVID A. BIES

10.1 Introduction

This chapter presents information of importance to the use of porous materials for the purpose of attenuating sound propagating in ducts or for improving sound-transmission loss through acoustical barriers. Information relevant to the use of porous materials for the reduction of standing waves in enclosures is found in Chaps. 9 and 15. The use of porous materials in mufflers and ducts is discussed in Chaps. 12 and 15.

10.2 Types of Porous Materials

Porous materials may be divided into two groups depending upon their thickness. If the wavelength of the sound which is of interest is long compared to the thickness of the material, then the material comprises a *sheet;* viscous effects and area density control its behavior. If the wavelength of the sound which is of interest is of the order of or shorter than the thickness of the material, then the latter is characterized as *bulk* material; viscous and thermal effects and solid-material density control its behavior.

Sheets of porous material backed by some form of cavity which may be filled with a bulk material or with air are extensively used for attenuating sound that is being propagated along ducts in which there is high-velocity air flow. Sheet materials may be comprised of bonded, felted, woven, or contained fibers or they may be plates perforated with small holes. Common constructions are sintered metal sheets or thin layers of glass or mineral-wool fibers contained by woven glass cloth.

Bulk materials may be composed of fibers or foams. In general, fibrous materials are more common and will be emphasized in these discussions. Fibrous bulk materials may be grouped either as unbonded fibers, blankets, or as boards. In the blanket, the binder partially cements fine lightly packed fibers together to produce a soft pliable material, while in the board, the binder rigidly cements generally larger and more densely packed fibers together to produce a material which is quite stiff.

10.3   Physical Properties

In the following discussion the definitions given will be those of the American Society for Testing and Materials.[1] Accordingly, the *flow resistance* of a material is defined as the quotient obtained when the air-pressure difference across the specimen is divided by the volume velocity of airflow through the specimen. When mks units are used, the unit obtained is the mks acoustic ohm. In general, we will not use *flow resistance* in this chapter, but rather the quantities *specific flow resistance* and *flow resistivity*, defined below. The term *flow resistance* will be used only where any of these three quantities could be used without confusion.

10.3.1   Specific Flow Resistance and Flow Resistivity

Specific Flow Resistance. The specific (unit area) flow resistance of any layer of porous material is defined as

$$R_f = \frac{\Delta p}{u} \qquad \text{mks rayls} \qquad (10.1)$$

where Δp = applied air-pressure differential measured between the two sides of the layer, N/m^2

u = particle velocity through and perpendicular to the two faces of the layer, m/sec *

Flow Resistivity. For bulk materials the flow resistivity (specific flow resistance per unit thickness of material) is

$$R_1 = \frac{R_f}{l} = \frac{\Delta p}{lu} \qquad \text{mks rayls/m} \qquad (10.2)$$

where l = thickness of the material, m

* Conversion factors are given in Appendix C.

Over some restricted range of u from zero to some small value, the quantity R_f is sensibly constant; above the constant or linear range R_f increases rapidly with increasing values of u. In bulk materials the range of linear flow resistance seems to be broader with coarse or large-diameter fibers than with fine or small-diameter fibers. The opposite behavior is observed for sintered sheet materials.

Measurement of R_f. The flow resistance of a sample material is usually measured with a steady airflow. An apparatus like that shown in Fig. 10.1 may be used.[1, 2] To determine the linear value of R_f, measurements are made at a number of different velocities and the data are extrapolated to zero velocity.

In using the flow-resistance apparatus of Fig. 10.1, the sample is carefully and securely mounted in the sample holder. The upper and lower surfaces must be clearly defined by the screens C and D. The thickness l of the sample is read from a scale G. Air is drawn through the sample, and its volume velocity is determined from the airflow meter. The linear velocity is

$$u = \frac{U}{S} \quad \text{m/sec} \tag{10.3}$$

where U = volume velocity, m³/sec
S = area of the sample, m²
The pressure Δp across the two faces of the sample is determined by a

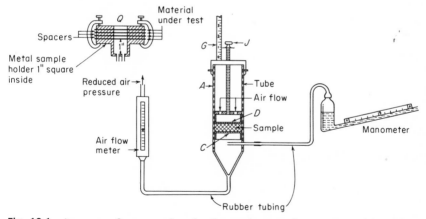

Fig. 10.1 Apparatus for measuring the flow resistance of porous materials. The sample of area S is held in cylinder A between two wire screens C and D. The reduced air pressure is created by drawing water from a large tank with a water control valve at the bottom. When water is removed from the tank steadily, air is sucked through the sample at a rate equal to the rate of water removed. The pressure drop across the sample is measured with a slant manometer. Note that to convert from inches of water on the manometer to N/m² multiply by 249, to convert from sample area in square inches to square meters multiply by 6.45×10^{-4}, and to convert from inches to meters divide by 39.37. (See also Appendix C.)

TABLE 10.1   Specific (Unit-area) Flow Resistance of Some Wire-mesh Screen, R_f

No. of wires/linear in.	Approx. no. of wires/ linear cm	Wire diameter		Flow resistance, mks rayls (N-sec/m³)
		mils (in. $\times$ 10^{-3})	microns (10^{-6} m)	
30	12	13	330	5.7
50	20	8.7	221	5.9
100	39	4.5	114	9.0
120	47	3.6	91	13.5
200	79	2.25	57	24.6

gauge, for example, a slant manometer with a range of 0 to 1 cm H_2O. The specific flow resistance R_f and the flow resistivity R_1 are then calculated using Eqs. (10.1) to (10.3).

Specific Flow Resistance of Sheet Materials. Sintered materials are commercially available with controlled specific flow resistances lying in the range from about 100 to 1,000 mks rayls. They are often used where sound-pressure levels are rather high, as in air inlets to compressors or similar air-handling devices. Some materials are available that remain linear in applications for levels as high as 135 dB *re* 2×10^{-5} N/m².

In a typical application, such as in an air-handling duct, a sheet of material is backed with an air-filled cavity. In this case the real part of the normal specific acoustic impedance * (measured perpendicular to the duct liner) is approximately equal to the specific flow resistance of the sheet. The reactive part is determined by the inertia of the gas in the sheet material and the depth of the backing cavity.

Wire-mesh screens and woven glass-fiber cloths may be used as containment devices or as protective facings for bulk materials. Tables 10.1 and 10.2 list the values of specific flow resistance of wire-mesh screens and some currently available glass-fiber cloths. For some cloths, the specific flow resistance of a small sample varied as much as ±20% from the value averaged over a large sample.

Flow Resistivity R_1 of Bulk Materials. Bulk materials by their nature are subject to considerable variation in thickness d and bulk density ρ_m. Both of these parameters affect flow resistance. Values of flow resistance for some typical commercially available homogeneous porous materials are presented in Figs. 10.2 and 10.3.

Flow resistance also depends upon fiber diameter. The specific flow resistivity of bulk glass-fiber products is related to the fiber diameter and bulk density by the following relation [3]

$$\frac{R_1 d^2}{\rho_m^{1.53}} = K \tag{10.4}$$

* See appendix to Chap. 1.

where $K = 8.75 \times 10^{-7}$ when the fiber diameter is in inches, the bulk density in lb/ft^3, and the flow resistivity in cgs rayls/in. Alternatively K has the value 3.18×10^3 when the fiber diameter is in microns (10^{-6} m), the bulk density in kg/m^3, and the flow resistivity in mks rayls/m

Equation (10.4) is plotted in Fig. 10.4 for practical ranges of fiber diameter and bulk density.

Nichols [4] has found a similar expression to hold for the materials which he tested, as follows

$$R_1 d^2 / \rho_m^{1+x} \approx \text{constant} \tag{10.5}$$

where x lies between 0.3 and 1.0. If the fibers are randomly oriented, x approaches unity, but if the fibers are layered crosswise to the direction of flow, x tends to 0.3. The empirical equation (10.4) is seen to be in good agreement with the general result given by Eq. (10.5).

10.3.2 Fiber Diameter   It is of value to know what ranges of fiber diameter are commercially available. Among sintered metal sheet materials the fiber diameter ranges between approximately 10 and 150 microns, and in a given product it is usually quite uniform. For glass-fiber products the fiber diameter ranges between 0.5 and 15 microns. The fiber diameters in a given product may be quite uniform or non-uniform depending upon the process by which the fibers were manufactured.

TABLE 10.2 Specific (Unit-area) Flow Resistance of Glass-fiber Cloths, R_f (Averaged over a Large Sample)

Manufacturer *	Cloth no.	Surface density		Construction, ends × picks	Flow resistance, mks rayls (N-sec/m^3)
		oz/yd^2	gm/m^2		
1,2,3	120	3.16	96	60 × 58	300
1,2,3	126	5.37	164	34 × 32	45
1,2,3	138	6.70	204	64 × 60	2,200
1,2,3	181	8.90	272	57 × 54	380
3	1044	19.2	585	14 × 14	36
2	1544	17.7	535	14 × 14	19
3	3862	12.3	375	20 × 38	350
1	1658	1.87	57	24 × 24	10
1	1562	1.94	59	30 × 16	<5
1	1500	9.60	293	16 × 14	13
1	1582	14.5	442	60 × 56	400
1	1584	24.6	750	42 × 36	220
1	1589	12.0	366	13 × 12	11

* Code numbers for manufacturers are as follows: (1) Burlington Glass Fabrics Company, (2) J. P. Schwebel and Company, and (3) United Merchants Industrial Fabrics.

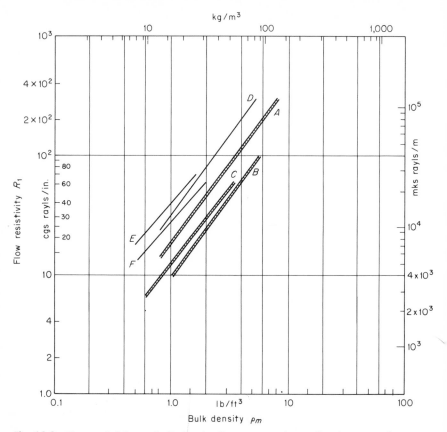

Fig. 10.2   Flow resistivity vs. bulk density for some products of various manufacturers. Gustin Bacon Company: (*A*) Ultralite No. 200, (*B*) Ultralite No. 100, (*C*) Ultralite No. 50 and No. 75; Certain-teed, St. Gobain Insulation Company: (*D*) Ultrafine No. 1001, (*E*) Ultrafine No. 751, (*F*) Ultrafine No. 501. (*Measurements courtesy of Bolt Beranek and Newman Inc.*)

The five basic processes by which glass fibers are produced are listed in Table 10.3.[5]   These processes determine the mean fiber diameter, fiber-diameter distribution, and the possible presence of unfiberized material called "shot."   Shot adds to the bulk density without contributing significantly to the flow resistance.

10.3.3 Porosity   The porosity Y of a porous material is defined as the ratio of the volume of the voids V_a in a material to the total volume V_m

$$Y = \frac{V_a}{V_m} \qquad (10.6)$$

The porosity of an open-cell material may be measured by means of a simple apparatus described in Ref. 2.

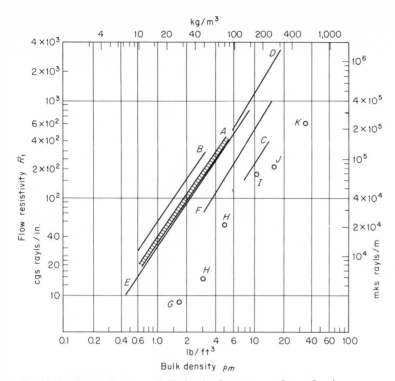

Fig. 10.3   Flow resistivity vs. bulk density for some products of various manufacturers. Johns-Manville Sales Corporation: (A) Microlite B-305, (B) Microlite B-310/0, (C) Spintex (400 series), (D) Thermoflex RF; (E) Babcock & Wilcox Kaowool Blanket B; (F) HITCO Refrasil Batt B-100; United States Gypsum Company: (G) Thermafiber Insulating Blanket, (H) Thermafiber Insulating Felt; Celotex Corporation: (I) Acoustiform-Mat Ceiling Board, (J) Glazed Ceramic Ceiling Board, (K) Natural Fissured Celotone. (*Measurements A, B, E, and F by Bolt Beranek and Newman Inc.; others courtesy of manufacturers.*)

TABLE 10.3   Glass-fiber Manufacturing Processes and Fiber Diameters

Process [5]	Fiber-diameter range,* microns	Comment
Steam-blown process †	·5–15	Fiber diameter not well controlled. Contains some shot
Spinning process	3–5	Contains little shot
Textile process	9–11	Fiber diameter well controlled. No shot
Flame-attenuated process	0.5–10	No shot
Rotary process	2–12	No shot

* Fiber diameters are dependent upon the temperature and composition of the melt, the design and speed of rotating parts, and the production rate.[6]

† This process is being discontinued.

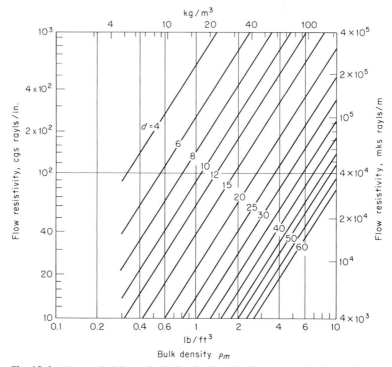

Fig. 10.4   Flow resistivity vs. bulk density showing the parametric dependence on fiber diameter.   The relationship shown was provided by the Owens-Corning Fiberglas Corporation and describes its materials in the ranges shown in the figure.   The fiber diameter d is in units of 10^{-5} in.

Calculation of Porosity.   The porosity of a fibrous material may be calculated where the fiber density is known, provided that the weight and volume of the binder, if any, can also be estimated.   For a material with negligible binder by weight, we obtain

$$Y = 1 - \frac{\rho_m}{\rho_f} \tag{10.7}$$

where ρ_m = bulk density of sample, kg/m³
       ρ_f = density of fibers, kg/m³

Example 10.1   A sample of glass fibers weighs 25 kg/m³.   What is the porosity?   Use of Eq. (10.7) and the fact that glass fibers have a density of about 2.5×10^3 kg/m³ leads to the following solution:

$$Y = 1 - \frac{25}{2.5 \times 10^3} = 0.99$$

10.3.4   Structure Factor   The detailed inner structure of a porous material has an effect upon its acoustical behavior in addition to the effect accounted for by the specific flow resistance. This effect has been described by Zwikker and Kosten [7] and amounts to an effective increase in the density of the gas in the material voids. The increase results from the tortuous paths that the gas particles must take in the material during the passage of a sound wave. A dimensionless quantity called the structure factor s has been introduced into the theory of propagation in porous materials to take account of this effect.

There does not appear to be any way of measuring the structure factor of a material other than to investigate the propagation of sound through the material and, from such measurements, to infer a value for the structure factor. An approximate relation between structure factor and porosity is shown in Fig. 10.5 for homogeneous materials made of fibers or granules, with interconnecting pores.

10.4 Propagation of Sound Waves in Porous Materials

The propagation of sound through a medium is controlled by the compressibility and inertia of the medium. If losses are important, the concepts of compressibility and inertia may be generalized to include them. Thus the propagation of sound through a porous material may be described in terms of an effective gas density and effective gas compressibility.[7,8]

10.4.1 Effective Gas Compressibility The compressibility of a gas contained within the interstices of a porous material is strongly influ-

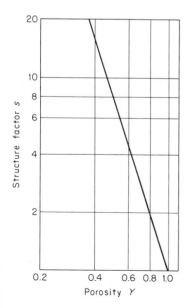

Fig. 10.5 Approximate relation between porosity Y and structure factor s for homogeneous materials made of fibers or granules with interconnecting pores and few "blind alleys."

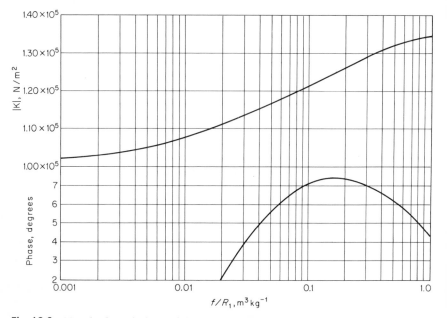

Fig. 10.6 Magnitude and phase of the complex effective compressibility of air contained in a fibrous porous material. One atmosphere pressure is assumed. The phase angle is small enough to be neglected in calculations. The abscissa is frequency f in hertz divided by flow resistivity R_1 in mks rayls/m.

enced by the presence of the material. In general, the thermal capacity of the porous material in a unit volume is very much greater than the thermal capacity of the contained gas. As a consequence, the temperature of the porous material remains essentially constant during a compression and rarefaction cycle of the gas as a sound wave traverses through it, even though the gas is heated and cooled during the same cycle. At very low frequencies the compression and rarefaction process takes place slowly enough that the thermal conduction process is able to maintain the gas temperature constant also by carrying heat back and forth from the gas to the porous material. In this case the compression process is essentially isothermal. At very high frequencies, the compression and rarefaction process takes place so rapidly that the thermal conduction process is not able to carry heat energy back and forth from the bulk of the gas to the porous material. In this case the process is essentially adiabatic. In many materials isothermal behavior extends up to about 100 Hz, while adiabatic behavior lies above 1,000 Hz.

For analytical purposes, a porous material with its contained gas may be replaced by a fictitious gas of effective compressibility determined by the gas and the porous material. The complex compressibility K of the contained gas may be defined as

$$K = -V_a \frac{dP}{dV} \tag{10.8}$$

where V_a is the volume of the contained gas and dP/dV is the incremental change in pressure associated with an incremental change in volume of the gas, and K has a magnitude and phase angle $K = |K| \underline{/\theta}$.

Equation (10.8) may be combined with Eq. (10.6) to obtain an expression for the effective compressibility $-V_m \frac{dP}{dV}$ of the contained gas

$$\frac{K}{Y} = -V_m \frac{dP}{dV} \tag{10.9}$$

From Eq. (10.9) we see that the effective compressibility of an equivalent gas to replace the gas–porous material system is simply the compressibility of the contained gas divided by the porosity of the medium.

Based on a study of the acoustic properties of a great many commercially available glass and mineral wool fibrous materials, the expressions plotted in Fig. 10.6 for the magnitude and phase of the complex compressibility K of air contained in the interstices of a fibrous porous material have been determined.[9] It is seen that the compressibility, for air at 1 atmosphere and room temperature, depends upon frequency divided by flow resistivity.

Figure 10.6 and other data [8,10] show that the phase angle of the complex compressibility remains relatively small. Thus we may ignore it in any calculation involving K and we need only consider the magnitude as given in the figure.

10.4.2 Effective Gas Density The effective density of the fictitious gas, which has been postulated to replace the contained gas in a porous material system, must include viscous as well as inertial effects and is, therefore, a complex number. The complex effective density ρ' can be written for two important cases [8]

$$\rho' \approx \frac{\rho s}{f_1} \left(f_2 - \frac{jR_2}{\rho s \omega} \right) \tag{10.10}$$

where f_1 and f_2 are functions which for semirigid materials are

$$f_1 = f_2 = 1 \tag{10.11}$$

and for very soft materials are

$$f_1 = 1 + \left(\frac{R_2}{\rho_m \omega} \right)^2 \tag{10.12}$$

$$f_2 = 1 + \left(Y + \frac{\rho_m}{\rho s}\right)\left(\frac{R_2}{\rho_m \omega}\right)^2 \tag{10.13}$$

where ρ_m = bulk density of the porous material, kg/m³
ρ = density of the gas (air), kg/m³
s = structure factor (dimensionless)
Y = porosity (dimensionless)
ω = frequency, radians/sec (frequency multiplied by 2π)
R_2 = approximately 1.2 times the flow resistivity R_1 in mks rayls/m
These equations were derived under the assumption

$$\frac{\rho(s-1)}{\rho_m} \ll 1 \tag{10.14}$$

Cottonlike materials would be treated as very soft, while materials which are not fairly soft would be treated as semirigid in the above formulation. Most commercially available materials will generally fall into one class or the other. Comparison of Eqs. (10.11) to (10.13) shows that at high frequencies very soft materials behave like semirigid materials, since at high frequencies inertial effects dominate the behavior of both types of material.

10.4.3 Characteristic Impedance and Complex Propagation Constant

The complex propagation constant is the product of j (equal to $\sqrt{-1}$) and the angular frequency ω divided by the complex speed of sound in the material.* The complex speed of sound in the porous medium may be written as $\sqrt{K/Y\rho'}$. Thus we may write the complex propagation constant b as

$$b = j\omega \sqrt{\frac{Y\rho'}{K}} \tag{10.15}$$

The complex propagation constant may also be written as

$$b = j\frac{\omega}{c_M} + \alpha = j\frac{2\pi}{\lambda_M} + \alpha \tag{10.16}$$

where c_M = speed of wave propagation in the material, m/sec
λ_M = wavelength in material, m/sec
α = attenuation constant, nepers/m in the material (to convert nepers to decibels, multiply nepers by 8.69)
The characteristic acoustic impedance Z_0 (the ratio of p/u at a plane in a very long tube containing the acoustical material) of the contained-gas porous-material system is the product of the effective complex density

---

* The concept of speed of sound is extended here, by means of the complex notation, to include dissipation during propagation.

ρ' and the complex speed of sound c_M in the material. Since the latter may be written in terms of the complex effective compressibility and density, we may write for the characteristic impedance

$$Z_0 = \sqrt{\frac{\rho' K}{Y}} \quad \text{N-sec/m}^3 \tag{10.17}$$

For later convenience the characteristic impedance may also be written in terms of the complex propagation constant as

$$Z_0 = -\frac{jKb}{\omega Y} \quad \text{N-sec/m}^3 \tag{10.18}$$

10.4.4 Estimation of Characteristic Impedance, Wavelength, and Attenuation in Fibrous Materials from Empirical Data   Values of the characteristic impedance Z_0 and propagation constant b may be presented as universal functions of the dimensionless parameter $\rho f/R_1$ where ρ is the gas density, f is the frequency, and R_1 is the flow resistivity.[9] A summary of the principal results valid for semirigid materials is given in Table 10.4.

TABLE 10.4 Empirical Power Law Approximations for the Complex Characteristic Impedance Z_0 and Complex Propagation Constant b of Semirigid Materials [9]

Characteristic Impedance
$Z_0 = R + jX$ $R = \rho c[1 + 0.0571(\rho f/R_1)^{-0.754}]$ $X = -\rho c[0.0870(\rho f/R_1)^{-0.732}]$

Propagation Constant
$b = \alpha + j\,(2\pi/\lambda_M) = \alpha + j\beta$ $\alpha = (\omega/c)[0.189(\rho f/R_1)^{-0.595}]$ $\beta = (\omega/c)[1 + 0.0978(\rho f/R_1)^{-0.700}]$ $0.01 \leq \rho f/R_1 \leq 1$

For convenience, the real and imaginary parts of the characteristic impedance, the attenuation coefficient, and the wavelength in the material, which have been calculated using the equations in Table 10.4, are plotted in Figs. 10.7 to 10.9.[9] The range over which the results are valid is indicated by the ranges in the plots; extrapolation is advised against. A comparison with experimental data of estimates of wavelength and

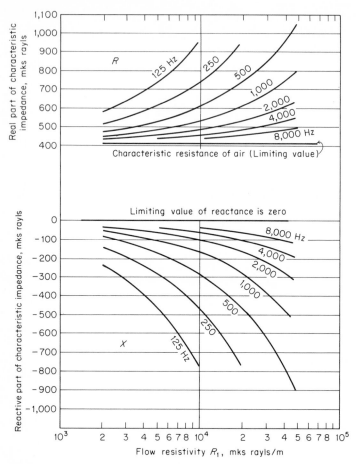

Fig. 10.7   Characteristic impedance of sound waves propagating in a semirigid fibrous sound-absorbing material.   The gas is air at 20°C. (*From Ref. 9.*)

attenuation based on the curves in Figs. 10.8 and 10.9 is shown in Fig. 10.13.   The comparison with the semirigid material is quite good, but the comparison with the nonrigid material is only fair.

10.4.5 Calculation of Characteristic Impedance, Wavelength, and Attenuation   Graphical solutions to Eqs. (10.12), (10.13), (10.16), and (10.17) have been prepared and are shown in Figs. 10.10 to 10.12. These graphs are used to calculate wavelength, attenuation, and specific acoustic impedance of a sound wave propagating in a porous medium.

Example 10.2   Let us assume that we wish to calculate the attenuation, wave-length, and characteristic acoustic impedance of a plane sound wave propagating in a soft fibrous porous material described as follows:

Trade name = PF-105, manufactured by Owens-Corning Fiberglas Corporation prior to 1959

Bulk density $\rho_m = 9.6$ kg/m^3

Gas in material, air, density $\rho_0 = 1.18$ kg/m^3

Fiber diameter $d = 1.0$ micron

Porosity $Y = 0.99$

Structure factor $s = 1.0$

Fiber density $\rho_f = 2.5 \times 10^3$ kg/m^3

Flow resistivity $R_1 = 4.1 \times 10^4$ mks rayls/m

Effective flow resistivity $R_2 \approx 4.9 \times 10^4$ mks rayls/m

First, determine the magnitude of the effective compressibility of the air contained in the interstices of the porous material. To do this, we calculate the parameter f/R_1 (where f is the frequency in hertz and R_1 is the flow resistivity in mks rayls/m) and enter Fig. 10.6 to determine $|K|$. Computed values of the frequency parameter f/R_1 and the corresponding values of $|K|$ in N/m^2 are shown in Table 10.5 in rows 2 and 3, respectively.

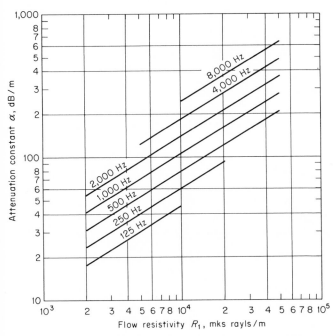

Fig. 10.8   Attenuation of sound propagating in a semirigid fibrous sound-absorbing material. The gas is air at 20°C. (*From Ref. 9.*)

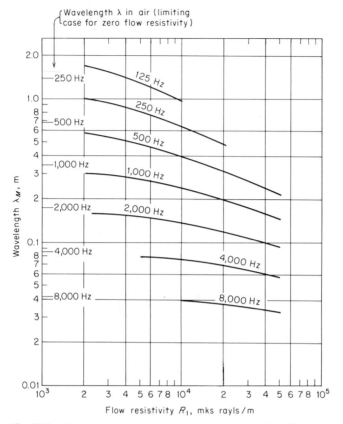

Fig. 10.9   Wavelength of sound propagating in a semirigid fibrous sound-absorbing material.   The gas is air at 20°C.   (*From Ref. 9.*)

Because the material under consideration is assumed to be very soft, we must determine values for f_1 and f_2 as given by Eqs. (10.12) and (10.13).   For this purpose, we compute the parameters $Y + (\rho_m/\rho s)$ and $f\rho_m/R_2$ and enter Fig. 10.10.   We see from the legend of Fig. 10.10 that f_1 may be determined by using the curve for $Y + (\rho_m/\rho s) = 1$.

For this example, at 100 Hz

$$\frac{f\rho_m}{R_2} = \frac{100 \times 9.6}{4.9 \times 10^4} = 0.020$$

and

$$Y + \frac{\rho_m}{\rho s} = 0.99 + \frac{9.6}{1.18 \times 1.0} = 9.1$$

Computed values of the parameter $f\rho_m/R_2$ and corresponding values of f_1 and f_2 are shown in rows 4, 5, and 6 of Table 10.5.

In order to obtain the dimensionless parameters of Fig. 10.11, we compute the parameters $\rho s f f_2/R_2$ and $f\sqrt{\rho s Y f_2/|K|f_1}$. For this example, at 100 Hz

$$\frac{\rho s f f_2}{R_2} = \frac{1.18 \times 1.0 \times 100 \times 550}{4.9 \times 10^4} = 1.32$$

$$f\sqrt{\frac{\rho s Y f_2}{|K|f_1}} = 100 \sqrt{\frac{1.18 \times 1.0 \times 0.99 \times 550}{1.03 \times 10^5 \times 65}} = 0.98 \text{ m}^{-1}$$

Computed values of the above parameters are presented in Table 10.5 in rows 7 and 8. The quantities shown in rows 9 and 10 in the same table are read from Fig. 10.11. Finally, the values of α and λ_M shown in rows 11 and 12 of Table 10.5 are computed using the data in rows 8 to 10. For this example at 100 Hz

$$\alpha = \frac{\alpha}{f}\sqrt{\frac{|K|f_1}{\rho s Y f_2}} \times f\sqrt{\frac{\rho s Y f_2}{|K|f_1}} = 3.4 \times 0.98 = 3.33 \text{ dB/m}$$

$$\lambda_M = \lambda_M f\sqrt{\frac{\rho s Y f_2}{|K|f_1}} \div f\sqrt{\frac{\rho s Y f_2}{|K|f_1}} = \frac{1.00}{0.98} = 1.02 \text{ m}$$

The calculated values of α_M and λ_M for PF-105 material are plotted in Fig. 10.13. The case of a semirigid fibrous porous material has also been worked out and is also shown in Fig. 10.13.

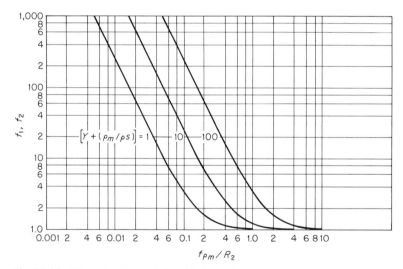

Fig. 10.10 Means for determining the dimensionless quantities f_1 and f_2 used in the calculation of propagation constant and characteristic impedance of soft porous blankets. The quantities are the porosity Y, the bulk density ρ_m in kg/m³, the gas density ρ in kg/m³, the structure factor s, the effective flow resistivity $R_2 \approx 1.2R_1$ where R_1 is the flow resistivity in mks rayls/m and f is the frequency in hertz. To obtain f_1, use the curve for $Y + (\rho_m/\rho s) = 1$. To obtain f_2 use the appropriate curve.

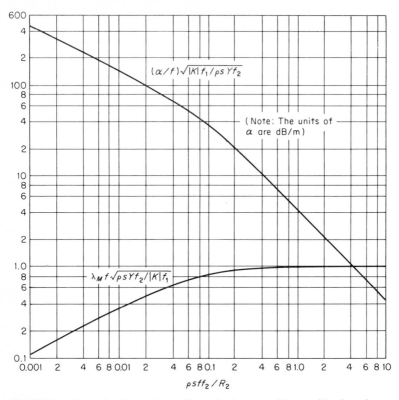

Fig. 10.11 Means for determining dimensionless quantities used in the calculation of wavelength λ_m and attenuation constant α for acoustic propagation in porous materials. The quantities are f_1 and f_2 from Fig. 10.10, the magnitude of the compressibility $|K|$ from Fig. 10.6, the structure factor s, the porosity Y, the gas density ρ in kg/m^3, the effective flow resistivity R_2 approximately equal to 1.2 times the flow resistivity R_1 in mks rayls/m, and the frequency f in hertz.

To compute the characteristic impedance Z_0 of the material, we make use of Fig. 10.12. The quantities listed in row 7 of Table 10.5 are used to enter the figure and read the quantities listed in rows 13 and 14. The magnitude of the impedance $|Z_0|$ may be calculated using the data in rows 1, 3, 8, and 13 of Table 10.5. For example, at 100 Hz according to Eq. (10.18), we have

$$|Z_0| = \frac{|K|}{Y} \frac{|b|}{\omega} = \frac{|K|}{Y} \times \left| f \sqrt{\frac{\rho s Y f_2}{|K| f_1}} \right| \times \frac{|b|}{\omega} \sqrt{\frac{|K| f_1}{\rho s Y f_2}} \times \left| \frac{1}{f} \right|$$

Using Table 10.5 yields

$$|Z_0| = \frac{1.03 \times 10^5}{0.99} \frac{0.98 \times 1.00}{100} = 1{,}020 \text{ mks rayls}$$

Computed values are listed in row 15 of Table 10.5.

TABLE 10.5   Calculation of the Characteristic Impedance Z_0, Wavelength λ_M, and Attenuation α for a Soft Fibrous Porous Material PF-105

(1) Frequency f, Hz	100	300	600	1,000	3,000	6,000
(2) f/R_1, m³-kg⁻¹	0.0024	0.0073	0.0146	0.0244	0.0732	0.146
(3) $\lvert K \rvert$, N-m⁻²	1.03×10^5	1.06×10^5	1.09×10^5	1.12×10^5	1.19×10^5	1.24×10^5
(4) $f\rho_m/R_2$	0.020	0.059	0.12	0.20	0.59	1.17
(5) f_1	65	8.3	2.8	1.6	1.08	1.00
(6) f_2	550	63	16	6.2	1.6	1.1
(7) $\rho s f f_2/R_2$	1.32	0.455	0.231	0.149	0.116	0.159
(8) $f\sqrt{\rho s Y f_2/\lvert K \rvert f_1}$, m⁻¹	0.98	2.74	4.70	6.35	11.4	19.3
(9) $(\alpha/f)\sqrt{\lvert K \rvert f_1/\rho s Y f_2}$	3.4	8.2	18	27	32	24
(10) $\lambda_M f\sqrt{\rho_0 s Y f_2/\lvert K \rvert f_1}$	1.00	1.00	0.93	0.89	0.85	0.90
(11) α, dB/m	3.33	22.5	84.6	172	365	463
(12) λ_M, m	1.02	0.365	0.20	0.14	0.075	0.047
(13) $(\lvert b \rvert/\omega)\sqrt{\lvert K \rvert f_1/\rho s Y f_2}$	1.00	1.00	1.12	1.29	1.30	1.20
(14) θ phase of Z_0, degrees	−3.5	−8.5	−18	−24	−27	−22
(15) $\lvert Z_0 \rvert$, mks rayls	1,020	978	947	927	594	483

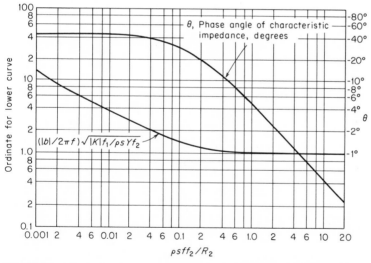

Fig. 10.12   Means for determining the magnitude $\lvert Z \rvert$ and phase θ of the characteristic impedance of acoustic waves propagating in porous media. The propagation constant $\lvert b \rvert$ is used for calculating the magnitude of the characteristic impedance.   The quantities are f_1 and f_2 from Fig. 10.10, the magnitude of the compressibility $\lvert K \rvert$ from Fig. 10.6, the structure factor s, the porosity Y, the gas density ρ in kg/m³, the effective flow resistivity R_2 approximately equal to 1.2 times the flow resistivity R_1 in mks rayls/m, and the frequency f in hertz.

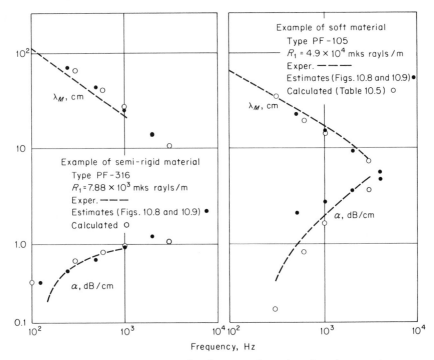

Fig. 10.13 Comparison of calculated and measured wavelength and attenuation constant for two porous materials, one semirigid, the other soft.

10.4.6 Comparison of Measured and Calculated Values of Attenuation and Wavelength   Comparison may be made between the results of calculations for wavelength and attenuation and the results of measurements in corresponding materials. In Fig. 10.13, we show a comparison between the measured [11] and calculated values for the materials indicated. The comparison between the measured and calculated wavelengths is satisfactory. At high frequencies, the calculated attenuation is closer to the measured attenuation than at low frequencies. The comparisons are probably satisfactory for most engineering purposes.

10.5 General Considerations

In order to carry out the design of noise control measures for a particular problem, the acoustical engineer must consider not only the basic properties of the materials already discussed, but also a number of practical aspects of the problem such as (1) space limitations, (2) weight limitations, (3) exposure to people or moving objects, (4) need for ready cleaning or painting, (5) weather exposure, (6) resistance to contamination, (7) gas-flow temperature, and (8) gas-flow velocity. For very intense acoustic field applications, one must also consider (9) linearity of response and (10) acoustic fatigue.

10.5.1 Choice of Materials   Materials that are satisfactory for out-of-door use as in installations for quieting transformers, turbines, motors, cooling towers, and roof ventilators include: (1) glass-fiber blankets, (2) sintered metals, (3) glass-fiber bulk materials, (4) mineral wools, and (5) metal wools. Such materials may be expected to give satisfactory performance over a period of many years although installed in installations which are exposed to the weather. This statement assumes that the installation does not permit them to become soaked with water, either during use, or during rest periods that might be accompanied by freezing. Acoustical materials which may be satisfactory for indoor use, besides those listed for out-of-door use, include: (1) natural fibers (kapok, paper, cotton, etc.), (2) wood fibers, (3) cellulose fibers, and (4) hair felt and wool felt.

Examples of some of the materials that are currently available for use where temperature must be considered are tabulated in Tables 10.6 and 10.7. The approximate maximum allowable temperature to which the material should be exposed is indicated.

In Table 10.6 materials are given that can be used to wrap high-temperature pipes and ducts or as inner fillers in wall panels for acoustical

TABLE 10.6   Porous Materials for Pipe and Duct Wrappings and Fillers in Wall Structures

Fibrous materials	Product and manufacturer key	Maximum allowable temperature, °F
Rockwool fibers..	....	150–300
Asbestos felts..	....	300
Bonded glass fibers.......................................	3A, 1A, 1B	350–400
Wood felts..	....	600
Asbestos fibers, rockwool...............................	....	700–900
Unbonded glass fibers....................................	1C	1,000
Mineral wool and asbestos and rockwool combinations...	2A	1,200
Felted block..	....	1,400
Mineral wool...	2B	1,700
Vitreous fiber – silica....................................	4A	1,800
Refractory fiber	....	2,000
Alumina-silica fiber.......................................	5A	2,000

Product and Manufacturer:
1. Owens-Corning: (A) Aerocor, (B) PF Fiberglas, (C) TWL Fiberglas
2. Keene Corporation: (A) Spun Felt white wool, (B) Spun Felt black wool
3. Johns-Manville: (A) Microlite
4. HITCO: (A) Refrasil
5. Babcock and Wilcox: (A) Kaowool

TABLE 10.7   Porous Materials for Use in Hot Gas Streams

Fibrous materials	Product and manufacturer key	Maximum allowable temperature, °F
Some mineral wools....................	....	125–150
Wool felts.................................	....	150–200
Hair felts and nylon fibers	....	200–250
Bonded glass fibers....................	3A, 1A, 7A, 6A	350–400
Asbestos fibers...........................	....	800
Unbonded glass fibers................	1B, 3B	1,000–1,100
Mineral wool felted block	2A	1,200
Basalt wool...............................	....	1,400
Mineral wool.............................	2B	1,700
Vitreous fiber – silica..................	4A	1,800
Alumina-silica............................	5A	2,000
Alumina-silica............................	3C	2,300
Refractor-silica fiber	3D	2,700
Other materials		
Haydite block (cracks under high transient temperatures)	....	900
Porous firebricks (including ceramic).................................	....	1,600–3,000
Gravel......................................	....	3,000

Product and Manufacturer:
1. Owens-Corning: (A) Aerocor, (B) TWL Fiberglas
2. Keene Corporation: (A) Spun Felt white wool, (B) Spun Felt black wool
3. Johns-Manville: (A) Microlite, (B) Unbonded B Fiber, (C) Thermoflex, Cerafiber, (D) Dyna-Flex, Dyna-quartz
4. HITCO: (A) Refrasil
5. Babcock and Wilcox: (A) Kaowool
6. Certain-teed St. Gobain Insulation Corp.: (A) Ultrafine
7. Gustin Bacon: (A) Ultralite

enclosures. Not all these materials are satisfactory for use in noise control problems where the acoustical material is to be used as an absorbing medium in the path of gas streams – mainly because the flow resistances of some of them are too high.

In Table 10.7 some porous materials that can be used satisfactorily in the presence of hot gases and that have desirable flow resistances are listed. The fibrous materials are normally encased behind perforated metal sheets forming the exterior of parallel baffles or acoustical liners for tuned ducts. Perforated bricks or blocks are sometimes used to line exhaust mufflers or to form parallel baffles. These bricks usually do not have as good acoustical properties as the fibrous materials and in many instances are more expensive.

Noise-induced fatigue may evidence itself in either sheet or bulk materials. In a sheet material, acoustic fatigue may evidence itself by cracking and disintegration of the material. In bulk materials, it may occur as the disintegration of the fibers or it may, in extreme cases, occur as spontaneous combustion of the porous material binder. In either case, the material will lose its form and usefulness.

10.5.2 Protective Facings

A porous material may be covered with a protective facing which may be a sheet material providing much or little of the required total flow resistance. There are many reasons for considering such layered construction and some of these will be considered here.

A very common situation is for an acoustical material to be located where it will be damaged by contact with people or moving objects, or where it may need frequent cleaning or painting. In these cases it is customary to cover the material with perforated metal or asbestos sheets, expanded metal grilles, rigidized metal sheets, or woven plastic (Saran) cloth. Unless the percentage open area of the covering sheet is 20% or more, the reduction of sound absorption at high frequencies may be significant.[12]

10.5.3 High-speed Gas Flows

Noise control problems often involve the use of acoustical materials in high-velocity gas streams; for example, in the exhausts of jet engines, diesel engines, compressors, high-velocity ventilating ducts, etc. The high-velocity gases flowing past the material results in a tearing off of particles of the material and is a cumulative process which eventually may lead to complete deterioration. In addition, turbulence in high-velocity flows subjects the materials to vibration which may lead to further deterioration. In many cases, high temperatures and high-velocity gas streams are present simultaneously so that the acoustical engineer must take into account the effects due to their interaction. One solution to these problems is to install the acoustical material behind some type of protective facing which may vary in complexity depending upon the requirements of the system.

Some quantitative information on the allowable limits of structures for noise control in the presence of high-speed gas flows is available on the basis of field experience, and this information is tabulated in Table 10.8. We believe that these values are useful in the design of acoustical structures and in most instances are believed to be conservative.

The gas velocities given in Table 10.8 represent average values for smooth, diffuse gas flow (no flow separation) at grazing incidence only. If high-velocity gradients (turbulence) exist near the surfaces of the protective facings, such as might be encountered in 90° bends or in the vicinity of sharp edges, or if there are sharp constrictions, local gas

TABLE 10.8   Protective Facings for Acoustical Linings Subjected to High-velocity Gas Streams*

Materials	Maximum allowable velocity in straight runs	Materials	Maximum allowable velocity in straight runs
A. Ventilating ducts		*B*. Large panels (e.g., test cell applications)	
1. Uncoated lining, see Note 1	25 ft/sec 8 m/sec	1. Same construction as A4	75 ft/sec 25 m/sec
2. Coated lining, · see Note 2	30 ft/sec 10 m/sec	2. 20 gauge perforated metal, minimum 25% open area, Wire screen Glass fiber cloth	180 ft/sec 60 m/sec
3. Blanket 20 gauge perforated metal, minimum 25% open area	75 ft/sec 25 m/sec	3. Three layers 20 gauge perforated metal, minimum 25% open area, shown separated by 2 corrugated perforated sheets Wire screen Glass fiber cloth Blanket	300 ft/sec 100 m/sec
4. Blanket Perforated facing as in *A*3 Glass fiber cloth	125 ft/sec 40 m/sec		

Notes: 1. Examples of suitable materials: Johns-Manville, Microtex, Micro-bar, Micro-Coustic.
2. Examples of suitable materials: Johns-Manville, Microlite, Tenu-Mat, Microtex; Gustin Bacon, Ultra * Liner; Owens-Corning, Fiberglas.
* For additional information see Ref. 13. Note that the air is assumed to be flowing over the perforated facing on the left of each sketch.

velocities might increase to values several times the calculated average speeds. It is generally wiser not to place acoustical structures where gas turbulence is high, because the probability of erosion is greatly increased. If acoustical structures are used in turbulent gas streams, extreme care should be exercised to protect the porous materials inside as much as possible.

The thicknesses of the perforated, protective, facing materials shown in Table 10.8 are governed by both gas temperature and gas velocities. For ventilating systems the thickness of the perforated facing is generally 20 gauge. For large panels, such as might be used in engine test-cell installations, the thickness ranges from about 20 gauge at normal room temperature and a maximum velocity of 75 fps to about 10 gauge for temperatures of 1,000°F at 300 fps velocities. The increased thickness in the metal is usually required because of the expected higher amount of turbulence associated with applications involving high temperatures. In these large panels the perforated facings should be at least 25% open. In the case of item *B*3 in Table 10.8, the facing consists of three perforated metal sheets separated by two orthogonal corrugated sheets, each about 30% open. The information given in Table 10.8 applies to acoustical panels that are installed in sections about 3 ft in length. If smaller sections of perhaps half this length are used, the velocity limits given may be somewhat increased.

REFERENCES

1. ASTM C522−69, "Standard Method of Test for Airflow Resistance of Acoustical Materials," American Society for Testing and Materials, 1916 Race Street, Philadelphia, Pa. 19103.
2. L. L. Beranek, *Acoustic Measurements*, pp. 836–869, Wiley, 1949.
3. Owens-Corning, private communication.
4. R. H. Nichols, Jr., "Flow-resistance Characteristics of Fibrous Acoustical Materials," *J. Acoust. Soc. Am.*, vol. 19, no. 5, pp. 866–871, 1947.
5. R. D. Dickenson, "Fiber Glass−There's a Difference," *Metal Building Rev.*, pp. 5–8, June, 1967.
6. Johns-Manville Corporation, private communications.
7. C. Zwikker and C. W. Kosten, "Sound Absorbing Materials," Elsevier, 1949.
8. L. L. Beranek, "Acoustical Properties of Homogeneous, Isotropic Rigid Tiles and Flexible Blankets," *J. Acoust. Soc. Am.*, vol. 19, no. 4, pp. 556–568, 1947.
9. M. E. Delany and E. N. Bazley, *Nat. Phys. Lab. Aerodyn. Div. Rept.* NPL AERO REPORT Ac37, March, 1969.
10. D. A. Bies, "Acoustic Properties of Steel Wool," *J. Acoust. Soc. Am.*, vol. 35, no. 4, pp. 495–499, 1963.
11. L. L. Beranek and S. Labate, in L. L. Beranek (ed.), *Noise Reduction*, Chap. 12, pp. 270–271, McGraw-Hill, 1960.
12. U. Ingard, "Perforated Facing and Sound Absorption," *J. Acoust. Soc. Am.*, vol. 26, no. 2, pp. 151–154, 1954.
13. Guy J. Sanders, "Silencers: Their Design and Application," *Sound and Vibration*, February, 1968.

Chapter Eleven

Interaction of Sound Waves with Solid Structures

ISTVÁN L. VÉR and CURTIS I. HOLMER

The response of structures to dynamic forces is the subject of structural dynamics and acoustics. Structural dynamics is concerned predominantly with dynamic stresses severe enough to endanger the structural integrity of a member, while acoustics deals with low-level dynamically excited waves in the structures which are still strong enough to radiate disturbing sound or to cause unacceptable vibration.

A dynamic forcing source feeds power into a structural system. In the steady state, the resulting vibration field builds up to that level at which the power from the source balances the power losses in the system. These system losses include the power dissipated, the power transmitted to neighboring systems through coupling, and the power radiated into the surrounding air.

The subject matter of this chapter is specific to noise control problems and thus is restricted to the audible frequency range and to air as the surrounding medium, though many of the concepts are directly applicable to other fluid media or to higher frequencies.

A typical noise control problem is illustrated by Fig. 11.1. A resiliently supported floor slab is excited by a tapping machine in the room to

270

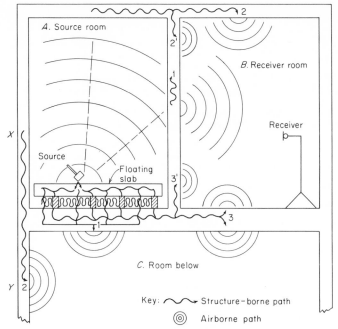

Fig. 11.1   Sound-transmission paths between an impact source in room A and a receiver in room B.   Also shown are the paths to room C, below.

the left, while a microphone stands in the room to the right.   A part of the induced vibrational energy is dissipated in the floating slab, a part is radiated as sound into the source room, and the remainder is transmitted through the resilient layer into the building structure.

The radiated sound energy builds up a reverberant sound field in the source room which in turn excites the walls.   The vibrations in that wall common to the two rooms, indicated by 1 in Fig. 11.1, radiate sound directly into the receiver room.   The vibrations in the other surfaces of the source room travel to the six surfaces of the receiving room as shown by waves 2, 2', 3, and 3', which then radiate sound into the receiving room.

To reduce the noise level in the receiving room, the acoustical engineer must estimate the power transmitted by each of the paths and then must design appropriate sound- and vibration-isolation measures to reduce the radiated sound by the desired amount.

11.1   Types of Wave Motion in Solids

Because they can store energy in shear as well as in compression there are a number of different types of waves possible in solids, including compressional waves, flexural waves, shear waves, and torsional waves. These types of waves result from different ways of stressing a solid. By contrast, only compressional waves are of practical importance in air since fluids can store energy only in compression. For a wave to propagate in solids, liquids, or gases, the medium must be capable of storing energy alternately in kinetic and potential forms. Kinetic energy is stored in any portion of a medium that has mass and is in motion. Potential energy is stored in any part of a medium that has undergone elastic deformation. All solids have mass and are elastic (i.e., they can store energy in compression and shear) and so are capable of storing both types of energy. Compressional waves and shear waves represent "pure" forms of propagation of deformations. Flexure and torsion may be locally represented as combinations of compression-tension and shear in the solid and give rise to bending and torsional waves, respectively.[1-6]

Speeds of Sound. Formulas for the speed of sound in compressional (longitudinal) waves in a bar, an infinite plate, and an infinite solid are shown in Table 11.1. Formulas are also shown for shear waves, torsional waves in bars, and bending waves in plates and bars.

The propagation speed of a bending wave is *frequency dependent*, so that a complex (e.g., any nonsinusoidal) waveform changes its shape with distance of travel.

The attenuation of bending waves in structures during propagation, owing to losses, is taken into account through a complex bending stiffness,[7,8] defined by

$$\bar{B} = B(1 - j\eta) \qquad \text{N-m} \tag{11.1}$$

where $\bar{B}$ = complex bending stiffness with magnitude B, N-m
η = loss factor of the structure (dimensionless)

The loss factor describes the fraction of stored energy which is lost in each "cycle" of energy storage.[8] (See Chap. 14.)

11.2 Acoustical Behavior of Infinite Panels

If a stone is tossed on the surface of a thin layer of ice covering a lake, a person a distance away listening to the sound created by the bounces of the stone will observe that the sound is "frequency modulated": the first sound to arrive has a high pitch that drops rapidly as the slower-moving low-frequency components arrive. The observer will also note very little low-frequency energy in the perceived sound. Two important characteristics of wave motion on infinite homogeneous plates are apparent. First, the propagation speed of free bending waves increases

TABLE 11.1   Speeds of Sound in Solids

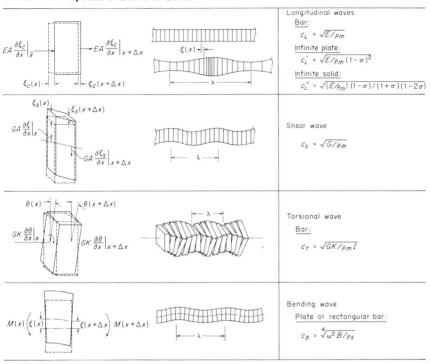

	Longitudinal waves
	Bar:
	$c_L = \sqrt{E/\rho_m}$
	Infinite plate:
	$c_L' = \sqrt{E/\rho_m (1-\sigma)^2}$
	Infinite solid:
	$c_L'' = \sqrt{(E/\rho_m)\,(1-\sigma)/(1+\sigma)(1-2\sigma)}$
	Shear wave
	$c_S = \sqrt{G/\rho_m}$
	Torsional wave
	Bar:
	$c_T = \sqrt{GK/\rho_m I}$
	Bending wave
	Plate or rectangular bar:
	$c_b = \sqrt[4]{\omega^2 B/\rho_S}$

Young's modulus E relates the stress S (force per unit area) to the strain (change in length per unit length), N/m^2. *Poisson's ratio* σ is the ratio of the transverse expansion per unit length of a circular bar to its shortening per unit length, under a compressive stress, dimensionless. It equals about 0.3 for structural materials and nearly 0.5 for rubber-like materials. The *density* of the material is ρ_m, kg/m^3. The *shear modulus* G is the ratio of shearing stress to shearing strain, N/m^2. I is the *polar moment* of inertia, m^4. The *torsional stiffness factor* K relates a twist to the shearing strain produced, m^4. The *bending stiffness* per unit width B equals $(Et^3/12)$, for a homogeneous bar or plate, N-m, where t is the thickness of the bar (or plate) in the direction of bending, m.

with increasing frequency, and second, the lack of low frequencies in the perceived sound suggests that the infinite plate does not radiate sound well below a certain frequency.

11.2.1   Undamped Infinite Plate with Enforced Motion   Let us assume a thin plate of infinite size, which has no internal damping, properly driven mechanically to carry a plane bending wave of constant amplitude and propagation speed c_p. At a certain arbitrarily chosen time such a plate will be deformed as shown in Fig. 11.2. A sound wave is *radiated* outward in the air in such a direction that

$$\lambda = \lambda_p \sin \theta \qquad \text{m} \qquad (11.2)$$

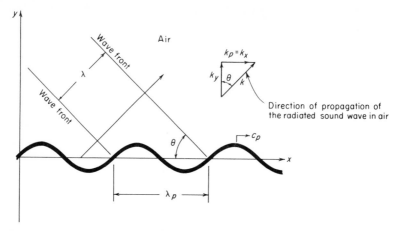

Fig. 11.2 Sound radiation from an infinite mechanically driven plate in which there is a prescribed plane bending wave of propagation speed c_p and wavelength λ_p, where $\lambda_p = \sqrt{1.8 t c'_L / f}$ and t = thickness of plate.

The mechanical wave propagates with this spatial pattern along the plate with speed c_p. The wavenumber k_p is

$$k_p \equiv \frac{2\pi f}{c_p} = \frac{2\pi}{\lambda_p} = 2\pi \sqrt{\frac{f}{1.8 t c'_L}} \qquad \text{m}^{-1} \qquad (11.3)$$

The radiated sound wave propagated in the air has a wavenumber $k = 2\pi/\lambda = 2\pi/\lambda_p \sin \theta = 2\pi f/c$.

Because the sound pressure in the air above or below the plate is caused by the transverse velocity of the plate vibration, the sound field parallel to the plate has the same spatial periodicity as the plate. The y component of the acoustic wavenumber k_y must assume a value such that $\bar{k} = \bar{k}_p + \bar{k}_y$ (see Fig. 11.2). Consequently, the magnitude as well as the direction of the acoustical wavenumber k is defined by the plate wavenumber k_p and frequency.

There are three frequency regions of interest: first, *above critical frequency*, where the propagation speed of the bending wave in the plate is greater than the propagation speed of sound in the air ($k_p < k$); second, *at critical frequency*, where the propagation speed of the bending wave in the plate equals the speed of sound in the air ($k_p = k$); and third, *below critical frequency*, where the propagation speed of the bending wave in the plate is less than the speed of sound in the air ($k_p > k$).

The *critical frequency* $f_c = \omega_c/2\pi$ occurs when $c_b = c$, and $k = k_p$, so that *

---

* The first expression in Eq. 11.4 is generally valid; the following two expressions are valid only for homogeneous isotropic plates.

$$f_c = \frac{c^2}{2\pi} \sqrt{\frac{\rho_s}{B}} = \frac{c^2}{1.8t} \sqrt{\frac{\rho_m}{E}} = \frac{c^2}{1.8c_L t} \tag{11.4}$$

where the symbols are defined in Table 11.1.

Radiation Ratio. Let us define a *radiation ratio* as the acoustic power radiated by the plate into the half space, divided by the acoustic power that an infinite piston (all parts vibrating in phase) would radiate into the same half space if it were vibrating with the same rms velocity as the plate.

Radiation Ratio above Critical Frequency $(f > f_c)$. For an infinite plate mechanically driven at a frequency above the critical frequency of the plate the bending wave travels faster than the speed of sound in air $(k > k_p)$. For this case it can be shown that the radiation ratio σ_{rad} is

$$\sigma_{\text{rad}} = \frac{W_A}{v_{\text{rms}}^2 \rho c} = \frac{1}{\sqrt{1 - \left(\frac{k_p}{k}\right)^2}} \qquad k > k_p \tag{11.5}$$

where v_{rms}^2 = mean-square velocity of the plate, (m/sec)²

W_A = acoustic power radiated, watts

Equation (11.5) is plotted in Fig. 11.3, as the curve for $\eta = 0$, for $k/k_p = c_p/c$ greater than unity.

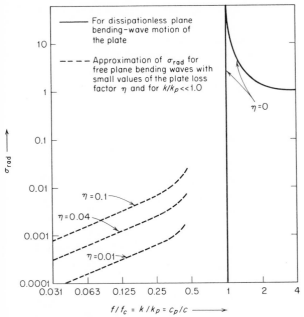

Fig. 11.3 Radiation ratio σ_{rad}, as a function of the normalized wavenumber k/k_p.

Radiation Ratio at Critical Frequency ($f = f_c$). At the critical frequency ($k = k_p$), given by Eq. (11.4), the radiated sound wave runs exactly parallel to the plate surface ($\theta = 90°$) and the acoustical wavelength exactly matches the bending wavelength in the plate, so that the power radiated by the various parts of the plate along the propagation direction "pile up." If the plate velocity could be kept constant the pressure as well as the radiation ratio would approach infinity (see Fig. 11.3).

Radiation Ratio below Critical Frequency ($f < f_c$). For a bending-wave speed that is less than the speed of sound in air (e.g., "subsonic" or "acoustically slow," $k < k_p$), the perpendicular velocity of the elements of the plate and the pressure at the same points on the plate surface are always 90° out of phase so that the plate cannot radiate any acoustic power to the surrounding medium; thus $W_A = 0$ and σ_{rad} is zero.

11.2.2 Damped Infinite Plate Supporting Free Bending Waves

In most practical examples of vibrating equipment, where one is concerned with the sound radiation from plates supporting free bending waves, no external energy is supplied to the plates except at a distant point of excitation. In these cases, the propagation of stresses and moments in the plate material is the only source of power. Accordingly, any power dissipated in the plate material or radiated as sound to the surrounding medium along the path of travel of the free bending wave must necessarily reduce the amplitude of the wave. Hence, both the bending wave and the radiated sound wave are attenuated with increasing distance from the source.

Let us assign to the damped plate a *composite loss factor* η (dimensionless) as defined by Eq. (11.1).

The approximate acoustic powers $W_A(x)$ radiated by unit area of the plate at location x, vibrating at $(v_0)^2_{\text{rms}}$ and for the radiation ratio σ_{rad} are given below:

Power Radiated for $k \gg k_p (f \gg f_c)$ *and* $\eta \ll 1$.

$$W_A(x) \approx (v_0)^2_{\text{rms}} \rho c e^{-\eta k_p x/2} \qquad \text{watts/m}^2 \tag{11.6}$$

and

$$\sigma_{\text{rad}} \approx 1$$

Equation (11.6) shows that the power radiated by the unit area of the plate decreases with increasing distance from the source of vibration but that the radiation ratio remains $\sigma_{\text{rad}} \approx 1$.

Power Radiated for $k \ll k_p (f \ll f_c)$ *and* $\eta \ll 1$.

$$W_A(x) \approx \left(\frac{1}{4} \eta \frac{k}{k_p} \right) (v_0)^2_{\text{rms}} \rho c e^{-\eta k_p x/2} \qquad \text{watts/m}^2 \tag{11.7}$$

and

$$\sigma_{\text{rad}} \approx \frac{1}{4} \eta \frac{k}{k_p} \tag{11.8}$$

Equation (11.8), which approximates the radiation ratio below the critical frequency, is plotted in Fig. 11.3 (dotted lines) for several small values of η.

We see that for subsonic speed of the plate wave ($k < k_p$) the damping changes the radiation behavior of the infinite plate. There is a small (but nonzero) power radiated from each part of the plate and a correspondingly low (but nonzero) value for the radiation ratio as given in Eq. (11.8).

11.2.3 Sound Radiation for Point * or Line Force Excitation

If a thin infinite plate is excited by a point or line force, then far away from the excitation point, the plate carries free bending waves. Because the plate is forced at the excitation point, a near field is created (in addition to the free bending wave field) whose strength decreases exponentially with increasing distance from the excitation point. Below the critical frequency the free-bending-wave field radiates sound very inefficiently so that almost the entire sound radiation is produced in the near field.

Point Source Excitation. Assume a point force excitation F_{rms}. Below the critical frequency, the power radiated into the air from one side of an infinite homogeneous isotropic plate is [6]

$$W_P = \frac{F_{\text{rms}}^2}{2\pi} \frac{\rho}{\rho_s^2 c} \quad \text{watts} \quad f \ll f_c \tag{11.9}$$

where ρ = density of air, kg/m³
ρ_s = mass per unit area of the plate, kg/m²
c = speed of sound in air, m/sec

Thus, the sound power radiated by a thin infinite plate excited by a point force at a frequency less than the critical frequency of the plate is independent of the stiffness of the plate and the frequency.

It is often convenient to measure the rms velocity v_{rms} of the plate at the excitation point rather than to measure the force. The radiated power for given velocity is

$$W_P = \frac{8}{\pi^3} v_{\text{rms}}^2 \rho c^3 \frac{1}{f_c^2} \quad \text{watts} \quad f \ll f_c \tag{11.10}$$

where $f_c = c/\lambda_c$ = critical frequency of the plate, Hz

Equating the power radiated from the region near the point source

* See Sec. 11.5 for point excitation of a finite panel.

with that radiated from a circular portion of radius r of an infinite large piston moving with velocity v_{rms}, namely, $v_{rms}^2 \rho c S$, gives

$$v_{rms}^2 \rho c \, \frac{8}{\pi^3} \, \lambda_c^2 = v_{rms}^2 \rho c \pi r^2 \qquad \text{watts} \qquad (11.11)$$

Solving for r yields the radius of the equivalent rigid piston

$$r = \sqrt{\frac{8}{\pi^3}} \, \lambda_c = 0.286 \lambda_c \qquad \text{m} \qquad (11.12)$$

Thus, a mechanical point source, driving a thin, homogeneous, isotropic plate well below its critical frequency radiates as much acoustical power as a circular portion of an infinite rigid piston having a radius of $r \approx \lambda_c/4$ moving with the velocity of the plate at the excitation point.

Line Source Excitation. Assume a line force excitation F_l along a length l which is large compared to the wavelength of the sound in air. Below critical frequency, the power radiated into the air from one side of the plate is (note that F_l is force per unit length)[4]

$$W_l = F_l^2 l \, \frac{\rho}{\rho_s^2 \omega} \qquad \text{watts} \qquad f \ll f_c \qquad (11.13)$$

where $\rho_s =$ surface mass of the plate, kg/m²
$\quad\quad\; l =$ length of the line source, m
The radiated sound power decreases with increasing frequency.

If the line is excited by a constant velocity source, the radiated power is

$$W_l = 0.64 v_{rms}^2 \lambda_c l \rho c \qquad \text{watts} \qquad f \ll f_c \qquad (11.14)$$

where $\lambda_c = c/f_c =$ wavelength at critical frequency, m
Because the power radiated by a portion of a large piston of area S is $v_{rms}^2 \rho c S$, the radiation of a line-excited, homogeneous, infinite plate can, from Eq. (11.14), be seen to be equivalent to that of a portion of a very large piston of area $S = 0.64 \lambda_c l$ moving with the same velocity as the excitation line and having a radiation ratio of unity.

If the frequency of excitation is above the critical frequency of the panel, the mechanical power introduced to the plate is partly dissipated and partly radiated as sound. In this case, the function $\rho c/(\rho c + \rho_s \omega \eta)$ gives the fraction of the mechanical power that gets radiated.

Example 11.1 Determine the acoustic power that a 1,000-Hz sinusoidal point force with an rms amplitude of 1 N (1 N = 0.22 lb$_f$) radiates from one side of a 0.25-cm-thick infinite, undamped steel plate. The physical parameters of the plate and the surrounding air are $\rho_s = 19$ kg/m², $c_L = 5,100$ m/sec, $f_c = 5,100$ Hz, $\rho = 1.2$ kg/m³, and $c = 344$ m/sec.

SOLUTION: The radiated power is given by Eq. (11.9)

$$W_{\text{point}} = \frac{1}{2\pi} \frac{F_{\text{rms}}^2 \rho}{\rho_s^2 c} = \frac{1}{2\pi} \frac{1.2}{(19)^2 \times 344} = 1.54 \times 10^{-6} \text{ watt}$$

corresponding to a sound-power level of 62 dB *re* 10^{-12} watt.

Example 11.2 Determine how inhomogeneities affect the sound radiation from an infinite plate carrying free plane bending waves which propagate at subsonic speed. Define an inhomogeneity as a point of zero velocity, e.g., a rigid rod that connects this point to an infinitely rigid structure.

The zero velocity at the chosen point of the plate can be thought of as the superposition of a traveling free bending wave and a point application of a transverse velocity equal in magnitude but 180° out of phase with the transverse velocity caused by the traveling wave. We have learned that plane free bending waves of subsonic propagation speed do not radiate, so that all of the radiation must come from the region near the hypothetical point source. Assume the same 0.25-cm steel plate used in the first example, and an rms transverse velocity of the plate due to the free bending wave equal to 10^{-3} m/sec.

SOLUTION: The acoustic power radiated from the near field introduced by the discontinuity is calculated from Eq. (11.10)

$$W_P = v_{\text{rms}}^2 \rho c \, \frac{8}{\pi^3} \lambda_c^2 = 10^{-6} \times 4.1 \times 10^2 \times \frac{8}{\pi^3} \times 4.5 \times 10^{-3}$$

$$= 4.75 \times 10^{-7} \text{ watt}$$

corresponding to a power level of $L_W = 57$ dB *re* 10^{-12} watt.

Example 11.3 Calculate the sound power radiated by a 0.25-cm infinite steel plate when it is forced in phase along a line of length $l = 2$ m to move with a transverse rms velocity of 10^{-3} m/sec. Assume that the frequency of the forcing is $f = 1,000$ Hz, so that both $f \ll f_c$ and $\lambda \ll l$ are satisfied.

SOLUTION: The radiated power as given by Eq. (11.14) is

$$W_l = 0.64 \lambda_c l \rho c v_{\text{rms}}^2 = 0.64 \times 6.75 \times 10^{-2} \times 2 \times 410 \times 10^{-6}$$

$$= 3.5 \times 10^{-5} \text{ watt}$$

corresponding to a sound-power level of 75 dB *re* 10^{-12} watt.

11.2.4   Sound-field Excitation (Distributed Coherent Excitation)

Plane Airborne Sound-wave Excitation. A very common acoustical problem is the excitation of a panel or wall from one side by an airborne sound wave and the radiation of sound from the other side as shown in Fig. 11.4. An infinite homogeneous panel with internal damping separates two air-filled half spaces, each with density ρ and speed of sound c. A plane sound wave is incident from one half space on the panel at angle θ. Part of the energy in the incident wave is reflected back into the same half space, and part excites the panel into vibration. This vibration in turn creates a plane wave, called the transmitted wave, in the second half space.

Even below the critical frequency, the wavelength of the forced wave in the plate must equal the separation of the airborne sound wavefronts as projected on the plate. Since sin θ must always be smaller than or equal to unity, the wavenumber of a forced bending wave on the plate is always less than or equal to the wavenumber in the air ($k_p \leq k$). Hence, the propagation speed of the waves forced on the plate by a sound field is always supersonic or sonic ($c_p \geq c$).

By analogy with Eq. (11.5), we find that the *radiation ratio for a plate forced by a plane sound wave is always real and equal to or greater than unity.*

$$\sigma_{\text{rad}} = \frac{1}{\sqrt{1 - \sin^2 \theta}} \tag{11.15}$$

Clearly, the radiation ratio is independent of frequency and is dependent only on angle of incidence. The consequence of a real, frequency-independent radiation ratio is that the airborne-sound-driven plate can radiate sound efficiently at all frequencies. This is the reason why *an infinite plate forced by an incident sound wave at any frequency, even below its critical fre-*

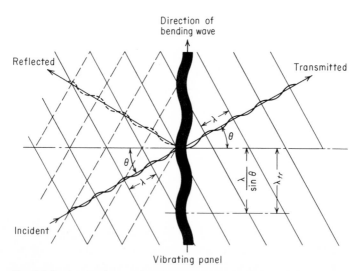

Fig. 11.4 A sound wave incident on an infinite panel. The angle θ is between the normal to the wavefront and the normal to the plate. The wavelength of the forced wave in the plate λ_{tr} is equal to the trace wavelength of the sound wave $\lambda/\sin \theta$. When the trace wavelength of a plane sound wave is such that $\lambda/\sin \theta$ is equal to λ_b, the wavelength of a free bending wave, the intensity of the transmitted wave approaches the intensity of the incident wave and wave coincidence is said to occur. The coincidence angle is θ_{co}.

quency, will radiate sound, although a free bending wave below critical frequency will not.

11.2.5 Transmission Coefficient; Plane Sound Wave Incident at Angle θ

The flow of power is the easiest form in which to express the transmission of sound through a structure. A transmission coefficient τ is defined as the ratio of transmitted acoustic power to the incident acoustic power. For a given angle of plane-wave incidence it can be shown to be [3,4,9,10]

$$\tau(\theta) = \frac{I_{\text{trans}}}{I_{\text{inc}}} = \frac{|p_t^2|}{|p_i^2|}$$

$$= \left\{ \left[1 + \eta \left(\frac{\omega \rho_s}{2\rho c} \cos\theta \right) \left(\frac{\omega^2 B}{c^4 \rho_s} \sin^4\theta \right) \right]^2 \right.$$

$$\left. + \left[\left(\frac{\omega \rho_s}{2\rho c} \cos\theta \right) \left(1 - \frac{\omega^2 B}{c^4 \rho_s} \sin^4\theta \right) \right]^2 \right\}^{-1} \qquad (11.16)$$

where $\tau(\theta) =$ transmission coefficient defined in Eq. (11.16)

$I =$ sound intensity, watts/m^2

$p =$ sound pressure, N/m^2

$\eta =$ composite plate loss factor (dimensionless)

$\rho_s = \rho_m t =$ plate surface density, kg/m^2

$\rho_m =$ density of the plate material, kg/m^3

$t =$ thickness of plate, m

$\rho =$ density of air, kg/m^3

$c =$ speed of sound in air, m/sec

$\theta =$ angle shown in Fig. 11.4

$B =$ plate bending stiffness per unit width, N-m

The subscripts trans and inc stand for the transmitted and incident quantities respectively.

Although an infinitely large panel does not appear to be a model closely representative of many practical problems, it provides useful predictions of the effects of the more important variables involved in airborne sound transmission through practical panels or walls.

11.2.6 Transmission Coefficient; Reverberant Sound Field

A forcing sound field consisting of a plane wave arriving from one particular angle is not a typical problem. The sound field in a room is better modeled by a diffuse sound field, which is an ensemble of plane sound waves of the same average intensity traveling with equal probability in all directions. A region of unit area on the plate will be exposed, at any instant, to plane waves incident equally from all areas on a hemisphere whose center is the area on the plate. The intensity incident on this unit area from any particular angle will be the intensity of the plane wave at that angle I_{inc} multiplied by the cosine of the angle of incidence. (See Chap. 9.) The total transmitted intensity is then

$$I_{trans} = \int_{\Omega} \tau(\theta)\, I_{inc} \cos \theta \, d\Omega \tag{11.17}$$

The integration is over a hemisphere of solid angle Ω, where $d\Omega = \sin \theta \, d\theta \, d\phi$.

Because I_{inc} is the same for all plane waves and τ is independent of the polar angle ϕ, an average transmission coefficient may be defined by

$$\bar{\tau} = \frac{\displaystyle\int_0^{\theta_{lim}} \tau(\theta) \cos \theta \sin \theta \, d\theta}{\displaystyle\int_0^{\theta_{lim}} \cos \theta \sin \theta \, d\theta} \tag{11.18}$$

where θ_{lim} is the limiting angle of incidence of the sound field. For random incidence, θ_{lim} is taken as $\pi/2$ or $90°$.

11.2.7 Definition of Transmission Loss R   To rate the sound-insulating properties of a wall, the difference between incident intensity level and transmitted intensity level is desired. This difference is called the transmission loss R in decibels, and is related to the transmission coefficient $\bar{\tau}$ by

$$R = 10 \log \frac{1}{\bar{\tau}} \quad \text{dB} \tag{11.19}$$

Let us interpret the result of combining Eqs. (11.16), (11.18), and (11.19) to obtain the transmission loss for several limiting cases.

11.2.8 $R(\theta)$ at Low Frequencies; Mass Law at Angle θ   At frequencies much less than the critical frequency (see Eq. 11.4) we have

$$\frac{\omega^2 B}{c^4 \rho_s} \ll 1 \tag{11.20}$$

With this approximation, the transmission loss obtained by combining Eqs. (11.16) and (11.19) is

$$R(\theta) \approx 10 \log \left[1 + \left(\frac{\omega \rho_s}{2\rho c} \cos \theta \right)^2 \right] \quad \text{dB} \tag{11.21}$$

Equation (11.21) is referred to as the *limp-wall mass law transmission loss* because the only characteristic of the wall involved is its surface mass. Thus under the mass law assumption of Eq. (11.20), $R(\theta)$ increases by 6 dB for each doubling of surface mass.

11.2.9 $R_0 = R(\theta = 0)$; **Normal-incidence Mass Law**   For normal incidence, $\theta = 0$

$$R_0 = 10 \log \left[1 + \left(\frac{\omega \rho_s}{2\rho c} \right)^2 \right] \quad \text{dB} \quad (11.22)$$

11.2.10 R_{random} **at Low Frequencies; Random-incidence Mass Law**
With a perfectly diffuse sound field and $R_0 > 15$ dB, the argument of
Eq. (11.21) may be averaged over a range of θ from 0 to 90° to yield the
random-incidence transmission loss [10, 12]

$$R_{\text{random}} = R_0 - 10 \log (0.23 \, R_0) \quad \text{dB} \quad (11.23)$$

11.2.11 R_{field} **at Low Frequencies; Field-incidence Mass Law**   It
has become common practice to use the field-incidence mass law, which
is sometimes defined [3] as (for $R_0 \geqslant 15$ dB),

$$R_{\text{field}} = R_0 - 5 \quad \text{dB} \quad (11.24)$$

This result approximates a diffuse field with a limiting angle θ_{lim} of
about 78° [see Eq. (11.18)].[11, 12]

The three types of mass law transmission loss valid for frequencies
below critical frequency are plotted vs. the product $f\rho_s$ in Fig. 11.5.

11.2.12   Transmission Factor at Wave Coincidence, $\tau(\theta_{\text{co}})$ The
terms in Eq. (11.16) involving the bending stiffness per unit width have

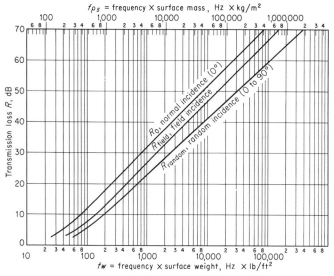

Fig. 11.5   Theoretical transmission-loss curves for mass-controlled
limp panels. Field incidence assumes a sound field which allows all
angles of incidence up to 78° from normal.

significance when $\omega^2 B/c^4 \rho_s$ is equal to or greater than unity. The propagation speed of bending waves c_b (see Table 11.1) also involves the term $(\omega^2 B/\rho_s)^{1/4}$.

We have already defined the *critical frequency* f_c as that frequency where $c = c_b$ [see Eq. (11.4)]. At each frequency *above* the critical frequency there will be some angle of wave incidence for which the trace-wave speed of the incident sound field $(c/\sin \theta)$ will equal the bending-wave speed c_b. This coincidence angle θ_{co} is

$$\sin \theta_{co} = \frac{c}{c_b} = \frac{\lambda}{\lambda_b} \tag{11.25}$$

where λ_b = wavelength of the free bending wave in the panel, m
λ = wavelength of the incident sound wave, m

For excitation by a plane wave at the critical angle θ_{co} a traveling bending wave is created at the surface of the panel of trace wavelength (i.e., in Fig. 11.4 $\theta = \theta_{co}$ and $\lambda_b = \lambda_{tr}$)

$$\lambda_{tr} = \frac{\lambda}{\sin \theta_{co}} \tag{11.26}$$

The plate is highly excited and a transmitted pressure wave of nearly the same intensity, and at the same angle as the incident wave, is radiated.

The transmission coefficient at each frequency for this particular angle of incidence $\theta = \theta_{co}$ is found from Eq. (11.16) by noting that at coincidence, $(\omega^2 B/c^4 \rho_s) \sin^4 \theta_{co} = 1$. Thus

$$\tau(\theta_{co}) = \left[1 + \eta \left(\frac{\omega \rho_s}{2\rho c} \cos \theta_{co} \right) \right]^{-2} \tag{11.27}$$

where η = composite loss factor for the panel

In Eq. (11.27) the quantity in parentheses is a measure of the sound-transmission loss owing to reflection of the incident wave from the panel. If the panel is lossless ($\eta = 0$), the velocity of the panel matches the normal component of the particle velocity of the sound field and the sound transmission is perfect. The addition of damping reduces the amplitude of the excited wave in the panel so that the transverse velocity of the panel no longer matches that of the sound field, and some of the incident power is reflected.

Figure 11.6 is a plot of the transmission coefficient given by Eq. (11.16) as a function of angle of incidence for several values of $(\omega/\omega_c)^2$. As an example, for the surface mass times frequency product $\rho_s f = 52,000$ Hz-kg/m², $(\omega/\omega_c)^2$ is 0.304 for steel and is 2.45 for aluminum. Note the high transmission factor at $\theta = \theta_{co}$.

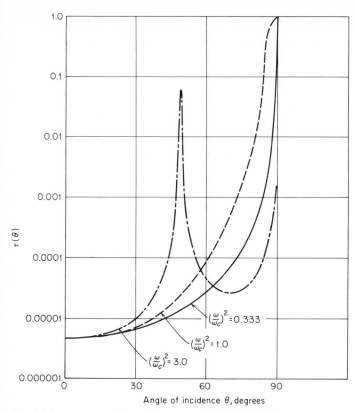

Fig. 11.6   Transmission coefficient as a function of angle for a few selected values of the parameters ω/ω_c, where ω_c is the critical angular frequency. Shown [13] are the cases of well below, $(\omega/\omega_c)^2 = 0.333$, well above, $(\omega/\omega_c)^2 = 3$, and at the critical frequency, $\omega/\omega_c = 1$, for plates whose characteristics are $\rho_s f = 5.2 \times 10^4$ Hz-kg/m^2 and loss factor $\eta = 0.01$. (Note that for a particular plate $\tau(0)$ would have different values at the three frequencies.)

11.2.13   Field-incidence Transmission Loss R_{field} Equations

(11.16) and (11.18) must be solved by numerical integration. The results of such an integration of the transmission coefficient between the angles 0 and 78°, and application of Eq. (11.19) to give the *field-incidence trans-mission loss*, are presented in Fig. 11.7 for all values of f/f_c. The ordinate is the difference between the field-incidence transmission loss R_{field} and the normal-incidence mass law transmission loss at the critical frequency $R_0(f_c)$. The value of $R_0(f_c)$ is easily determined from Eq. (11.22) or from Fig. 11.5 when the mass per unit area and the critical frequency of the panel are known.   Note that predicted transmission losses of less

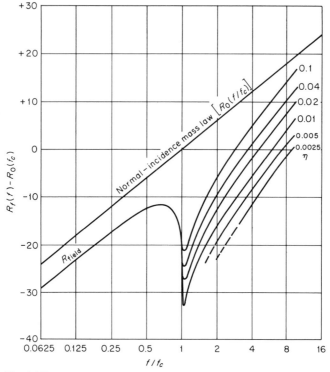

Fig. 11.7 Field-incidence forced wave transmission loss. The ordinate is the difference between the field-incidence transmission loss at the frequency f and the normal-incidence transmission loss *at the critical frequency* ($f/f_c = 1$). Note that for a predicted transmission loss of less than 15 dB, or for the dashed areas on the figure, the transmission loss depends on both the surface weight and the loss factor, and the curves provide only a lower-bound estimate to the actual transmission loss. Use of curve: (1) determine wf_c from Table 11.4, (2) determine f_c, (3) determine $R_0(f_c)$ from Fig. 11.5 or Table 11.4, (4) read $R_f(f) - R_0(f_c)$ from Fig. 11.7 at the required η, and (5) $R_f(f) = [R_f(f) - R_0(f_c)] + R_0(f_c)$.

than about 15 dB for $f \ll f_c$, or less than 25 dB for $f \approx f_c$ from Fig. 11.7 are not accurate.*

Example 11.4 Calculate the normal-incidence mass law for an aluminum panel weighing 10 lb/ft² at a frequency of 500 Hz. Also determine the random-incidence and field-incidence mass laws. What is R_{field} at 2,800 Hz when $\eta = 10^{-2}$?

SOLUTION: The normal-incidence mass law is given by Eq. (11.22) and the upper curve of Fig. 11.5. We have $fw = 500 \times 10 = 5,000$ Hz-lb/ft². From Fig.

* A similar family of curves given in Ref. 3 is in error in the frequency range above critical frequency.

11.5, $R_0 = 45.5$ dB.   The random-incidence mass law is given by Eq. (11.23) and the lower curve of Fig. 11.5; i.e., $R_{random} = 35$ dB.   The field-incidence mass law is given by Eq. (11.24) and the middle curve of Fig. 11.5; i.e., $R_{field} = 45.5 - 5 = 40.5$ dB.

From Table 11.4, $wf_c = 7,000$; $f_c = 7,000/10 = 700$ Hz and $R_0(f_c) = 48$ dB from Fig. 11.5.   Evaluating Fig. 11.7 at $f/f_c = 2,800/700 = 4$ and $\eta = 0.01$, we get $R_f(f) - R_0(f_c) = -6$ dB yielding $R_f(f) = [R_f(f) - R_0(f_c)] + R_0(f) = -6 + 48 = 42$ dB.

11.3   Acoustical Behavior of Finite Panels

Wave motion in finite panels differs from that in infinite panels owing to the presence of edges which produce reflected waves.   Interference between incident and reflected traveling bending waves produces standing-wave patterns, which may result in transverse panel motions of large amplitude.

11.3.1   Resonances   A resonance occurs when a wave travels a closed path in a system and arrives at its starting point in phase with itself.   For a rectangular panel mounted so that both the displacement and the curvature of the panel at the edges are zero, i.e., the panel is *simply supported*, the condition for bending-wave resonance is [7]

$$\left(\frac{1}{\lambda_{mn}}\right)^2 = \left(\frac{m}{2l_x}\right)^2 + \left(\frac{n}{2l_y}\right)^2 \qquad (11.28)$$

where m and $n =$ integers, both different from zero

$l_x, l_y =$ panel-edge lengths in the x and y directions, respectively, m

$\lambda_{mn} =$ wavelength of the bending wave on the panel, m

Substituting c_b/f_{mn} for λ_{mn} in the above equation and using Table 11.1 yields the frequency at which the (m, n)th resonance occurs, assuming no effects from the surrounding medium

$$f_{mn} = \frac{\pi}{2} \left(\frac{B}{\rho_s}\right)^{1/2} \left[\left(\frac{m}{l_x}\right)^2 + \left(\frac{n}{l_y}\right)^2\right] \qquad m, n = 1, 2, \ldots \quad (11.29)$$

where $B =$ bending stiffness of the panel per unit width, N-m

$\rho_s =$ mass per unit area of the panel, kg/m^2

Neither m nor n can equal zero because the displacement at all edges is restricted to zero.   Thus, if there is a displacement in one coordinate direction, there must also be a displacement in all other directions.

The transverse displacement of the panel at these resonance frequencies is

$$\xi(x, y) = \sin\frac{m\pi x}{l_x} \sin\frac{n\pi y}{l_y} \qquad m, n = 1, 2, \ldots \quad (11.30)$$

The spatial deformation shape for a particular m, n is called a *normal mode of vibration*. For each pattern described by Eq. (11.30) there is a gridwork of lines of zero displacement, called *nodes*, which subdivide the panel into a number of smaller rectangular vibrating areas. For the (m, n)th mode, there are $m - 1$ such lines equally spaced along the x direction, and $n - 1$ in the y direction. Each of these small areas displaces the surrounding fluid as it vibrates, and the fluid motion from each area interacts in a complex manner with that from neighboring areas so that the radiated power is in general not a simple function of the average panel velocity.

A means is needed for determining those characteristics that control some "average" quantity, such as the radiated power in a frequency band. One means is to represent the power radiated as equal to the power radiated from a representative mode multiplied by the number of modes of vibration in the band.

The *wavenumber* of a mode designated by the integers m, n is found from Eq. (11.28) to be

$$k_{mn} = \frac{2\pi}{\lambda_{mn}} = \sqrt{\left(\frac{m\pi}{l_x}\right)^2 + \left(\frac{n\pi}{l_y}\right)^2} \qquad (11.31)$$

The x and y components of the *wavenumber vector* $\bar{k}_{mn} = \bar{k}_{mx} + \bar{k}_{ny}$ are

$$k_{mx} = \frac{m\pi}{l_x} \qquad k_{ny} = \frac{n\pi}{l_y} \qquad (11.32)$$

All the points that correspond to different resonances in a given plate may now be plotted as shown in Fig. 11.8. Each mode of plate vibration is characterized by a point in wavenumber space ($k_{mx} = m\pi/l_x$; $k_{ny} =$

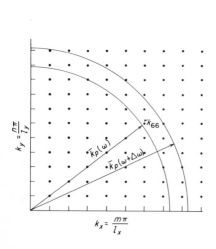

$$k_x = \frac{m\pi}{l_x}$$

Fig. 11.8 Typical mode lattice (resonance mode wavenumbers) in wavenumber space for a simply supported, rectangular panel (with $l = 0.66\, l_y$). The length of the vector from the origin to a particular point equals the mode wavenumber $k_{m,n}$ for a particular resonance in a given plate. The frequency of that resonance, in Hz, equals the length of that vector divided by $2\pi/c_b$. The points which are located k_{mn} from the origin [see Eq. (11.31)] satisfy the conditions for bending-wave resonance at those frequencies where the plate free bending wavenumber $\bar{k}_p(\omega)$ [see Eq. (11.33)] equals the resonance mode wavenumber k_{mn}.

$n\pi/l_y$) and the wavenumber vector of the mode is represented by the vector from the origin to the respective point. The orientation of the wavenumber vector to a particular resonance point in wavenumber space gives the direction of propagation of one of the traveling waves in the panel which, when combined with its reflections from the panel boundaries, produces the mode shape corresponding to that point. For example, the point nearest the end of $\bar{k}_p(\omega)$ in Fig. 11.8 corresponds to the 6, 6 mode. This mode will be strongly excited by a point force source (whose driving frequency is $f_{(6,6)}$) placed at any point on the plate halfway between the nodal lines.

The resonances at a particular frequency are those points where the mode wavenumber $k_{m,n}$ equals the bending wavenumber $k_p(\omega)$ at that frequency. These mode points lie on an arc of radius

$$k_p(\omega) = \frac{\omega}{c_b} = \omega^{1/2} \left(\frac{\rho_s}{B}\right)^{1/4} \qquad (11.33)$$

where $k_p(\omega) =$ infinite panel bending wavenumber.*

Modal Density. The modal density is the number of modes with resonance frequencies in an angular frequency band of unit width ($\Delta\omega = 1$ radian/sec). The number of resonant modes in a band of frequencies $\Delta\omega$ between ω and $\omega + \Delta\omega$ equals the number of modal points lying in the area between the arcs of radius $k_p(\omega)$ and $k_p(\omega + \Delta\omega)$, and is given by

$$n(\omega)\Delta\omega = \frac{l_x l_y}{4\pi} \left[k_p^2(\omega + \Delta\omega) - k_p^2(\omega) \right] \qquad (11.34)$$

When the panel wavenumber k_p is much greater than the first mode wavenumber k_{11}, Eq. (11.34) reduces to †

$$n(\omega) \approx \frac{l_x l_y}{4\pi} \sqrt{\frac{\rho_s}{B}} = \frac{1}{2\pi} \left(\frac{\sqrt{12}S}{2tc_L}\right) \qquad (11.35)$$

where $B =$ bending stiffness per unit width, N-m
   $\rho_s =$ mass per unit area, kg/m²
   $l_x l_y = S =$ area of plate, m²
   $t =$ thickness of plate, m
   $c_L =$ longitudinal wave speed in plate, m/sec

* The mode wavenumber k_{mn} identifies the spatial deformation of the panel. This mode of vibration may be forced at any frequency, but will only respond "freely," i.e., at resonance, at a frequency where the mode wavenumber equals the "free" bending wavenumber $k_p(\omega)$ for the panel.

† The first expression in Eq. (11.35) is generally valid; the second is limited to homogeneous isotropic plates only.

This equation is a good approximation for the modal density of any panel of area S of arbitrary shape and boundary conditions, provided the bending wavelength of the mode is much smaller than the shorter plate dimension.[4] This is because the edges can only affect the mode shape of a panel in the vicinity (i.e., within $0.5\lambda_b$) of the edge of the panel. Note that the number of modes in *unit frequency bandwidth* (Hz), e.g., $n(f)$, is given by the quantity in the parentheses of Eq. (11.35), i.e., is 2π larger than $n(\omega)$. *The modal density is large for a thin large panel and is independent of frequency.*

11.3.2 Radiation of Sound from Resonant Modes

We have already shown that there is a substantial difference in the efficiency of radiation of sound depending upon whether the wavenumber k_p in a plate is much greater or smaller than the wavenumber k for sound in air at the same frequency. Figure 11.9 shows that, for an infinite homogeneous panel, k_p increases as the square root of frequency while k increases as the first power of frequency. At one frequency only, the critical frequency, the two have the same value on a given panel. We shall now see how the radiation ratio of the finite plate differs from that of the infinite plate.

First consider a high-order mode such that both k_{mx} and k_{ny} are greater than k. This implies that the nodal (zero displacement) lines in both directions are separated by less than half a wavelength in air. The relative phase relations for various areas of the panel for a particular mode ($m = 5$; $n = 4$) are indicated by Fig. 11.10. Because adjacent subsections are separated by much less than a wavelength in the surrounding air, the air displaced outward by one subsection moves to occupy the space left by the motion of the adjacent subsections, without being compressed,

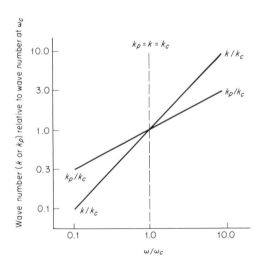

Fig. 11.9 Relation between wavenumber for bending waves in an infinite panel $k_p(\omega)$ and wavenumber k in the surrounding medium, that is, k/k_c and k_p/k_c, are plotted vs. ω/ω_c, where ω_c is the critical frequency of the panel.

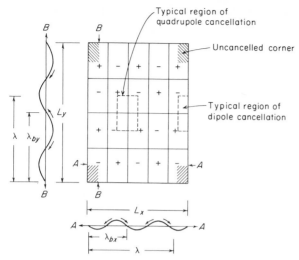

Fig. 11.10 Displacement pattern for the $m = 5$, $n = 4$ mode on a panel with simply supported edges. The relative phases of the antinodes are indicated by $+$ and $-$. Arrows indicate air movement during that half-period of vibration. The uncancelled corner areas are widely separated compared to a wavelength in air and do not cancel each other. This mode is called a *corner mode*.

and very little power is radiated. At the edges of the panel, the "cancellation" is not quite so effective, because the subsection is bounded by cancelling areas on only three sides. The corners of the panel have cancelling subsections on only two sides, and so they radiate the most effectively of all areas.

Multipole Radiation. A more analytical description of the radiation may be developed from the multipole description of sound sources,[2,6] as defined in Table 11.2. We see that the net power radiated by a dipole is less than that radiated by one of the constituent monopoles [2] by a factor of $\frac{1}{3}(kd)^2$, and considerably less power is radiated by a quadrupole.[14]

Application of Table 11.2 to Fig. 11.10 indicates that the central regions of the panel radiate very little power because they combine as quadrupole sources. The edges of the panel form a line of dipole sources. Four monopoles remain at the corners of the panel. Thus, for classification purposes this mode when radiating at resonance is called a *corner mode*, because only the panel corners radiate power effectively.

If the panel is much less than a wavelength in size ($kl_x < 1$, $kl_y < 1$), then the corner monopoles interact with each other. The radiation will be monopole in character if the corners are in phase (e.g., m, n odd), dipolelike if one adjacent pair is in phase, but out of phase with respect

TABLE 11.2   Properties of Multipole Sources [2]

Source type	Radiated pressure (Note 1)	Radiated power
1. Monopole or simple source	$\|p\|^2 = \dfrac{S_0}{r^2}$	$W_{rad} = W_0 = \dfrac{\rho c k^2}{4\pi} A^2 \langle v^2 \rangle$ A = source area $\langle v^2 \rangle$ = mean-square source velocity
2. Dipole source $d \ll \lambda$, or $kd < 1$	$\|p\|^2 = D_0 \dfrac{\cos^2 \alpha}{r^2} (kd)^2$ $\times \left[1 + \dfrac{1}{(kr)^2} \right]$	$W_{rad} = \tfrac{1}{3}(kd)^2 W_0$
3. Quadrupole source $kd_x < 1$ $kd_y < 1$	$\|p\|^2 = \dfrac{Q_0(\theta, \phi)}{r^2} (kd_x)^2 (kd_y)^2$ $\times \left[1 + \dfrac{3}{(kr)^2} + \dfrac{9}{(kr)^4} \right]$	$W_{rad} = \tfrac{1}{15}(kd_x)^2 (kd_y)^2 W_0$

Note 1. S_0, D_0, Q_0 are source constants which depend only on W_0; Q_0 also depends on the orientation to the measurement point.

to the other (m odd, n even as shown in Fig. 11.10, or m even, n odd), or quadrupolelike when all adjacent corners are out of phase with each other (n, m even). When the panel is much greater than a wavelength in size, the corner monopoles are uncoupled from each other, and their radiated powers combine additively.

The case of one of the wavenumber components of the plate vibration being less than k is shown in Figs. 11.11 and 11.12. For this case the separation between subareas in the y direction is greater than an acoustic wavelength in the y direction so that the central areas form long, narrow dipole radiators. There is little cancellation along the entire y-edge length. Since the edges must be more than an acoustic wavelength apart when this mode is at resonance, they radiate independently as monopoles. This type of vibration is classified as an *edge mode*. Edge modes

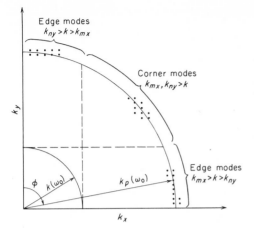

Fig. 11.11   Mode lattice showing the relative magnitude of k_{mx} and k_{ny} for different modes at a frequency ω_0 well below the critical frequency such that $k_p > k$. The angle ϕ gives the direction of wave propagation relative to the y edge of the plate for one component of the traveling wave making up the resonance wave.

typically radiate more power than corner modes in the frequency range where they occur.

As the exciting frequency approaches the critical frequency, k approaches k_p in magnitude so that the edge modes form an increasing percentage of the total mode population, while at the same time the effective radiating width of the average edge mode is decreasing. In addition, the effectiveness of the cancellation in the central area of the

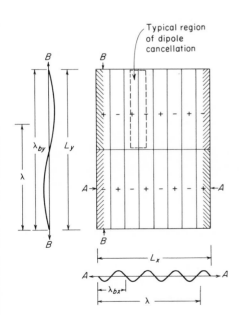

Fig. 11.12   Displacement pattern for $m = 8$, $n = 2$ on a panel with simply supported edges. The uncancelled edge areas are widely separated, compared to a wavelength in air, and do not cancel each other. This form of cancellation yields an "edge mode."

panel is decreasing as the separation between subsections approaches a half acoustic wavelength ($\lambda/2$) in size.

At and above the critical frequency, the cancellation breaks down completely and the entire panel radiates. These modes are therefore called *surface modes*. The situation approximates that for the infinite panel except that for a single mode the directivity of the radiation is modified from the single direction in the infinite case to radiation into four lobes that are oriented in the direction of $\vec{k}_{mn}$ and its mirror images, with respect to the edges of the plate. If the plate is sufficiently large so that there are many modes in any frequency band (e.g., $k_p^2 S \gg 1$), then we recognize that there will be many allowed directions for $\vec{k}_{mn}$. The resulting directivity pattern will approximate a cone with an apex centered on the panel and sides inclined at the angle $\theta = \sin^{-1}[k_p(\omega)/k]$.

Sound Radiation. It was hypothesized above that the power radiated from a set of resonant modes might be found from the product of the power radiated by an "average" mode and the number of modes in the frequency band $n(\omega)\Delta\omega$. If the modes are excited by a random noise in a narrow bandwidth $\Delta\omega$ centered on frequency ω, and if the space-time average transverse velocities of the modes within the band are equal, the sound power radiated is given by

$$W_{\text{rad}} = \rho c S \sigma_{\text{rad}} \langle v^2(\omega) \rangle \qquad (11.36)$$

where σ_{rad} = average radiation ratio for the panel at the frequency ω (dimensionless)

$\langle v^2(\omega) \rangle$ = space-time average mean-square velocity of the panel at the frequency ω in the band $\Delta\omega$, m²/sec²

$S = l_x l_y$ = panel area, m²

Radiation Ratio Curves.[14-16] The average radiation ratio vs. frequency for a simply supported rectangular panel is given in Table 11.3. An approximation to these expressions, accurate to within several decibels, is depicted in Fig. 11.13. The radiation ratio in the low-frequency region (corner modes and single modes controlling) is overestimated by the horizontal line at $\sigma_{\text{rad}} = \lambda_c^2/S$. In the region of edge mode control σ_{rad} is approximated by a line with a slope of 1.8 dB/octave (6 dB/decade) which intercepts the flat portion of σ_{rad} at the frequency $100\,(\lambda_c/P)(c/P)$, where S is the panel surface area (one side only) and P is its perimeter. The region of edge mode control does not begin, however, until $f \approx 3c/P$. The horizontal portion of σ_{rad} is extended to this frequency, and then faired into the curve for edge mode control. Between $f_c/4$ and f_c the edge mode control curve rises smoothly to the peak value of approximately $(P/2\lambda_c)^{1/2}$, and then descends smoothly to $\sigma_{\text{rad}} = 1$, above the critical frequency.

It is clear that *the radiation from a panel with resonance modes is character-*

TABLE 11.3   Average Radiation Ratio for a Simply Supported (Rectangular) * Panel (One Side Only)[14]

Frequency region	Radiation ratio σ_{rad}
I.A. Well below coincidence $(k \ll k_P)$ $l_x, l_y \ll \lambda$ $(kl_x, kl_y < 1)$ 1. Up to f_{11} †	$\sigma_{\text{rad}} \approx \dfrac{4S}{c^2} f^2$
2. Just above f_{11}	σ_{rad} fluctuates widely (see modal radiation ratio)
I.B. Below coincidence $(k < k_p)$, multimodal region $(k_P l_x, k_P l_y > 2)$, corner modes and edge modes occur when frequency is such that $f \geq c/P$	$\sigma_{\text{rad}} \approx \dfrac{\lambda_c^2}{S} g_1(\alpha) + \dfrac{P\lambda_c}{S} g_2(\alpha)$ $P = 2(l_x + l_y),\ S = l_x l_y$ (for rectangular panel) $g_1(\alpha) = \begin{cases} \dfrac{8}{\pi^4} \dfrac{(1 - 2\alpha^2)}{\alpha(1 - \alpha^2)^{1/2}} & f < f_c/2 \\ 0 & f > f_c/2 \end{cases}$ $g_2(\alpha) = \dfrac{1}{4\pi^2} \left[\dfrac{(1 - \alpha^2) \ln\left(\dfrac{1 + \alpha}{1 - \alpha}\right) + 2\alpha}{(1 - \alpha^2)^{3/2}} \right]$ $\alpha = \left(\dfrac{f}{f_c}\right)^{1/2}$
II. At the critical frequency	$\sigma_{\text{rad}} \approx \left(\dfrac{l_x}{\lambda_c}\right)^{1/2} + \left(\dfrac{l_y}{\lambda_c}\right)^{1/2}$
III. Above the critical frequency	$\sigma_{\text{rad}} \approx \left(1 - \dfrac{f_c}{f}\right)^{-1/2}$

* The values of σ_{rad} are determined from the single-mode radiation ratio $\sigma_{mn_{\text{rad}}}$ for a rectangular panel, averaged over a narrow frequency band. However, the values are expected to be reasonably accurate for any simply supported panel of area S and perimeter P.

† $f_{11}/f_c = c^2\beta/2S;\ \beta = \dfrac{1}{2}\left(\dfrac{l_x}{l_y} + \dfrac{l_y}{l_x}\right)\gamma$, and $\gamma \begin{cases} = 1 \text{ for simply supported edges} \\ > 1 \text{ for clamped edges} \\ < 1 \text{ for free edges} \end{cases}$

istically different from the radiation from an infinite panel. The power radiated from the panel is no longer simply related to the panel velocity, but depends to a great extent on the details of the panel displacement. In addition, we note that while the dipole and quadrupole canceled areas in the middle of the panel will not radiate significant sound *power* below the critical frequency, they produce very intense sound *pressures* in the "near field" (within about $\lambda/2$ from the surface of the panel).

Effect of Edge Conditions. For all nondissipative edge conditions with zero edge displacement, the greatest radiation is from a clamped edge,[14]

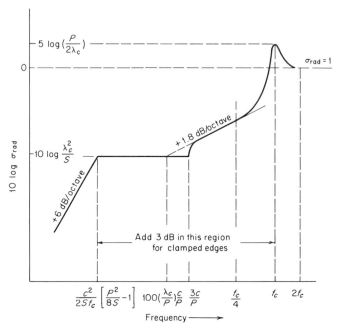

Fig. 11.13 Design curve for approximating the radiation ratio σ_{rad} of a finite panel of perimeter P and area S with simply supported or clamped edges. (Note that $\lambda_c = c/f_c$ = wavelength in the panel or in air at the critical frequency.)

$[\xi = 0, (\partial \xi / \partial x) = 0]$, where σ_{rad} for corner and edge modes is increased by a factor of 2 (i.e., 3 dB) over the simply supported case. The range of the radiation ratio for all stiffness controlled edges lies between that for the simply supported and clamped edges.[15]

11.3.3 Statistical Analyses

Since most practical structures are too complex to permit the use of deterministic procedures, acoustical engineers have learned to apply statistical methods in their analysis. In this section the statistical viewpoint is explained by progressing from the simple power balance to Statistical Energy Analysis.

Power Balance. To get a feel for this technique, let us start with the case of a resiliently suspended plate that has no connections to neighboring structures. Assume a steady-state source of power supplied by some kind of vibrator. The power supplied by the source must equal the sum of the power lost through internal damping and the power radiated as sound into the surrounding air. The steps in determining the response are *

---

* See also discussion in Sec. 11.5.

1. The total energy E stored in the plate equals the sum of the kinetic and potential energies for the plate (which are equal on the average); that is, $E = 2 \times (0.5M\langle v^2 \rangle)$, where $M =$ total mass of the plate (in kg) and $\langle v^2 \rangle$ is the mean-square transverse vibrational velocity averaged over the plate. This relation can also be written $E = \rho_s S \langle v^2 \rangle$, where $\rho_s =$ mass per unit area of the plate (kg/m^2) and $S =$ area of plate (one side) (m^2).

2. The energy dissipated in the plate in one cycle of vibration is equal to the total energy stored E times the dissipation loss factor η_d. Thus, at angular frequency $\omega W^d = E\omega\eta_d$ (watts).

3. The acoustic power radiated into free space is simply equal to the mean-square velocity, times the characteristic resistance for air, times the radiation efficiency. Thus, $W^{\mathrm{rad}} = \langle v^2 \rangle 2S\rho c \sigma_{\mathrm{rad}}$. The factor 2 is added, because the plate has two sides, each of which radiates sound.

If the plate has rigid (structureborne) connections to neighboring structures, energy is lost to the plate across these boundaries. This loss can be taken into account by increasing the dissipation loss factor η_d so that it becomes η_T, where $\eta_T > \eta_d$.

Finally, if we divide W^{rad} by W_{in}, we get the ratio of power transfer from the source to the air. Thus, $W^{\mathrm{rad}}/W_{\mathrm{in}} = 2\rho c \sigma_{\mathrm{rad}}/(\rho_s \omega \eta_T + 2\rho c \sigma_{\mathrm{rad}})$. If the losses in the plate are large, so that most of the power is dissipated there, then $W^{\mathrm{rad}} = W_{\mathrm{in}} (2\rho c \sigma_{\mathrm{rad}}/\rho_s \omega \eta_T)$. If, on the other hand, the losses in the plate are very small, $W^{\mathrm{rad}} \approx W_{\mathrm{in}}$!

Statistical Energy Analysis (SEA). Statistical Energy Analysis (SEA) is a point of view in dealing with the vibration of complex resonant structures. It permits calculation of the energy flow between connected resonant structures, such as plates, beams, etc., and between plates and the reverberant sound field in an enclosure.[17–21]

System of Modal Groups. In respect to the energy E stored in a structure, it may be thought of as a *system* of resonant modes or resonators. First let us consider the power flow between two groups of resonant modes of two coupled structures having their modal resonance frequencies within a narrow frequency band $\Delta\omega$ (see Fig. 11.14).

We assume that each resonant mode of the first system (box 1 in Fig. 11.14) has the same energy. Also, assume that the coupling of the indi-

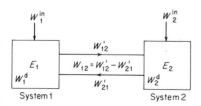

Fig. 11.14 Block diagram illustrating the power flow between two non-dissipatively coupled, incoherent systems.

vidual resonant modes of the first system with each resonance mode of the second system is approximately the same.

If we further assume that the waves carrying the energy in one system are uncorrelated with the waves carrying the energy gained through coupling to the other system, we can separate the power flow as *

$$W'_{12} = E_1 \omega \eta_{12} \qquad (11.37a)$$

$$W'_{21} = E_2 \omega \eta_{21} \qquad (11.37b)$$

where W'_{12} = power system 1 transmits to system 2, watts
   W'_{21} = power system 2 transmits to system 1, watts
   E_1 = total energy in system 1, m-kg/sec
   E_2 = total energy in system 2, m-kg/sec
   ω = center frequency of the band, radians/sec
   η_{12} = coupling loss factor from system 1 to system 2, as defined in Eq. (11.37a)
   η_{21} = coupling-loss factor from system 2 to system 1, as defined in Eq. (11.37b)

The net power flow between the two systems is, accordingly,

$$W_{12} = W'_{12} - W'_{21} = E_1 \omega \eta_{12} - E_2 \omega \eta_{21} \qquad (11.38)$$

Modal Energy E_m.   Let us define modal energy as

$$E_m = \frac{E(\Delta \omega)}{n(\omega) \Delta \omega} \qquad \text{watt-sec/Hz} \qquad (11.39)$$

where $E(\Delta \omega)$ = total energy in the system in angular frequency band $\Delta \omega$
   $n(\omega)$ = modal density = number of modes in unit bandwidth ($\Delta \omega = 1$) centered on ω, the angular frequency
   $\Delta \omega$ = bandwidth, radians/sec

If the previously made assumptions about the equal distribution of energy in the modes and the same coupling-loss factor are valid, then it may be shown that

$$\frac{\eta_{21}}{\eta_{12}} = \frac{n_1(\omega)}{n_2(\omega)} \qquad (11.40)$$

where $n_1(\omega)$ = modal density of system 1 at frequency ω, sec
   $n_2(\omega)$ = modal density of system 2 at frequency ω, sec

Equation (11.40) implies that for equal total energies in the two systems, $E_1 = E_2$, the system that has the *lower* modal density [lower $n(\omega)$] transfers more energy to the second system than is transferred from the second to the first system.

---

* Each equation is written for a narrow frequency band $\Delta \omega$.

Combining Eqs. (11.38) and (11.40) yields *

$$W_{12} = \omega \eta_{12} n_1(\omega) \left[E_{m_1} - E_{m_2} \right] \Delta\omega \qquad \text{watts} \qquad (11.41)$$

where W_{12} = net power flow between system 1 and 2 in the band $\Delta\omega$, centered at ω, watts

E_{m_1}, E_{m_2} = modal energies for systems 1 and 2, respectively [see Eq. (11.39)], watt-sec/Hz

This equation is positive if the first term in the brackets is greater than the second.

The principle of the SEA method is given by Eq. (11.41), which is a simple algebraic equation with energy as the independent dynamic variable. It states that *the net power flow between two coupled systems in a narrow frequency band centered at frequency ω is proportional to the difference in the modal energies of the two systems at the same frequency. The flow is from the system with the higher modal energy to that with the lower modal energy.*

It may help to understand Eq. (11.41) if we use the thermodynamical analogy of heat transfer between two connected bodies of different temperature, where the heat flow is from the body of higher temperature to that of lower temperature and the net heat flow is proportional to the difference in temperature of the two bodies. Consequently, the modal energy E_m is analogous to temperature, and the net power flow W_{12} is analogous to heat flow. The case of equal modal energies of the two systems where the net power flow is zero is analogous to equal temperature of the two bodies.

Equal Energy of Modes of Vibration. Equal energy of the modes within a group will usually exist if the wave field of the structure is diffuse. Also, since the frequency-adjacent resonance modes of a structure are coupled to each other by scattering and damping, there is always a tendency for the modal energy of resonant modes to equalize within a narrow frequency band even if the wave field is not diffuse.

Noncorrelation between Waves in the Two Systems. In sound-transmission problems usually only one system is excited. The power W'_{12} transmitted to the nonexcited system builds up a semidiffuse vibration field in that system. Accordingly, the waves that carry the transmitted power W'_{21} back to the excited system are almost always sufficiently delayed and randomized in phase with respect to the waves carrying the incident power W'_{12} that there is little correlation between the two wave fields.

Realization of Equal Coupling-loss Factor. Equality of the coupling-loss factors between individual modes within a group is a matter of

---

* Note that Eq. (11.40) is a necessary requirement if Eq. (11.41) is to obey the consistency relationship $W_{12} = -W_{21}$.

grouping modes of similar nature. If the coupled system is a plate in a reverberant sound field, the material presented earlier in this section about the classification of "acoustically slow edge and corner modes," and "acoustically fast surface modes," provides the necessary tools to perform such a grouping since the radiation ratio, as we will show later, is closely connected with the coupling-loss factor.

Composite Structures. Composite structures generally consist of a number of elements such as plates, beams, and stiffeners, etc. We may divide a complex structure into its simpler members provided the wavelength of the structureborne vibration is small compared with the characteristic dimensions of the elements. Where this is true, the modal density of a complex structure is approximately that of the sum of the modal densities of its elements. If the power input, the various coupling-loss factors, and the power dissipated in each element are known, the power-balance equations will yield the vibrational energy in the respective elements of the structure.

The dissipative loss factor for an element of a structure is obtained by separating that element from the rest of the structure and measuring its decay rate as discussed in Chap. 14. The coupling-loss factor can be determined experimentally from Eq. (11.41); however, the procedure is difficult. Theoretical solutions are available for the coupling-loss factors of a few simple structural connections.[21] When the coupling-loss factor between a sound field and a simple structure is desired, such a loss factor can be calculated from Eq. (11.40) if the radiation ratio of the structure is known as shown in the next section.

Power Balance in a Two-structure System. The power balance of the simple two-structure system of Fig. 11.14 is given by the following two algebraic equations

$$W_1^{in} = W_1^d + W_{12} \qquad \text{watts} \qquad (11.42)$$

$$W_2^{in} + W_{12} = W_2^d \qquad \text{watts} \qquad (11.43)$$

where W_1^{in} = input power to system 1, watts
W_1^d = power dissipated in system 1, watts
$W_{12} = W_{12}' - W_{21}'$ = net power lost by system 1 through coupling *
to system 2, watts
W_2^{in} = input power to system 2, watts
W_2^d = power dissipated by system 2, watts

The power dissipated in a system is related to the energy stored by that system, E_i, through the dissipative loss factor η_i; namely

$$W_i^d = E_i \omega \eta_i \qquad (11.44)$$

* As in our previous analysis, we assume that the coupling is nondissipative.

where E_i = energy stored in system i, N-m

Assuming that the second system does not have direct power input ($W_2^{in} = 0$), the combination of Eqs. (11.39) to (11.41), (11.43), and (11.44) (with $i = 2$) yields the ratio of the energies stored in the two respective systems

$$\frac{E_2}{E_1} = \frac{n_2}{n_1} \frac{\eta_{21}}{\eta_{21} + \eta_2} \qquad (11.45)$$

If the coupling-loss factor is very large compared to the loss factor in system 2, i.e., if $\eta_{21} \gg \eta_2$, Eq. (11.45) yields the equality of the modal energies ($E_1/n_1\Delta\omega = E_2/n_2\Delta\omega$).

Diffuse Sound Field Driving a Freely Hung Panel. Let us now examine the special case of the excitation of a homogeneous panel (system 2) which hangs freely, exposed to the diffuse sound field of a reverberant room (system 1).

For this case the total energies for each system are given by

$$E_1 = DV = \frac{\langle p^2 \rangle}{\rho c^2} V \qquad (11.46)$$

$$E_2 = \langle v^2 \rangle \rho_s S \qquad (11.47)$$

where D = average energy density in the reverberant room [see Eq. (9.8)], watt-sec

$\langle p^2 \rangle$ = mean-square sound pressure (space-time average), N^2/m^4

V = room volume, m^3

$\langle v^2 \rangle$ = mean-square plate-vibration velocity (space-time average), m^2/sec^2

S = plate surface area (one side), m^2

ρ_s = mass per unit area of the panel, kg/m^2

To find the coupling-loss factor η_{21} we must first recognize that W'_{21} equals the power that the plate, having been excited into vibration, radiates back into the room. Thus, $W'_{21} = W_{rad}$. From Eq. (11.36), (11.44) (with $i = 2$), and item 1 of the discussion under "Power Balance," we obtain

$$W_{rad} = W'_{21} = 2\langle v^2 \rangle \rho c \sigma_{rad} S \equiv E_2 \omega \eta_{21} = \langle v^2 \rangle \rho_s S \omega \eta_{21} \qquad (11.48)$$

where σ_{rad} = radiation ratio for the plate, dimensionless

W_{rad} = acoustical power radiated by both sides of the plate, which accounts for the factor 2, watts

Solving for η_{21} yields

$$\eta_{21} = \frac{2\rho c \sigma_{rad}}{\rho_s \omega} \qquad (11.49)$$

The modal density of the reverberant sound field in the room $n_1(\omega)$ and that of the thin homogeneous plate $n_2(\omega)$ are given in Table 11.5 near the end of the chapter, namely

$$n_1(\omega) = \frac{\omega^2 V}{2\pi^2 c^3} \tag{11.50}$$

$$n_2(\omega) = \frac{\sqrt{12}S}{4\pi c_L t} \tag{11.51}$$

where c_L = propagation speed of longitudinal waves in the plate material, m/sec

t = plate thickness, m

Inserting Eqs. (11.46) to (11.51) into Eq. (11.45) yields the desired relation between the sound pressure and plate velocity

$$\langle v^2 \rangle = \langle p^2 \rangle \frac{\sqrt{12}\pi c^2}{2\rho c t c_L \rho_s \omega^2} \frac{1}{1 + \rho_s \omega \eta_2 / 2\rho c \sigma_{\mathrm{rad}}} \tag{11.52}$$

The mean-square (space-time average) acceleration of the panel is simply

$$\langle a^2 \rangle = \omega^2 \langle v^2 \rangle = \langle p^2 \rangle \frac{\sqrt{12}\pi c^2}{2\rho c t c_L \rho_s} \frac{1}{1 + \rho_s \omega \eta_2 / 2\rho c \sigma_{\mathrm{rad}}} \tag{11.53}$$

It can be shown that as long as the power dissipated in the plate is small compared with the sound power radiated by that plate ($\rho_s \omega \eta_2 \ll 2\rho c \sigma_{\mathrm{rad}}$), the equality of the modal energies of the sound field and the plate yields the proper plate velocity and acceleration. Also, under this condition the ratio of the mean-square plate acceleration to the mean-square sound pressure is independent of frequency.

In general, *the plate response is always smaller than that calculated by the equality of the modal energies by the last factor on the right of Eq. (11.52) or (11.53), which equals the ratio of the power loss by acoustical radiation to the total power loss.* In dealing with the excitation of structures by a sound field, the concept of equal modal energy often enables one to give a simple estimate for the upper bound of the structure's response.

Example 11.5   Calculate the rms velocity and acceleration of a 0.005-m-(1/10-in.-)thick homogeneous aluminum panel resiliently suspended in a reverberant room. The space-average sound-pressure level $L_p = 100$ dB ($\sqrt{\langle p^2 \rangle} = 2$ N/m^2) as measured in a one-third-octave band centered at a frequency $f = \omega/2\pi = 1{,}000$ Hz. The appropriate constants of the panel and surrounding media are $\rho_s = 13.5$ kg/m^2; $c_L = 5.2 \times 10^3$ m/sec; $t = 5 \times 10^{-3}$ m; $\rho = 1.2$ kg/m^3; $c = 344$ m/sec; and $\eta_2 = 10^{-4}$.

SOLUTION: First, calculate the factor

$$\frac{\rho_s \omega \eta_2}{2\rho c \sigma_{\text{rad}}} = \frac{13.5 \times 2\pi \times 10^3 \times 10^{-4}}{2 \times 1.2 \times 344 \times \sigma_{\text{rad}}} = \frac{8.5}{820\sigma_{\text{rad}}} \ll 1$$

According to the above inequality the mean-square acceleration of the panel given by Eq. (11.53) simplifies to

$$\langle a^2 \rangle \approx \langle p^2 \rangle \frac{\sqrt{12}\pi c^2}{2\rho c t c_L \rho_s} = 19 \text{ m}^2/\text{sec}^4$$

or $a_{\text{rms}} = \sqrt{\langle a^2 \rangle} = 4.34$ m/sec², an acceleration level of 113 dB re 10^{-5} m/sec². The mean-square velocity is

$$\langle v^2 \rangle = \frac{\langle a^2 \rangle}{\omega^2} = \frac{19}{4\pi^2 10^6} = 4.76 \times 10^{-7} \text{ m}^2/\text{sec}^2$$

or $v_{\text{rms}} = \sqrt{\langle v^2 \rangle} = 6.9 \times 10^{-4}$ m/sec, a velocity level of 97 dB re 10^{-8} m/sec.

11.3.4 Sound-transmission Loss of a Simple Homogeneous Structure by the SEA Method   The SEA method may be used to analyze the transmission of sound between two rooms coupled to each other by a common, thin homogeneous wall.[20]   System 1 is the ensemble of modes of the diffuse, reverberant sound field in the source room, resonant within the frequency band $\Delta\omega$.   System 2 is an appropriately chosen group of modes of the wall.   System 3 is the ensemble of modes of the diffuse reverberant sound field in the receiving room, resonant within the frequency band $\Delta\omega$.   A loudspeaker in the source room is the only source of power, and the power dissipated in each system is assumed to be large compared with the power lost to the other two systems through the coupling (see Fig. 11.15).

The procedure is as follows:

1. Relate W_1^{in} to the power lost by sound absorption in the room [see Eqs. (2.4) and (9.5)].   This yields $\langle p_1^2 \rangle$, the space-average mean-square sound pressure in the source room.

Loudspeaker
input power

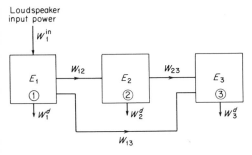

Fig. 11.15 Block diagram illustrating the power flow in a three-way coupled system. W_{13} represents the transmission of sound by those modes whose resonance frequency lies outside the source band. The "nonresonant" modes are important below the critical frequency.

2. Calculate $E_1 = \langle p_1^2 \rangle V_1 / \rho c^2$, where V_1 is the volume of the source room.

3. The reverberant sound power incident on the dividing wall of area S_2 is $W_{\text{inc}} = E_1 c S_2 / 4 V_1$.

4. The power balance of the wall for resonant modes within the bandwidth $\Delta \omega$ is next determined, that is, $W_{12} = W_2^d + W_{23}$, so that

$$W_{12} = [\text{Eq. (11.44)}] + 0.5 \, [\text{Eq. (11.48)}]$$

This sum equals Eq. (11.41). Because η_{12} of Eq. (11.41) is not well known, it is replaced by using Eq. (11.40) and the definition of loss factor η_{21}, which yields $\eta_{21} = \rho c \sigma_{\text{rad}} / \rho_s \omega$.

5. The vibrational energy of the wall is $E_2 = \langle v^2 \rangle \rho_s S_2$.

6. Combining steps 3, 4, and 5 yields the mean-square wall velocity $\langle v^2 \rangle$ as a function of mean-square source room pressure $\langle p_1^2 \rangle$.

7. The power radiated into the receiving room is

$$W_{23} = \rho c S_2 \sigma_{\text{rad}} \langle v^2 \rangle$$

8. Finally, the transmission coefficient τ_r is found by dividing step 7 by step 3.

9. The *resonance transmission loss*, defined as $R_r = 10 \log (1/\tau_r)$, is computed from step 8, using Eq. (11.4) and assuming that $\rho_s \omega \eta_2 \gg 2 \rho c \sigma_{\text{rad}}$, to yield

$$R_r = 20 \log \left(\frac{\rho_s \omega}{2 \rho c} \right) + 10 \log \left(\frac{f}{f_c} \frac{2}{\pi} \frac{\eta_2}{\sigma_{\text{rad}}^2} \right) \quad \text{dB} \quad\quad (11.54)$$

The first term in Eq. (11.54) is approximately the normal-incidence mass law transmission loss R_0, so that Eq. (11.54) becomes

$$R_r = R_0 + 10 \log \left(\frac{f}{f_c} \frac{2}{\pi} \frac{\eta_2}{\sigma_{\text{rad}}^2} \right) \quad \text{dB} \quad\quad (11.55)$$

where f_c = critical frequency [see Eq. (11.4)], Hz

η_2 = total loss factor of the wall, dimensionless

σ_{rad} = radiation ratio for the wall, dimensionless

Thus we have obtained the transmission loss between two rooms separated by a common wall, using the SEA method.

Below the critical frequency and when the dimensions of the wall are large compared with the acoustical wavelength, the radiation factor σ_{rad} can be taken from Table 11.3 or from Fig. 11.13.

It is important to note that if the sound-transmission loss of an equivalent *infinite* wall is compared with the data measured and predicted by the SEA method it is found that *above the critical frequency,* the transmission loss R for the infinite wall yields the same results as Eq. (11.55), which takes into account only the resonance transmission of a finite wall.

Below the critical frequency the sound-transmission loss of a finite panel is more controlled by the contribution of those modes that have their resonance frequencies outside of the frequency band of the excitation signal than by those with resonance frequencies within that band. [See the discussion following Eq. (11.15).] Since only the contributions of the latter are included in the previous SEA calculation, Eq. (11.55) usually overestimates the sound-transmission loss of a finite panel below the critical frequency. Figure 11.16 shows that below the critical frequency for a $\frac{1}{8}$-in.-thick aluminum panel the sound-transmission loss of the resonant modes alone (curve a) is approximately 10 dB higher than that measured on the actual panel (curve d).

A composite transmission factor which approximately takes into account both the forced and resonance waves is closely approximated by

$$\frac{1}{\tau} = \frac{W_{\text{inc}}}{W_{\text{forced}} + W_{\text{res}}} = \frac{\dfrac{\langle p^2 \rangle}{4\rho c} S_2}{\langle p^2 \rangle \dfrac{\pi \rho c S_2}{\rho_s^2 \omega^2} + \langle p^2 \rangle \dfrac{\sqrt{12}\pi c^3 \rho \sigma_{\text{rad}}^2 S_2}{2\omega^3 \rho_s^2 c_L t \eta_2}} \qquad (11.56)$$

At low frequencies where the first term in the denominator dominates, the transmission factor becomes

$$\frac{1}{\tau} \approx \frac{1}{\pi} \left(\frac{\rho_s \omega}{2\rho c} \right)^2 \qquad (11.57)$$

and the sound-transmission loss is [see Eq. (11.24)]

$$R = 10 \log \left(\frac{1}{\tau} \right) \approx R_0 - 5 = R_{\text{field}} \qquad \text{dB} \qquad (11.58)$$

At high frequencies where the second term in the denominator becomes dominant, the sound-transmission loss is given by Eq. (11.55).

11.3.5 Alternative Method for Calculating Sound-transmission Loss
An alternate technique, useful in preliminary design, is illustrated in Fig. 11.16. In essence, it considers the loss factor of the material to be determined completely by the material selection and substitutes a "plateau" or horizontal line for the peak and valley of the forced wave analysis in the region of the critical frequency.[3, 23] Its use will be demonstrated in Example 11.6.

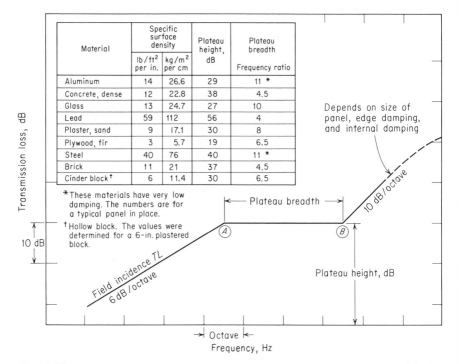

Material	Specific surface density		Plateau height, dB	Plateau breadth
	lb/ft² per in.	kg/m² per cm		Frequency ratio
Aluminum	14	26.6	29	11 *
Concrete, dense	12	22.8	38	4.5
Glass	13	24.7	27	10
Lead	59	112	56	4
Plaster, sand	9	17.1	30	8
Plywood, fir	3	5.7	19	6.5
Steel	40	76	40	11 *
Brick	11	21	37	4.5
Cinder block†	6	11.4	30	6.5

*These materials have very low damping. The numbers are for a typical panel in place.

†Hollow block. The values were determined for a 6-in. plastered block.

Fig. 11.16 Approximate design chart for estimating the sound-transmission loss of single panels. The chart assumes a reverberant sound field on the source side, and approximates the behavior around the critical frequency with a horizontal line or plateau. The part of the curve to the left of A is determined from the field-incidence mass law curve (Fig. 11.5). The plateau height and length of the line from A to B is determined from the table. The part above B is an extrapolation. This chart is fairly accurate for large panels. Length and width of the panel should be at least 20 times the panel thickness.[3]

11.3.6 Properties of Common Materials

To complete the usefulness of these design techniques, some constants and typical physical properties of a number of common materials are given in Table 11.4.[3] Conversion factors are given in Appendix B.

Examples 11.6–11.9 Calculate the transmission loss of a $\frac{1}{8}$-in.-thick, 5×6.5 ft aluminum panel by the alternate (plateau) method.

SOLUTION: From Table 11.4, the product of surface mass and critical frequency is $\rho_s f_c = 34{,}700$ Hz-kg/m². The $\frac{1}{8}$-in.-thick aluminum plate has a surface density of 8.5 kg/m². From Fig. 11.5 we find that the normal-incidence transmission loss at the critical frequency is $R_0(f_c) = 48.5$ dB, and that the field-incidence transmission loss at 1,000 Hz is $R_F(8{,}500$ Hz-kg/m²$) = 31.5$ dB.

Example 11.6 The procedure by the plateau method is as follows:[3, 23]

1. Using semilog paper (with coordinates dB vs. log frequency), plot the

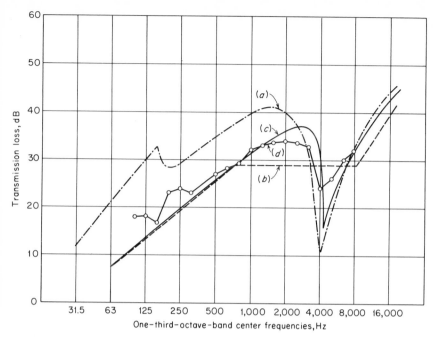

Fig. 11.17 Comparison of experimental and theoretical transmission loss of a 5 ft × 6.5 ft × 1/8 in. aluminum panel. The theoretical calculations are based on (a) resonance mode calculation, (b) plateau calculation, and (c) forced wave calculation. Curve (d) shows the experimental results. (*After Price and Crocker.*[20])

field-incidence mass law transmission loss as a line with a 6-dB/octave slope through the point 31.5 dB at 1,000 Hz.

2. From Fig. 11.16, the plateau height for aluminum is 29 dB. Plotting the plateau gives the intercept of the plateau with the field-incidence mass law curve at approximately 750 Hz.

3. From Fig. 11.16, the plateau width is a frequency ratio of 11. The upper frequency limit for the plateau is, therefore, 11 × 750 = 8,250 Hz.

4. From the point 29 dB, 8,250 Hz, draw line sloping upward at 10 dB/ octave. This completes the plateau-method estimate as shown in Fig. 11.17 (curve *b*).

Example 11.7 Calculate the radiation ratio for the resonant modes using the data of the previous example. The perimeter P of the panel is 23 ft and the area S is 33 ft^2. The wavelength at the critical frequency is $\lambda_c = c/f_c = 1,130/4,000 = 0.28$ ft.

SOLUTION: From the abscissa of Fig. 11.13 compute the six frequencies and the two noted values of σ_{rad} in decibels.

$$f_1 = \frac{c^2}{2Sf_c}\left(\frac{P^2}{8S}-1\right) = \frac{(1,130)^2}{2\times33\times4,000}\left(\frac{(23)^2}{8\times33}-1\right) = 4.8 \text{ Hz}$$

$$10 \log\left[\sigma_{\mathrm{rad}}(f_1)\right] = 10 \log_{10}\frac{\lambda_c^2}{S} = 10 \log_{10}\frac{(0.23)^2}{33} = -26 \text{ dB}$$

TABLE 11.4 Speed of Longitudinal Waves, Density, Internal Damping Factors, and Products of Surface Density and Critical Frequency for Common Building Materials

Material	c_L, m/sec	ρ_m, lb/ft³	ρ_m, kg/m³	Product of surface density and critical frequency			Normal-incidence transmission loss at critical frequency $R_0(f_c)$, dB	Internal damping factor at 1,000 Hz η *
				wf_c, Hz-lb/ft²	$M_s f_c$, Hz-slug/ft²	$M_s f_c$, Hz-kg/m²		
Aluminum	5,150	170	2,700	7,000	217	34,700	48.5	10^{-4}–10^{-2} †
Brick	......	120–140	1,900–2,300	7,000–12,000	217–373	34,700–58,600	48.5–53	0.01
Concrete, dense poured	3,400	150	2,300	9,000	279	43,000	50.5	0.005–0.02
Concrete (clinker) slab plastered on both sides, 2 in. thick	......	100	1,500	10,000	310	48,800	51.5	0.005–0.02
Masonry block:								
Hollow cinder (nominal 6 in. thick)	......	50	750	4,750	148	23,200	45.0	0.005–0.02
Hollow cinder, ⅝ in. sand plaster each side (nominal 6 in. thick)	......	60	900	5,220	162	25,500	46.0	0.005–0.02
Hollow dense concrete (nominal 6 in. thick)	......	70	1,100	4,720	147	23,000	45.0	0.007–0.02
Hollow dense concrete, sand-filled voids (nominal 6 in. thick)	......	108	1,700	8,650	269	42,200	50.0	Varies with frequency
Solid dense concrete (nominal 4 in. thick)	......	110	1,700	11,100	345	54,100	52.5	0.012

Fir timber	3,800	40	550	1,000	31	4,880	31.5	0.04
Glass	5,200	156	2,500	7,800	242	38,000	49.5	0.001–0.01†
Lead:								
Chemical or tellurium	1,200	700	11,000	124,000 (approx.)	3,850	605,000	73.5	0.015
Antimonial (hard)	1,200	700	11,000	104,000	3,240	508,000	71.5	0.002
Plaster, solid, on metal or gypsum lath	......	108	1,700	5,000	165	24,500	45.5	0.005–0.01
Plexiglas or Lucite	1,800	70	1,150	7,250	225	35,400	49.0	0.002
Steel	5,050	480	7,700	20,000	621	97,500	57.5	10^{-4}–10^{-2}†
Gypsum board (⅛ to 2 in.)	6,800	43	650	4,500	140	20,000	45.0	0.01–0.03
Plywood (¼ to 1¼ in.)	......	40	600	2,600	81	12,700	40.0	0.01–0.04
Wood waste materials bonded with plastic, 5 lb/ft²	......	48	750	15,000	466	73,200	55.0	0.005–0.01

* The range in values of η are based on limited data. The lower values are typical for the material alone while the higher values are the maximum observed on panels in place.

† The loss factor for structures of these materials is very sensitive to construction techniques and edge conditions.

$$f_2 = 100 \frac{\lambda_c}{P} \frac{c}{P} = \frac{100 \times 28 \times 1{,}130}{23^2} = 60 \text{ Hz}$$

$$f_3 = \frac{3c}{P} = \frac{3 \times 1{,}130}{23} = 150 \text{ Hz}$$

$$f_4 = \frac{f_c}{4} = 1{,}000 \text{ Hz}$$

$$f_5 = f_c = 4{,}000 \text{ Hz}$$

$$10 \log [\sigma_{\text{rad}}(f_5)] = 10 \log \sqrt{\frac{P}{2\lambda_c}} = 8 \text{ dB}$$

$$f_6 = 2f_c = 8{,}000 \text{ Hz}$$

Because f_1 is off the frequency scale of Fig. 11.18, plot $10 \log \sigma_{\text{rad}}$ beginning at 16 Hz with $10 \log \sigma_{\text{rad}} = -26$ dB, and extend a line to $f_3 = 150$ Hz. Returning to $f_2 = 60$ Hz, extend a dashed line with a slope of 1.8 dB/octave (6 dB/decade) from $10 \log \sigma_{\text{rad}}$ (f_2). At $f_3 = 150$ Hz, draw a smooth curve connecting to the sloped line, and extend the line to $f_4 = 1{,}000$ Hz. At $f_5 = 4{,}000$ Hz, plot the peak value of $10 \log \sigma_{\text{rad}} = 8$ dB, and smoothly join this point with the curve of $10 \log \sigma_{\text{rad}}$ at $1{,}000$ Hz, and the value 0 dB at $f_6 = 8{,}000$ Hz.

Since σ_{rad} for a panel with clamped edges is desired, add 3 dB to the curve in the frequency range between f_c and f_1. Figure 11.18 shows the results of the computation of radiation ratio for the sample.

For comparison, the exact expressions for σ_{rad} from Table 11.3 are also plotted in this figure, together with the results of a measurement[20] of σ_{rad} for the panel under discussion.

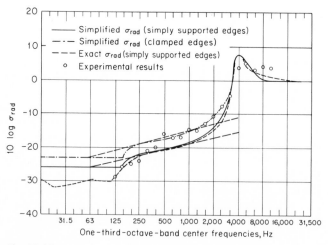

Fig. 11.18 Comparison of experimental[20] and theoretical radiation ratio (σ_{rad}) for a 5 ft $\times$ 6.5 ft $\times$ 1/8 in. aluminum panel. The theoretical calculations are discussed in Ex. 11.7.

Example 11.8   Estimate the transmission loss of resonant modes.   The resonance transmission loss is given by Eq. (11.55), or, rewritten

$$R_0(f_c) + 10 \log \left[\left(\frac{f}{f_c} \right)^3 \frac{2\eta_2}{\pi} \right] - 20 \log \sigma_{\text{rad}}$$

SOLUTION: The second term above will have a slope of 9 dB/octave (since it is proportional to f^3), and takes the value $10 \log [(2 \times 0.01)/\pi] = -22$ dB at $f = f_c$.

Sketch a line with a 9-dB/octave slope through $R_0(f_c)$. To this line add the quantity -22 dB $- 2 \times (10 \log \sigma_{\text{rad}})$. For example, at 125 Hz, $10 \log \sigma_{\text{rad}} = -23$ dB, so, $-22 - 2(-23 \text{ dB}) = 24$ dB. The curve with 9-dB/octave slope through $R_0(f_c)$ takes the value 6 dB at this point so that the total resonance transmission loss is $6 + 24 = 30$ dB. The complete curve is shown in Fig. 11.7.

Example 11.9   Estimate the field-incidence forced wave transmission loss. From Table 11.5 we have that $f_c = 34{,}700/8.5 \sim 4{,}000$ Hz and $R_0(f_c) = 48.5$ dB. From Fig. 11.7 the field-incidence transmission loss at any frequency is

$$R_f(f) = R_0(f_c) + [R_f(f) - R_0(f_c)]$$
$$= 48.5 + [R_f(f) - R_0(f_c)] \quad \text{dB}$$

At 500 Hz, for example, $f/f_c = \frac{1}{8}$ and $[R_f(f) - R_0(f_c)] = -23.5$ dB. $R_f(f) = 48.5 - 23.5 = 25$ dB. The solid curve c in Fig. 11.17 shows the result of this calculation.

From Fig. 11.18, which compares the σ_{rad}'s, we see that the SEA approach describes radiation from resonant modes with creditable accuracy. On the other hand, the resonant mode description does not accurately describe the experimentally observed sound-transmission loss except in the frequency region well above the critical frequency. The plateau method is reasonably accurate in predicting the observed transmission loss considering its intended use as a preliminary design aid. The forced wave prediction provides the greatest accuracy of all three techniques for this sample.

11.4   Sound Transmission through Complex Partitions

11.4.1   Composite Barrier (Window in Wall)   One form of a complex barrier consists of two or more elements with different transmission losses, such as a window or a door in a wall. The sound power transmitted by two or more elements with a common average incident intensity is

$$W_{\text{trans}} = \bar{I}_{\text{inc}} \Sigma \bar{\tau}_i S_i \quad \text{watts} \quad (11.59)$$

where $\bar{\tau}_i =$ average transmission coefficient, for the ith element (dimensionless)

$S_i =$ area of the ith element, m²

$\bar{I}_{\text{inc}} =$ average intensity incident on the element, watts/m²

Wave effects and structural interactions have been ignored.

The composite transmission loss for a two-element system is, therefore,

$$R_c = 10 \log \frac{\sum_i S_i}{\sum_i \tau_i S_i} = 10 \log_{10} \frac{S_1 + S_2}{\tau_1 S_1 + \tau_2 S_2} \quad \text{dB} \quad (11.60)$$

This equation is plotted in normalized form in Fig. 11.19.

11.4.2 Layered Media   Panels that are composed of two or more solid homogeneous layers are frequently used as partitions. If the layers are joined along their interface so that no slipping occurs (continuous stress at the interface) then the composite structure bends around a common neutral axis.

Two-ply Laminate.[13]   For a two-layer panel, the mass per unit area ρ_s is

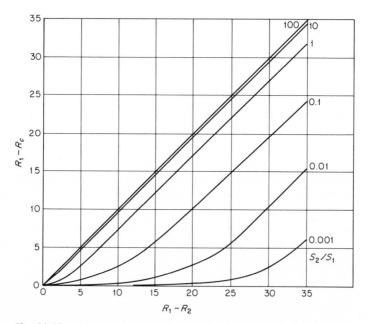

Fig. 11.19 Transmission loss of a two-element composite barrier as a function of the relative transmission loss of the components. R_1 is the larger transmission loss, R_2 is the smaller transmission loss, and R_c is the transmission loss of the composite. S_2/S_1 is the ratio of the areas of the respective components.

$\rho_s = \rho_1 t_1 + \rho_2 t_2$, where ρ_i = density of layer i, kg/m³, and t_i = thickness of layer i, m.

The approximate bending stiffness per unit width B of a two-layer plate is [22]

$$B \approx \frac{E_1 t_1}{12}\left[t_1^2 + \left(y - \frac{t_1}{2}\right)^2\right] + \frac{E_2 t_2}{12}\left[t_2^2 + \left(y - \frac{2t_1 + t_2}{2}\right)^2\right] \qquad \text{N-m} \quad (11.61)$$

where E_i = Young's modulus of the ith layer, N/m²

$\quad y$ = distance between the neutral axis of the composite bar, and the more distant face of layer 1, i.e., $y = [E_1 t_1 + E_2(2t_1 + t_2)]/2(E_1 + E_2)$, m

With the bending stiffness B and mass per unit area ρ_s, the critical frequency of the composite panel is $f_c = (c^2/2\pi)\sqrt{\rho_s/B}$, as given in Eq. (11.4).

The loss factor may be estimated or, in the case where one layer is very lossy, it may be calculated from Chap. 14. Figure 11.7 is used to calculate the transmission loss as discussed in Sec. 11.2.

If the added layer is relatively limp ($E_2 t_2 \ll E_1 t_1$), the mass per unit area of the composite can be increased without significantly altering the bending stiffness. This increases the "mass law" transmission loss, and simultaneously increases the critical frequency.

Three-ply Laminate. The three-layer panel usually exhibits shearing deformation in the middle layer. The reaction forces which are developed depend on the ratio of the shearing stress in the core to the extensional stress in the outer layers. This ratio, in turn, depends on the wavelength of the deformation of the panel. The dynamic bending stiffness is variable and dependent on the way in which the panel is forced.[24,25,26]

The effective bending stiffness for free bending wave propagation in a three-ply laminate is frequency dependent. At low frequencies the core acts as an ideal spacer which distributes the stress between the two outer layers and couples them together, producing maximum bending stiffness. At intermediate frequencies the bending stiffness decreases and is controlled primarily by the propagation of shear waves in the core. At high frequencies the bending stiffness is asymptotic to a constant value approximately equal to the sum of the bending stiffnesses of the two faces.

A sandwich panel with identical faces (e.g., $E_1 = E_3$, $t_1 = t_3$) is one form of a three-layer panel that is commonly encountered. For this case the low-frequency (coupled) bending stiffness per unit width is

$$B_c \approx \frac{E_1 t_1 (t_1 + t_2)^2}{2} \qquad \text{N-m} \qquad (11.62)$$

In the transitional region we obtain

$$B_T \approx \left(\frac{G_2}{2\pi f}\right)^2 \frac{\rho_s}{\rho_2^2} \quad \text{N-m} \tag{11.63}$$

where G_2 = shear modulus of the core, N/m²
ρ_s = mass per unit area of the panel, kg/m²
ρ_2 = density of the core, kg/m³
At high frequencies the bending stiffness per unit width is asymptotic to

$$B_1 + B_3 = 2B_1 = \frac{E_1 t_1^3}{6} \quad \text{N-m} \tag{11.64}$$

If the shear-wave speed in the core material is very much less than the speed of sound in air ($c_s < 0.1c$), then the critical frequency is given approximately by [26]

$$f_c' \approx \frac{c^2}{2\pi} \sqrt{\frac{6\rho_s}{E_1 t_1^3}} \quad \text{Hz} \tag{11.65}$$

High critical frequencies can be achieved in sandwich panels with solid rubber or high-density foam-plastic cores (where $c_s \ll c$).

When the shear-wave velocity is greater than about one-tenth the velocity of sound in air ($c_s > 0.1c$), a lower limit for the critical frequency is given by

$$f_c'' = \frac{c^2}{2\pi} \left[\frac{2\rho_s}{E_1 t_1 (t_1 + t_2)^2}\right]^{1/2} \quad \text{Hz} \tag{11.66}$$

Above this frequency (for $c_s > 0.5c$) the transmission-loss–vs.–frequency curve does not follow the same pattern as that for a panel with constant bending stiffness. The transmission loss may be predicted accurately by using a wavenumber-dependent bending stiffness in Eq. (11.16) and using numerical integration as for the homogeneous panel.[24]

We have assumed that the core of the panel may be treated as incompressible, which is valid for most solid materials, such as rubber, when constrained between two stiff layers. However, for porous cores, such as honeycombs or foams, the core is compressible and relative transverse motion of the panel faces occurs, leading to *dilational resonances.*[28] The first dilational resonance in which the faces act as simple masses connected by a springlike core is given by [13]

$$f_d = \frac{1}{2\pi} \left[\frac{4E_{c_2}}{t_2(2\rho_1 t_1 + \rho_2 t_2/3)}\right]^{1/2} \quad \text{Hz} \tag{11.67}$$

where E_{c_2} = effective Young's modulus in compression for the core (for large, thin panels, equal to the bulk modulus), N/m²
ρ_1 = density of face, kg/m³

Near and above this frequency, the transmission loss will be considerably reduced.

11.4.3 Orthotropic Panels   In orthotropic panels either the Young's modulus E or the moment of inertia I is not constant in all directions in the panel. Many reinforced materials are good examples of the former, while a corrugated panel is an example of the latter.

The major deviation in transmission loss for large panels is observed near and above the lower critical frequency (e.g., the critical frequency corresponding to the stiffest direction in the panel). Furthermore, if the ratio of the dynamic bending stiffness in the stiffest and limpest directions is less than about 1.4, the effects will be small, and the transmission loss will be closely approximated by using an average bending stiffness per unit width in the forced wave theory. If the ratio is greater than 1.4, the transmission-loss curve is bounded by transmission-loss curves corresponding to the range of allowed critical frequencies, as indicated in Fig. 11.20, and usually is in the region between the mean of the envelope and the lower bound. The lower and upper critical frequencies are determined by

$$f_{cl} = \frac{c^2}{2\pi} \left(\frac{\rho_s}{B_x}\right)^{1/2}$$

(11.68)

$$f_{cu} = \frac{c^2}{2\pi} \left(\frac{\rho_s}{B_y}\right)^{1/2}$$

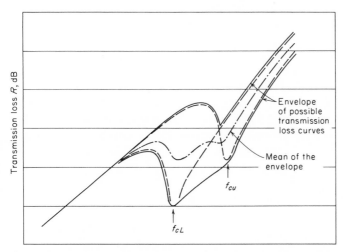

Fig. 11.20 Envelope of possible transmission loss of an orthotropic panel. The ratio of critical frequencies is given by $f_{cL}/f_{cu} = \sqrt{B_y/B_x}$ where B_y is the highest bending stiffness per unit width in the panel and B_x is the lowest. The actual panel transmission loss should lie between the mean of the envelope and the lower bound.

where B_x, B_y = highest and lowest bending stiffness per unit width for different directions in the panel, respectively, N-m

The effect of orthotropicity is to smooth out the coincidence dip, by reducing the depth of the dip and broadening it.

11.4.4   Ribbed Panels   The effects of ribbing on the dynamics of panel motion are to alter the static bending stiffness, or the average mass per unit area of the composite panel, or both.[14]   The ribs will act to subdivide the panel into subpanels.   The first resonance of a subpanel (f_{11}) can be evaluated by assuming that the ribs provide simply supported edge conditions for the subpanels using Eq. (11.29) with $m = n = 1$, and with B and ρ_s as the subpanel properties.   Well below the first resonance of the largest subpanel, no significant wave motion can occur in these subpanels and the entire panel behaves approximately as an orthotropic plate.[29]   For the purpose of computing its transmission loss in this frequency region, the panel acts with an average mass per unit area ρ_s' given by

$$\rho_s' = \rho_1 t_1 + \frac{\rho_r L}{S} \qquad \text{kg/m}^2 \qquad (11.69)$$

where ρ_s' = average mass per unit area of the panel, kg/m²
   $\rho_1 t_1$ = mass per unit area of the face material, kg/m²
   $\rho_r L$ = total mass of the ribs, kg
   S = total area of the panel, m²
The bending stiffness per unit width equals the static bending stiffness of the panel, which usually differs in two perpendicular directions.   Thus orthotropic results will apply.   For most panels, the first subpanel resonance lies below the lower critical frequency f_{cl} of the orthotropic plate approximation, so that the transmission loss below f_{11} is adequately predicted by the field-incidence mass law using ρ_s' from Eq. (11.69).

At and above the first subpanel resonance f_{11}, very stiff or massive ribs form definite edge conditions for the subpanels.   This effectively increases the perimeter of the panel, which increases the radiation ratio σ_{rad}[14] in the frequency region of edge mode control.   The effective perimeter is

$$P' = 2(l + w + L) \qquad \text{m} \qquad (11.70)$$

where l, w = length and width of the panel
   L = total length of the applied ribs
The modal density is approximately equal to the sum of the modal densities of the panel face and the ribs.[14]   The total increase in σ_{rad} is frequently sufficient that the resonance transmission loss R_r will be important for low-order resonances of the subpanels.   On the other hand, at frequencies of high-order subpanel resonances, the forced wave trans-

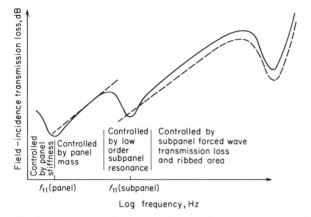

Field –incidence transmission loss, dB

Controlled by panel stiffness

Controlled by panel mass

Controlled by low order subpanel resonance

Controlled by subpanel forced wave transmission loss and ribbed area

f_{11}(panel) f_{11}(subpanel)

Log frequency, Hz

Fig. 11.21   Schematic transmission loss for a ribbed panel. The first subpanel resonance is f_{11}. The upper dashed curve (left) represents the field-incidence mass law of the composite panel. The lower dashed curve (right) represents the forced wave transmission loss of the subpanels. The observed transmission loss in the upper frequency range is increased because of the reduced radiating area attributable to the rib coverage.

mission of the subpanels will dominate the results. If the ribs occupy a significant area of the panel, they reduce the effective radiating area for the subpanels which increases the effective transmission loss of the subpanels. Figure 11.21 provides a schematic illustration of this discussion.

11.4.5 Double Walls   A double wall is a barrier composed of two separate panels separated by an airspace. There are two major paths by which sound energy is transmitted through a double wall: *the first* involves radiation from the first panel into the airspace, where it excites the second panel, and this panel then radiates the energy into the receiving room; *the second* involves structureborne transmission of vibrational energy from the first panel to the second panel through mechanical links between the panels, and the second panel radiates the transmitted energy. This latter path is called the *structureborne flanking path*.

Because of these factors, a comprehensive theory of the sound-transmission loss through double walls is the subject of current research. Our intent in this section is to provide a qualitative introductory description of the phenomenon, based in part on Refs. 3, 12, 55–58.

High-frequency Transmission Loss.   For two panels that are coupled only by an airspace with a depth greater than a half wavelength of sound in air, the transmission loss at high frequencies is approximately[3]

$$R_{\text{field}} = R_{1\text{field}} + R_{2\text{field}} - 10 \log_{10} \left(\frac{1}{4} + \frac{S_w}{S_2 \bar{\alpha}_2} \right) \qquad (11.71)$$

where $R_{1\text{field}}$ = field-incidence transmission loss of the first panel, dB
$R_{2\text{field}}$ = field-incidence transmission loss of panel 2, dB
S_w = panel area (one side), m^2
$S_2\bar{\alpha}_2$ = absorption in the cavity between the panels, metric sabins; that is, S_2 is in m^2

Effect of Absorption in the Cavity.   Equation (11.71) is restricted by the assumptions of no mechanical coupling between the faces, and to those frequency regions where the sound field in the cavity may be considered to be diffuse.   To satisfy this latter assumption, the cavity must preferably be 1 or more wavelengths deep and have little absorption. We see that if the "absorption" in the cavity is only that due to transmission through the panels, then the transmission loss of the total wall is at best 3 dB higher than the transmission loss of the better panel. This oversimplified theory also suggests that if the effective area of absorption in the cavity is approximately equal to the area of the panel, then the transmission loss may approach the sum of the transmission losses of the component panels.

For a practical double wall in this high-frequency range, a relatively small amount of absorption (such as provided by a fibrous blanket at the perimeter) will ensure that the transmission by the acoustic path is less than that by the flanking path, except perhaps near the critical frequency.[57]   The absorption attenuates the sound waves propagating in the cavity parallel to the panel faces and so reduces the excitation of the second panel.   This is especially important because most of the energy is transmitted by sound waves of nearly grazing incidence.

Effect of Wave Coincidence.   In the high-frequency region, the least transmission loss for a double wall with identical panels occurs in the frequency range near the critical frequency of the panel.   If there is no absorptive material in the cavity, the transmission loss near the critical frequency can be estimated by assuming that it is somewhat less than twice the forced wave transmission loss of a single panel.   In this frequency range the addition of absorption between the panels dramatically improves the transmission loss, perhaps to the point where the coincidence dip is only apparent as a plateau in the transmission-loss curve. For a double wall composed of panels with considerably different critical frequencies and for no absorption in the cavity, the dip due to the wave coincidence will be widened and made less pronounced in comparison with a wall with identical faces.

Low- and Mid-frequency Transmission Loss.   At very low frequencies, where the cavity is much less than a wavelength in depth, the contained air behaves as a soft spring which couples the two panels together.   Deflection of one panel will compress the air and displace the second panel to some extent.   If the motion of the individual panels is mass controlled, then it is reasonable to assume that the wall obeys the mass law, where the mass is the total mass of the wall.   This estimate is reasonably valid for very light double walls with a thin cavity, at low frequencies.   It also predicts a second phenomenon, namely, the *double-wall resonance*.   The double-wall resonance is due to the panels acting as two masses connected by a spring (air).   The frequency of resonance for identical panels is [3]

$$f_0 = \frac{1}{2\pi} \sqrt{\frac{2\rho c^2}{d\rho_s}} \qquad \text{Hz} \qquad (11.72)$$

where $d =$ cavity depth, m

   $\rho_s =$ mass per unit area of one panel, kg/m²

   Panel stiffness has been ignored in Eq. (11.72), which is valid for a double wall where f_0 is well above the first panel resonance.

   For most double walls, with moderately heavy panels and typical cavity depths, the double-wall resonance is below the frequency range of interest.

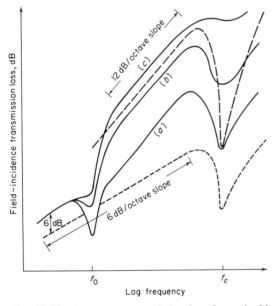

Fig. 11.22   Schematic transmission loss for a double wall with identical panels, and no flanking transmission, indicating the controlling factors.   The lower (dashed curve) is the field-incidence transmission loss of a single panel, while the upper dashed curve is twice this (in decibels).   The double-wall resonance frequency is f_0 and the critical frequency of the panel is f_c.   Curve a is for no added absorptive material in the cavity, and curve b is for some.   Curve c is for a cavity uniformly filled with absorptive material.

Above the double-wall resonance, the transmission-loss curve rises with a slope of about 18 dB/octave, until it is limited by either the acoustic path or by the structureborne flanking path. Figure 11.22 illustrates the range of transmission loss which may be achieved for double walls with no ties between the panels.

Effect of Flanking Transmission.[55] In general, a structureborne flanking path can be approximated as a point or line source to the second (radiating) panel of the wall. This forcing excites free bending waves in the panel which radiate sound in a manner already described in Secs. 11.2 and 11.3. The result is that the transmission-loss curve usually lies a fixed number of decibels above that for a single panel. This curve will have a slope of about 6 dB/octave, rather than 12 dB/octave. For panels of similar weight the limits of the transmission-loss curve with mechanical flanking paths are between 6 dB above the single-panel curve, when flanking is complete, up to the smaller of (1) the maximum achievable by the acoustic path or (2) about 30 dB improvement when the flanking transmission is reduced to a practical minimum.

11.4.6 Improvement in Sound-transmission Loss ΔR Obtained by Adding a "Resilient Skin" to a Heavy Partition

The sound-transmission loss of a single homogeneous wall increases by about 6 dB for each doubling of its mass per unit area. Beyond a certain point, however, it becomes uneconomical to increase the transmission loss by further increasing the mass. A more practical way is to add a thin resilient plate, separated by an airspace, to the original, considerably heavier, partition.

The added plate may be attached to the heavy wall either at individual points (e.g., with resilient clips) or along lines (e.g., rigidly attached to studs which are intrinsically stiff). In order to prevent the buildup of resonances in the airspace between the wall and the skin and also to extract energy from the near field of the thin plate, the airspace is usually filled with a sound-absorbing material. Figure 11.23 shows the three most frequently used constructions.

The vibrational energy from the wall can reach the resilient skin either through compression of the air in the airspace or through the attachment points.

Assuming that the vibration of the basic wall is not appreciably affected by the added thin plate, the vibration velocity of the resilient skin at points removed from any attachment point is controlled by the forces arising from the compression of the fluid in the airspace, yielding

$$\langle v_2^2 \rangle \approx \langle v_1^2 \rangle \left(\frac{f_0}{f}\right)^4 \quad \text{for } f > f_0 \qquad (11.73)$$

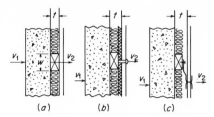

Fig. 11.23 Various attachments of a resilient skin to a massive wall. (a) Rigid line attachment (tightly nailed to studs); (b) rigid point attachment (skin not touching studs); (c) flexible point attachment (resilient clips).

where $\langle v_1^2 \rangle$ = space-average mean-square velocity of the basic wall, $(\mathrm{m/sec})^2$

$\langle v_2^2 \rangle$ = space-average mean-square velocity of the skin at locations far away from attachment points, $(\mathrm{m/sec})^2$

f_0 = resonance frequency, determined by the resiliency of the fluid in the air gap and the surface mass density of the skin given by Fig. 11.24 in both metric and English units

The improvement in sound-transmission loss for point attachment [31] of the resilient skin is *

$$\Delta R_{\text{point}} = -10 \log \left[\beta^2 \frac{8}{\pi^3} n'' \frac{c^2}{f_c^2} + \left(\frac{f_0}{f} \right)^4 \right] + 10 \log \sigma_{\text{rad}_1} \quad \text{dB} \quad (11.74)$$

and for line attachment *

$$\Delta R_{\text{line}} = -10 \log \left[0.64 n' \frac{c}{f_c} + \left(\frac{f_0}{f} \right)^4 \right] + 10 \log \sigma_{\text{rad}_1} \quad \text{dB} \quad (11.75)$$

where β = vibration-isolation factor of the flexible support ($\beta = 1$ for rigid support)

n'' = number of attachment points per unit area, m^{-2}

n' = number of studs per unit length, m^{-1}

f_c = critical frequency of the resilient skin given in Eq. (11.4) and in Fig. 11.25

f_0 = resonance frequency given in Eq. (11.72) and plotted in Fig. 11.24

σ_{rad_1} = radiation ratio of the heavy wall

Graphical Method to Estimate ΔR. A graphical method for estimating the incremental sound-transmission loss ΔR or, alternatively, to determine other constructional details such as the depth of airspace, surface mass and critical frequency of the skin material, stud-to-stud spacing or the permissible number of attachment points, is given in Figs. 11.24 to 11.27.

The characteristic shape of the curve of ΔR vs. frequency is shown in Fig. 11.24. Starting at f_0, ΔR increases with a slope of 40 dB/decade in

* Equations (11.74) and (11.75) are valid only in the frequency region $f < f_c$.

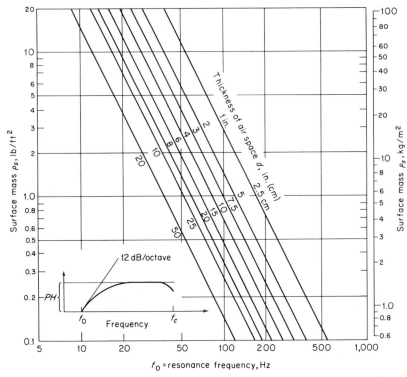

f_0 = resonance frequency, Hz

Fig. 11.24 Resonance frequency f_0 (for the case of Fig. 11.23) as function of the surface mass of the added resilient skin with the thickness of the airspace as parameter plotted for adiabatic compression ($\kappa = 1.4$). Multiply f_0 by 0.85 if the airspace is filled with glass fiber.

frequency (12 dB/octave) until it levels off at the plateau height (PH). It then is constant until the critical frequency f_c of the skin where ΔR decreases.

The resonance frequency f_0 is found from Fig. 11.24 as a function of the surface density of the skin material and the thickness of the airspace. The critical frequency f_c is found from Fig. 11.25. The values of PH for rigid point and line attachment are given in Figs. 11.26 and 11.27, respectively.

Example 11.10, Added Skin Calculate the curve of ΔR vs. frequency for a 0.5-in. gypsum board skin with rigid point attachment ($\beta = 1$) to 2×2 in. studs as shown in Fig. 11.23*b* spaced 24 in. on-center with a 15-in. nail spacing along the stud ($n'' = 0.4/\text{ft}^2$).

SOLUTION: From Fig. 11.25 the critical frequency is $f_c = 5,000$ Hz. Entering Fig. 11.26 at $n'' = 0.4/\text{ft}^2$ and $f_c = 2,500$ yields the plateau height, PH = 17 dB.

Finally, entering Fig. 11.24 at $w_{s_2} = 2$ lb/ft^2 and $d = 2$ in. yields the resonance frequency $f_0 = 90$ Hz.

The ΔR-vs.-frequency curve for this configuration will then slope upward at 12 dB/octave from the point $\Delta R = 0$ at 80 Hz and will level off at 17 dB.

If the skin were tightly attached to the stud (line attachment), the plateau height would be obtained from Fig. 11.27; the values $f_c = 2,500$ Hz and $1/n' = 2$ ft yield PH $= 8.5$ dB.

Example 11.11, Remedial Measures   Choose a suitable construction to achieve an improvement of $\Delta R = 15$ dB at 400 Hz.   Calculate the required stud-to-stud spacing if a 0.5-in. gypsum board, tightly nailed to 2 × 2 in. wood studs, is chosen as the resilient panel.

SOLUTION: Using the values for f_0, w_{s_2}, and f_c from the previous example, and entering Fig. 11.27 at $f_c = 2,500$ Hz and $\Delta R = 10$ dB, yields the required stud-to-stud spacing, $1/n' = 3$ ft, a value which is too great to be practical.   Instead choose *rigid point* attachment.   From Fig. 11.26 the number n'' of attachment points is 1.8 per square foot.   Thus, assuming a 24-in. stud-to-stud

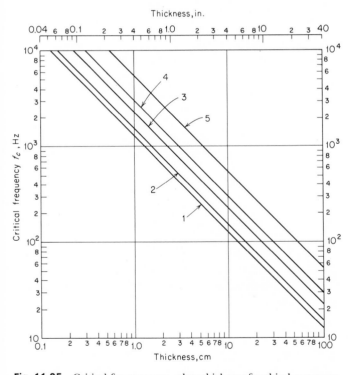

Fig. 11.25   Critical frequency vs. plate thickness for thin homogeneous, isotropic plates.   (1) aluminum or steel; (2) glass * or sand plaster; (3) dense concrete, plywood,* or brick; * (4) gypsum board; (5) lead, foamed lightweight concrete * (* indicates that f_c may vary substantially for different samples).

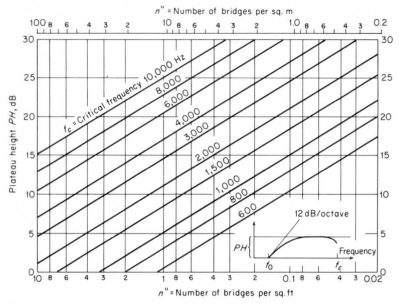

Fig. 11.26   Plateau height PH for rigid point attachment as a function of the number of bridges per unit area with the panel's critical frequency as parameter.

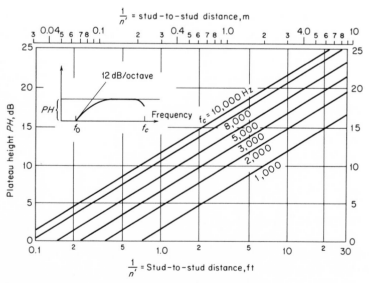

Fig. 11.27   Plateau height PH for rigid line attachment as a function of stud-to-stud spacing with the panel's critical frequency as parameter.

spacing, the distance between attachment points along the stud must be 14 in. or more.

11.4.7 Floating Floors

When either the sound-transmission loss or impact noise isolation (see Sec. 11.6 for impact noise isolation) or both of a structural floor has to be increased, a second, usually thinner, resiliently supported slab is floated on top of the structural slab, either on point or strip resilient mounts or on a continuous resilient, porous layer. A load-bearing resilient layer must also be porous and have sufficient flow resistance to prevent the propagation of sound in the airspace horizontally. Suitable materials are properly chosen glass-fiber and mineral-wool boards.

If the dead load of the floor is variable it is often advantageous to use a large number of appropriately placed resilient unit mounts such as rubber, cork, or precompressed glass fiber. The airspace between the unit mounts is filled with non-load-bearing flow-resistive material, usually mineral-wool or glass-fiber blanket, to eliminate sound propagation in the cavity parallel to the floated slabs. This filling may improve impact noise isolation by up to 20 dB.

Because of the resilient nature of the supports, the coupling of the vibration field of the floating slab (which is directly exposed to the sound field or to the striking hammer of the tapping machine) to the structural slab depends on frequency. The basic resonance frequency f_0 is approximately

$$f_0 \approx \frac{1}{2\pi} \sqrt{\frac{s}{\rho_{s_1}}} \qquad (11.76)$$

where ρ_{s_1} = mass per unit surface area of the floating slab, kg/m^2
 s = stiffness per unit area of the resilient support including the stiffness of trapped air, N/m

Below f_0 the two slabs are closely coupled and move in phase. The improvement in sound-transmission loss is small since the floating floor adds only relatively small mass to the structural floor.

In the vicinity of the resonance frequency f_0 the composite floor often performs less favorably than the structural floor alone (see Fig. 11.35), and the transmission loss depends primarily on the loss factor of the resilient mounts.

Above the resonance frequency f_0 the decoupling of the slabs increases with frequency.

Excitation by a Diffuse Sound Field. A diffuse sound field in the source room forces the floating slab to move. Because the forcing is due to the incident sound wave the trace velocity of the pressure wave on the slab is always acoustically fast or "supersonic" [Eq. (11.15)]. This acoustically fast forced wave is partially transmitted to the structural slab through the resilient supports which, in turn, forces the structural slab into acoustically fast motion. The acoustically fast forced wave of the structural slab then radiates sound efficiently ($\sigma_{\rm rad} = 1$) to the receiving room. Due to the finite size of the slabs the forced waves, when encountering boundaries or discontinuities, create a reverberant, free-bending-wave field in both slabs.

Locally Reacting Floating Slab.   When the forced wave encounters a boundary or discontinuity, free bending waves are created. If the floating slab is highly damped, the free bending waves are quickly attenuated so that the motion of the floating slab is controlled by the forced waves. In this case the acoustically fast forced waves in the floating slab are locally transmitted through the resilient mounts. Because of the resiliency of the mounts and the inertia of the floating slab, the pressure acting on the structural slab is smaller than the pressure acting on the floating slab by the factor $(f/f_0)^2$ for $f > f_0$. Accordingly, the improvement in sound-transmission loss over the single structural slab is

$$R = 20 \log \frac{p_1}{p_2} \approx 40 \log \frac{f}{f_0} \qquad \text{for } f > f_0 \qquad (11.77)$$

It has been shown that Eq. (11.77) is valid also for an undamped but infinitely large homogeneous floating floor where the forced wave never encounters boundaries.[43]

Resonantly Reacting Floating Slab.   In practice, free bending waves created at the boundaries travel back and forth in the floating slab many times at a propagation speed that depends on frequency and the physical parameters of the floating slab. Below the critical frequency the free bending waves are acoustically slow; above they are acoustically fast. The free bending waves form a diffuse reverberant vibration field on the floating slab which usually dominates the motion of the slab. The intensity of the reverberant vibration field is inversely proportional to the losses of the floating slab (internal losses and power lost to the structural slab). The structural slab responds to both types of forcing and a forced wave is produced in it which has the same wavenumber as the exciting wave. In addition, the forced waves at points of discontinuity excite free bending waves in the structural slab.

The amplitude of the diffuse reverberant field of the free bending waves in the structural slab is usually much larger than the amplitude of the forced waves. If the excitation frequency is above the critical frequency of the structural slab, both the forced and free bending waves radiate with the same efficiency; namely, $\sigma_{\rm rad} = 1$. Below the critical frequency the radiation ratio of the acoustically fast forced waves remains unity while that of the acoustically slow free bending waves decreases with decreasing frequency.

A calculation of the improvement in sound-transmission loss which takes into account only the contribution of the resonant nature of the floating slab has been made using statistical energy analysis, and is given here.

Assuming (1) no sound propagation in the airspace parallel to the slabs, (2) no correlation of the slab motion at different mounts, (3) light coupling of the two slabs above the resonance frequency so that the motion of slab 1, except the power loss through the mounts, is not substantially affected by the presence of the second slab, (4) real, frequency-independent point input admittances for both slabs, (5) both slabs of the same material, and (6) unity radiation ratios, the resonant sound-transmission loss is [30]

$$\Delta R_{\text{res}} \approx 10 \log \left(\frac{2.3}{2\pi} c_{L_1} t_1 \eta_1 n' \frac{f^3}{f_0^4} \right) \tag{11.78}$$

where c_{L_1} = propagation speed of longitudinal waves in the floating slab material, m/sec

t_1 = thickness of the floating slab, m

η_1 = dissipative loss factor of the floating slab

n' = number of mounts per unit slab area, m^{-2}

f_0 = basic resonance frequency of the system

If the loss factor η_1 is independent of frequency, then the ΔR_{res}-vs.-frequency curve follows a 30-dB/decade slope. The improvement depends only on the physical parameters of the floating slab and the mounts. Obviously, the floating slab should have as high a loss factor as practicable, which is the reason why the improvement for an asphalt-type floating slab is much greater than that for a lightly damped concrete slab of the same surface density.

Approximation of the Improvement in Sound-transmission Loss ΔR. The ΔR-vs.-frequency curve above the basic resonance frequency f_0 increases with a slope of 40 dB/decade [see Eq. (11.77)] if the floating slab is highly damped and if the acoustically fast transmitted forced waves control the sound radiation. In those cases where the free resonant bending waves control the sound transmission, Eq. (11.78) applies, which yields a ΔR-vs.-frequency curve that asymptotically approaches a 30-dB/decade slope.

11.4.8 On Flanking Transmission   The improvement in sound-transmission loss achieved by adding a second resiliently connected partition to an existing partition is frequently limited by flanking transmission through structural elements common to the source and receiving

room. Of course, a partition can also be flanked by gaps around its perimeter or cracks and such openings as back-to-back mounted bathroom cabinets, electrical outlet boxes, etc. Accordingly, the measured sound-transmission loss of effective composite structures such as double walls, floating floors, hung ceilings, etc., is often only a few decibels higher than that of the single partition alone. Flanking transmission paths must be eliminated or reduced in effectiveness to a minimum.

Figure 11.1 illustrates the direct path and two principal paths of structureborne flanking between the upper source room and the lower receiving room if they are separated by a floating floor. *Path 1* is the direct sound transmission through the floor. *Path 2* is the flanking transmission through the walls that enclose the upper and lower rooms. This flanking path is often responsible for the limitation on sound-isolation performance of a sound barrier. *Path 3* is the flanking transmission through the walls of the source room into the structural floor slab, which in turn radiates sound into the adjacent rooms.

The bending wave transmission (in dB) through a junction of single homogeneous partitions (path 2) is approximately equal to 12 dB plus 20 log of the ratio of the surface density of the flanked partition (e.g., structural floor slab) to the surface density of the flanking partitions. Using this relation, the flanking transmission loss can be calculated from

$$R_f = 10 \log \frac{W_{kd}^{\text{inc}}}{W_{ng}^{\text{rad}}} = R_{ng} + 10 \log \frac{S_{kd}}{S_{ng}} + 20 \log \frac{\rho_s^{kd}}{\rho_s^{nd}} + 12 \qquad \text{dB}$$

where W_{kd}^{inc} = the sound power incident on the flanked partition from the sound field in the source room, watts

W_{ng}^{rad} = the sound power reaching the receiving room through flanking path 2, watts

R_{ng} = the sound transmission loss of the flanking partition, dB

S_{kd} = the surface area of the flanked partition (the floor separating the source room and the room below), m²

S_{ng} = the surface area of all flanking partitions in the room below (four walls in this case), m²

ρ_s^{kd} = mass per unit area of the flanked partition, kg/m²

ρ_s^{ng} = mass per unit area of the flanking partitions, kg/m²

Because the addition of a resilient skin on a partition or a floating slab on a floor decreases the transmission of power through the direct path (path 1), the indirect paths (e.g., paths 2 and 3) usually remain unaffected. Accordingly, the sound-transmission loss is usually controlled by the flanking paths. In order to illustrate the severity of flanking, we have calculated the improvement in sound-transmission loss ΔR achieved by adding a 10-cm (4-in.) floating floor on the top of a 20-cm (10-in.)

structural slab in a typical mechanical equipment room where the four side walls of 10-cm (4-in.) dense concrete are common to the upper source room and the lower receiver room. Because of the flanking through the side walls (path 2), the addition of the 10-cm resiliently supported floating floor results in not more than a few decibels increase in actual transmission loss. However, the floating slab will significantly reduce the transmission of impact noise. We should also note that the vibration-isolation performance of a floating floor is usually not subject to flanking since the floating slab generally does not have rigid connections to the walls.[33]

11.4.9   Sound Transmission through Curved Structures [34-36]   Curving a flat panel increases its ability to sustain transverse loads. The curvature alters the bending stiffness of the panel, which, in turn, suggests that the dynamic response of the panel is also altered. Consider the dynamics of a uniform tube, or pipe. If a static transverse force is applied, the deformation is predicted by treating the pipe as a beam with a particular bending stiffness determined by the radius and wall thickness. When the stresses in the pipe wall are analyzed, we find that the force is sustained in much the same way as in a hollow beam. For a thin-walled pipe, there will be nearly uniform compressive stresses through the wall thickness on the side of the pipe where the force is applied, and nearly uniform tensile stresses through the wall thickness on the opposite side of the pipe.

If the force is now changed to an oscillating point force at some constant frequency, the stresses will vary in time. The local stress in the pipe wall near the excitation point will be essentially compressive when the force is pushing into the pipe, but will be primarily tensile when the force is pulling the pipe. The stresses which occur on the opposite side of the pipe cannot be so precisely determined. If the frequency of excitation is very low then the picture at any instant will closely resemble the static case, where the stress was opposite in sense on the opposite side of the pipe. However, the "information" concerning whether the force is pulling or pushing travels from one side of the pipe to the other at a finite speed.

The stress wave travels with the propagation speed of longitudinal waves c_L in the pipe material, which is given approximately by Table 11.1, where here we shall use the symbol ρ_p as the density of the pipe material, kg/m^3. For a particular frequency of excitation, the stress wave will arrive at the opposite side of the pipe out of phase with the driving force. At this frequency the traveling wave will arrive in phase with the driving force at the point of application, due to a delay of exactly one period. The frequency at which this occurs is called the *ring frequency*, given by

$$f_r = \frac{c_L}{2\pi R} \qquad\qquad (11.79)$$

where f_r = ring frequency, Hz

c_L = longitudinal wave velocity, m/sec

R = radius of the pipe, m

As a rule of thumb the ring frequency for most structural pipe material (i.e., steel, aluminum, glass) is given by $f_r \approx 2.9 \times 10^4/R_{in.}$, where $R_{in.}$ is the pipe radius in inches. At and above this frequency, the pipe can clearly no longer be treated as a simple beam. In fact, for an exciting force of frequency greater than f_r the pipe responds locally as if it were a flat plate with a thickness equal to the pipe wall thickness [34] and our previous analyses for flat panels apply.

Below the ring frequency, wave motion occurs in the pipe wall, but this wave is more complex than a bending wave. One type of wave which can occur is a helical wave, which has a wavefront that progresses in a spiral along the pipe wall.[35] These waves travel at speeds much higher than the bending-wave speed in a flat panel at this frequency and frequently much higher than the speed of sound in air, so that they may radiate sound in a coincident manner. Due to their complex propagation path, however, they are not well excited by external sound fields. Thus for an external sound field it is difficult to assess the effect of helical waves on the transmission of sound through the pipe walls.

The interior geometry of the pipe also influences the transmission loss. For example, when the wavelength of sound in the fluid in the pipe is larger than the pipe diameter, only a plane sound wave moving along the pipe axis is permitted.* Because a diffuse sound field will not occur inside the pipe in this frequency range the sound transmission is not easily evaluated. In fact, the meaning of the term *transmission loss* in this case is no longer completely clear, since presumably no sound power may be incident on the pipe wall. Thus the transmission loss for acoustic power flowing from inside the pipe to outside is different from that for outside to inside, because of the difference in the exciting sound fields.

Figure 11.28 shows the measured and calculated radiation ratio for a particular finite cylinder. The marked increase in radiation at the ring frequency is apparent. For this particular cylinder, helical wave propagation was not expected to occur at frequencies below 630 Hz because of the short length of the pipe. The sharp increase in radiation ratio occurring at this frequency results from the presence of acoustically fast helical waves above 630 Hz. The more complex problem of predicting the excitation of curved structures by sound fields is a problem of current research.

* See also Sec. 8.7.

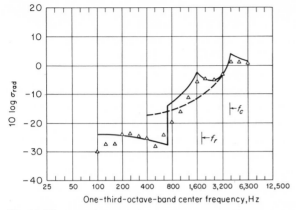

Fig. 11.28   Experimental ($\triangle$) and theoretical (solid line) radiation ratio[36] (σ_{rad}) for a short cylinder.[36]   At 630 Hz the first resonance for helical waves occurs, causing a step change in σ_{rad}.   The ring frequency and critical frequency are identified by f_r and f_c.   Also shown is the theoretical radiation ratio (dashed line) for an equivalent flat plate.

11.5   Point Force Excitation of a Finite Plate

Practical structures involve plates of finite size.   Their vibrational behavior is a logical extension of that of infinite plates.   In an infinite plate the bending waves excited by a point force spread indefinitely.   In a finite plate the same radially spreading bending waves encounter the boundaries of the plate, where a part of the energy is reflected.   The reflection process builds a reverberant vibration field which may be considered as separate from the direct wave.   If the plate is lightly damped and the power loss across the plate boundaries is small, the vibration field of the plate, except in the immediate vicinity of the excitation point, is dominated by this reverberant field.   The vibration field has highly defined spatial characteristics dictated by the geometry of the plate.[37]

Let us imagine a sinusoidal point force acting on the plate at a point. If the frequency is swept continuously, the velocity at the point will go through a succession of maxima and minima whose frequencies and magnitudes depend on the plate geometry and the location of the source. The normal modes of vibration of the plate result from positive reinforcement of the traveling waves, which cause a definite pattern of vibration on the plate surface.   Because there are many possible modes of vibration, each with its normal frequency, and because the geometry is specific to a plate, one expects to observe some trend in the fluctuations of the velocity.   One characteristic is that the average frequency separation between peaks in velocity is constant and is given by the reciprocal of the modal density, $1/n(f)$ [see Eq. (11.35) ff.].

The mechanical power introduced into a thin plate by a point force of rms value F_{rms} is

$$W_{in} = F^2_{rms} \, \text{Re}[Y] \qquad \text{watts}$$

where Y is the point input admittance of the plate defined as the ratio of the plate velocity to the applied force at the excitation point. If the force and velocity are not in phase, Y is a complex number. The point input impedance $Z = 1/Y$ is given in Table 11.6. Fortunately, *both the space average (i.e., the force source moved around) and the frequency average (i.e., the frequency warbled) of the point input admittance of a finite plate are the same as those of an infinite plate.* Thus, either for excitation of the plate with a band of noise at a fixed point or for excitation of the plate by a large number of randomly spaced point forces over its surface, the average input admittance can be well approximated by the point admittance of the equivalent infinite plate which is [38]

$$Y_\infty = \langle Y \rangle_{x,y} = \langle Y \rangle_f = \frac{1}{2.3 \rho_p c_L t^2} \qquad \text{m-N/sec}$$

for thin, homogeneous, isotropic plates. For definition of symbols see Table 11.6. With the above value of the point input admittance the mechanical power introduced by the point force of rms amplitude F_{rms} is given by

$$W_{in} = \frac{F^2_{rms}}{2.3 \rho_p c_L t^2} \qquad \text{watts} \qquad (11.80)$$

The balance of power in the steady state requires that this power either be dissipated or radiated by the plate as sound into the surrounding medium. Let us consider the radiated power.

The total acoustic power radiated from one side of the finite plate, W_{rad}, is the sum of the power radiated by the near field at the excitation point and that radiated by the reverberant bending-wave field. If the power lost by radiation of sound to the surrounding medium is small compared with the power lost by dissipation in the plate and edge losses ($2\rho c \sigma_{rad} \ll \rho_s \omega \eta_p$), then

$$W_{rad} \approx \frac{1}{2\pi} \frac{F^2_{rms} \rho}{\rho_s^2 c} + \frac{F^2_{rms} \rho c \sigma_{rad}}{2.3 \rho_s^2 c_L t \omega \eta_p} \qquad \text{watts} \qquad (11.81)$$

where ρ, ρ_s, c, c_L and t are defined in Table 11.5

F_{rms} = rms value of the point force, N

σ_{rad} = radiation ratio of the plate (one side), dimensionless

η_p = composite loss factor of the plate, dimensionless

If we wish to reduce the sound power radiated by the plate, it is of prime importance to know whether the radiation is controlled by the near field or by the reverberant bending-wave field. Let us examine the relative importance of these two terms on the right side of Eq. (11.81).

The frequency f_g above which the radiation from the plate is predominantly from the reverberant field is

$$f_g = 0.78 \frac{\sigma_{\text{rad}}}{\eta_p} f_c \qquad \text{Hz} \qquad f_g \ll f_c \tag{11.82}$$

The radiation ratio σ_{rad} increases sharply when the frequency approaches the critical frequency, and there is almost always a frequency region $f > f_g$ where the radiated power is controlled by the resonant motion of the plate.

In the frequency region $f > f_g$ the radiated sound-power level is approximated by an expression taken from the last term of Eq. (11.81)

$$L_W = 10 \log \frac{W_{\text{rad}}}{W_{\text{ref}}} = 10 \log \left(F_{\text{rms}}^2 \frac{\rho c \sigma_{\text{rad}}}{14.4 \rho_p^2 c_L \eta_p t^3 f} \right)$$
$$+ \; 120 \qquad \text{dB } re \; 10^{-12} \text{ watt} \tag{11.83}$$

In the frequency region $f < f_g$ the total radiated power is nearly that radiated by the near field yielding

$$W_{\text{rad}} = \frac{1}{2\pi} \frac{F_{\text{rms}}^2 \rho}{\rho_p^2 c t^2} \tag{11.84}$$

or

$$L_W = 10 \log \left[F_{\text{rms}}^2 \rho / (2\pi \rho_p^2 t^2 c) \right] + 120 \qquad \text{dB } re \; 10^{-12} \text{ watt}$$

In this frequency region ($f < f_g$) *the radiated power cannot be reduced by increasing the internal damping.*

Note that for $f > f_g$ *the radiated power decreases with the third power of the plate thickness. Accordingly, each doubling of the plate thickness decreases the sound-power level by 9 dB.*

Equation (11.83) can be rearranged so that for concrete (see Table 11.4)

$$L_W = 20 \log F_{\text{rms}} + 10 \log \sigma_{\text{rad}} - 10 \log \eta_p + \theta$$

$$\text{where } \theta = \begin{cases} \theta_{\text{N}} = 32 - 30 \log t - 10 \log f, \text{ dB (with } t \text{ in meters and } F_{\text{rms}} \\ \qquad \text{in newtons)} \\ \theta_{\text{lb}} = 93 - 30 \log t - 10 \log f, \text{ dB (with } t \text{ in inches and } F_{\text{rms}} \\ \qquad \text{in lb}_f) \end{cases}$$

Because concrete slabs are so frequently encountered in design the quantities θ_{N} (with t in meters) and θ_{lb} (with t in inches) are plotted in Fig. 11.29 to facilitate computations.

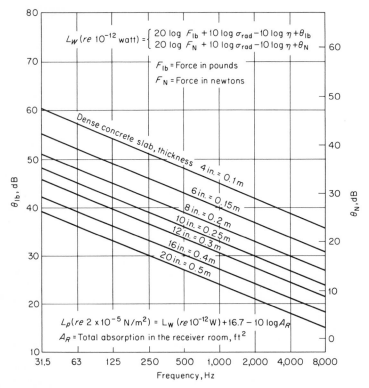

Fig. 11.29 Graph of θ_N or θ_{lb} needed to calculate the power level L_W of the sound radiated by dense concrete slabs for point force excitation of F_N or F_{lb} in the frequency region where the resonance waves control the radiation. If the panel radiates the sound into a room with a total sound absorption, $A_R = S\bar{\alpha}_{\text{Sab}}$ ft^2 [see Eqs. (9.1) and (9.2)], the sound pressure level in the room equals L_p, dB. [See Eq. (11.81) ff. for definition of symbols.]

11.6   Impact Noise

There are many practical cases where the excitation of a structure can be represented reasonably well by the periodic impact of a mass on its surface. Footfalls in dwellings, punch presses, and forge hammers fall into this category.

11.6.1   Standard Tapping Machine [52]

A standard tapping machine is used to rate the impact noise isolation of floors in dwellings. This machine consists of five hammers equally spaced along a line, the distance between the two end hammers being about 40 cm. The hammers successively impact on the surface of the floor to be tested at a rate of 10 times per second. Each hammer has a mass of 0.5 kg and falls with a velocity equivalent to a free-drop height of 4 cm. The area of the striking surface of the hammer is approximately 7 cm²; the striking surface is rounded as though it were part of a spherical surface of 50 cm radius.

The impact noise isolation capability of a floor is rated by placing the standard tapping machine on the floor to be tested and measuring the one-third-octave-band sound-pressure level L_p' averaged in space in the room below. The one-third-octave-band measurements then are corrected to correspond to octave bands by the addition of 5 dB to the one-third-octave-band level; namely, $L_p = L_p' + 5$ dB. The normalized impact sound level is, by definition,[52]

$$L_n \equiv L_p - 10 \log \left(\frac{A_0}{S\bar{\alpha}_{\text{Sab}}} \right) \qquad \text{dB } re \ 2 \times 10^{-5} \text{ N/m}^2 \qquad (11.85)$$

where L_p = equivalent octave-band sound-pressure level as measured, dB

$S\bar{\alpha}_{\text{Sab}}$ = total absorption in the receiving room (see Chap. 9), m²

$A_0 = 10$ m² is the reference value of absorption

The physical formulation of the problem of impact noise is that of the excitation of a plate by periodic force impulses. Such periodic forces can be represented by a Fourier series consisting of an infinite number of discrete-frequency components, each with amplitude F_n given by

$$F_n = \frac{2}{T_r} \int_0^{T_r} F(t) \cos \frac{2\pi n}{T_r} t \, dt \qquad \text{N} \qquad (11.86)$$

where $T_r = 1/f_r = 0.1$ sec = time interval between hammer strikes

$n = 1, 2, 3, \ldots$

Curve a in Fig. 11.30 shows the time function of the force $F(t)$, and curve b, the amplitude of its Fourier components.

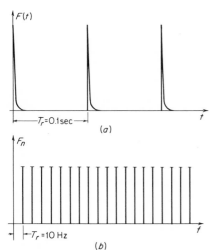

Fig. 11.30 Time function (a) and Fourier components (b) of the force that a standard tapping machine exerts on a massive rigid floor.

It is an experimental fact that when the hammer strikes a hard concrete slab the duration of the force impulse is small compared even with the period of the highest frequency of interest in impact testing. For less stiff structures, like wooden floors, this assumption is not valid and the exact shape of $F(t)$ has to be determined and used in Eq. (11.86). For a thick concrete slab the effective length of the force impulse is short enough so that $\cos[(2\pi n/T_r)t] \approx 1$, and all components have the same amplitude. Because the integral in Eq. (11.86) is the momentum of a single hammer blow equal to mv_0 (in kg-m/sec), the amplitudes of the Fourier components of the force, for a repetition frequency f_r, are

$$F_n = 2f_r mv_0 \qquad \text{N} \qquad (11.87)$$

The velocity of the hammer at the instant of impact is

$$v_0 = \sqrt{2gh} \qquad \text{m/sec} \qquad (11.88)$$

where h = falling height of the hammer, m

g = acceleration of gravity (9.8 m/sec^2)

Let us define a mean-square-force spectrum density S_{f_0} which when multiplied by the bandwidth will yield the value of the mean-square force in the same bandwidth

$$S_{f_0} = \frac{T_r}{2} F_n^2 = 4f_r m^2 gh \qquad \text{N}^2/\text{Hz} \qquad (11.89)$$

For the standard tapping machine the numerical value of S_{f_0} is 4 N^2/Hz.

Accordingly, the mean-square force in an octave band $\Delta f_{\text{oct}} = f/\sqrt{2}$ is

$$F_{\text{rms}}^2(\text{oct}) = \frac{4}{\sqrt{2}} f \qquad \text{N}^2 \qquad (11.90)$$

The octave-band sound-power level radiated by the impacted slab (which is assumed to be isotropic and homogeneous) into the room below is calculated by inserting Eq. (11.90) into Eq. (11.83) which yields

$$L_w(\text{oct}) \approx 10 \log_{10}\left(\frac{\rho c \sigma_{\text{rad}}}{5.1\rho_p^2 c_L \eta_p t^3}\right) + 120 \qquad \text{dB } re \ 10^{-12} \text{ watt} \qquad (11.91)$$

where ρ = density of air, kg/m^3

c = speed of sound in air, m/sec

σ_{rad} = radiation factor of the slab

ρ_p = density of the slab material, kg/m^3

c_L = propagation speed of longitudinal waves in the slab material, m/sec

η_p = composite loss factor of the slab

t = thickness of the slab, m

Note that *the sound-power level is independent of the center frequency of the octave,* that *doubling the slab thickness decreases the level of the noise radiated into the room below by 9 dB,* and that the *sound-power level decreases with increasing loss factor.*

11.6.2   Improvement of Impact Noise Isolation by an Elastic Surface Layer
Experience has shown that the impact noise level of even an 8- to 10-in. thick dense concrete slab is too high to be acceptable. A further increase of thickness to reduce impact noise is not economical.

Impact noise may be reduced effectively by an elastic surface layer, much softer than the surface of the slab, applied to the structural slab. The resilient layer changes the shape of the force pulse and the amount of mechanical power introduced into the slab by the impacting hammer, as shown in Fig. 11.31.

We would expect, if the elastic layer is linear and nondissipative, that the velocity will be at its maximum v_0 at the instant of impact $t = 0$. It will then decrease to zero and the mass will rebound to nearly the same velocity (it is assumed the hammer is not permitted to bounce a second time) according to the function shown by curve a of Fig. 11.31. The force function is shown by curve b.

The improvement in impact noise isolation achieved by the addition of the soft surface layer is defined in terms of the logarithmic ratio [53]

$$\Delta L_n = 20 \log \frac{F}{F'} = 20 \log \left(\frac{4/\pi}{\left| \dfrac{\sin x}{x} + \dfrac{\sin y}{y} \right|} \right) \qquad \text{dB} \qquad (11.92)$$

$$\text{where } x = \frac{\pi}{2} \left(1 - n \frac{f_r}{f_0} \right)$$

$$y = \frac{\pi}{2} \left(1 + n \frac{f_r}{f_0} \right) \qquad n = 1, 2, 3, \ldots \qquad (11.93)$$

$$f_0 = \frac{1}{2\pi} \sqrt{\frac{A_h}{m}} \sqrt{\frac{E}{t}} \qquad \text{Hz} \qquad (11.94)$$

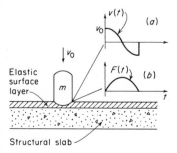

Fig. 11.31 Velocity and force pulse of a single hammer blow on an elastic surface layer over a rigid slab. (*a*) Velocity pulse; (*b*) force pulse.

F' and F = forces acting on the slab with and without the resilient sur-
    face layer, respectively, N
A_h = striking area of the hammer, m²
m = mass of the hammer, kg
E = dynamic Young's modulus of the elastic material, N/m
t = thickness of the layer, m

The characteristic frequency f_0 of an elastic surface layer, for the standard tapping machine, is plotted in Fig. 11.32 as a function of E/t.

Equation (11.92), which assumes no damping, is plotted in Fig. 11.33
as a function of the normalized frequency f/f_0.   Below $f/f_0 = 1$, the improvement is zero.   Above $f/f_0 = 1$ the improvement increases with an
asymptotic slope of 40 dB/decade.

Figures 11.32 and 11.33 (use the 40 dB/decade asymptote) permit one
to select an elastic surface layer to achieve a specified ΔL_n.

Example 11.12   The required improvement in impact noise isolation should
be 20 dB at 300 Hz.   Design a resilient covering for the concrete slab.
SOLUTION: From Fig. 11.33 we obtain $f/f_0 \approx 3$, which gives $f_0 = 100$ Hz.
Entering Fig. 11.32 with this value of f_0 yields $E/t = 2.8 \times 10^8$ N/m³ (or
$E/t \approx 1,000$ psi/in.).   Any material having this ratio of Young's modulus to

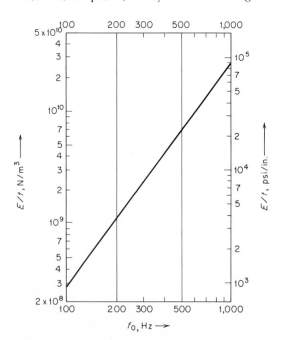

Fig. 11.32   Chart for the selection of an elastic surface layer where
f_0 = characteristic frequency, E = Young's modulus, and t = thickness of the layer.

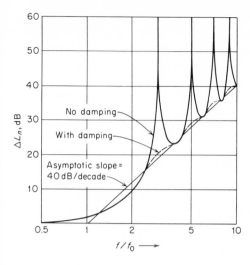

Fig. 11.33 Improvement in impact noise isolation ΔL_n vs. normalized frequency for a resilient surface layer (select f_0 to yield desired improvement).

thickness will provide the required improvement.   If we wish to select a 0.31-cm ($\frac{1}{8}$-in.) thick layer, the dynamic modulus of the material should be 8.7 × 10⁵ N/m² (8,000 psi).   Since the dynamic modulus of most elastic materials is about twice the statistically measured Young's modulus,[27,41] a material with $E \leqslant 4.35 \times 10^5$ N/m² (4,000 psi) should be selected.

Frequently used materials for the elastic surface layer are rubberlike materials, vinyl-cork tile, or carpet.   The impact isolation improvement curve (ΔL_n vs. frequency) has been measured and reported[32,42] for a large variety of elastic surface layer configurations.

The expected normalized impact sound level in the room below (see Fig. 11.1) for a composite floor (with a heavy structural slab) is that of the bare concrete structural floor minus the improvement caused by the elastic surface layer: i.e.,

$$L_{n_{\text{comp}}} = L_{n_{\text{bare}}} - \Delta L_n \qquad (11.95)$$

where $L_{n_{\text{bare}}}(\text{oct}) = 116 + 10 \log \left(\dfrac{\rho c \sigma_{\text{rad}}}{5.1 \rho_p^2 c_L \eta_p t^3} \right)$ dB re 2 × 10⁻⁵ N/m²

$$(11.96)$$

for homogeneous isotropic slabs.

Improvement through Floating Floors.   It is often more practical to use a floating floor above a structural slab than a soft resilient surface layer.   The advantages are that (1) both the impact noise isolation and the airborne-sound-transmission loss of the composite floor are improved, and (2) the walking surface is hard.   For analysis, floating floors can be categorized as either (1) locally reacting or (2) resonantly reacting, as defined below.

Locally Reacting Floating Floors.   A locally reacting floor is one where the impact force of the hammer on the upper slab (slab 1) is transmitted to the structural slab (slab 2) primarily in the immediate vicinity of the excitation point, and where there is no spatially homogeneous reverberant vibration field on slab 1.   In this case the bending waves in the floating slab are highly damped.   If the Fourier amplitude of the force acting on plate 1 is given by Eq. (11.87), the reduction in transmitted sound level is [43]

$$\Delta L_n = 20 \log \left[1 + \left(\frac{f}{f_0} \right)^2 \right] \approx 40 \log \frac{f}{f_0} \qquad (11.97)$$

where

$$f_0 = \frac{1}{2\pi} \sqrt{\frac{s'}{\rho_{s_1}}} \qquad (11.98)$$

and ρ_{s_1} = mass per unit area of the floating slab, kg/m²
   s' = dynamic stiffness per unit area of the resilient layer between slab 1 and slab 2 including the trapped air, N/m

Resonantly Reacting Floating Floors.   If the floating slab is thick, rigid, and lightly damped, the impact force of the hammers excites a more-or-less spatially homogeneous reverberant bending-wave field.

   The improvement in impact noise isolation at higher frequencies, where the power dissipated in slab 1 exceeds the power transmitted to slab 2, can be approximated by [30,53]

$$\Delta L_n \approx 10 \log \frac{2.3 \rho_{s_1}^2 \omega^3 \eta_1 c_{L_1} t_1}{n' s^2} \qquad (11.99)$$

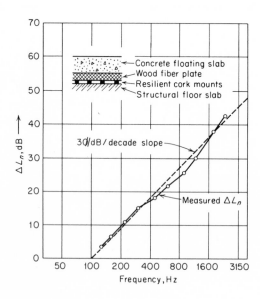

Fig. 11.34   Measured improvement in impact noise isolation ΔL_n for a resonantly reacting floating slab. (*Measured data by Gösele.*[47])

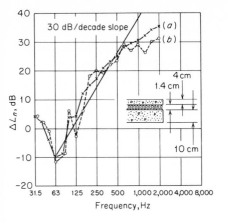

Fig. 11.35 Improvement in impact noise isolation ΔL_n for a resonantly reacting floating floor for excitation by (a) a standard tapping machine, and (b) high-heeled shoes. Note the negative ΔL_n in the vicinity of the resonance frequency. (*After Josse and Drouin.*[45])

where t_1 = thickness of the floating slab, m

c_{L_1} = propagation speed of longitudinal waves in the floating slab, m/sec

ρ_{s_1} = mass per unit area of the floating slab, kg/m^2

η_1 = loss factor of the floating slab

n' = number of resilient mounts per unit area of slab, 1/m^2

s = stiffness of a mount, N/m

Equation (11.99) indicates that, in contrast to the locally reacting case where ΔL_n increases at a rate of 40 dB for each decade increase in frequency, the increase is only 30 dB/decade if the loss factor of the floating slab η_1 is frequency independent. Another difference is the marked dependence of ΔL_n on this loss factor. The loss factor is determined both by the energy dissipated in the slab material itself and by the energy dissipated in the resilient mounts.

Figure 11.34 shows that the measured [47] improvement in impact noise isolation for a resonantly reacting floating slab follows the 30 dB/decade slope predicted by Eq. (11.99). Figure 11.35 shows the improvement for another floating floor system under impacting by a standard tapping machine and high-heel shoes respectively.[45] The negative improvement in the vicinity of the resonance frequency f_0 can be observed.

11.6.3 Impact Noise Isolation vs. Sound-transmission Loss

Whether a floor is excited by the hammers of a tapping machine or by an airborne-sound field in the source room, it will in both cases radiate sound into the receiving room. There is a close relation between the airborne-sound-transmission loss R and the normalized impact noise level L_n for a given floor.

In the case of acoustical excitation the sound power transmitted to the receiver room comprises the contribution of forced waves and of resonance waves. The forced waves usually dominate below the critical frequency of the slab and the resonance waves above. The sound power transmitted by exciting the slab by a standard tapping machine is made up of the contributions of the near-field component and of the reverberant component.

The relation between sound-transmission loss R and normalized impact noise level, assuming measurement in octave bands, is [54]

$$L_n + R = 84 + 10 \log \left[\frac{S_{f_0} f}{\sqrt{2}} \left(\frac{\dfrac{\rho}{2\pi \rho_s^2 c} + \dfrac{\rho c \sigma_{\text{rad}}}{2.3 \rho_s^2 c_L \omega \eta_p t}}{\dfrac{\pi \rho c}{\omega^2 \rho_s^2} + \dfrac{\pi \sqrt{12} c^3 \rho \sigma_{\text{rad}}^2}{2 \rho_s^2 c_L \omega^3 \eta_p t}} \right) \right] \quad \text{dB} \quad (11.100)$$

where ρ_s = mass per unit area of the slab, kg/m^2
c_L = propagation speed of longitudinal waves in the slab, m/sec
σ_{rad} = radiation factor of the slab
t = thickness of the slab, m
η_p = composite loss factor of the slab
S_{f_0} = mean-square-force spectrum density as given in Eq. (11.89), N^2/Hz

In the special case of a thick, lightly damped slab,

$$L_n + R = 43 + 30 \log f - 10 \log (\sigma_{\text{rad}}) - \Delta L_n \quad (11.101)$$

where ΔL_n represents the effect of the surface layer only. For a bare structural slab, by definition, $\Delta L_n = 0$.

Equation (11.101) states that the *sum of the airborne-sound-transmission loss and the normalized impact noise level is independent of the physical characteristics of the structural slab above the critical frequency of the slab where* $\sigma_{\text{rad}} \approx 1$.[46]

Below the coincidence frequency where the forced waves control the airborne-sound-transmission loss but the impact noise isolation is still controlled by the resonant vibration of the impacted slab Eq. (11.100) yields

$$R + L_n = 39.5 + 20 \log f - \Delta L_n - 10 \log \frac{\eta_p}{f_c \sigma_{\text{rad}}} \quad \text{dB} \quad (11.102)$$

where ΔL_n = effect of the surface layer only (zero for structural slab), dB
f_c = critical frequency of the structural slab, Hz

In this case, the sum $(R + L_n)$ dB depends on the physical characteristics of the slab and frequency.

These equations present powerful tools for determining the extent of flanking transmission in such situations where both R and L_n have been measured. *Any difference between the calculated and measured values of the sum $R + L_n$ suggests flanking.*

11.7 Power Transmission between Structural Elements

In the preceding sections, the power lost to a connected structure was usually considered only as an additional mechanism that increases the loss factor of the excited structure. In many practical problems, however, the power transmitted to a neighboring structure is the prime reason for a noise reduction program.

The power-balance equation states that the power introduced into the directly excited structure is either dissipated in it or is transmitted to neighboring structures. Accordingly, *if, in a noise reduction problem, the power is to be confined to the excited structure, the power dissipated in the structure must exceed the power transmitted to the neighboring structures. Required is a high loss factor of the excited structure and a construction that minimizes the power transmission to neighboring structures.* Methods to achieve high damping are the subject of Chap. 14.

11.7.1 Reduction of Power Transmission through a Change in Cross-sectional Area The simplest construction that causes a partial reflection of an incident compression or bending wave is a sudden change in cross-sectional area as shown schematically in Fig. 11.36. Assume a primary wave v_{inc} of perpendicular incidence arriving from the left. When it

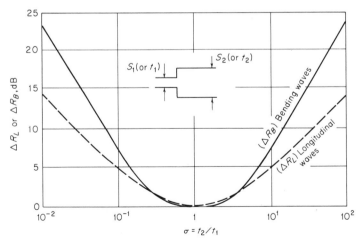

Fig. 11.36 Attenuation at discontinuity in cross section, as function of thickness ratio. (*After Cremer and Heckl.*[4])

reaches the location of the cross-section change, a part of the energy is reflected v_r, and a part is transmitted v_t.

Transmission of Compression Waves. For compression or longitudinal waves, the boundary conditions are the equality of the velocity and equality of the force on both sides of the junction; namely,

$$v_{\text{inc}} + v_r = v_t \quad \text{m/sec} \tag{11.103}$$

$$F_1 = F_2 \quad \text{N} \tag{11.104}$$

It can be shown that *the reflection loss ΔR_L, defined as the logarithmic ratio of the incident to the transmitted power* (for both sections of the same material), is [4,49]

$$\Delta R_L = 20 \log \left(\frac{\sigma^{1/2} + \sigma^{-1/2}}{2} \right) \quad \text{dB} \tag{11.105}$$

where $\sigma = S_2/S_1 = $ ratio of the cross-sectional areas (see Fig. 11.36).

The reflection loss as a function of cross-sectional-area ratio is plotted as the dashed line in Fig. 11.36. Since Eq. (11.105) is symmetrical for S_1 and S_2, the reflection loss is independent of the direction of the incident wave. This equation is also valid for plates where $\sigma = t_2/t_1$ is the ratio of the thicknesses.

We note that a $1:10$ change in cross-sectional area yields only 4.8-dB reflection loss. To achieve a 10-dB reflection loss, a 1:40 change in cross-sectional area would be necessary!

Transmission of Bending Waves. Since the energy in a bending wave is transmitted not only by forces but also by moments, the boundary conditions at the junction can be described by four equations requiring the equality of the transversal forces, the transversal velocities, the moments, and angular velocities on both sides of the junction. The reflection loss for bending waves of perpendicular incidence at low frequencies is independent of frequency and is given by [4,49]

$$\Delta R_B = 20 \log \left[\frac{\left(\sigma + \dfrac{1}{\sigma} \right)^{5/4} + \left(\sigma + \dfrac{1}{\sigma} \right)^{3/4}}{1 + \dfrac{1}{2} \left(\sigma + \dfrac{1}{\sigma} \right)^2 + \left(\sigma + \dfrac{1}{\sigma} \right)^{1/2}} \right] \quad \text{dB} \tag{11.106}$$

The equation is also plotted on Fig. 11.36 (solid line).

We conclude from Fig. 11.36 that a change in cross-sectional area is not a practical way to achieve high reflection loss in load-bearing structures.

11.7.2 Transmission of Free Bending Waves at an L Junction

Structural elements that necessitate a change in the direction of a bending wave play an important role in architectural structures. For example, when a free bending wave reaches an L junction, such as shown in Fig. 11.37, a part of the incident energy is transmitted and the rest is reflected. For low frequencies and for perpendicular incidence both the transmitted and reflected energy is predominantly in the bending waves. In this frequency range the reflection loss (logarithmic ratio of incident to transmitted power) for perpendicularly incident waves in plates and beams of the same material is given by [4,49]

$$\Delta R_{BB} = 20 \log \left[\frac{\left(\sigma + \dfrac{1}{\sigma} \right)^{5/4}}{\sqrt{2}} \right] \quad \text{dB} \qquad (11.107)$$

This equation is plotted in Fig. 11.37. Because ΔR_{BB} is symmetric in σ, the reflection factor does not depend upon whether the prime bending wave is incident from the thicker or from the thinner beam or plate. Note that the lowest reflection loss of 3 dB occurs for equal thicknesses ($\sigma = 1$). The reflection loss increases rapidly with increasing or decreasing ratio of thicknesses on both sides of this minimum. If the two plates or beams constituting the junction are of different material, replace the ratio $\sigma = t_2/t_1$ by

$$\sigma = \left(\frac{B_2 c_{B_1}}{B_1 c_{B_2}} \right)^{2/5} \qquad (11.108)$$

where B and $c_B =$ bending stiffness and propagation speed of free bending waves, respectively (see Table 11.1)

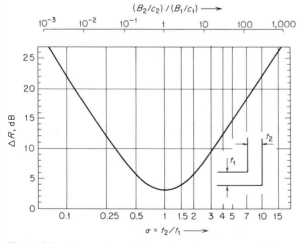

Fig. 11.37 Attenuation of flexural waves at corners (in absence of longitudinal wave interactions), as function of thickness ratio. (*After Cremer and Heckl.*[4])

At higher frequencies the incident bending wave also excites longitudinal waves in the second structure.[4, 49]

11.7.3   Transmission of Bending Waves through Cross and T Junctions

Other architectural structures that may provide a substantial reflection of an incident bending wave are the *cross junction* of walls as shown schematically in Fig. 11.38 and the *T junction* in Fig. 11.39. If a bending wave of perpendicular incidence reaches the cross junction from plate 1 it is partially reflected and partially transmitted to the other plates. The transmitted power splits into a number of different wave types, namely, bending waves in plate 3 and longitudinal and bending waves in plates 2 and 4. Because of the symmetry in the geometry, plates 2 and 4 will have the same excitation.[4]

The reflection loss (defined as the logarithmic ratio of the power in the incident to that in the transmitted bending wave) is given as a function of the ratio of the plate thicknesses for plates or beams of the same material and is shown in Figs. 11.38 and 11.39. When the plates are made of different material the ratio $\sigma = t_2/t_1$ is given by Eq. (11.108). The amplitudes of the bending waves transmitted without a change in direction are restrained by the perpendicular plate, and the reflection loss in this direction ΔR_{13} increases monotonically with increasing thickness of the restraining plate, t_2. Since this plate effectively stops the vertical motion of the horizontal plate at the junction, even for very thin vertical walls, ΔR_{13} remains level at 3 dB, indicating that only the power carried by the

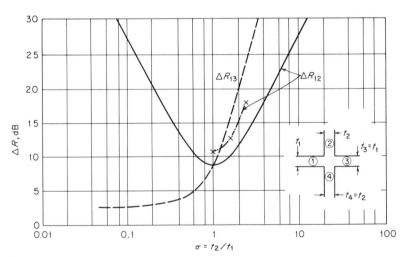

Fig. 11.38   Attenuation of flexural waves at plate intersections (in absence of longitudinal wave interactions), as function of thickness ratio. (*After Cremer and Heckl.*[4]) x— · —x is ΔR_{12} for random incidence computed by Kihlman.[50]

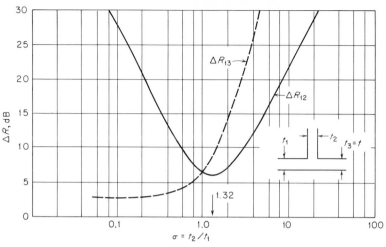

Fig. 11.39 Attenuation of flexural waves at plate junctures (in absence of longitudinal wave interactions), as function of thickness ratio. (*After Cremer and Heckl.*[4])

bending moment can pass the junction. For those bending waves that change direction at the junction, the reflection loss becomes a minimum ($\Delta R_{12} = 9$ dB) at a thickness ratio of $\sigma = t_2/t_1 = 1$ for the cross junction and 6.5 dB for the T junction at a thickness ratio of $\sigma = t_2/t_1 = 1.32$. The reflection loss then increases symmetrically for increasing or decreasing thickness ratio (t_2/t_1).

The transmission of free bending waves at cross junctions for random incidence has been computed and the reflection loss for a number of combinations of dense and lightweight concrete plates determined.[50] The results for ΔR_{12} are plotted in Fig. 11.38 (as x's) which indicate that ΔR_{12} for random incidence is somewhat higher than that for normal incidence. It was also found that for random incidence ΔR_{12} is independent of frequency, but ΔR_{13} decreases with increasing frequency.

11.7.4 Power Transmission from a Beam to a Plate

The structural parts of a modern building frequently include columns and structural floor slabs. Consequently, the power transmission from a beam to a plate, which models this situation, is of practical interest. Let us first consider the reflection loss for longitudinal and bending waves incident *from* the beam onto an infinite homogeneous plate.

Reflection Loss for Longitudinal Waves in a Beam. When a longitudinal wave in the beam reaches the plate, its energy is partly reflected back up the beam and is partly transmitted to the plate in the form of a bending wave. The reflection loss (equal to the logarithmic ratio of the incident to the transmitted power) is given by[51]

TABLE 11.5 Point Input Impedance of Some Simple Infinite Homogeneous Systems and Modal Density of Equivalent Finite Systems

System	Picture	Driving point impedance	Modal density of finite system, $n(\omega)$	Auxiliary expressions and notes
Thin bar in compression		$Z = S_b\sqrt{E\rho_b}$ $= \rho_b S_b c_L$	$\dfrac{l}{\pi}\sqrt{\dfrac{\rho_b}{E}} = \dfrac{l}{\pi c_L}$	$c_L = \sqrt{\dfrac{E}{\rho_b}}$
Semi-infinite Thin bar in bending		$Z = \tfrac{1}{2}\rho_b S_b c_B(1+j)$	$\dfrac{l}{2\pi\sqrt{\omega\kappa_b c_L}}$	$c_B = \sqrt[4]{\dfrac{EI\omega^2}{\rho_b(1-\sigma^2)S_b}}$
Infinite		$Z = 2\rho_b S_b c_B(1+j)$		$\kappa_b c_L = \sqrt{\dfrac{EI}{\rho_b S_b}}$
Infinite		$Z = 8\sqrt{B'\rho_p t} \approx 2.3\rho_p c_L t^2$		$B' = \dfrac{Et^3}{12(1-\sigma^2)}$
Thin plate			$\dfrac{S}{4\pi\kappa_p c_L} = \dfrac{\sqrt{12}\,S}{4\pi c_L t}$	
Semi-infinite		$Z = 3.5\sqrt{B'\rho_p t} \approx \rho_p c_L t^2$		$\kappa_p c_L = \sqrt{\dfrac{Et^2}{12\rho_p(1-\sigma^2)}}$
for $\omega > \omega_r$		$Z_\infty \approx 2.3\rho_c c_L t^2$	$\approx \dfrac{S\sqrt{12}}{4\rho c_L t}$	
Cylindrical shell		$R_e\left\{\dfrac{1}{Z_\infty}\right\} \approx \dfrac{\sqrt{12}}{8\rho_c c_L t^2}\left(\dfrac{\omega}{\omega_r}\right)^{2/3}$		$\omega_r = \dfrac{c_L}{\pi D}$
for $\omega < \omega_r$			$\approx \left(\dfrac{\omega}{\omega_r}\right)^{2/3}\dfrac{S\sqrt{12}}{4\pi c_L t}$	
Membrane			$\dfrac{S\omega}{2\pi c_m^2}$	$c_m = \sqrt{\dfrac{F'}{\rho t}}$
String		$Z = 2\sqrt{F\rho_l}$	$\dfrac{l}{\pi c_s}$	$c_s = \sqrt{\dfrac{F}{\rho_l}}$
Air volume in a large enclosure			$\dfrac{\omega^2 V}{2\pi^2 c^3}$	$V = $ room volume
Small monopole source		$Z = \rho c\left(\dfrac{k^2 r^2 + jkr}{1 + k^2 r^2}\right)$ (free-field value)		Radiated power $W = \dfrac{k^2\rho c U_{\text{rms}}}{4\pi}\beta$ for $kr \ll 1$ $\beta = 1$ for source in free field $\beta = 2$ for source at a wall, center $\beta = 4$ for source at an edge, center $\beta = 8$ for source at a corner

348

TABLE 11.5   (Continued)

Glossary of Symbols

B' = bending stiffness per unit width, N-m
c = propagation speed of sound in air, m/sec
c_B = propagation speed of free bending waves, m/sec
c_L = propagation speed of longitudinal waves in the solid material, m/sec
c_m = propagation speed of waves in a membrane, m/sec
c_s = propagation speeds of waves in a string, m/sec
D = cylinder diameter, m
E = Young's modulus, N/m²
t = thickness, m
I = centroidal moment of inertia of the cross section, m⁴
$j = \sqrt{-1}$
k = wavenumber in the fluid medium (air), m⁻¹
l = length, m
p = sound pressure, N/m²
r = radius of the small monopole source, m
S = surface area, m²
S_b = cross-sectional area of the beam, m²
F = tension force in the string or force in general, N
F' = tension force per unit length in the membrane, N/m
U_{rms} = rms volume velocity of the monopole source, m³/sec
V = volume of the room, m³
ρ_l = mass per unit length of string, kg/m
ρ = density of the fluid medium, kg/m³
ρ_b = density of the material of the bar, kg/m³
ρ_c = density of the material of the cylinder wall, kg/m³
ρ_s = mass per unit length of the string, kg/m
ρ_p = density of the plate material, kg/m³
κ_b = radius of gyration of the bar cross section, m
κ_p = radius of gyration of the plate cross section, m
σ = Poisson's ratio, dimensionless

Conversion Relationships

$n(f) = 2\pi n(\omega)$

$R_e \left\{ \dfrac{1}{Z_\infty} \right\} \approx n(\omega) \dfrac{\pi}{2M}$ sec/N-m

$n(f)$ = number of resonant modes per 1-Hz bandwidth centered on frequency f
$n(\omega)$ = number of resonant modes per unit radian bandwidth centered at angular frequency ω
M = mass of the finite system, kg
Z_∞ = driving point impedance of the equivalent infinite system, N-m/sec

$$\Delta R_L = -10 \log \left(1 - \left| \frac{Y_b - Y_p}{Y_b + Y_p} \right|^2 \right) \qquad \text{dB} \qquad (11.109)$$

where $Y_b = 1/Z_b$ = admittance of the beam for compressional waves, m/N-sec
$Y_p = 1/Z_p$ = point admittance of the infinite plate, m/N-sec
Both Y_p and Y_b are real and frequency independent and can be found for infinite beams and plates from Table 11.5 by taking the reciprocal of the impedances, i.e., $Y_b = 1/\rho_b c_{L_b} S_b$ and $Y_p = 1/2.3 \rho_p c_{L_p} t^2$.

Matching between Beam and Plate. Inspecting Eq. (11.109), we note that the reflection factor is zero (all the incident energy is transmitted to the plate) when $Y_b = Y_p$. Equating the values for Y_b and Y_p given above yields the requirements for matching.

$$S_b = 2.3 \frac{\rho_p c_{L_p} t^2}{\rho_b c_{L_b}} \quad \text{m}^2 \quad (11.110)$$

If the column and the slab are of the same material $\rho_p c_{L_p} = \rho_b c_{L_b}$, the matching requirements simplify to

$$S_b = 2.3 t^2 \quad \text{m}^2 \quad (11.111)$$

This equation says that for perfect power transfer from a beam of square cross section to a large plate, the cross dimension of the beam must be 1.52 times the thickness of the plate and for a beam of circular cross section, the radius must be 0.86 times the plate thickness. Actually, this is well within the range of slab thicknesses to column-cross-section ratios that are commonly found in architectural structures. Thus we may as a first conservative estimate assume that in architectural structures columns and slabs are approximately matched and all the power carried by the longitudinal waves in the column is transmitted to the slab. The reflection factor for different geometries of steel beam and plate

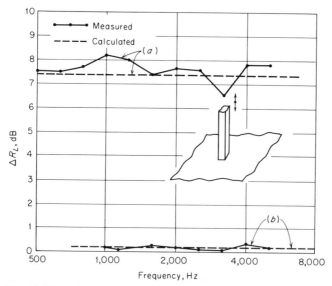

Fig. 11.40 Reflection loss for longitudinal waves ΔR_L for a steel beam plate system. (a) Plate thickness 2 mm, beam cross section 10 × 20 mm; (b) plate thickness 4 mm, beam cross section 5 × 10 mm. (*After Paul.*[51])

connections has been measured.[51]   The results (*a*) for a substantial mismatch and (*b*) for a near matching are plotted in Fig. 11.40.   The agreement between experimental and theoretical results is good.

Reflection Loss for Bending Waves in a Beam.   When a beam carries a
free bending wave, a part of the energy carried by the wave is transmitted to the plate by the effective bending moment and excites a radially
spreading free bending wave in the plate.   A part of the incident energy
is reflected from the junction.   Here the reflection loss is determined by
the respective moment impedances [4,51] of the plate and the beam

$$\Delta R_b = -10 \log \left(1 - \left| \frac{Y_b^M - Y_p^M}{Y_b^M + Y_p^M} \right|^2 \right) \quad \text{dB} \quad\quad (11.112)$$

where the moment admittances Y_b^M and Y_p^M are given by [4,51]

$$Y_b^M = \left(\frac{2}{1+j} \right) \frac{k_b^2}{\rho_b S_b c_{B_b}} \quad\quad (11.113)$$

and

$$Y_p^M = \frac{\omega}{16 B_p} \left[1 + j \frac{4}{\pi} \ln \left(\frac{1.1}{k_p a} \right) \right] \quad\quad (11.114)$$

where k = wavenumber, 1/m
  B_p = bending stiffness per unit width of the plate
  a = effective distance of the pair of point forces making up the
      moment on the plate, m.   For a rectangular and circular
      beam cross section a is respectively $a_r = d/3$ and $a_c = 0.59r$, m
  d = side dimension of the rectangular beam cross section (in the
      direction of bending), m
  r = radius of the circular beam cross section, m

Matching for Bending Waves.   Since the moment impedances of the
plate and beam are both frequency dependent, the reflection factor r_B
can vanish only at a single frequency.   The criteria for the vanishing reflection factor which indicates perfect matching is achieved when both
the real and imaginary parts of the moment impedances of the beam and
plate are equal, which requires that

$$\lambda_b = 0.39 \frac{B_b}{B_p} \quad \text{m} \quad\quad (11.115)$$

$$\lambda_p = 2.6a \quad \text{m} \quad\quad (11.116)$$

where B_b = bending stiffness of the beam
  λ_b = bending wavelength in the beam, m
  λ_p = bending wavelength in the plate, m

The reflection loss obtained [51] for the bending-wave excitation of a

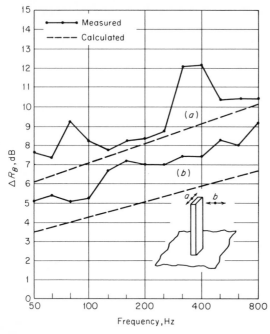

Fig. 11.41   Reflection loss for bending waves ΔR_B for a steel beam plate system for two different directions of bending of the beam (a and b).   Plate thickness 2mm.   Beam cross section 10×20 mm. (*After Paul.*[51])

steel rod of 1×2 cm cross-sectional area attached to a 0.2-cm-thick semi-infinite steel plate for two perpendicular directions of bending of the rod is plotted in Fig. 11.41.   The agreement between theory and experiment is satisfactory.

11.7.5   Reduction of Power Transmission between Plates by a Thin Resilient Layer   In architectural structures it is customary to provide a so-called vibration break by inserting a thin layer of resilient material between structural elements.   The geometry of such a vibration break is shown in Fig. 11.42.   Often this construction serves also as an expansion joint.

Reflection Loss for Compression Waves.   Let us assume in Fig. 11.42 a primary wave of perpendicular incidence arriving from the left.   At the junction a part of the energy is reflected and a part is transmitted to the right side structure as a compressional wave.   The reflection loss is given by [4,49]

$$\Delta R_L = 10 \log \left[1 + \left(\frac{\omega Z_1}{2 s_i} \right)^2 \right] \quad \text{dB} \quad (11.117)$$

where $Z_1 =$ impedance of the solid structure for compressional waves, N-sec/m

$\quad s_i =$ stiffness of the resilient layer in compression, N/m

Above a certain frequency ($\omega = 2s_i/Z_1$) the ΔR_L-vs.-frequency curve increases with a 20-dB/decade slope with increasing frequency. Below this frequency the resilient layer transmits the incident wave almost entirely. As an example [49] the ΔR_L-vs.-frequency curve for a 3-cm-thick layer of cork inserted between two 10-cm-thick concrete slabs (or columns) is shown in Fig. 11.42.

In order to achieve a high reflection loss, the resilient layer must be as soft as is permitted by the load-bearing requirements ($s_i/\omega \ll Z_1$). However, the stiffness of the layer cannot be reduced indefinitely by increasing the thickness. For frequencies where the thickness of the layer is comparable with the wavelength of compressional waves in the resilient material, the layer can no longer be considered as a simple spring characterized by its stiffness alone. The reflection loss in this frequency region is given by [4,49]

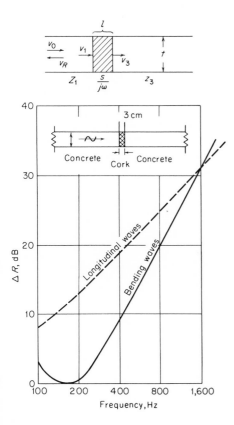

Fig. 11.42 Attenuation due to an elastic interlayer, as function of frequency. (*After Cremer and Heckl.*[4])

$$\Delta R_L = 10 \log \left[\frac{1}{\cos^2 k_i l + \frac{1}{4} \left(\frac{Z}{Z_i} + \frac{Z_i}{Z} \right)^2 \sin^2 k_i l} \right] \quad \text{dB} \quad (11.118)$$

where k_i = wavenumber for compressional waves in the resilient material, $1/m$
$Z = Z_1 = Z_3$ = impedance of the equivalent infinite structure for compression waves, N-sec/m
Z_i = impedance of the equivalent infinite (length) resilient material for compression waves, N-sec/m
l = length of the resilient layer, m

As expected, $\Delta R_L = 0$ for $Z_i = Z$. The maximum of ΔR_L coincides with the minimum of the denominator which is reached when $k_i l = (2n + 1)$ $(\pi/2)$ where the reflection loss becomes

$$\Delta R_{L_{max}} = 20 \log \left(\frac{2ZZ_i}{Z_i^2 + Z^2} \right) \quad \text{for } l = (2n + 1) \frac{\lambda_i}{2} \quad (11.119)$$

For $k_i l = n\pi$ ($l = n\lambda_i/2$), the denominator of Eq. (11.119) becomes unity independent of the magnitude of Z, yielding

$$\Delta R_{L_{min}} = 0 \quad \text{for } l = \frac{n\lambda_i}{2} \quad (11.120)$$

Finally for $k_i l \ll 1$ and $Z_i \ll Z$, Eq. (11.118) simplifies to Eq. (11.117).

Reflection Loss for Bending Waves. The geometry in Fig. 11.42 suggests that for bending waves of perpendicular incidence the moments and forces acting on both sides of the junction must be equal. However, the transverse velocity and angular velocity on both sides of the junction are different because of shear and compressional deformation of the resilient layer. Owing to the more complicated nature of the bending waves, the calculation of the reflection factor is much more complicated than that for compression waves. The reflection factor for this configuration has also been calculated.[49] It is found that the elastic layer behaves quite differently for bending waves than it does for compressional waves. The most striking difference is the *complete transmission of the incident bending wave at a certain frequency and a complete reflection of it at another, higher frequency.* Unfortunately, it turns out that the frequency of complete transmission for architectural structures of interest usually occurs in the audio frequency range. As an example, Fig. 11.42 shows the reflection factor for bending waves as a function of frequency for a layer of 3-cm-thick cork between 10-cm-thick concrete slabs. The transmission is complete at a frequency of 170 Hz and then decreases with increasing frequency. The reflection loss for bending waves can be approximated by[4,49]

$$\Delta R_B = 10 \log \left[1 + \left(\frac{1 - \dfrac{E}{E_i} \dfrac{2\pi^2 l t^2}{\lambda_{B_1}}}{2} \right) \right] \quad \text{dB} \qquad (11.121)$$

where E = Young's modulus of the structural material
E_i = Young's modulus of the elastic material
l = length of the elastic layer, m
t = thickness of the structure, m
λ_{B_1} = wavelength of bending waves in the structure, m
The bending wavelength in the structure for which the elastic layer provides a complete transmission of bending waves is given by

$$\lambda_{B_{\text{trans}}} = \pi^3 \sqrt{\frac{E}{E_i}} \, t^2 l \qquad \text{m} \qquad (11.122)$$

If one wishes to reduce the frequency of complete transmission, the ratio E/E_i and the length of the resilient layer must be large. However, the length of the resilient material should always be small compared with the wavelength of bending waves in the elastic layer to avoid resonances.

A complete reflection of the incident bending wave occurs when the bending wavelength in the plate

$$\lambda_{B_S} = \pi t = \frac{c_b}{f_s} \qquad \text{m} \qquad (11.123)$$

Consequently, the *frequency where complete reflection occurs is independent of the dynamic properties and the length of the elastic layer* and is given by the thickness and by the dynamic properties of the plate or beam.

REFERENCES

1. L. D. Landau and E. M. Lifshitz, *Theory of Elasticity*, Pergamon, 1959.
2. P. M. Morse and K. U. Ingard, *Theoretical Acoustics*, McGraw-Hill, 1968.
3. L. L. Beranek (ed.), *Noise Reduction*, McGraw-Hill, 1960.
4. L. Cremer and M. Heckl, *Körperschall*, Springer-Verlag, 1967.
5. L. L. Beranek, *Acoustic Measurements*, Wiley, 1949.
6. L. E. Kinsler and A. R. Frey, *Fundamentals of Acoustics*, 2d ed., Wiley, 1962.
7. J. C. Snowdon, *Vibration and Shock in Damped Mechanical Systems*, Wiley, 1968.
8. E. E. Ungar and E. M. Kerwin, Jr., "Loss Factors of Viscoelastic Systems in Terms of Energy Concepts," *J. Acoust. Soc. Am.*, vol. 34, no. 7, pp. 954–957, 1962.
9. A. London, "Transmission of Reverberant Sound through Single Walls," National Bureau of Standards, RP1998, vol. 42, pp. 605–615, 1949.
10. H. Feshbach, "Transmission Loss of Infinite Single Plates for Random Incidence," *Bolt Beranek and Newman Inc. Rept.* 1, pp. 1–10, October, 1954.
11. A. deBruijn, "Influence of Diffusivity on the Transmission Loss of a Single Leaf Wall," *J. Acoust. Soc. Am.*, vol. 47, no. 3, pt. II, pp. 667–675, 1970.
12. L. L. Beranek, "The Transmission and Radiation of Acoustic Waves by Structures" (the 45th Thomas Hawksley Lecture of the British Institution of Mechanical Engineers), *J. Inst. Mech. Engrs.*, vol. 6, pp. 162–169, 1959.

13. C. I. Holmer, "Sound Transmission through Structures: A Review," M. S. thesis, John Carroll University, Cleveland, 1969.
14. G. Maidanik, "Response of Ribbed Panels to Reverberant Acoustic Fields," *J. Acoust. Soc. Am.*, vol. 34, no. 6, pp. 809–826, 1962.
15. P. W. Smith, Jr., "Response and Radiation of Structural Modes Excited by Sound," *J. Acoust. Soc. Am.*, vol. 34, no. 5, pp. 640–647, 1962.
16. P. W. Smith, Jr., "Coupling of Sound and Panel Vibration below the Critical Frequency," *J. Acoust. Soc. Am.*, vol. 36, no. 8, pp. 1516–1520, 1964.
17. R. H. Lyon and G. Maidanik, "Power Flow between Linearly Coupled Oscillators," *J. Acoust. Soc. Am.*, vol. 34, no. 5, pp. 623–639, 1962.
18. E. Skudrzyk, "Vibrations of a System with a Finite or Infinite Number of Resonances," *J. Acoust. Soc. Am.*, vol. 30, no. 12, pp. 1140–1152, 1958.
19. E. E. Ungar, "Statistical Energy Analysis of Vibrating Systems," *J. Eng. Ind.*, vol. 87, *Trans. ASME*, ser. B, pp. 629–632, 1967.
20. M. J. Crocker and A. J. Price, "Sound Transmission Using Statistical Energy Analysis," *J. Sound Vibration*, vol. 9, no. 3, pp. 469–486, 1969.
21. J. E. Manning and P. J. Remington, "Statistical Energy Methods," *Bolt Beranek and Newman Inc. Rept.* 2064, December, 1970.
22. W. Flügge, *Handbook of Engineering Mechanics,* pp. 35–39, McGraw-Hill, 1962.
23. B. G. Watters, "Transmission Loss of Some Masonry Walls," *J. Acoust. Soc. Am.*, vol. 31, no. 7, pp. 898–911, 1959.
24. C. I. Holmer, "Sound Transmission through Sandwich Panels," paper 64, presented at the 75th Meeting of the Acoustical Society of America, Ottawa, Canada, May, 1968.
25. D. Ross and E. M. Kerwin, "Damping Flexural Vibrations in Plates by Free and Constrained Viscoelastic Layers," final report Phases I–III, *Bolt Beranek and Newman Inc. Rept.* 632, Appendix A, PB 153794.
26. G. Kurtze and B. G. Watters, "New Wall Design for High Transmission Loss or High Damping," *J. Acoust. Soc. Am.*, vol. 31, no. 6, pp. 739–748, 1959.
27. K. Gösele, "Die Bestimmung der Dynamischen Steifigkeit von Trittschall-Dämmstoffen," *Boden, Wand und Decke,* Heft 4 and 5, 1960.
28. R. D. Ford, P. Lord, and A. W. Walker, "Sound Transmission through Sandwich Constructions," *J. Sound Vibration*, vol. 5, no. 1, pp. 9–21, 1967.
29. G. Maidanik, "The Influence of Fluid Loading on the Radiation from Orthotropic Plates," *J. Sound Vibration*, vol. 3, no. 3, pp. 288–299, 1966.
30. I. L Vér, "Acoustical and Vibrational Performance of Floating Floors," *Bolt Beranek and Newman Inc. Rept.* 72, October, 1969.
31. M. Heckl, "Untersuchungen über die Luftschalldämmung von Doppelwänden mit Schallbrücken," *Proc. Intern. Congr. Acoust., 3d, Stuttgart, 1959,* Elsevier, Vol. II, pp. 1010–1014, 1961.
32. K. Gösele, "Trittschall-Entstehung und Dämmung," *VDI-Berichte,* Band 8, pp. 23–28, 1956.
33. K. Gösele, "Untersuchungen zur Schall-Längsleitung in Bauten," *Schallschutz in Gebäuden,* Heft 56, p. 25, 1968.
34. M. C. Junger and P. W. Smith, Jr., letter to the editor on "The TL of Curved Structures," *Acustica,* vol. 5, no. 1, pp. 47–48, 1955.
35. P. W. Smith, Jr., "Sound Transmission through Thin Cylindrical Shells," *J. Acoust. Soc. Am.,* vol. 29, no. 6, pp. 721–729, 1957.
36. J. E. Manning and G. Maidanik, "Radiation Properties of Cylindrical Shells," *J. Acoust. Soc. Am.,* vol. 36, no. 9, pp. 1691–1698, 1964.
37. C. H. Agren and K. A. Stetson, "Measuring the Wood Resonances of Treble Violin Plates by Hologram Interferometry," paper EE3, presented at the 77th Meeting of the Acoustical Society of America, Philadelphia, April, 1969.

38. R. H. Lyon, "Statistical Analysis of Power Injection and Response in Structures and Rooms," *J. Acoust. Soc. Am.*, vol. 45, no. 3, pp. 545–565, 1969.

39. "Impact Sound Transmission and Airborne Sound Transmission Loss Tests on Histress Flexicore Slabs," Test Report of the Cedar Knolls Acoustical Laboratories, 1966, Test 6612-12.

40. V. I. Zaborov, G. S. Rosin, and L. P. Tyumentseva, "Reduction of Impact Noise by Flooring Materials" (English trans.), *Soviet Phys.-Acoust.*, vol. 12, no. 3, pp. 263–265, 1967.

41. I. L. Vér, "Measurement of the Dynamic Stiffness and Loss Factor of Elastic Mounts as a Function of Static Load," paper 7C1, presented at the 78th Meeting of the Acoustical Society of America, San Diego, November, 1969.

42. K. Gösele, "A Simple Method to Calculate the Impact Noise Isolation of Floors," *Boden Wand und Decke*, vol. 11, 1964.

43. L. Cremer, "Theorie des Kolpfschalles bei Decken mit Schwimmenden Estrich," *Acustica*, vol. 2, no. 4, pp. 167–178, 1952.

44. F. Bruckmayer, "Zur Akustik von Verkehrsbauwerken," *Der Bauingenieur*, vol. 42, no. 6, pp. 201–208, 1967.

45. R. Josse and C. Drouin, "Étude des impacts lourds à l'interieur des bâtiments d'habitation," Rapport de fin d'étude, Centre Scientifique et Technique du Bâtiment, Paris, Feb. 1, 1969, DS No. 1, 1.24.69.

46. M. Heckl and E. J. Rathe, "Relationship between the Transmission Loss and the Impact-noise Isolation of Floor Structures," *J. Acoust. Soc. Am.*, vol. 35, no. 11, pp. 1825–1830, 1963.

47. K. Gösele, "Neue Wege zur Entwicklung von Trittschall-Dämmstoffen," *Gesundheits-Ingenieur*, vol. 75, Heft 1/2, 3/4, 1954.

48. B. G. Watters, "Impact Noise Characteristics of Female Hard-heeled Foot Traffic," *J. Acoust. Soc. Am.*, vol. 37, no. 4, pp. 619–630, 1965.

49. L. Cremer, "The Propagation of Structure-borne Sound," *Dept. Sci. Ind. Res. Rept.* 1, ser. B, 1950, Germany.

50. T. Kihlman, "Transmission of Structure-borne Sound in Buildings," *Nat. Swedish Inst. Building Res., Stockholm, Rept.* 9, 1967.

51. M. Paul, "Die Messung von Transmissionsgraden bei Schallübertritt von Stäben auf Platten," *Acustica*, vol. 20, pp. 36–40, 1968.

52. ISO Recommendation R-140, 1960-E, "Field and Laboratory Measurements of Airborne and Impact Sound Transmission," Ref. No. ISO/R-140-1960(E).

53. I. L. Vér, "Impact Noise Isolation of Composite Floors," paper B3, presented at the 79th Meeting of the Acoustical Society of America, Atlantic City, April, 1970.

54. I. L. Vér, "Connection between the Impact Noise Level and Sound Transmission Loss," paper B4, presented at the 79th Meeting of the Acoustical Society of America. Atlantic City, April, 1970.

55. V. I. Zaborov, "Sound Insulation of Double Walls Joined at the Edges" (Russian translation), *Soviet Phys.-Acoust.*, vol. 11, no. 2, pp. 135–140, 1965.

56. P. H. White and A. Powell, "Transmission of Random Sound and Vibration through a Rectangular Double Wall," *J. Acoust. Soc. Am.*, vol. 40, no. 4, pp. 821–832, 1966.

57. R. D. Ford, P. Lord, and P. C. Williams, "The Influence of Absorbent Linings on the Transmission Loss of Double-Leaf Partitions," *J. Sound & Vib.*, vol. 5, no. 1, pp. 22–28, 1967.

58. K. A. Mulholland, H. D. Parbrook, and A. Cummings, "The Transmission Loss of Double Panels," *J. Sound Vibration*, vol. 6, no. 3, pp. 324–334, 1967.

Appendix to Chapter Eleven

SOUND-TRANSMISSION CLASS (STC) AND IMPACT ISOLATION CLASS (IIC)

Determination of Sound-transmission Class (STC)   The purpose of the STC classification is to provide a single-number rating that can be used for comparing partitions for general building-design purposes. The procedure described here has been adopted formally by the American Society for Testing and Materials.[1]   The procedure adopted by the International Standards Organization yields identical results.[2]   A general discussion of sound-transmission loss is given in Sections 11.2 to 11.4 and in Ref. 3.

To determine the sound-transmission class (STC) of a test specimen, the sound-transmission losses R for the specimen are measured [4] in series of 16 one-third-octave test bands and these data are compared with the points on a reference STC contour, as described below.

Graphical Determination of STC.   If the measured one-third-octave-band transmission losses for the test specimen are plotted on the specified graph paper (see Fig. A.11.1 legend) as a function of frequency, the sound-transmission class may be determined by comparison with a standard transparent overlay on which the STC contour of Fig. A.11.1 is drawn.   The STC contour is shifted vertically relative to the test curve until some of the measured transmission-

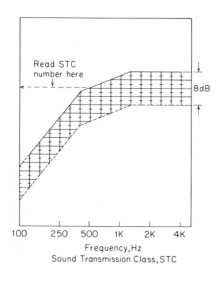

Fig. A.11.1   Form of overlay for STC determination.   See text for details of positioning of the solid curve above.   The STC equals the R value on the ordinate of the transmission-loss curve that lies beneath the arrow above.   Use this overlay with one-third-octave sound-transmission-loss data plotted on graph paper that meets these specifications: 5 cm = 25 dB; 5 cm = 1 decade of frequency.

loss R values for the test specimen fall below those of the STC contour and the following two conditions are fulfilled: the maximum deficiency in any single one-third-octave band shall not exceed 8 dB and the sum of the deficiencies (that is, the deviations below the contour) shall not be greater than 32 dB. When the contour is adjusted to the highest value (in integral decibels) that meets the above requirements, the sound-transmission class for the specimen is the R value corresponding to the intersection of the contour and the 500-Hz ordinate (read on the vertical scale of the R-vs.-frequency plot of the test specimen).

Presentation of Results. It is recommended that the test data (for one-third-octave bands) be plotted on the specified graph paper together with the corresponding STC contour obtained as described above. In this way attention is drawn to the frequency regions that limit the sound-transmission class of the test specimen. Whenever possible, the ordinate scale should start at 0 dB.

Limitation. The STC rating [5] is useful when designing walls that provide insulation against the sounds of speech, music, radio, or television. It is not valid for noise sources with spectra that differ markedly from the above, e.g., industrial processes, aircraft, motor vehicles, power transformers, and the like.

Determination of Impact Isolation Class (IIC)   The impact isolation class (IIC) is a single-number rating that provides a means for comparing the acoustical performance of floor-ceiling assemblies when excited by impacts produced by a standard tapping machine.[6]

The *impact-sound-transmission levels* resulting from the operation of a standard tapping machine on a test floor are characterized by the band spectrum of the space-average sound-pressure levels produced in the receiving room beneath the test floor. The band sound-pressure levels are measured with an indicating device that reads rms sound pressure in decibels $re\ 2 \times 10^{-5}$ N/m².

The impact-sound-transmission levels L_{IST} are generally normalized to remove the effects of varying degrees of sound absorption that might occur from one test receiving room to another. The relation between L_{IST} and the measured band level is

$$L_{\text{IST}} = L_2 + 10 \log_{10} (A/A_0)$$

where L_2 = space-average, one-third-octave-band sound-pressure level measured in the receiving room, dB

$A = \Sigma S_i \alpha_i$ = actual total sound absorption in the receiving room at the center frequency of the band, m² (see Chap. 9)

$A_0 = 10$ m² = reference sound absorption

Frequency Range of Measurements. The space-average sound-pressure levels should be determined in 16 contiguous frequency bands, each one-third octave wide, and covering the range of mean frequencies from 100 to 3,150 Hz. If possible, the measurements should be made at higher and lower frequencies although those values would not enter into the IIC rating.

Statement of Results. The measured one-third-octave-band sound-pressure levels should be normalized to a standard reference absorption of $A_0 = 10$ m², and the resulting L_{IST} should be plotted, for all frequencies of measurement, as a curve on the specified graph paper (see Fig. A.11.2 legend).

Improvement of Impact Sound Insulation.   For measuring the improvement of impact sound insulation, such as the improvement owing to floor coverings or floating floors,[7] it is often possible to measure the impact-sound-transmission level before and after treatment, particularly in laboratory tests.   In such cases, the improvement (lowering of levels) owing to the treatment may be denoted by ΔL_{IST} and should be given in the form of a curve on the specified graph paper.

Impact Isolation Class (IIC).   The purpose of this single-number rating is to permit easy comparison of the impact noise isolation performance of floor-ceiling assemblies for general guidance in building design.   The procedure for assigning the single-number rating IIC is to plot the measured impact-sound-transmission levels L_{IST} on the specified graph paper.   The transparent overlay of Fig. A.11.2 is then placed on the graph, aligned with the frequency scale, and adjusted so that all data points lie on or below the broken-line contour.   This procedure ensures initially that single deviations are less than or equal to a maximum of 8 dB.   Then the deviations above the solid-line contour are summed.   The total must not exceed 32 dB; if greater, the overlay is adjusted upward until the total equals 32 dB.   The IIC value, *read from the overlay*, is that overlay value which lies over (corresponds to) the impact-sound-pressure level of 60 dB on the graph scale.[3]

Other single-number rating procedures in the literature include impact noise rating (INR)[8] and impact sound insulation class I_i.[9]

The significance of impact-sound-transmission levels produced by the stan-

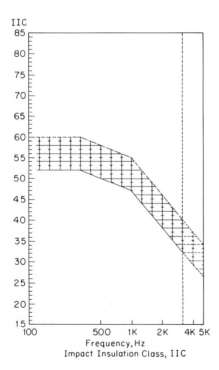

Fig. A.11.2   Form of overlay for IIC determination.   See text for details of positioning of the solid curve above.   The IIC equals the value on the above ordinate that lies over the $L_{ISP} = 60$ dB value on the impact-sound-transmission curve beneath the overlay.   Use this overlay with one-third-octave impact-sound-transmission data plotted on graph paper that meets these specifications: 5 cm = 25 dB; 5 cm = 1 decade of frequency.

dard tapping machine and the corresponding single-number rating IIC is the subject of current debates in several national and international standard organizations.[10] The reader is advised to keep abreast of documents of ASTM, ANSI, and ISO for information on current standards for tapping machines and single-number rating procedures.

REFERENCES FOR APPENDIX TO CHAPTER ELEVEN

A.1 ASTM E90−70, "Recommended Practice for Laboratory Measurement of Airborne Sound Transmission Loss of Building Partitions," American Society for Testing and Materials, 1916 Race St., Philadelphia, Pa. 19103.

A.2 ISO/R-717-1968, "Rating of Sound Insulation for Dwellings." Available through American National Standards Institute, 1430 Broadway, New York, N.Y. 10018.

A.3 Raymond D. Berendt, George E. Winzer, and Courtney B. Burroughs, "A Guide to Airborne, Impact and Structure Borne Noise-Control in Multifamily Dwellings," FT/TS-24, U.S. Department of Housing and Urban Development, Washington, D.C. 20410, January, 1968.

A.4 The test procedures are specified in either the "Recommended Practice for Laboratory Measurement of Airborne Sound Transmission Loss of Building Partitions" (ASTM E90−70), or the "Tentative Recommended Practice for Measurement of Airborne Sound Insulation in Buildings" (ASTM E336−67T), or ISO Recommendation R-140 (January, 1960), "Field and Laboratory Measurements of Airborne and Impact Sound Transmission."

A.5 A large number of measured values of STC may be found in "Solution to Noise Problems in Apartments and Hotels," Owens-Corning Fiberglas Corporation, 1969; in "Sweet's Catalog," Sweet's Construction Division, McGraw-Hill Information Systems Company; in "Sound Isolation of Wall, Floor and Door Constructions," *Natl. Bur. Std. Nomograph* 77, 1964; and in Ref. A.3.

A.6 The standard tapping machine and the measurement procedure for laboratory and field evaluation of the impact sound transmission loss are described in ISO Recommendation R-140, Ref. No. ISO/R-140-1960(E). Practical measurement details are given in Ref. A.3.

A.7 For the measurement of the improvement of insulation due to a floating floor or a floor covering, ISO/R-140 recommends that a standard reinforced concrete floor of thickness 12 ± 2 cm should be used. A recent draft of a supplementary ISO standard, dealing specifically with determining the effectiveness of floor coverings, recommends a 14-cm slab, weighing approximately 350 kg/m², and approximately 10 m² in surface area.

A.8 "Impact Noise Control in Multifamily Dwellings," *FAA Rept.* 750, Federal Housing Administration, Washington, D.C., January, 1963.

A.9 ISO Recommendation R-717-1968, "Rating of Sound Insulation for Dwellings."

A.10 See, for example, T. Mariner and H. W. W. Hehmann, "Impact-noise Rating of Various Floors," *J. Acoust. Soc. Am.*, vol. 41, pp. 206–214, 1967.

Chapter Twelve

Mufflers

T. F. W. EMBLETON [*]

12.1 Introduction

A muffler may be described as any section of a duct or pipe that has been shaped or treated with the intention of reducing the transmission of sound, while at the same time allowing the free flow of a gas. A muffler is an acoustical filter, and like an electrical filter its performance varies with frequency.

12.1.1 Criteria

The properly designed muffler for any particular application should satisfy the often-conflicting demands of at least five criteria simultaneously:

1. The *acoustical* criterion which specifies the minimum noise reduction required from the muffler as a function of frequency. The operating conditions must be known because large steady-flow velocities or large alternating velocities (high sound-pressure levels) may alter its acoustical performance.

2. The *aerodynamic* criterion which specifies the maximum acceptable

* Division of Physics, National Research Council, Ottawa, Ontario.

average pressure drop through the muffler at a given temperature and mass flow.

3. The *geometrical* criterion which specifies the maximum allowable volume and restrictions on shape.

4. The *mechanical* criterion which may specify the materials from which it is constructed, or the design of the muffler, so that it is durable and requires little maintenance. This is especially important in cases involving high temperature or corrosive gases, or where the gaseous flow is carrying solid particles in suspension that might become deposited and reduce the muffler's effectiveness. In special cases, e.g., in the food-processing industry, considerations of hygiene may require that it be readily cleanable.

5. The *economic* criterion is vital in the marketplace. A muffler usually must be as inexpensive as possible, where in any particular situation not only the initial purchase price but also the operating costs must be considered.

12.1.2 Terminology and Definitions of Performance

Dissipative Muffler. A dissipative muffler is one whose acoustical performance is determined mainly by the presence of sound-absorbing, i.e., flow-resistive, material. Dissipative mufflers usually have relatively wideband noise reduction characteristics, and hence are most useful for noise control problems associated with noise spectra such as fan noise, jet-engine noise, and speech. They may also be useful if the center frequency of a narrowband noise varies over a wide range as the operating conditions of the source fluctuate.

Reactive Muffler. The performance of a reactive muffler is determined mainly by its geometrical shape. The one or more chambers, resonators, or finite sections of pipe which collectively make up a reactive muffler provide an impedance mismatch for the acoustic energy traveling along the duct. This impedance mismatch results in a reflection of part of the acoustic energy back toward the source of sound, or back and forth among the chambers, preventing that part from being transmitted past the muffler. In practice, this is often accompanied by enhanced alternating sound pressures and particle velocities within the muffler, which can result in a significant absorption of energy by even a small amount of sound-absorbing material, or by nonlinear behavior of the gas.

Insertion Loss L_{IL}. Insertion loss is generally defined as the difference, in decibels, between two sound-pressure levels (or power levels or intensity levels) which are measured at the same point in space before and after a muffler is inserted between the measurement point and the noise source.

Transmission Loss L_{TL}.   Transmission loss of a muffler is defined as 10 times the logarithm, to base 10, of the ratio of sound power incident on the muffler to the sound power transmitted by the muffler.   The units are decibels.   Transmission loss is a useful analytic concept, but its measurement is hampered by the lack of an acoustic wattmeter.   Sound power can be derived from measurements of sound pressure only under certain limited circumstances.

Noise Reduction L_{NR}.   Noise Reduction, *SPL difference,* and *end differences* all have the same meaning.   The term *Noise Reduction* * will be used throughout this chapter to refer to the difference between the sound-pressure levels measured at the input of a muffler and at its output.

Attenuation L_A.   Attenuation is the decrease of sound power in decibels between two points in an acoustical system.   It is a useful quantity in describing wave propagation in lined ducts where the acoustically effective material is continuously and uniformly distributed along the direction in which energy flows.   In this case the attenuation is measured by determining the decrease in sound-pressure level per unit length of duct—measured inside the muffler, but not too near the ends of the duct—and multiplying by the total length.

It is important to note that insertion loss, transmission loss, and Noise Reduction are not uniquely related to the physical properties of a muffler. Each depends also on the source or termination impedance, or both, which create losses or reflections of energy, known as *end effects.*   Thus, each is a different measure of the interaction between the muffler and its acoustical environment.   In noise control engineering, insertion loss is perhaps the most useful measure of acoustical effectiveness.   One generally measures or calculates the sound-pressure level at some point for a system which has no acoustical treatment.   The difference between this and the acceptable sound-pressure level is just the insertion loss required for a muffler.

Wavelength.   In the discussion of the performance of individual mufflers an important parameter is often the ratio of some length dimension of the muffler to the wavelength of sound at the frequency of interest.   The wavelength depends on the temperature of the gas within the muffler and may be calculated as

$$\lambda = \frac{345}{f}\sqrt{\frac{\theta_C + 273}{295}} \qquad \mathrm{m} \qquad (12.1a)$$

or

$$\lambda = \frac{1,130}{f}\sqrt{\frac{\theta_F + 460}{530}} \qquad \mathrm{ft} \qquad (12.1b)$$

* Noise Reduction is capitalized to avoid confusion with the nonspecific term *noise reduction.*

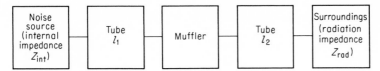

Fig. 12.1 Block diagram of a muffler system, showing the noise source, the connecting tubes l_1 and l_2, the muffler, and the surroundings into which the sound is radiated.

where f = frequency, Hz
  θ_C = muffler temperature, °C
  θ_F = muffler temperature, °F

12.2 Acoustical Performance of Reactive Mufflers

A block diagram of a typical muffler system is shown in Fig. 12.1.   A source of sound pressure p, with internal impedance Z_{int},* is joined by a tube l_1 to the muffler, which in turn is joined by a tube l_2 to the surroundings.   Radiation from l_2 to the surroundings is characterized by a radiation impedance Z_{rad}.

In most mufflers the alternating (acoustic) flow is superimposed on a steady, dc flow.   Several authors [1,2] have shown that the effectiveness of some muffler shapes changes with the presence of a steady flow by an amount which is dependent on the magnitude of this flow.   In this section we restrict ourselves to those levels of alternating flow — roughly, below L_p = 120 dB — for which the acoustical performance is not dependent on L_p (see Sec. 12.7 for these nonlinear effects).   In order to focus attention more sharply on the features of acoustical performance that are related to the various shapes and sizes of cavities and pipe sections, we shall assume that the inlet and outlet tubes, l_1 and l_2 in Fig. 12.1, are each infinitely long, or themselves contain mufflers with impedances equal to ρc.

12.2.1 Expanded Cross-section Muffler   The simplest type of muffler is the simple expansion chamber, shown in Fig. 12.2, having a length l and an abrupt change in cross-sectional area at each end.   Its behavior can be described in terms of two parameters m and kl

$$m = \frac{\text{cross-sectional area of chamber}}{\text{cross-sectional area of duct}} = \frac{S_2}{S_1} \qquad (12.2)$$

* Impedance here means "specific acoustic impedance," equal to the complex ratio of the rms sound pressure to the rms particle velocity.   An understanding of "complex impedance" is not essential to this chapter.   Further information on impedance is given in the appendix to Chap. 1.

and

$$kl = \frac{2\pi l}{\lambda} \qquad (12.3)$$

where λ is the wavelength of sound at the temperature of the gas in the muffler; see Eq. (12.1). Its transmission loss in the absence of a steady air flow is given by the equation describing the family of curves in Fig. 12.3 [3,4]

$$L_{TL} = 10 \log \left[1 + \frac{1}{4} \left(m - \frac{1}{m} \right)^2 \sin^2 kl \right] \qquad dB \qquad (12.4)$$

We see that L_{TL} is a periodic function in kl, repeating every π radians (180°). This result is valid at frequencies below that for which 0.8 times the wavelength is equal to the largest transverse dimension.

The performance shown by Fig. 12.3 may be interpreted in terms of the wave system existing inside the muffler. At very low frequencies ($kl \to 0$) or whenever the length of the muffler equals $\lambda/2$, λ, $3\lambda/2$, etc., a standing wave system is produced with enhanced sound pressures at the end walls of the cavity. This has the effect of increasing the characteristic impedance of the duct of cross section S_2 from $\rho c/S_2$ to $m\rho c/S_2$ which, from Eq. (12.2), is exactly the value for the inlet and outlet pipes, i.e., $\rho c/S_1$. Thus, at these resonance frequencies the muffler is a perfect impedance match for the pipe and its L_{TL} is 0 dB. At intermediate frequencies (and wavelengths) reflected waves inside the muffler interfere destructively with incident waves at the inlet and thus provide a mismatch of impedance between the muffler and the inlet pipe—leading to the reflection of sound energy back along the inlet pipe toward the source of sound.

At higher frequencies, where the wavelength is equal or less than the transverse dimension of the chamber, L_{TL} is dependent on other parameters. For example, when the diameter d_2 (corresponding to the area S_2) equals λ, a significant reduction of L_{TL} is found for a very large value of m.[3,4] However, for $m = 4$ and $\lambda = d_2$ in a muffler consisting of a series of coupled expansion chambers, the transverse wavemotion may be utilized [5] to enhance L_{TL}. Theoretically, there should be a maximum of L_{TL} when the difference in the diameters of expansion chamber and inlet

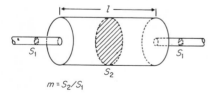

$m = S_2/S_1$

Fig. 12.2 Diagram of a single-expansion-chamber muffler.

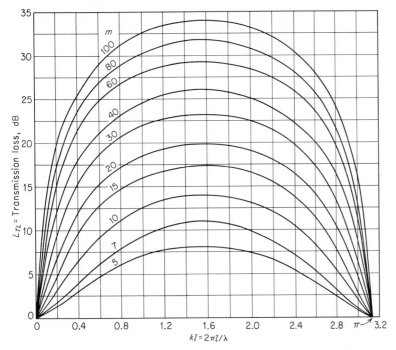

Fig. 12.3 Transmission loss L_{TL} of an expansion chamber of length l and $S_2/S_1 = m$ (see Fig. 12.2). The cross section of the muffler need not be round, but its greatest transverse dimension should be less than $0.8\ \lambda$ (approximately) for the graph to be valid. For values of kl between π and 2π, subtract π and use the scale given along the abscissa. Similarly, for values between 2π and 3π, subtract 2π; etc. Note that when $kl = \pi$, then $l = \lambda/2$; when $kl = 2\pi$, then $l = \lambda$; etc.

pipe is an odd number of half wavelengths,[6] i.e., when $d_2 - d_1 = \lambda/2$, $3\lambda/2$, $5\lambda/2$, etc.

The L_{TL} of a simple expansion chamber is not affected by the presence of a superimposed steady flow,[2] at least up to a velocity of 35 m/sec. At very high velocities the flow noise may become large enough to render the muffler ineffective.

The foregoing discussion may be extended to include two or more expansion chambers in series, either with or without external connecting tubes, as shown in Fig. 12.4. It is seen[3] that L_{TL} increases as the number of chambers is increased, although the addition of a third chamber represents only a small improvement over two chambers. From (d) we see that L_{TL} increases as the length of the tube separating the two cavities increases.

12.2.2 Pipe-resonator Muffler
Another design for joining a single expansion chamber to an inlet and outlet tube, or for the connecting

tube between two chambers, is to permit the tubes to project into the chamber from one or both ends, as shown in Figs. 12.5 and 12.6.   With this design higher L_{TL} over the whole frequency range is obtained than with the same sizes of expansion chambers without projecting pipes.

The pipe resonator cannot provide an exact impedance match $\rho c/S_1$ simultaneously to both the inlet and outlet tubes, except when the orifices of the inlet and outlet tube are at similar locations in relation to the standing-wave pattern in the resonator.   Thus in the absence of the tube connecting the two cavities, L_{TL} would be 0 dB at about 280 Hz, i.e., when $l = \lambda/2$, as shown in Fig. 12.5a.   With the connecting tube terminating at the center of each cavity as shown in (b), i.e., at the "$\lambda/4$ point" at this

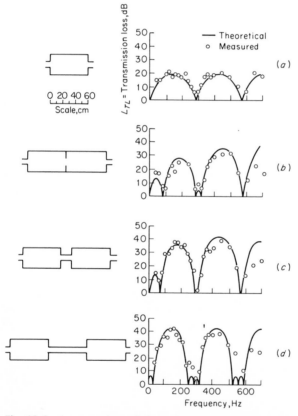

Fig. 12.4   A comparison of theoretical and experimental results for $m = 16$, without flow, for two expansion chambers in series compared to one expansion chamber.   The L_{TL} is greater for two chambers than one, and also increases as the length of the connecting tube increases.   Note that the "passbands," where $L_{\text{TL}} \approx 0$, are increased in width when the connecting tube has a length of $\lambda/2$, λ, etc.   (*After Davis, Stokes, Moore, and Stevens.*[3])

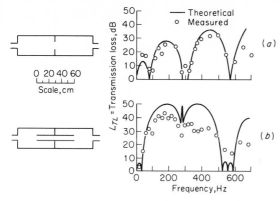

Fig. 12.5 A comparison of theoretical and experimental values of L_{TL} for (a) two chambers in series and (b) two pipe resonators in series, where the pipes terminate in the center of each chamber. The transmission loss of the system remains large at the frequency of the "$\lambda/2$ resonance" of each chamber, since the connecting-pipe ends are then at "$\lambda/4$ points," i.e., points of low impedance compared with that of the inlet duct. (*After Davis, Stokes, Moore, and Stevens.*[3])

frequency, there is a maximum mismatch of impedance and L_{TL} remains large.

When both inlet and outlet tubes extend into the chamber, the length of one or both may be varied in order to obtain either a large increase in a narrow frequency range or a small increase in transmission loss (compared with a simple expansion chamber of the same size) over a broad range of frequencies. Figure 12.6 shows the measured transmission loss of two such mufflers.

Unlike a simple muffler of expanded cross section, the transmission loss of a pipe resonator is reduced considerably by a steady flow, though this reduction is mainly confined to the frequency range near $l = \lambda/4$. Figure 12.7 gives L_{TL} as a function of frequency for a particular pipe resonator with and without airflow, and also the value of L_{TL} as a function of airflow near the frequency where $l = \lambda/4$. For this pipe resonator an air velocity of less than 15 m/sec has little effect on L_{TL}; at higher velocities L_{TL} decreases rapidly to a value of 6 to 10 dB for flows of over 30 m/sec. In industrial and truck applications, gas velocities in the inlet tube may be 50 to 120 m/sec.

Depending on many details of construction, including, for example, pipe resonator or simple chamber, resistance in a muffler may decrease the transmission loss. The decrease of L_{TL} in the presence of a steady flow is explained qualitatively by the addition of acoustic resistance (due to the flow) to the impedance of the resonator. The L_{TL} of a pipe resonator at its resonance frequency is [2]

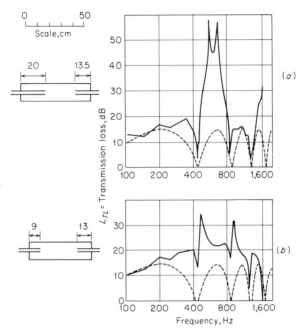

Fig. 12.6   An example of how L_{TL} of a pipe resonator may be enhanced by a large amount in a narrow range of frequencies or by a small amount over a broad range of frequencies, by varying the distance to which the inlet and outlet pipes extend into the chamber.   The L_{TL} of an expansion chamber of the same size, without projecting pipes, is shown by the dotted curves.   (*After Gösele.*[2])

$$L_{TL} = 10 \log_{10} \left(1 + \frac{\rho c}{2 R_S m} \right)^2 \qquad (12.5)$$

where ρ = density of gas, kg/m³

c = speed of sound in gas, m/sec

R_S = specific acoustic resistance of the resonator (its reactance is zero at resonance), N-sec/m³ (mks rayls)

$m = S_2/S_1$ (see Fig 12.2)

The performance can be interpreted by assuming R_S increases strongly with flow velocity, and that in addition there is a small resistance that is independent of flow.   (See Sec. 12.7 for a more detailed discussion of the relation between flow velocity and impedance.)

12.2.3   Volume-resonator Muffler

A volume-resonator muffler differs from those just discussed in that no gas flows through the chamber and the muffler connects to the main duct through one or more small openings or tubes (see Fig. 12.8).   The resonance frequencies are determined by the dimensions of the openings and by the volume of the branched cavity. In the literature these mufflers are also called *side-branch resonators* or *Helmholtz resonators*. It is assumed that the linear dimensions of the cavity are less than one-tenth wavelength at all frequencies under consideration; if this condition is violated, the wave-motion in the resonator must be taken into account and the situation becomes similar to that considered for pipe resonators (though usually with increased acoustic resistance).

The equation for the transmission loss of a volume-resonator muffler is given by [7]

$$L_{\mathrm{TL}} = 10 \log_{10} \left[1 + \frac{\alpha + 0.25}{\alpha^2 + \beta^2 (f/f_0 - f_0/f)^2} \right] \quad \mathrm{dB} \quad (12.6)$$

where α = resonator resistance (dimensionless) = $S_1 R_s / A_0 \rho c$
  β = resonator reactance (dimensionless) = $S_1 c / 2\pi f_0 V$
  S_1 = area of main duct, m²
  R_s = flow resistance in resonator tubes, mks rayls
  V = volume of resonator, m³
  A_0 = total aperture area, m²
  f_0 = resonance frequency, Hz
  ρ = density of gas, kg/m³
  c = speed of sound, m/sec

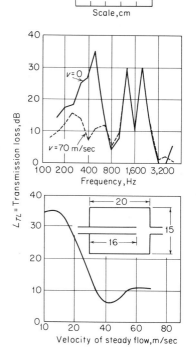

Fig. 12.7 The L_{TL} of a pipe resonator is reduced by the presence of a steady airflow. The reduction occurs mainly at the lower frequencies. The magnitude of this reduction at the frequency near $l = \lambda/4$ is shown as a function of flow velocity. (*After Gösele.*[2])

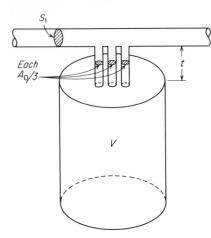

Fig. 12.8 Diagram of a volume (side-branch) resonator. The volume V is joined to a duct of cross-sectional area S_1 by three tubes in parallel, each of length t and area $A_0/3$. The parameters V, A_0, and t are all important in determining the properties of the resonator.

The shapes of the transmission-loss curves calculated by Eq. (12.6) are shown in Figs. 12.9 and 12.10, as a function of f/f_0 for two different ratios of α to β, namely, 1.0 and 0.5. At resonance ($f = f_0$) L_{TL} is a function of α only, and can be written

$$L_{\mathrm{TL}}\ (\text{at } f = f_0) = 20 \log_{10} \left(\frac{\alpha + 0.5}{\alpha} \right) \qquad \text{dB} \qquad (12.7)$$

For α much less than 0.25 and for frequencies much lower or higher than f_0, Eq. (12.6) becomes

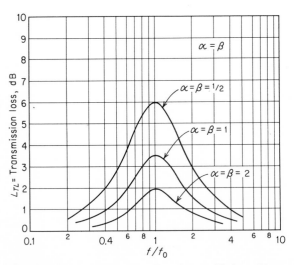

Fig. 12.9 Transmission loss of a volume (side-branch) resonator. For this chart, the resistance parameter α is equal to the reactance parameter β. (*After Ingard.* [7])

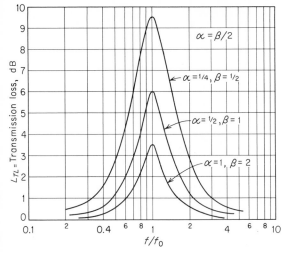

Fig. 12.10   Transmission loss of a volume (side-branch) resonator. For this chart, the resistance parameter α is equal to half the reactance parameter β. (*After Ingard.*[7])

$$L_{TL} \text{ (for negligible } \alpha) = 10 \log_{10} \left[1 + \frac{1}{4\beta^2 (f/f_0 - f_0/f)^2} \right] \quad \text{dB} \quad (12.8)$$

A procedure for choosing the appropriate resonator dimensions for a particular application is itemized as follows:

1. Starting with the desired L_{TL} as a function of frequency, select, from Figs. 12.9 and 12.10, the values of α, β, and f_0; alternatively use Eqs. (12.7) and (12.8) or Eq. (12.6) to investigate other combinations of α and β. (The assumption that the resonator dimensions are small compared to a wavelength requires that β should exceed $\frac{1}{2}$.)

2. Using the known main-duct area S_1 and the resonance frequency f_0, determine the resonator volume from

$$V = \frac{S_1 c}{2\pi f_0 \beta} \quad \text{m}^3 \quad (12.9)$$

3. Calculate the "length" quantity

$$q = \frac{2\pi f_0 S_1}{c\beta} \quad \text{m} \quad (12.10)$$

Choose an appropriate neck length t (in m) and number of tubes n. Compute the dimensionless quantity q/nt. Then enter Fig. 12.11 and from it obtain the value of the dimensionless quantity A_0/qt from the value of q/nt just computed.   Determine A_0.

If the resulting area of the apertures A_0 is unsatisfactory, vary the choices of n and t.

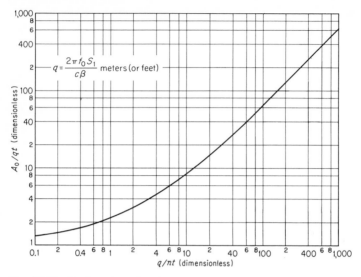

Fig. 12.11 Design chart for determining resonator aperture area A_0. This chart gives A_0/qt as a function of q/nt.

4. Choose a wire or cloth screen or other flow-resistive material such as sintered metal to cover or fill the aperture area A_0 in accordance with the desired resistance factor α. The acoustic resistance R_s needed is

$$R_s = \frac{A_0}{S_1} \alpha \rho c \quad \text{N-sec/m}^3 \text{ (mks rayls)} \tag{12.11}$$

If the acoustic resistance is smaller than that calculated from Eq. (12.11) the transmission loss may be greater than predicted from the chosen value of α. However, there must be sufficient acoustic resistance to ensure that the resonator absorbs energy rather than stores it in order to prevent the resonator from "ringing."

At the resonance frequency f_0 the total reactance of the resonator is zero; this occurs when the reactance of the resonator openings is equal to the reactance of the volume (these are always of opposite sign). The equation for f_0 is

$$f_0 = \frac{c}{2\pi} \sqrt{\frac{A_0}{Vt'}} \quad \text{Hz} \tag{12.12}$$

where t' = equivalent neck length = $t + 0.8 \sqrt{A_0/n}$, m

When discussing volume resonators it is sometimes convenient to write formulas in terms of a single parameter for the openings, which takes into account both the total area A_0 and the effective length t'. This parameter is called the *conductivity* c_0 and is equal to A_0/t'.

Typical measurements are shown in Fig. 12.12;[3] these data also illustrate the modification in performance that takes place when the volume of the resonator is sufficiently large that wavemotion becomes important.   Figure 12.12a shows the L_{TL} curve for a simple volume resonator having a resonant frequency of about 340 Hz.   In (b) the development of a region of low L_{TL} near 580 Hz is shown, at which frequency the length of the cavity equals one wavelength and the standing-wave pattern presents a high impedance to the sound waves traveling along the duct.   At half this frequency, 290 Hz, the cavity is half a wavelength long and the impedance due to wavemotion is a minimum at the location of the opening.

Figure 12.12c and d illustrates that once wavemotion is present in the cavity, this wavemotion becomes the dominant mechanism determining the L_{TL} curves.   The difference between (c) and (d) lies only in the position of the openings: in (c) the impedance at the opening is high (low

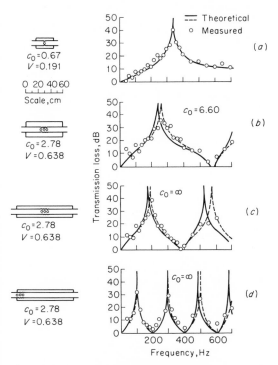

Fig. 12.12 Comparison of theoretical and experimental values of L_{TL} for a simple volume (side-branch) resonator, and the changes that occur with wavemotion.   When wavemotion is present it is the dominant factor in determining the L_{TL} of the muffler; the location of the openings in the duct wall is now more important than their area.   (*After Davis, Stokes, Moore, and Stevens.*[3])

L_{TL}) whenever the length of the cavity is an integral number of wavelengths; in (*d*) this condition occurs for integral numbers of half wavelengths.

When the sound waves are superimposed on a steady gas flow the transmission loss of a volume resonator decreases as the flow velocity increases. This effect occurs mainly near the resonance frequency, i.e., where L_{TL} is greatest. Figure 12.13 shows L_{TL} for one particular volume resonator [2] with and without a steady flow.

12.2.4 Conical-connector Muffler   A common shape of muffler is the conical connector of length l_s which joins a duct of cross-sectional area S_1 to another duct of cross-sectional area S_2.[8] The transmission loss is given in Fig. 12.14 as a function of l_s/λ. The parameter m is the area ratio S_2/S_1. The transmission loss results from an impedance mismatch at the junction of the two ducts, which causes the reflection of sound energy back toward the sound source.

12.2.5 Finite Length of Tailpipe   In this section we have assumed that the duct tubes on either side of the muffler, see Fig. 12.1, are infinitely long. We shall now consider the insertion loss of a muffler consisting of a single expansion chamber, terminated by a tailpipe of finite length. The results for one particular geometry are shown in Fig. 12.15.[9] At some resonance frequency f_0 the reactances of tailpipe and

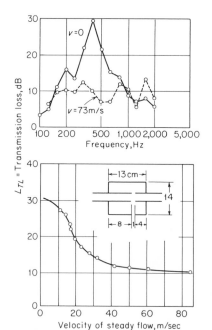

Fig. 12.13 The transmission loss of a volume (side-branch) resonator is decreased by steady gas flow. The effect is most marked near the resonance frequency, where L_{TL} is otherwise greatest. In the lower graph, L_{TL} at the resonance frequency is shown as a function of flow velocity for one typical resonator. (*After Gösele.*[2])

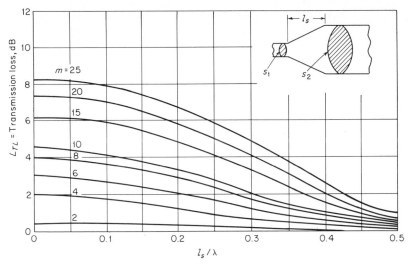

Fig. 12.14   The transmission loss of conical connectors as a function of the non-dimensional parameters l_s/λ and $m = S_2/S_1$. As shown, l_s is the length of the conical section measured along the axis of the duct.

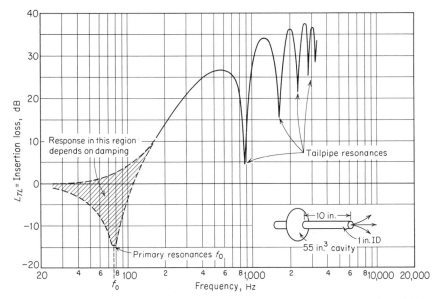

Fig. 12.15   Insertion loss of a particular cavity muffler terminated by a tailpipe of finite length (straight-through resonator) as determined by an electrical analog. (*After Watters, Hoover, and Franken.*[9])

cavity cancel each other and the insertion loss then depends only on the resistive impedance of the muffler. At higher frequencies there are resonances in the tailpipe, due to its finite length, which occur whenever its length equals an integral multiple of $\lambda/2$.

The effect of airflow (Fig. 12.16) is primarily to improve the insertion loss at the resonance frequency f_0. The resonance (and negative insertion loss) becomes less marked as the flow velocity is increased and disappears entirely at a velocity of about 70 m/sec (see Ref. 2).

12.2.6   Commercial Nondissipative Muffler Designs   Three commercial designs of nondissipative mufflers are shown in Figs. 12.17 and 12.18. The "straight-through" resonator is an economy replacement design. The "two-pass" and "in-and-out same end" mufflers are practical and efficient designs that illustrate a combination of side-branch, straight-through, wave-cancellation, and nonlinear effects which yield high transmission loss over a wide frequency range. Commercial designs are generally developed empirically to fit particular engines and usually call for specific lengths of pipes before and after the muffler in order to minimize loss of engine power and maximize the insertion loss in those parts of the frequency range where the loudness contributions of the source are the greatest.

12.3   Lined Ducts and Parallel Baffles

A lined duct is an air passage with at least one interior surface covered with a porous acoustical material (see Fig. 12.19). Parallel baffles are a grouping of side-by-side ducts that generally have a rectangular cross

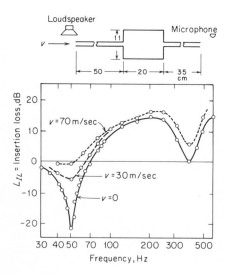

Fig. 12.16   Effect of airflow on insertion loss. Near the resonance the insertion loss is increased by steady airflow; if the flow velocity is large enough, the resonance (and negative insertion loss) may be eliminated. (*After Gösele.*[2])

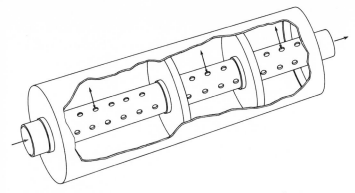

Fig. 12.17 Artist's sketch of a low-priced, low-performance, straight-through type of muffler.

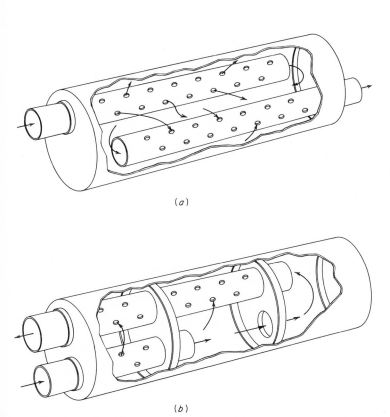

(*a*)

(*b*)

Fig. 12.18 Artist's sketches of two practical, high-performance mufflers for trucks or automobiles. (*Designs courtesy Nelson Muffler Corporation.*)

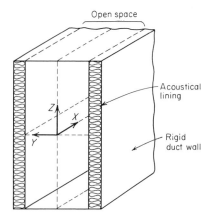

Open space

Acoustical lining

Rigid duct wall

Fig. 12.19 A sketch of a lined duct, showing the nomenclature used in describing the direction of a sound wave in the duct.

section and may be considered as a special type of lined duct (see Fig. 12.20). Prefabricated mufflers for ventilation systems, see Sec. 12.8, are usually of this type. A special type of multiple-channel baffle system is shown in Fig. 12.21.

The transmission loss of lined ducts arises primarily from the attenuation of sound by the absorbing lining. When the total cross-sectional area of the lining becomes as great as the cross-sectional area of the free air passage, or when there is a significant change in area at the ends of the ducts, reflected waves are produced which cause increased losses above the attenuation caused by the lining. For plane waves * traveling along the longitudinal axis of a duct, L_{TL} can be calculated. In practice, the noise input may consist of many waves that impinge on the inlet to the muffler, both parallel to the longitudinal axis of the muffler and from other directions. However, the sound power entering the inlet from the oblique waves is usually not significantly larger than that from waves incident parallel to the longitudinal axis. When the duct is wide compared with a wavelength the open area presented to a wave incident at an angle θ to the axis is decreased by a factor $\cos \theta$. When the duct is narrow, angle of incidence becomes irrelevant. In either case, both on entering the inlet of the duct or muffler and during propagation inside it, the sound waves undergo processes of reflection, scattering, and diffraction which, after a short distance of travel, cause the amplitudes and directions of propagation of individual waves to become changed into a form determined more by the duct than by the origin of the waves.

A calculation of L_{TL} of a dissipative muffler which considers only waves whose direction of travel is parallel to the longitudinal axis of the duct

---

* A truly plane wave cannot exist in a duct. Because of losses at the boundary, the wavefront near the side walls is bent toward them. Furthermore, in the presence of a steady gas flow the wavefronts may become more (or less) bent toward the walls depending on whether the sound is being propagated in the same direction as (or against) the steady flow.

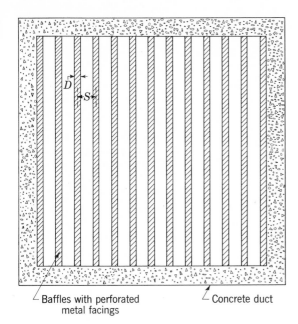

⌐ Baffles with perforated   ⌐ Concrete duct

  metal facings

Fig. 12.20 Elevation view of the end of a large duct containing parallel baffles. In engine test cells, each baffle is constructed of two parallel sheets of perforated metal, filled with glass or mineral wool. Special constructions and materials are needed for high temperatures and gas flows.

has the virtue of simplicity, but results in a "conservative" (low) estimate of L_{TL} if the inlet and outlet ducts are short. Such a simple calculation is reasonably accurate, however, when the inlet and outlet ducts are reasonably straight and acoustically untreated for a distance equal to several duct widths on either end of the muffler.

The calculation of L_{TL} of lined ducts involves four steps which are described in the following sections.

12.3.1 Attenuation in Lined Ducts of Plane Waves Traveling along the "x" Axis

The attenuation of a sound wave traveling along the x axis of a lined duct is dependent primarily on the duct length, the thickness of the lining, the flow resistance of the lining, the width of the air passage, the wavelength of the sound, and superimposed airflow.[10-21] Cremer's theory [11,12] contained in Chap. 15, is probably the most accurate available for the calculation of duct attenuation.* However, a series of simplified charts are given here from Ingard [13] that are valid over a fairly wide frequency range and that show explicitly the relations among the

* Reference 12 contains a condensed English version of the original German work given in Ref. 11.

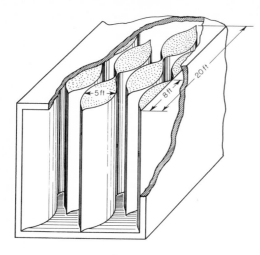

Fig. 12.21 Isometric drawing for a Soundstream® muffler. The columns are separated so that the lateral distance between them is 4 ft 10 in.

various parameters and attenuation. The reader should learn to use both.

The attenuation characteristics of rectangular ducts that have flow-resistive linings mounted directly on two walls (see Fig. 12.19) with no facing or a facing with relatively large (20% or more) open area are given in Fig. 12.22. The other two walls are assumed to be rigid, hard surfaces.

The attenuation parameter Al_y is the attenuation per length of duct equal to the duct width l_y. The attenuation for a duct of length l_e is, therefore, $Al_e = Al_y(l_e/l_y)$. The frequency parameter l_y/λ is the ratio of the width between the linings to the wavelength of sound in air at the frequency (and temperature) of interest. To facilitate the use of this graph a frequency scale for an air duct 1 ft wide at an ambient temperature of 25°C (77°F) has been added. To use this chart for a duct 2 ft wide, for example, the frequency scale shown for a 1-ft duct must be halved.

The flow resistance R_1t of the duct linings for which these data are most nearly applicable is given on the graph in a dimensionless form. The attenuation is not very sensitive to changes in the flow-resistance parameter, and this graph may be applied over an appreciable range of values of the flow-resistance parameter, say, from one-half to twice the nominal value given.

If the duct is lined on four sides, the total attenuation may be obtained by adding, arithmetically, the attenuation of the "sides" to the attenuation of the "top and bottom" of the duct. A square duct lined on four sides will have about twice the attenuation of a duct lined on only two opposite sides. If the duct is lined on only one side, the attenuation is approximately equal to the attenuation calculated for an equivalent duct which is lined on two sides and which is twice as wide as the actual duct.

The attenuation for a sound wave traveling along the longitudinal axis of a set of parallel baffles, such as that shown in Fig. 12.20, is determined from Fig. 12.22 by considering each channel as a rigid rectangular duct of width S lined on two sides with a layer of acoustical material of thickness $D/2$. Obviously, this approach will only be valid if the wave enters the baffles in the x direction, so that all channels receive the wave in the same phase. However, after the wave has traveled a distance equal to about $\lambda/2$, the attenuation is that for x-direction propagation. The difference is an end effect.

The attenuation for parallel baffles decreases at high frequencies because of the "beaming" effect. A design that provides parallel channels of constant width, but without visual openness, is shown in Fig. 12.21. Typical attenuation curves are shown in Fig. 12.23. The increase in attenuation at high frequencies is apparent.

At very low frequencies, $l_y/\lambda < 0.1$, a rough approximation for attenuation [19] may be used even though the flow resistance is other than

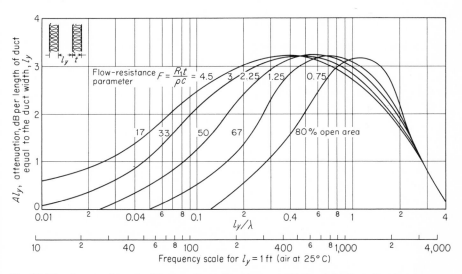

Fig. 12.22 Attenuation of axial waves (traveling along x axis) for rigid ducts lined on two sides with a porous material. The flow-resistance parameter F is nondimensional so that t (the thickness of the lining), R_1 (the flow resistance per unit thickness), and ρc (the characteristic impedance of gas) must be expressed in consistent units. A is the attenuation in decibels per unit length of duct. Al_y, the ordinate scale, is the attenuation per length of duct equal to the duct width l_y. To find the attenuation for a duct length l_e multiply the ordinate by l_e/l_y. To achieve the attenuation shown, for each percentage open area indicated, the flow resistance of the lining should be chosen to yield the proper flow-resistance parameter F.

that shown in Fig. 12.22.   Sabine found empirically that at very low frequencies the attenuation of a lined duct could be expressed as

$$A' = 12.6 \ (\bar{\alpha}_{Sab})^{1.4} \frac{P}{S} \quad dB/ft \quad (12.13)$$

where A' = attenuation, dB/ft
  $\bar{\alpha}_{Sab}$ = Sabine sound-absorption coefficient for the duct lining (dimensionless) [see Eqs. (9.1) and (9.2)]
  P = acoustically lined perimeter of duct, in.
  S = cross-sectional open area of duct, in.²
This formula becomes increasingly inaccurate as l_y/λ increases, and its use must be restricted to values of $l_y/\lambda < 0.1$ (low frequencies).[15]

Cremer (see Chap. 15) has shown that when the impedance of the lining is optimum the maximum attainable value of the attenuation parameter Al_y tends to a value of 19 dB (for a length equal to a duct width) at low frequencies and $7\lambda/l_y$ dB at higher frequencies.

12.3.2   Effect of a Steady Airflow   When the sound waves in a lined duct propagate through a steady gas flow, the effect is usually to increase the attenuation of the sound if the sound is traveling in an opposite direction to the flow, and to decrease the attenuation if the sound energy is traveling in the same direction as the steady flow.[1]   Several mechanisms contribute to this phenomenon,[21, 22] the most obvious being the convection effect of the flow itself.   The sound waves are carried along in the steady flow and the time available for the pressure alternations of the sound to interact with the acoustical lining of the duct is multiplied by a factor of $1/(1 + M)$, where M is the Mach number of the flow in the

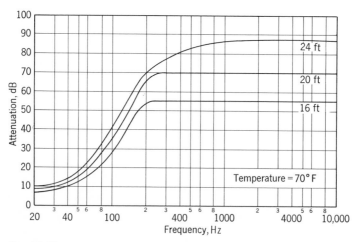

Fig. 12.23   Performance curves for three lengths of standard Soundstream® absorbing units, constructed as shown in Fig. 12.21.

direction of the sound propagation. If the velocity of the steady flow is equal to the speed of sound, the attenuation of sound traveling upstream against the flow is very large; M equals -1 for this situation.

Another mechanism affecting attenuation with airflow is the change in the absorptive properties of the duct lining as a function of steady flow. For materials having large numbers of small interstices, the increase in acoustic resistance is found to be nearly proportional to the magnitude of v, but independent of its direction. The change in attenuation of the duct depends on the relation between attenuation per unit distance and the flow resistance of the lining. Taking both these mechanisms into account we therefore find the attenuation per unit length of the duct to be (where γ is an experimental constant dependent on many factors)

$$\text{Attenuation} = \frac{A(1 + \gamma|M|)}{1 + M} \qquad (12.14)$$

where A is the attenuation without flow, $A\gamma|M|$ is that part of the attenuation which depends on the magnitude of the velocity, and the denominator represents the convective effect. For small values of M, Eq. (12.14) can be expanded

$$\text{Attenuation downstream} = A[1 + (\gamma - 1)M]$$

and $\qquad (12.15)$

$$\text{Attenuation upstream} = A[1 + (\gamma + 1)|M|]$$

Assuming that γ is positive, Eq. (12.15) shows that the attenuation upstream is always increased by these two mechanisms; downstream the attenuation may be increased (for $\gamma > 1$) or decreased ($\gamma < 1$).

12.3.3 Transmission Loss of Finite Lined Ducts and of Expansion Chambers

The transmission loss of a finite length of a lined duct can be found by considering it as a lined expansion chamber. In this way one can account for the reflective effects of the change in cross section at the beginning and end of the lining. In most cases L_{TL} is found to be greater than the sum of the attenuation and the reflection effects considered separately.

The effects of adding dissipation to an expansion chamber can be derived [3,4] from the analysis that led to Eq. (12.4) by changing the phase-propagation exponent from kl_e to $kl_e - j\sigma l_e/2$. The expression for the transmission loss is

$$L_{TL} = 10 \log_{10} \left\{ \left[\cosh \frac{\sigma l_e}{2} + \frac{1}{2}\left(m + \frac{1}{m}\right) \sinh \frac{\sigma l_e}{2} \right]^2 \cos^2 kl_e \right.$$

$$\left. + \left[\sinh \frac{\sigma l_e}{2} + \frac{1}{2}\left(m + \frac{1}{m}\right) \cosh \frac{\sigma l_e}{2} \right]^2 \sin^2 kl_e \right\} \qquad \text{dB} \qquad (12.16)$$

where σ = energy attenuation per unit length for lined duct neglecting
end losses (to obtain σ from dB/m, divide by 4.34)

m = ratio of cross-sectional area of expanded or lined section to
cross-sectional area of inlet and outlet sections of duct. The
presence of the sound-absorbing lining is ignored in determining m. Implicitly, therefore, we have assumed that the
thickness of the lining is small compared with the wavelength
of sound λ_m in the material comprising the lining

$k = \omega/c = 2\pi f/c$

l_e = length of expanded or lined section, m

Substitution of $1/m$ for m in no way affects Eq. (12.16). Thus a "con-

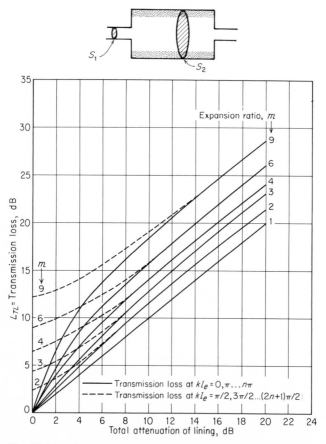

Fig. 12.24 The transmission loss of a lined expansion chamber
(see sketch) as a function of the area expansion ratio $m = S_2/S_1$ and
the total attenuation of the lining ($4.34\sigma l_e$ dB). The solid curves
show L_{TL} for $kl_e = 0$, π, 2π, etc., and the dotted curves show L_{TL}
for $kl_e = \pi/2$, $3\pi/2$, $5\pi/2$, etc.

traction" chamber has the same L_{TL} as an expansion chamber; the difference lies in the static pressure drop between the inlet and outlet ends of the chamber. For a "nonexpansion" chamber, $m = 1$, one obtains from Eq. (12.16) the correct valué for L_{TL} of a lined duct

$$L_{TL} = \text{total attenuation} = 10 \log_{10} e^{\sigma l_e}$$

$$= 4.34 \, \sigma l_e \quad \text{dB} \quad (12.17)$$

Because m and $1/m$ are interchangeable in the equations, and the high frequency limit of validity of the plane-wave theory is set by the transverse dimensions of the duct, which must be less than 0.8λ, it follows that this theory can be used to higher frequencies in contraction than in expansion chambers.

When finding L_{TL} of lined ducts, it is usually not necessary to evaluate Eq. (12.16) at all values of the frequency parameter kl_e. It is usually sufficient to find L_{TL} at the minima ($kl_e = 0$) and the maxima ($kl_e = \pi/2$). L_{TL} at $kl_e = 0, \pi, \ldots, n\pi$ and at $kl_e = \pi/2, 3\pi/2, \ldots, (2n + 1)\pi/2$ is given as a function of the total attenuation of the lining and the expansion ratio m in Fig. 12.24.

For large attenuations, L_{TL} at $kl_e = 0$ approaches L_{TL} at $kl_e = \pi/2$. Then L_{TL} is equal to the attenuation plus $20 \log_{10} [(1 + m)(1 + 1/m)/4]$. Figure 12.24 may be used to estimate L_{TL} for mufflers with an arbitrary expansion ratio and known attenuation characteristics. L_{TL} for other frequencies (values of kl_e) will lie between the value at $kl_e = \pi/2$ and the value at $kl_e = 0$.

Figure 12.25 shows L_{TL}'s of lined and unlined (lowest curve) expansion chambers for $m = 9$ and $m = 25$. The total attenuation for sound waves propagating through the duct of length l_e is given as a parameter. There is a large increase in L_{TL} at $kl_e = 0$ as the total attenuation is increased from 0 up to 6 dB, a small increase at $kl_e = \pi/2$, and for large σl_e (e.g., equal to 6 dB) a relatively small dependence on kl_e.*

The addition of dissipation to an expansion chamber (1) increases the frequency range in which a certain L_{TL} may be achieved, (2) lowers the expansion ratio required to achieve a given L_{TL} in a specified frequency range, and (3) decreases the length of muffler needed to obtain a given L_{TL} for a fixed expansion ratio.

12.3.4 Transmission Loss of Transverse Waves in Lined Ducts   It
is useful to consider briefly the enhanced attenuation, per unit length of duct, for very high frequency (short-wavelength) waves traveling obliquely to the axis of the duct.

---

* Observe from Fig. 12.25 (for $m = 9$) that when the attenuation (created by adding a dissipative lining to the expansion chamber) is only 2 dB, the transmission loss, at $kl_e = 0$, is 6 dB. Thus the increase in L_{TL} is three times as great as the increase in attenuation.

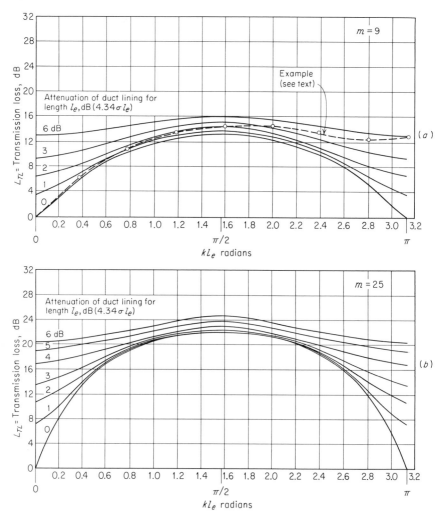

Fig. 12.25   The transmission loss of lined expansion chambers for area expansion ratios of 9 and 25 as a function of nondimensional frequency. The parameter is the total attenuation of the lining, $4.34\sigma l_e$ dB. Here, $k = 2\pi/\lambda$, l_e = length of the expansion chamber in meters (or feet), and $m = S_2/S_1$.

In practice, even a high-frequency sound wave traveling inside a duct is not specularly reflected at the walls of the duct, but is reflected or scattered in various directions with a distribution of amplitudes. This sound energy, when it next impinges on the walls of the duct, has various angles of incidence associated with the various amplitudes. As this process repeats itself, the relative amplitudes and directions of propagation become increasingly influenced by the shape of the duct and the absorption and scattering properties of its walls.   It is only near the inlet and outlet of the duct that the distribution is determined mainly by the sound field external to the duct.   In general the waves have both grazing z and normal y components, and the noise reduction characteristics described above are applicable to each component.   In particular, the more oblique components are attenuated more rapidly than waves traveling nearly parallel to the axis, because they interact with the walls for a longer time in traveling a given distance through the muffler.   However, as a result of the interactions between the sound waves and the surfaces of the muffler, the sound energy is continually being redistributed among the various possible directions of propagation.   The result is dynamic equilibrium among the component waves being propagated in the various directions, with a net transfer of energy from the more axial components to the more oblique components (which have a larger attenuation per unit length of duct).   This accounts for the fact that measured attenuations are usually greater than those given by the theoretical predictions of Fig. 12.22.

One would expect the influence of a steady flow to be as marked for waves traveling obliquely down the duct as it is for plane axial waves. All the effects described earlier still exist, and in addition the convective effect of the steady flow changes the angle of incidence of the oblique waves at the surfaces of the duct walls—waves traveling downstream become more axial to the duct and those propagating upstream become more transverse.

12.4   Lined and Unlined Bends

In air-handling systems, bends in ducts are usually necessary for physical reasons.   If the duct is lined with sound-absorbing material, a bend will contribute to the total transmission loss of the duct.   The acoustical effectiveness of a lined bend can be illustrated qualitatively by Fig. 12.26a. At plane B just before the bend, mainly axial waves exist because any oblique components which may have been present at plane A have been selectively suppressed by the lining.   At the bend, mainly diffracted waves are transmitted into the vertical duct because the end wall either absorbs or reflects the axial waves.   Between planes C and D, the oblique components are selectively suppressed and mainly axial waves exist by the time the energy arrives at plane D.   The usual definition of Noise Reduction does not apply because it requires that a difference between the noise levels at planes B and C be used to determine the effect of the bend. The very important fact that the Noise Reduction between planes C and D is increased because the bend transfers energy from near-axial into more oblique waves is not considered.   The most useful way to evaluate a bend in a lined duct is to measure the insertion loss at some point, such as plane D, well beyond the bend.   One can obtain an equivalent result by measuring the sound-pressure levels along the center line of the whole duct and plotting sound-pressure levels as a function of distance.[23]   The plot is a straight line in front of and also beyond the bend, but there is a transition in the vicinity of the bend.   The "height" of this transition is the insertion loss of the bend.

Unlined Bends.   The insertion loss of unlined bends is negligible, since, in the absence of any dissipative mechanism or any sharp mismatch of impedance which would reflect sound energy, there is little to prevent the sound energy from flowing around the bend; see Fig. 12.26*b*.

Lined Bends.   The insertion loss of a lined bend is obtained by three mechanisms, partial absorption of near-axial waves at the wall of the bend, partial reflection of these near-axial waves back toward the source, and transfer of sound energy into transverse modes that are attenuated

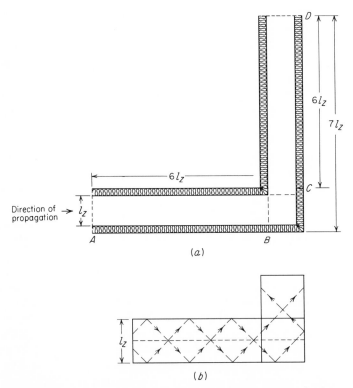

Fig. 12.26 Sketches used in the discussion of the noise reduction characteristics of 90° bends in ducts: (*a*) a typical lined bend; (*b*) a typical unlined bend showing how oblique waves in the horizontal portion of the duct are transmitted around the bend.

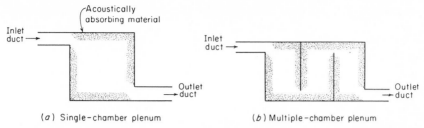

(a) Single-chamber plenum   (b) Multiple-chamber plenum

Fig. 12.27 Sketch showing (a) a single-chamber plenum and (b) a multiple-chamber plenum. The inlet and outlet ducts are staggered to minimize the direct transmission of sound at high frequencies. The amount of sound absorption in each chamber should be as large as possible.

by the lining beyond the bend. At low frequencies, the insertion loss of a lined bend, apart from the absorption due to the lining, is the same as the insertion loss of an unlined bend.[20, 24-27] At frequencies well above $l_z/\lambda = 0.5$ the insertion loss of a lined bend is increased considerably due to the conversion of sound energy from waves propagating in a generally axial direction into highly oblique waves. Field measurements indicate that the insertion loss of a lined bend at high frequencies is usually about 10 dB greater than that of the same length of lined duct. This is valid provided the lining extends 2 to $4l_z$ on each side of the bend.

Very little data exist concerning the insertion loss of bends at angles greater or less than 90°.[23, 28] As a rough approximation, one may assume that insertion loss is proportional to the angle. For example, one might estimate the insertion loss of a 30° bend to be one-third that of a 90° bend. It appears that the random-incidence Noise Reduction (and probably also insertion loss) for a 180° bend (duct foldback) is about $1\frac{1}{2}$ times the Noise Reduction for a 90° bend.[20]

12.5   Plenum Chambers

To perform well as a noise reducing element the interior surfaces of a plenum chamber should be covered with sound-absorbing material, and the inlet and outlet ducts should not be located directly opposite each other (see Fig. 12.27). Because the linear dimensions of a plenum are usually large compared with a wavelength of the sound field, its acoustical performance is closely related to that of a large enclosure; see Chap. 9. Plenum theory can also be applied to an acoustically treated corridor with doors into adjacent rooms staggered so as to minimize sound transmission from room to room.

A plenum having n chambers does not have n times the transmission loss of a single chamber. However, assuming one knows the transmis-

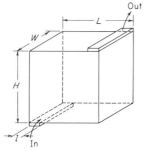

Fig. 12.28 Diagram of a single-plenum chamber showing nomenclature used in the text.

sion loss of a single chamber and also that of a double chamber, it is possible to predict the transmission loss of n chambers as[29]

$$L_{\mathrm{TL}_n} = (n - 1)L_{\mathrm{TL}_{\mathrm{double}}}$$

$$- (n - 2)L_{\mathrm{TL}_{\mathrm{single}}} \qquad \mathrm{dB} \qquad (12.18)$$

The geometry for a single plenum is given in Fig. 12.28, and its transmission loss can be found approximately by noting that the energy density at any point within a reverberant chamber can be written as the sum of the direct field and the reverberant field. The sound power W_{out} leaving the outlet from the direct field is $W_{\mathrm{in}}S \cos \phi/2\pi q^2$,* and from the reverberant field is $W_{\mathrm{in}}S (1 - \bar{\alpha})/a$, where

W_{in} = sound power radiated into the plenum from the inlet duct, watts. It is assumed that the inlet to the plenum behaves, acoustically, like a perfectly absorbing surface.

W_{out} = sound power transmitted from the outlet, watts

S = area of outlet, ft² (or m²)

$\bar{\alpha}$ = random-incidence absorption coefficient of plenum lining, dimensionless

a = total absorption of lining, sabins (or m²) (total lined area in chamber times random-incidence absorption coefficient)

q = slant distance between inlet and outlet openings, ft (or m)

$\cos \phi = H/q$

Thus the transmission loss of the single plenum is given by

$$L_{\mathrm{TL}} = -10 \log_{10} \left(\frac{W_{\mathrm{out}}}{W_{\mathrm{in}}}\right) = -10 \log_{10} \left[S \left(\frac{\cos \phi}{2\pi q^2} + \frac{1 - \bar{\alpha}}{a}\right)\right] \qquad \mathrm{dB} \quad (12.19)$$

This formula agrees with measurements at high frequencies and for values of l/L (see Fig. 12.28) that are not too large.[29] The calculated values are lower than measurements by 5 to 10 dB at low frequencies, which is not surprising since the plenum dimensions are then comparable with the wavelength, while the theory assumes the dimensions to be much greater than the wavelength.

When the absorption coefficient of the lining is large the transmission

---

* The factor 2 in the denominator arises because the inlet duct is assumed to radiate into a hemisphere. If the adjacent walls of area $L \times W$ and $H \times W$ are nonabsorbing, this factor tends to unity.

loss of the plenum is limited by the direct transmission between inlet and outlet, Eq. (12.19). The transmission loss of a special plenum designed to eliminate this direct path as well as other paths with only a small number of reflections[29] is much more effective than the ordinary single-chamber plenum.

12.6 Aerodynamic Performance

12.6.1 Back Pressure
When a steady airflow passes through a duct or muffler, there is a steady pressure drop which is related to the magnitude of the flow and the geometry of the air passages. If there is an alternating (acoustic) flow superimposed on the steady flow it is the instantaneous velocity, the sum of steady and alternating, that determines the pressure drop. This *back pressure* is important because it may influence the mechanical performance of a system to which the muffler is attached.

Consider the simple muffler shown in Fig. 12.15 which consists of a chamber and a tailpipe. The steady pressure drop through such a muffler is made up of three principal components:

1. A pressure drop ΔP_c caused by expansion and contraction of flow in the chamber
2. A pressure drop ΔP_l caused by turbulent flow in the tailpipe
3. A pressure drop ΔP_x caused by expansion of the flow at the end of the tailpipe

Component 1. ΔP_c is a function of the chamber dimensions and the relative size of the inlet and exhaust ducts to the cavity. The following design features tend to reduce ΔP_c:

1. Avoid square corners at the junctions of the ducts with the cavity.
2. In straight-through muffler design, locate the inlet and exhaust ducts in line, and make the exhaust duct somewhat larger than the inlet to allow for expansion of the jet flow. This requirement is not compatible with good design for high transmission loss at high frequencies.
3. Avoid designs that require the flow to change direction unless necessary for reasons of space or large noise reduction. Bends increase the back pressure considerably compared with a linear design. If change of direction of flow is necessary, it should be done at reduced velocity, if possible.

Component 2. ΔP_l, the pressure drop owing to turbulent flow in the tailpipe, may be expressed as

$$\Delta P_l = \left(4 \frac{F_m l}{d}\right)\left(\frac{1}{2}\rho v^2\right) \quad \text{N/m}^2 \quad (12.20)$$

where l = tailpipe length, m

d = tailpipe diameter, m

ρ = average gas density, kg/m³

F_m = Fanning friction factor for tailpipe, dimensionless

v = instantaneous linear velocity in tailpipe, m/sec

For turbulent flow in commercial steel pipe, or other pipe of comparable smoothness (roughness), F_m is approximately 8×10^{-3} when $l/d \gg 1$.[30]

The instantaneous value of ΔP_l depends on the square of the total instantaneous particle velocity in the tailpipe. This means that the presence of an alternating particle velocity (owing to a sound wave) will tend to increase ΔP_l. In the case of a noise source such as a reciprocating engine, significant alternating velocities are generated at the fundamental firing frequency and its harmonics. Thus, it may be worthwhile to use a muffler with appreciable attenuation at the fundamental firing frequency and harmonics to reduce the alternating velocities and thereby reduce ΔP_l. Of course, care must be taken not to make l too large or d too small, since Eq. (12.20) indicates that these changes tend to increase ΔP_l.

Component 3. Experiments indicate that ΔP_x, the pressure drop owing to end expansion, is approximately

$$\Delta P_x \approx 0.4(\tfrac{1}{2}\rho v^2) \qquad \text{N/m}^2 \qquad (12.21)$$

where 0.4 is a dimensionless constant. Thus, reduction of any important alternating component will aid in reducing ΔP_x also. This behavior has been observed experimentally. A cavity muffler with a long tailpipe could improve the mechanical performance of a small reciprocating engine, compared with the unmuffled engine or with a muffler with a shorter tailpipe.

12.6.2 Noise Generation by Steady Airflow

Gas flow can produce noise. Attenuation and self-noise combine to determine the overall noise reduction of a muffler, and these factors are different functions of flow velocity.[31]

Consider a single muffler element inserted in a duct that extends infinitely far in either direction (Fig. 12.29). The duct carries a steady flow which is responsible for the generation, inside the muffler, of a noise power N; assume that half is transmitted downstream and half upstream.

Superimposed on the flow entering the muffler from the inlet duct is an input noise power W, of which a fraction τ is transmitted to the outlet duct. In general, N, W, and τ are each functions of frequency. The noise power entering the outlet duct is $\tau W + N/2$, and the insertion loss is therefore given by

$$L_{IL} = 10 \log_{10} \left(\frac{W}{\tau W + N/2} \right)$$

$$= 10 \log_{10} \left(\frac{1}{\tau} \right) - 10 \log_{10} \left(1 + \frac{N}{2\tau W} \right) \quad dB \quad (12.22)$$

where W = noise power incident on the muffler from the inlet duct
τ = fraction of the incident sound power transmitted to the outlet duct
N = noise power generated inside the muffler by the steady gas flow

The implications of Eq. (12.22) are

1. The insertion loss L_{IL} is always less than the L_{TL} of the muffler without self-noise (first term on the right side of the equation).

2. Regardless of the value of τ, the minimum sound power entering the outlet duct is $N/2$. Thus the insertion loss of the muffler is negative (the muffler is worse than no muffler) at those frequencies and flow velocities for which $N/2 > W$. The insertion loss is positive only when $W > (\tau W + N/2)$. Obviously, this term must be examined as a function of frequency.

3. The insertion loss L_{IL} approaches the transmission loss L_{TL} for large values of the input sound power, i.e., when $W \gg N/2\tau$.

In some cases, for example in a lined duct, a muffler cannot be considered as a lumped unit but rather as a continuous distribution of elements over an extended distance. If the acoustical elements are identical at all points along the treated section of duct, the noise attenuation may be described by a constant β per unit length and the noise generation by a second constant n per unit length. The insertion loss for a uniformly distributed muffler of length l is shown [31] in Fig. 12.30 for two values of β and three values of the ratio of the noise generation coefficient n to W as a parameter. It is seen that the value of β determines the initial part of the curves (and hence the insertion loss for short mufflers) while the value of n/W determines the maximum insertion loss for long mufflers.

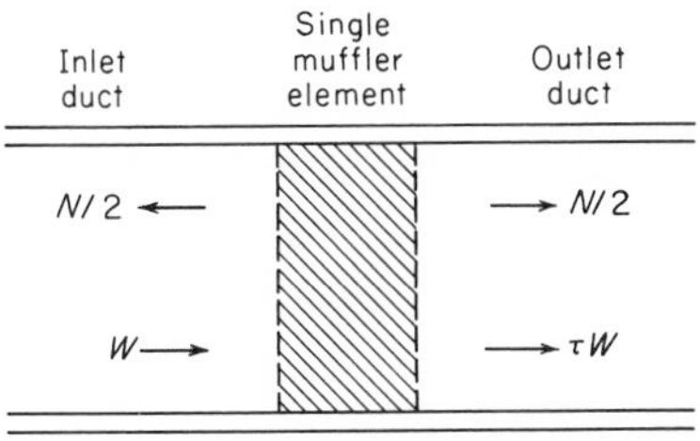

Fig. 12.29 Schematic diagram of a muffler having both attenuation and noise generation characteristics. A fraction τ of the incident noise power W is transmitted into the outlet duct. Steady flow through the muffler generates a noise power N, and it is assumed that half of this is propagated into the duct in each direction. The noise power at the outlet is the sum of the two.

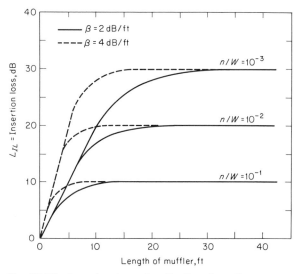

Fig. 12.30   Insertion loss of a distributed muffler (e.g., a lined duct) having both attenuation and noise generation characteristics. The initial slope of the curves, for short mufflers, is determined by the attenuation coefficient per unit length, β. The asymptotic value, for long mufflers, is determined by the ratio of the noise power n generated in the muffler per unit length to the initially incident noise power, W. (*After Ingard.*[31])

12.7   Mufflers for Large-amplitude Sound Waves

In the linear regime, dissipative mufflers absorb acoustic energy by converting it into heat.   In reactive mufflers most of the sound energy is reflected back toward the source rather than being absorbed within the muffler, though some small fraction of the energy is converted to heat by the same process as in a dissipative muffler.   When the amplitude of the incident sound waves becomes very large (sound-pressure levels in excess of 120 dB) the oscillatory flow becomes turbulent, and at any given point in the vicinity of fixed obstacles it does not repeat itself merely with a change of direction half a cycle later.   This again leads to the conversion of sound energy into heat, though the "rules" are not very well understood.

12.7.1   Nonlinearity and Acoustic Impedance of an Orifice   Using a hot-wire probe to measure the velocity of alternating flow through a small 7-mm-diameter hole in a plate 0.1 mm thick, the relation between the incident sound pressure and the peak orifice velocity over a range of incident sound-pressure levels from 110 to 162 dB has been deter-

mined.[32]   No steady flow was involved (see Fig. 12.31). At pressure amplitudes up to about 100 N/m² ($L_p = 131$ dB) the incident pressure amplitude p_0 is proportional to the velocity amplitude in the orifice, u_0, while at higher pressures the pressure is proportional to u_0^2. The waveform of the incident sound pressure is sinusoidal at all amplitudes, and at low levels the velocity waveform is also sinusoidal, indicating a linear relation.

At higher levels the velocity waveforms are significantly distorted, as revealed by the quadratic relationship shown in Fig. 12.31. In the quadratic region the incident sound-pressure amplitude and velocity in the orifice are related approximately by $p_0 = \rho u_0^2$, where ρ is the density of the gas — this implies a pressure-velocity ratio equal to ρu_0. Knowing the phase of the velocity relative to that of the incident sound pressure, both the in-phase (resistive) and the 90° difference-in-phase (reactive) components of the ratio of the rms sound pressure to the rms particle velocity (impedance) can be found, and these are shown in Fig. 12.32 as a function of the orifice particle velocity.

In the linear region (at low sound pressures) the flow near the orifice is nonturbulent and the acoustic pressure is mainly proportional to the acceleration of the flow in this region. At low frequencies the viscous drag is very small and can be neglected; both the pressure and velocity waveforms are sinusoidal but 90° out of phase (the reactive component

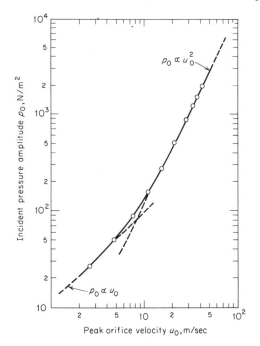

Fig. 12.31 Measured peak particle velocity in an orifice of 7-mm diameter, as a function of peak sound pressure of the incident waves, at 150 Hz. At low amplitudes the relation is linear; at larger amplitudes it is quadratic. (*After Ingard and Ising.*[32])

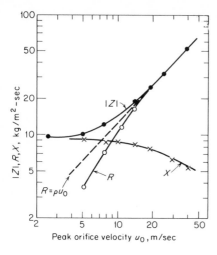

Fig. 12.32 Acoustic impedance of an orifice $|Z|$ and its resistive and reactive components R and X as a function of peak particle velocity amplitude u_0. Also shown is the line $R = \rho u_0$. The magnitude of the ratio of the incident rms sound pressure to the orifice rms particle velocity is $|Z|$; R is the component of Z that expresses the in-phase component of the ratio of p to u; and X is the component of Z that expresses the 90° difference-in-phase component of the ratio of p to u. $|Z| = \sqrt{R^2 + X^2}$. (*After Ingard and Ising.*[32])

dominates). The flow pattern is similar (apart from direction) at any instant on either side of the plate.

Now consider the nonlinear region (high sound pressures). Flow separation occurs at the outlet side of the orifice, thereby creating a central jet through the hole which is surrounded by a "doughnut-shaped" rotational flow (turbulence) in the shadow of the plate. The inflow remains essentially irrotational. The flow is now not symmetrical on the two sides of the plate. Half a cycle later the direction of the particle velocity in the orifice is reversed. The flow patterns are now interchanged. On both sides of the orifice one finds an outflowing, pulsating jet and a rotational "doughnut," but they occur alternately.

12.7.2 Resonant Absorbers for High Intensity Sound

Let us now consider the behavior of an array of closely spaced volume resonators which may, for example, be used as a resonator type of sound-absorption treatment on a plane wall or as the lining of an absorbing duct.[1,32,33] Such a construction typically consists of a plate perforated with an array of holes or slots spaced a certain distance away from a rigid, impervious wall, and this space is usually subdivided into compartments; see Fig. 12.33a. The ratio of the sound pressure in the duct p_1 to the average particle velocity through the perforated surface can be expressed as

$$\frac{p_1}{u_{\text{plate}}} = Z_b = \frac{Z}{\sigma} + Z_c \qquad (12.23)$$

where σ = ratio of the area of each orifice to that portion of the complete surface of the plate associated with it. σ is the percentage open area of the plate.

Z = orifice impedance, namely, the ratio of p_{duct} to u_{orifice}, i.e., p_1/u_0, N-sec/m³

Z_c = acoustic impedance of the air cavity behind each orifice as viewed from the perforation, i.e., ratio of p_{cavity} to u_{orifice} = p_c/u_0, N-sec/m³

The resistive part of the orifice impedance is ρu_0. Allowing for the fact that the area of the orifices may be a significant fraction of the total area, the absorption coefficient α (dimensionless) corresponding to the impedance of Eq. (12.23) is obtained as

$$\alpha = 4(u_0/c\sigma_1)/\{[(u_0/c\sigma_1) + 1]^2 + X^2\} \tag{12.24}$$

where $\sigma_1 = \sigma/(1 - \sigma^2)$

u_0 = particle velocity in an orifice, m/sec

c = speed of sound, m/sec

X = sum of the orifice reactance and the cavity impedance Z_c, N-sec/m³

Impedance

(a)

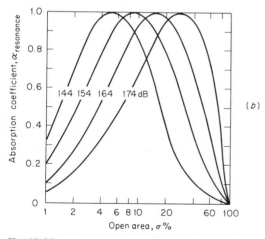

(b)

Fig. 12.33 Absorption of high-intensity sound by an array of volume resonators. (a) Diagram of the array of resonators; (b) absorption coefficient at resonance as a function of the fractional open area, σ, of the surface. The parameter is L_p in decibels of the sound pressure in the duct. (*After Ingard and Ising*.[32])

One notes that the absorption coefficient α is a function of u_0, and hence of the duct sound-pressure amplitude p_1. It is also a function of the fractional open area σ and of frequency, having a maximum value when the cavity is at resonance, at which time $X = 0$.

To obtain complete absorption at the resonance frequency, the value of σ is dependent on the magnitude of the duct sound pressure and is given approximately by the formula

$$\sigma \approx \sqrt{\frac{p_1}{\gamma p_s}} = \sqrt{\frac{p_1}{\rho c^2}} \tag{12.25}$$

where $p_1 =$ duct sound pressure, N/m²
  $p_s =$ static pressure, N/m²
  $\gamma =$ ratio of principal specific heats of the gas ($\gamma = 1.4$ for air)
For example, when $L_{p_1} = 154$ dB, i.e., $p_1/\gamma p_s = 0.01$, we obtain $\sigma = 10\%$; when $L_{p_1} = 160$ dB, $\sigma = 14\%$. The absorption coefficient at resonance as a function of σ and for several different levels of the incident waves is shown in Fig. 12.33b.

In designing a section of duct lined with resonators to attenuate sound waves of large amplitude as efficiently as possible, one must remember that the amplitudes of the waves decrease in passing through the muffler. Near the input end where the incident pressure is large, σ should have a certain value related to p_1 by Eq. (12.25). As sound energy is absorbed, p_1 decreases and the percentage open area σ should be reduced accordingly. This applies, of course, only for cases where the alternating pressures are large enough to produce nonlinear effects in the orifices.

12.8   Prefabricated Dissipative Mufflers

Prefabricated dissipative mufflers for ventilating systems are manufactured in a wide range of lengths, noise reduction characteristics, cross-sectional areas, and airflow performance.* The primary advantage of prefabricated units is that a noise reduction requirement can be satisfied in a much shorter length than is possible by use of lined ducts and bends. Pressure drop should be a major consideration in their selection, and it may be higher or lower than the pressure drop through the length of lined duct, with bends, required for equal noise reduction. The use of a prefabricated muffler may also be necessary if the noise reduction required is very large.

* Among the companies manufacturing such items are Industrial Acoustics Company, New York, N.Y., Koppers Company, Inc., Baltimore, Md., and Burgess-Manning Company, Dallas, Tex.

Most prefabricated mufflers are constructed on a module basis. The cross section of the smallest units commonly sold is 30 × 15 cm (12 × 6 in.) or 30-cm diameter. Parallel combinations of the basic rectangular module may be used in larger ducts. Rectangular units up to 0.6 × 0.9 m as well as units of circular cross section up to 1.5 m in diameter are also available. When prefabricated units are used in parallel, the noise reduction characteristics are approximately the same as for a single unit. Stacking units in parallel or increasing the module size decreases the pressure drop for a given volume of flow and also the internally generated noise power, since both of these effects depend mainly upon the flow velocity (see Sec. 12.6).

Figure 12.34 shows typical cross sections of several prefabricated mufflers. Figure 12.35 shows the insertion loss of various units, each 1.5 m (5 ft) long, measured in a long duct with and without a superimposed steady flow traveling in the same direction as the sound propagation. Other mufflers perform differently, providing a greater insertion loss at some frequencies and less at others. The insertion loss is reduced in the direction of the steady airflow. Similarly the insertion loss is increased in the upstream direction, though usually by a smaller amount.

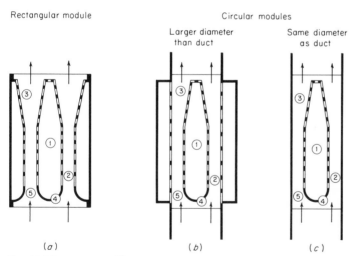

Rectangular module   Circular modules

Larger diameter   Same diameter
than duct   as duct

(a)   (b)   (c)

Fig. 12.34 Cross section of several typical prefabricated mufflers for use in air-handling systems: (a) Rectangular module which may be stacked in parallel to increase the cross-sectional area; (b) circular module in which the diameter of the muffler exceeds the diameter of the inlet and outlet ducts; and (c) circular module for use when the muffler diameter may not exceed the diameter of the duct. (1) Absorbing material; (2) air passage; (3) expansion passage at outlet, the taper is typically linear or exponential; (4) nose, usually solid; (5) bell mouth.

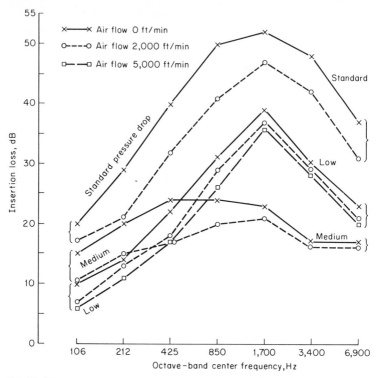

Fig. 12.35 Insertion loss of several dissipative mufflers as a function of frequency. The values given are for mufflers 1.5 m (5 ft) long, having three different internal designs, and, hence, different static pressure drops. For each muffler the effect of an airflow in the direction of sound propagation is also shown. The corresponding pressure drops are shown in Fig. 12.36. (These results are from the 1969 sales literature of two manufacturers.)

The designations of the curves indicate whether the design has a "standard," "medium," or "low" pressure drop for a given volume flow. Rectangular mufflers are supplied by several manufacturers in 0.9-, 1.5-, and 2.1-m lengths. To within a few decibels, the insertion loss is proportional to the length of the muffler.

Figure 12.36 shows the static pressure drop vs. steady-flow velocity for the same group of mufflers. The pressure drop is proportional to the square of the flow velocity, in agreement with Eq. (12.21). In general, larger values of insertion loss are obtained at the cost of increased pressure drop. The pressure drop increases by only about 25% as the length of the muffler is doubled, which indicates that the greater part of the pressure drop is attributable to the changes in flow velocity and direction at the inlet and outlet.

Figure 12.37 shows the self-noise power level (dB *re* 10^{-12} watt) generated internally by a steady flow velocity of 300 m/min (1,000 ft/min). The values relate to mufflers having a face area of 0.09 m² (1 ft²) but are independent of the length of the unit. Because the generated sound power is proportional to the cross-sectional area of the muffler, the curves of Fig. 12.37 may be modified to other areas by adding a term $10 \log_{10} S/S_0$, where S is the desired cross-sectional area in m² (ft²) and S_0 is 0.09 m² (1 ft²).

The internal noise generation is also a sharply varying function of the linear velocity through the muffler. One may adapt the values of Fig. 12.37 to other velocities by increasing the self-noise power level in each octave band by about $60 \log_{10} (V/V_0)$ dB, where V is the linear velocity in m/min (ft/min) and $V_0 = 300$ m/min (1,000 ft/min). This is about 18 dB per doubling of velocity. This correction for velocity may not be valid for all designs of mufflers. The difference in internally generated

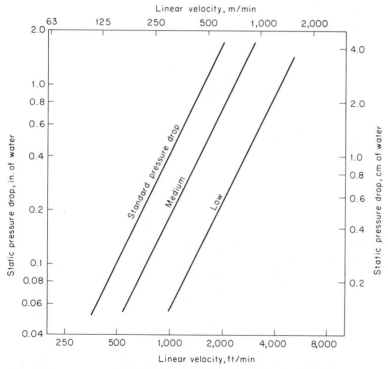

Fig. 12.36   Static pressure drops as a function of steady airflow velocity for the same dissipative mufflers as in Fig. 12.35. Note that pressure drop is proportional to the square of the air velocity, in accordance with Eq. (12.21).

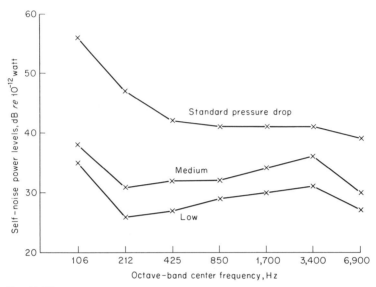

Fig. 12.37 Sound-power levels of noise generated within typical prefabricated mufflers by steady flow. The group of mufflers is the same as in Fig. 12.35. The values are given for a muffler having a face area of 1 ft²; correction to an area of S ft² is made by adding 10 $\log_{10}(S)$ to the sound-power level in each octave band. Similarly, the values in the figure are normalized to a steady flow velocity of 1,000 ft/min; correction to a different velocity V ft/min may be made by adding 60 $\log_{10}(V/1,000)$ to the levels in each octave band, i.e., 18 dB per doubling of velocity.

noise levels between mufflers is small and irrelevant except in a very few situations, such as television studios and concert halls, where background noise levels must be kept very low.

REFERENCES

1. E. Meyer, F. Mechel, and G. Kurtze, "Experiments on the Influence of Flow on Sound Attenuation in Absorbing Ducts," *J. Acoust. Soc. Am.*, vol. 30, pp. 165–174, 1958.
2. K. Gösele, "The Damping Behavior of Reflection-Mufflers with Air Flow" (in German), *VDI-Berichte*, vol. 88, pp. 123–130, 1965.
3. D. D. Davis, Jr., G. M. Stokes, D. Moore, and G. L. Stevens, Jr., "Theoretical and Experimental Investigation of Mufflers with Comments on Engine-exhaust Muffler Design," *NACA Rept.* 1192, 1954.
4. D. D. Davis, Jr., "Acoustical Filters and Mufflers," in C. M. Harris (ed.), *Handbook of Noise Control*, Chap. 21, McGraw-Hill, 1957.
5. G. J. Sanders, private communication, 1967.
6. W. P. Mason, *Electromechanical Transducers and Wave Filters*, Van Nostrand, 1946.
7. K. U. Ingard, "Side Branch Resonators in Ducts," Bolt Beranek and Newman Inc. (unpublished).
8. G. W. Stewart and R. B. Lindsay, *Acoustics*, Van Nostrand, 1930.

9. B. G. Watters, R. M. Hoover, and P. A. Franken, "Designing a Muffler for Small Engines," *Noise Control*, vol. 5, no. 2, pp. 18–22, March, 1959.

10. P. M. Morse, "Transmission of Sound Inside Pipes," *J. Acoust. Soc. Am.*, vol. 11, pp. 205–210, 1939.

11. L. Cremer, "Theory of Damping of Airborne Sound in a Rectangular Duct with Absorbing Walls and the Resulting Maximum Attenuation Constant" (in German), *Acustica*, vol. 3, pp. 249–263, 1953.

12. S. J. Lukasik, A. W. Nolle, and the Staff of Bolt Beranek and Newman Inc., "Physical Acoustics," vol. 1, suppl. 1 of *Handbook of Acoustic Noise Control*, pp. 252–261, *WADC Tech. Rept.* 52–204, April, 1955.

13. See Ref. 12, pp. 217–240.

14. See Ref. 12, pp. 240–249.

15. L. L. Beranek, "Sound Absorption in Rectangular Ducts," *J. Acoust. Soc. Am.*, vol. 12, pp. 228–231, 1940.

16. C. P. Brittain, C. R. Maguire, R. A. Scott, and A. J. King, "Attenuation of Sound in Lined Air Ducts," *Engineering*, p. 97, Jan. 30, 1948.

17. A. J. King, "Attenuation of Sound in Lined Air Ducts," *J. Acoust. Soc. Am.*, vol. 30, pp. 505–507, 1958.

18. I. Dyer, "Noise Attenuation of Dissipative Mufflers," *Noise Control*, vol. 2, no. 3, pp. 50–57 and 78, May, 1956.

19. H. J. Sabine, "The Absorption of Noise in Ventilating Ducts," *J. Acoust. Soc. Am.*, vol. 12, pp. 53–57, 1940.

20. N. Doelling, "Noise Control for Aircraft Engine Test Cells and Ground Run-up Suppressors," vol. 2, *Design and Planning for Noise Control, WADC Tech. Rept.* 58-202(2), November, 1961, prepared under contract by Bolt Beranek and Newman Inc.

21. F. Mechel, "Research on Sound Propagation in Sound-absorbent Ducts with Superimposed Air Streams," vol. 1, *Tech. Doc. Rept.* AMRL-TDR-62-140(1), December, 1962.

22. K. U. Ingard, private communication, 1967.

23. C. P. Brittain, C. R. Maguire, R. A. Scott, and A. J. King, "Attenuation of Sound in Lined Air Ducts," *Engineering*, pp. 97–98, Jan. 30, and pp. 145–147, Feb. 13, 1948.

24. J. W. Miles, "The Diffraction of Sound due to Right-angled Joints in Rectangular Tubes," *J. Acoust. Soc. Am.*, vol. 19, pp. 572–579, 1947.

25. W. K. R. Lippert, "The Measurement of Sound Reflection and Transmission at Right-angled Bends in Rectangular Tubes," *Acustica*, vol. 4, pp. 313–319, 1954.

26. American Society of Heating and Air-conditioning Engineers, "Sound Control," *Heating, Ventilating, and Air Conditioning Guide*, chap. 40, 1957.

27. B. G. Watters, S. Labate, and L. L. Beranek, "Acoustical Behavior of Some Engine Test Cell Structures," *J. Acoust. Soc. Am.*, vol. 27, pp. 449–456, 1955.

28. L. L. Beranek, *Acoustics*, McGraw-Hill, 1954.

29. R. J. Wells, "Acoustical Plenum Chambers," *Noise Control*, vol. 4, no. 4, pp. 9–15, July, 1958.

30. J. G. Knudsen and D. T. Katz, *Fluid Dynamics and Heat Transfer*, University of Michigan Press, 1954.

31. K. U. Ingard, "Attenuation and Regeneration of Sound in Ducts and Jet Diffusers," *J. Acoust. Soc. Am.*, vol. 31, pp. 1202–1212, 1959.

32. K. U. Ingard and H. Ising, "Acoustic Nonlinearity of an Orifice," *J. Acoust. Soc. Am.*, vol. 42, pp. 6–17, 1967.

33. L. L. Beranek, S. Labate, and K. U. Ingard, "Noise Control for NACA Supersonic Wind Tunnel," *J. Acoust. Soc. Am.*, vol. 27, pp. 85–98, 1955.

Chapter Thirteen
Isolation of Vibrations

DOUGLAS MUSTER * and ROBERT PLUNKETT †

Vibration isolation is a means of decreasing transmission of vibratory motions or forces from one structure to another. The term *isolation* as used here means interposing a relatively flexible element between the two structures. The vibration amplitude of the driven structure is often largely controlled by its inertia. If the isolating element is flexible enough, it will transmit little force to the second structure, except at frequencies in the vicinity of resonance. Adding damping in the vibrating system for the purpose of reducing the vibratory response at the resonance frequency may have the concomitant effect of decreasing the isolation that otherwise would be achieved at higher frequencies.

There are two types of vibration-isolating applications: (1) those in which one seeks to prevent transmission of vibratory forces from a machine to its foundation, and (2) those in which one desires to reduce the transmission of motion of a foundation (or a substructure) to a machine

* Brown and Root Professor of Engineering, Department of Mechanical Engineering, University of Houston, Houston, Tex.

† Professor, Department of Aerospace Engineering and Mechanics, University of Minnesota, Minneapolis, Minn.

(or other device) mounted on it. Reciprocating engines and other types of rotating equipment (such as electric motors, fans, turbines, etc.), mounted on vibration isolators, are examples of the first type. Radio equipment in an aircraft, mounted resiliently so that the motion it experiences will be less than that of the airframe, is an example of the second type.

An elastic device that reduces the transmission of vibratory force or displacement is called a *vibration isolator*. By contrast, *vibration dampers*, which have the property of turning motion into heat, serve to reduce the vibration response of a system only in the vicinity of a resonance frequency and are always placed in parallel with another element of the system. This chapter is concerned only with vibration isolation.

13.1   Vibration Isolation

Transmissibility: Force and Displacement. The effectiveness of an isolator is measured by its transmissibility ϵ, of which there are two types. *Force transmissibility* ϵ_f is defined as the ratio of the steady-state amplitude of the force transmitted through the isolator to the amplitude of the exciting force applied to a mass on it. *Displacement (or motion) transmissibility* ϵ_d is defined as the ratio of the displacement amplitude transmitted through the isolator to the exciting displacement amplitude applied to it. Obviously, in both cases the transmissibility depends not only upon the characteristics of the isolator but also upon the properties of its support and of the isolated item. If, for the present, we limit our discussion to the single-degree-of-freedom systems shown in Fig. 13.1, then we can show that the transmissibilities ϵ_f and ϵ_d are functions of the isolator properties and the mass of the isolated item only.

Undamped, Single-degree-of-freedom System. Consider the case of Fig. 13.1a, which shows a simple undamped mass-spring system, under the action of a periodic force $F_0 \cos \omega t$. Superposed on any other displacement, such as a static deflection due to its own weight, the mass m experiences a displacement x due to the force F. The equation which characterizes this motion is

$$m\ddot{x} + kx = F_0 \cos \omega t \qquad \text{N} \tag{13.1}$$

where k = stiffness of the isolator, assumed massless, N/m (lb/ft)

m = lumped mass mounted on isolator, kg (lb)

The steady-state solution of Eq. (13.1) is

$$x = \frac{F_0/k}{1 - (\omega/\omega_n)^2} \cos \omega t \qquad \text{m} \tag{13.2}$$

where $\omega_n^2 = k/m$. The force f_0 transmitted to the foundation is kx, and,

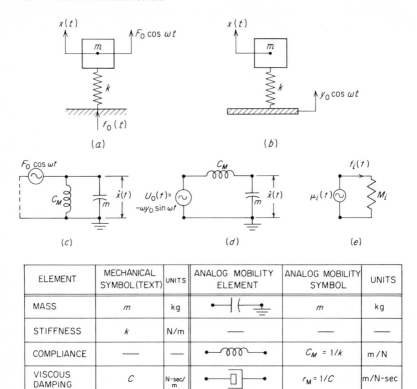

(a)

(b)

(c)   (d)   (e)

ELEMENT	MECHANICAL SYMBOL (TEXT)	UNITS	ANALOG MOBILITY ELEMENT	ANALOG MOBILITY SYMBOL	UNITS
MASS	m	kg		m	kg
STIFFNESS	k	N/m	—	—	—
COMPLIANCE	—	—		$C_M = 1/k$	m/N
VISCOUS DAMPING	C	N-sec/m		$r_M = 1/C$	m/N-sec
MOBILITY	—	—		M_i	m/N-sec

(f)

Fig. 13.1   Mechanical diagrams and symbols for an undamped, single-degree-of-freedom system: (a) force driven and (b) displacement driven. Analogous electrical-type schematic diagrams for the same system as above using the notation of table f with velocity $\dot{x}(t)$ equivalent to electrical voltage: (c) force driven and (d) velocity driven. (e) General relationship where $u_i(t)$ is the instantaneous velocity *across* (voltagelike) the ith element, $f_i(t)$ is the instantaneous force *through* (currentlike) the ith element, and M_i is the mobility (in m/N-sec). In the table (f) the mechanical symbols used in the text of Chap. 13 are shown in the second column. In the fourth and fifth columns the elements and symbols commonly used in books on acoustics are shown in the manner illustrated by (c) to (e). In the analogous schematic diagrams, one side of mass must always be at ground (reference-frame velocity) and elements with the same velocity are joined together.

by definition, force transmissibility is the magnitude of the ratio of transmitted force to applied force, or

$$\epsilon_f = \left| \frac{kx}{F_0 \cos \omega t} \right| \tag{13.3}$$

which by Eq. (13.2) becomes

$$\epsilon_f = \left| \frac{1}{1 - (\omega/\omega_n)^2} \right| \tag{13.4}$$

For the case in Fig. 13.1b, displacement transmissibility is defined as the magnitude of the ratio of the displacement of the mass m to the applied displacement y_0, or

$$\epsilon_d = \left| \frac{x}{y_0 \cos \omega t} \right| \tag{13.5}$$

The equation of motion for the mass m due to the displacement $y_0 \cos \omega t$ is

$$m\ddot{x} = k(y_0 \cos \omega t - x) \qquad \text{N} \tag{13.6}$$

The steady-state solution of this equation is

$$x = \frac{y_0}{1 - (\omega/\omega_n)^2} \cos \omega t \qquad \text{m} \tag{13.7}$$

which when substituted into Eq. (13.5) gives

$$\epsilon_d = \left| \frac{1}{1 - (\omega/\omega_n)^2} \right| \tag{13.8}$$

Thus, we see that the expressions for force and displacement transmissibility are identical, for this example.

An interesting aspect of Eqs. (13.4) and (13.8) is that the expression of which the absolute value is taken becomes negative for $\omega/\omega_n > 1$. This sign change is associated with a change in phase between the direction of the applied force and the motion of the mass; that is, for $\omega/\omega_n < 1$, m moves downward as $F_0 \cos \omega t$ is directed *downward;* for $\omega/\omega_n > 1$, m moves downward as $F_0 \cos \omega t$ is directed *upward.* This implies that the force transmitted to the foundation is either in phase or 180° out of phase with the applied force according as ω/ω_n is less or greater than unity. A plot of transmissibility as a function of frequency ratio (in terms of the absolute values, as defined) is shown in Fig. 13.2.

The principle underlying all vibration isolation is contained in Fig. 13.2. For very low frequency ratios ($\omega/\omega_n \approx 0$), the base in Fig. 13.1 senses almost the same force as if the mass m were fastened directly to it. As the frequency ω of the exciting force increases and approaches the natural frequency ω_n, the force sensed by the base increases, until at $\omega = \omega_n$ (since we have assumed there is no damping in the system) the force becomes infinitely large.

As the forcing angular frequency ω becomes greater than the natural angular frequency ω_n, the force applied to the base through the spring

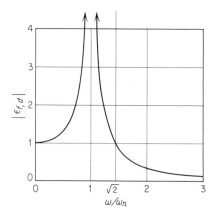

Fig. 13.2 Undamped force and displacement transmissibility as a function of frequency ratio.

decreases rapidly, until at $\omega = \sqrt{2}\,\omega_n$ the applied force and the force sensed by the base are the same. For exciting frequencies greater than $\sqrt{2}\,\omega_n$, the transmitted force is less than the applied force and, in general, the higher the frequency the more effective will be the action of the isolator. There are limits to this generalization which are discussed in a later section, but if the assumptions of the linear, lumped theory are valid for the conditions of a given problem, the greater the frequency ratio, the more effective the isolator.

From the curve in Fig. 13.2, it is clear that a resilient mounting can make the situation worse if its natural frequency is incorrectly selected. An isolator is effective for only $\omega > 1.4\,\omega_n$.

The natural frequency ω_n, which in the case of an undamped, single-degree-of-freedom system is also the *resonance* frequency, is given by

$$\omega_n = \sqrt{\frac{k}{m}} \qquad \text{radians/sec} \qquad (13.9)$$

or

$$f_n = \frac{1}{2\pi}\,\omega_n = \frac{1}{2\pi}\sqrt{\frac{k}{m}} = \frac{1}{2\pi}\sqrt{\frac{k'g'}{W}} \qquad \text{Hz} \qquad (13.10)$$

where $k =$ spring constant, N/m (lb/ft). (See Appendix B for a discussion of units. The various parameters are shown in two consistent systems of units, the mks and foot-slug-second systems. The slug is simply the weight of a mass in pounds divided by $g = 32.2$ ft/sec².)

$m =$ mass, kg (slugs)

$g =$ acceleration of gravity 9.8 m/sec² (32.2 ft/sec²)

$k' =$ spring constant, lb/in.

$W =$ weight of the mass m, lb

$g' =$ acceleration of gravity, 386 in./sec²

The static deflection δ_{st} (or δ'_{st}) of the isolator is given by $\delta_{st} = mg/k$ (or $\delta'_{st} = W/k'$), which when used in Eq. (13.10) yields

$$f_n = \frac{\sqrt{g}}{2\pi} \sqrt{\frac{1}{\delta_{st}}} \approx 0.5 \sqrt{\frac{1}{\delta_{st}}} \quad \text{Hz} \tag{13.11a}$$

where δ_{st} = deflection of the mass under its own weight, m

In British units

$$f_n = 3.13 \sqrt{\frac{1}{\delta'_{st}}} \quad \text{Hz} \tag{13.11b}$$

where δ'_{st} = deflection of the mass under its own weight, in.

Thus, f_n, the natural frequency of a mass on a simple isolator, is a function only of its static deflection.

Damped, Single-degree-of-freedom System. In Fig. 13.3, a damped, single-degree-of-freedom system is shown under the action of an applied force $F_0 \cos \omega t$. The applicable equation of motion is

$$m\ddot{x} + C\dot{x} + kx = F_0 \cos \omega t \quad \text{N} \tag{13.12}$$

where C = coefficient of viscous damping, N-sec/m

The solution of Eq. (13.12) is

$$x = e^{-(C/2m)t}(A \cos \omega_d t + B \sin \omega_d t) + X \quad \text{m} \tag{13.13}$$

where ω_d = damped natural frequency equal to $\sqrt{\omega_n^2 - (C/2m)^2}$

X = particular (steady-state) solution of Eq. (13.12)

It is clear that the first group of terms in Eq. (13.13) will become negligibly small for large values of time and that X (the steady-state solution) will then govern the motion of the mass m. It can be shown that

$$X = \frac{F_0/k}{\left[\left(1 - \frac{\omega^2}{\omega_n^2}\right)^2 + \left(2\frac{C}{C_c}\frac{\omega}{\omega_n}\right)^2\right]^{1/2}} \cos(\omega t - \phi) \tag{13.14}$$

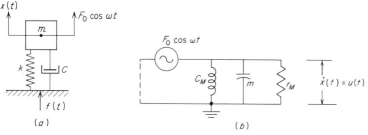

Fig. 13.3 (a) Mechanical diagram for a force-driven, damped single-degree-of-freedom system. (b) The same system drawn in analogous electrical-type schematic diagram, using the notation of Fig. 13.1f.

where $C_c = 2m\omega_n =$ coefficient of viscous damping at which critical damping occurs, N-sec/m

The phase angle ϕ is defined by

$$\tan \phi = \frac{2\,\dfrac{C}{C_c}\,\dfrac{\omega}{\omega_n}}{1 - \dfrac{\omega^2}{\omega_n^2}}$$

The force $f_0 \cos (\omega t - \phi)$ transmitted to the foundation is $C\dot{x} + kx$, and the magnitude of the force transmissibility ϵ_f of the system in Fig. 13.3 is

$$\epsilon_f = \left| \frac{f_0 \cos (\omega t - \phi)}{F_0 \cos \omega t} \right|$$

or

$$\epsilon_f = \left\{ \frac{1 + \left(2\,\dfrac{C}{C_c}\,\dfrac{\omega}{\omega_n} \right)^2}{\left[1 - \left(\dfrac{\omega}{\omega_n} \right)^2 \right]^2 + \left(2\,\dfrac{C}{C_c}\,\dfrac{\omega}{\omega_n} \right)^2} \right\}^{1/2} \tag{13.15}$$

A plot of this equation presents a rather accurate picture of the behavior of resilient mountings for a viscously damped single-degree-of-freedom system. Force transmissibility [Eq. (13.15)] as a function of frequency ratio for several damping ratios is shown in Fig. 13.4.

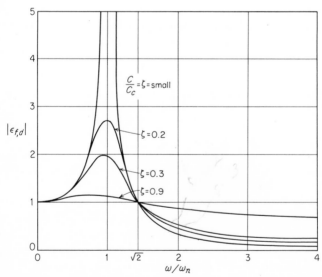

Fig. 13.4 Damped force and displacement transmissibility as a function of frequency ratio. Note that $\zeta = C/Cc$.

The shift downward in frequency of the resonance peak with damping is evident. Since we are plotting transmissibility vs. frequency, it is logical to ask at what frequency value is ϵ_f a maximum and how is this value related to other resonance frequencies?

From Eq. (13.15), it can be shown that the *frequency of maximum transmissibility* ω_ϵ is

$$\omega_\epsilon = \frac{\omega_n}{2\zeta} [(8\zeta^2 + 1)^{1/2} - 1]^{1/2} \qquad (13.16)$$

where $\zeta = C/C_c$, dimensionless
   $\omega_n = \sqrt{k/m}$, radians/sec

Equation (13.15) implies that the *frequency of maximum forced amplitude* ω_a is

$$\omega_a = \omega_n \sqrt{1 - 2\zeta^2} \qquad \text{radians/sec} \qquad (13.17)$$

Thus, the four important frequencies are

$$\omega_n = \sqrt{\frac{k}{m}} \qquad \text{(natural frequency)}$$

$$\omega_d = \omega_n \sqrt{1 - \zeta^2} \qquad \text{(damped natural frequency)}$$

$$\omega_a = \omega_n \sqrt{1 - 2\zeta^2} \qquad \text{(frequency of maximum forced amplitude)}$$

$$\omega_\epsilon = \frac{\omega_n}{2\zeta} \sqrt{\sqrt{8\zeta^2 + 1} - 1} \qquad \text{(frequency of maximum transmissibility)}$$

Commercial resilient mountings without external dashpots or friction devices have relatively little damping — normally, nearly always $\zeta < 0.25$ and in most cases $\zeta < 0.10$. For $\zeta = 0.10$ (that is, the damping is 10% of critical damping), the frequencies listed above take the values

$$\omega_d = 0.995 \ \omega_n$$

$$\omega_a = 0.990 \ \omega_n$$

$$\omega_\epsilon = 0.990 \ \omega_n$$

Thus, for all practical values of damping, the four frequencies can be taken as practically equal.

The curves of Fig. 13.4 show that for $\omega/\omega_n < \sqrt{2}$, the force transmitted through the mount to the base is actually magnified for any degree of damping. When $\omega/\omega_n = \sqrt{2}$, the transmitted force is equal to the applied force. In the region of isolation ($\omega/\omega_n > \sqrt{2}$), we can see from the curves of Fig. 13.4 that increased viscous damping adversely affects the amount of isolation achieved.

In the case of rotating machinery, a secondary problem makes it mandatory that there be some damping present.   As a rotating device is started or stopped, it will, in most cases, pass through a resonance range, i.e., a range of frequencies where the system response is magnified by a resilient mounting.   If the machine can be accelerated rapidly, it may pass through this region so quickly that the amplitude of the transmitted force does not have time to build up to the steady-state levels indicated by Fig. 13.4.   If, on the contrary, the machine accelerates slowly through the resonant range, then the transmitted force may become very large. In this case, a large amount of damping (say, $\zeta = 0.5$) may be required to prevent excessive vibration near resonance.   An external damper can be installed to accomplish this.   Fortunately, even with as large a damping as $\zeta = 0.5$, the amount of isolation achieved at higher frequencies is adequate in most cases.

Nonviscous Damping.   The previous analysis considered only viscous damping, that is, systems in which the component of force due to damping in Eq. (13.12) is directly proportional to velocity.   It has been shown that the effects at resonance due to other forms of damping can be represented in terms of an "equivalent viscous damping," using energy dissipation per cycle as the criterion of equivalence.[1]   However, in such cases, the frequency effects of damping are generally quite different from those shown in Fig. 13.4.

For hysteresis or structural damping, the damping term depends on displacement instead of velocity.   In this case the transmissibility at high frequencies is virtually independent of the damping.   The equation for ϵ_f is

$$\epsilon_f = \left[\frac{1 + \delta^2}{\left(1 - \frac{\omega^2}{\omega_n^2}\right)^2 + \delta^2} \right]^{1/2} \tag{13.18}$$

where $\delta = 1/Q$

$Q =$ amplification factor at resonance, equal to the ratio of the resonance frequency divided by the bandwidth at the half-power points; see Eq. (14.5)

Generally, δ is less than 0.2.   A few calculations will show that Eq. (13.18) is closely approximated by Eq. (13.4) up to the frequencies where ω is very much larger than ω_n.   At $\omega > 10\,\omega_n$, wave effects and other factors cause measured values of response to be significantly greater than those predicted by the approximations of Eqs. (13.14), (13.15), and (13.18).

Isolation at Acoustic Frequencies.   The theory so far presented is that which is customarily found in books and papers dealing with vibration isolation.[2,3]   With modifications to take care of the three-dimensional nature of the machine and the fact that several mounts are used, it gives satisfactory results at relatively low frequencies. ~~This~~ infrasonic frequency range ~~is where we are concerned~~ about physical damage or ~~fatigue failure.~~ Unfortunately, results calculated by Eq. (13.15) predict attenuations for the audio frequency range which are apt to be very much higher than those achieved in practice.

Since there is nothing wrong with the mathematical development, we must look to the assumptions to explain this discrepancy. Figure 13.5 shows a rigid mass mounted on an isolator which in turn rests on an absolutely rigid foundation. If the foundation were actually rigid, the isolation problem would be of no interest since the foundation could not be moved. In actuality, almost any foundation and any reasonable machine will have many resonances in the audio frequency range.

Let us consider first the foundation. Since for small motions it is a linear system, when a force $F = F_0 \cos \omega t$ is applied to it, the vibratory velocity at the point of application is given by

$$v = UF_0 \cos (\omega t - \phi)$$

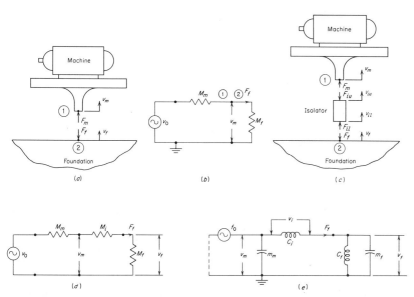

Fig. 13.5 Interconnection mobilities of a motor, isolator, and foundation: (a) mechanical diagram without isolator; (b) analogous schematic for this configuration with M_m = mobility of the machine, M_f = mobility of the frame, and v_0 = velocity the machine would have if it were freely suspended in space; (c) mechanical diagram with isolator; and (d) and (e) analogous schematic diagrams for this configuration with M_m, M_i, and M_f equal to the mobilities, in m/N-sec, for the machine, isolator, and foundation, respectively; m_M mass of machine in kilograms, C_i = compliance of the isolator in m/N, C_f = compliance of frame in m/N, and m_f = mass of frame in kilograms. Note also that $v_0 = M_m f_0$.

416 Noise and Vibration Control

where U and ϕ are functions of frequency ω which we shall discuss below. This equation may also be written in the form

$$v = UF_0 \cos \omega t \cos \phi + UF_0 \sin \omega t \sin \phi$$

If we use the Euler equation

$$e^{j\theta} = \cos \theta + j \sin \theta$$

we may redefine the force F as the real part of

$$F = F_0 e^{j\omega t}$$

and the velocity at the point of application of the force as the real part of

$$v = UF_0 e^{j(\omega t - \theta)}$$
$$= Ue^{-j\theta} F_0 e^{j\omega t}$$
$$= M_f F \tag{13.19}$$

where $M_f = $ *mobility*, defined as the complex number whose modulus $|M_f|$ is the rms magnitude of the velocity resulting from the application of a sinusoidal force whose rms magnitude is unity and whose argument θ is the phase angle between the velocity and the applied force

The internal mobility of the machine M_m can be defined in a similar manner. By superposition, the net motion of that point on the machine base where the machine is fastened to the foundation is the sum of the motion it would have due to internal forces only (that is, the point of attachment is free) and the motion due to the resisting force (Fig. 13.5a and b), or

$$v_m = v_0 + M_m F_m \tag{13.20}$$

It should be noted that v_0 is not necessarily a rigid-body motion of the machine and may be modified appreciably by internal resonances and antiresonances. It is not always easy to calculate but can be measured with reasonable accuracy if the machine is supported on mountings (isolators), which are very soft in comparison to M_m (that is, $M_m \ll M_i$, or in Fig. 13.5b, $M_f \gg M_m$).

Let us consider that point 1 on the machine is attached to point 2 on the foundation (Fig. 13.5a); then for motion of the combined system (which we denote by a tilde)

$$\tilde{v}_m = \tilde{v}_f$$
$$\tilde{F}_m = -\tilde{F}_f$$

From Eqs. (13.19) and (13.20)

$$\tilde{F}_f = \frac{v_0}{M_m + M_f} \qquad (13.21)$$

By the analogous circuit of Fig. 13.5b, the generator velocity v_0 is equal to the force "flowing through" the elements times the total mobility of the circuit.   If a massless isolator is interposed between the machine and its foundation (Fig. 13.5c and d), then for motion of the combined system

$$\tilde{v}_m = \tilde{v}_{iu}$$

$$\tilde{v}_f = \tilde{v}_{il}$$

$$\tilde{F}_m = -\tilde{F}_{iu} = +\tilde{F}_{il} = -\tilde{F}_f \qquad (13.22)$$

We define the force in and relative velocity across the isolator by, respectively,

$$\tilde{F}_i \equiv \tilde{F}_{iu} = -\tilde{F}_{il}$$

$$\tilde{v}_i \equiv \tilde{v}_{iu} - \tilde{v}_{il}$$

which, with Eqs. (13.19), (13.20), and (13.22), becomes

$$\tilde{v}_i = M_i \tilde{F}_i = v_0 + M_m \tilde{F}_m - M_f \tilde{F}_f$$

Because

$$F_i = \tilde{F}_f = -\tilde{F}_m$$

we obtain (see Fig. 13.5d)

$$\tilde{F}_f = \frac{v_0}{M_m + M_i + M_f} \qquad (13.23)$$

If we now redefine ϵ_f as the ratio of the force on the foundation with the isolator to that without

$$\epsilon_f = \left| \frac{F_{\text{wi}}}{F_{\text{wo}}} \right| = \left| \frac{M_m + M_f}{M_i + M_m + M_f} \right| \qquad (13.24)\,*$$

It will be found that this definition leads to the same result as the previous one if the previous assumptions are used.

It may also be shown that the displacement transmissibility ϵ_d will have the same value as the force transmissibility ϵ_f.

From Eq. (13.24) and Fig. 13.5d we find that an isolator does no good unless it is more flexible (has *greater* mobility, i.e., a greater value of M_i) than the sum of the mobilities of the machine and foundation.   The mobility of a simple structure may be calculated, and that of any structure may be measured.   At present, we have very few measured values, but those that are available indicate that M_m and M_f are very much larger

* The same result is obtained by Ungar and Dietrich [4] under more general assumptions.

than might be anticipated. It is rare to get more than 20 dB attenuation at acoustic frequencies ($\epsilon = 0.1$) with isolation mounts of reasonable stiffness, and it is not uncommon to get no attenuation at all. For this reason, very soft mounts ($f_n = 5$ to 6 Hz) are being offered, and special constructions, such as pneumatic mounts, are now marketed. In any case, we may see from Eq. (13.24) and Fig. 13.5d that if a mount is effective at all, a softer one (M_i larger) will be that much more effective.

Figure 13.5d can be expanded into Fig. 13.5e by noting that, if there is no dissipation, the mobilities are equal to

$$M_m = \frac{1}{j\omega m_m}$$

$$M_i = j\omega C_i$$

$$M_f = \frac{C_f/m_f}{j(\omega C_f - 1/\omega m_f)}$$

and $f_0 = v_0/M_m$

where the symbols are defined in Fig. 13.1f. The mobility of the foundation M_f is represented here by a different equivalent compliance C_f and mass m_f for each resonance condition of the foundation.

In Fig. 13.6 a predicted curve and some measured values of ϵ are shown for an experimental setup.[5] In Fig. 13.7 a measured value for M_f of a typical baseplate is plotted.[6] The high-frequency effects displayed in these figures can be taken into account as due to wave effects

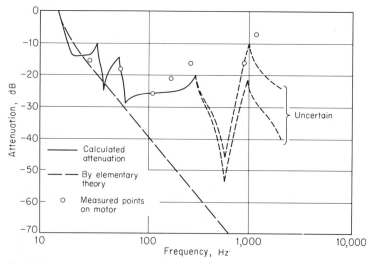

Fig. 13.6 Predicted and measured attenuation of an isolator.[5]

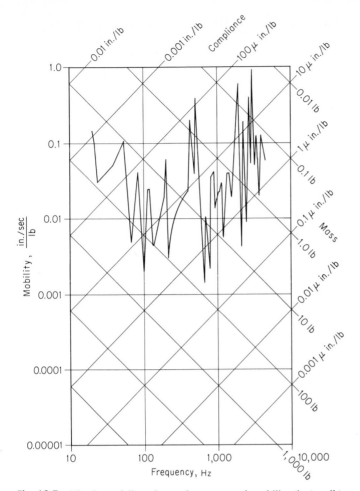

Fig. 13.7   The jagged line shows the measured mobility (in./sec-lb) as a function of frequency of a ribbed motor-generator baseplate 3/8 in. thick weighing 60 lb.   Note that the lines running from upper left to lower right on the graph give the mobility as a function of frequency for the masses indicated.   For example, 100 lb (a weight of 100 lb) at 100 Hz has a mobility of 0.0062 in./sec-lb.   The lines running from upper right to lower left give the mobility as a function of frequency for the compliances (reciprocals of the spring constants) indicated. For example, a spring with a compliance of 10^{-5} in./lb has, at 1,000 Hz, a mobility of 0.062 in./sec-lb.   This graph shows us that below 100 Hz this motor base behaves on the average like a mass of about 40 lb. Above 1,000 Hz it behaves on the average like a spring with a compliance of 3×10^{-6} in./lb.   (*After Plunkett.*[6])

in the resilient mounting in which the mass on the mounting and/or the mass of the mounting are significant contributors. In Fig. 13.8, the effects are shown of such waves for two values of structural loss factor ($\eta = 0$, 0.06) and three values of the ratio of the mass on the mounting to the mass of the resilient element of the mounting ($\mu = 10, 100, 1{,}000$).[4] Superposed on these computed curves is a plot of experimental data taken from a transmissibility test of a U.S. Navy resilient mounting.[7]

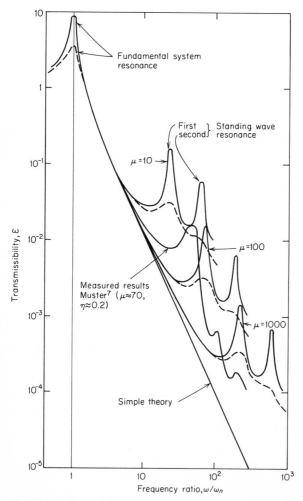

Fig. 13.8 Effect of the ratio μ of mass on an isolator to mass of the resilient element and of structural loss factor η on high-frequency transmissibility. The three solid calculated lines are for $\eta \approx 0$. The three dashed calculated lines are for $\eta \approx 0.06$.[4,7]

The computed and measured results are not directly comparable but the similarity is obvious.

Ungar and Dietrich [4] and Snowdon [8] have published detailed results of both analytical and experimental studies of the effects of isolator mass and damping on high-frequency performance of vibration isolators. The net effect of these phenomena is to increase significantly the transmission of forces or displacements through a resilient mounting over the levels that are predicted by the relatively simple classical theory. The first effects may occur at forcing frequencies as low as 10 to 30 times the natural frequency of the isolator mass system. It is clear that resilient mountings with good audio frequency isolation properties must be used in situations where both η and μ are large.

13.2   Vibration-isolation Systems

For systems in which more than a single degree of freedom must be considered in order to represent adequately the physical situation, there are no simple approaches. Fortunately, some multi-degree-of-freedom isolation problem procedures have been developed. For example, the SAE G-5 Committee on Aerospace Shock and Vibration has published a complete, stepwise procedure for the design of linear, center-of-gravity type vibration-isolation systems whose function is to protect fragile equipment.[9]   A monograph devoted to the theory and practice of cushioning systems has been published recently by The Shock and Vibration Information Center (DoD).[10]   However, the latter is not a design handbook. It contains no design procedures, but is intended to be a critical review of the current art in selecting, designing, analyzing, and using cushions.

A matrix approach to the fundamentals underlying vibration isolation for multi-degree-of-freedom systems has been developed by Smollen [11] which complements earlier analyses of Himelblau and Rubin,[12] who derived expressions for the effective stiffness and damping characteristics of a general isolation system and the associated equations of rigid-body motion using conventional techniques. They treat (see Fig. 13.9) the problem of a rigid body supported in a general way by isolators consisting of spring and damping elements. The solutions of specific examples are given by Smollen.[11]

13.3   Vibration Isolators and Isolator Materials

Four resilient materials are most commonly used in vibration isolators: metals in the form of various types of springs are used as well as mesh pads (or other shapes) of rubber, cork, and felt. Other materials, such as steel-mesh pads, pneumatic sacks, and even a gelatinous material similar to hectograph pad gel, have been used, but the majority of vibration isolators use one or more of the first four materials as the resilient element.

Metal Springs. Metal springs are by far the most commonly used, because the field of their use is as broad as that of machine design itself. They are used to isolate the most delicate scientific instruments from foundation vibrations, and yet masses up to 450 tons have been satisfactorily isolated with them. In theory, at least, the complete spectrum of frequencies can be isolated by metal springs. This is due, in part, to the large range of deflections which can be obtained by changing the dimensions and materials used in the design of the springs. This has been indicated in Fig. 13.10, which is a plot of Eq. (13.11b) with the nominal maximum permissible deflections of some elastic media superimposed upon it.

Metal springs have the advantages of ready interchangeability and several beneficial chemical characteristics (they resist corrosion by oil and water and are not affected by extremes of temperature). An advantage of industrial importance is that they can be produced in large quantities with only small variation in their individual characteristics. Inherently, metal springs have very little damping; the damping is about 0.001 of critical ($\zeta \approx 0.001$). However, external damping can be added to a system, if it is required by a particular application. A dashpot can be inserted in parallel with the metal spring (Fig. 13.3). Recoil mechanisms are sometimes built employing this principle—a metal spring for elasticity, a separate oil dashpot for a large amount of viscous damping.

Some resilient mountings are fabricated with a metal spring inside a rubber sack with a calibrated orifice that regulates the flow of air in and out of the sack and, thus, furnishes viscous damping. Coulomb (fric-

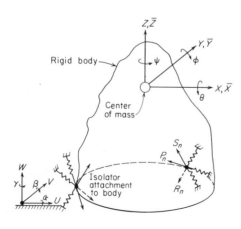

Fig. 13.9 General isolator support of rigid body (damping elements not shown).

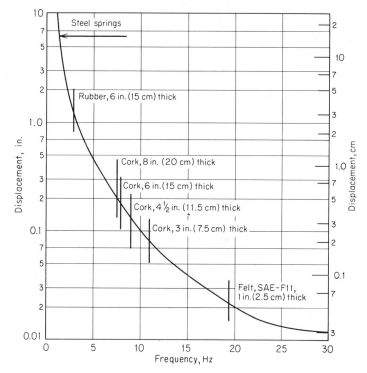

Fig. 13.10 Relation between static deflection and natural frequency. Typical natural frequencies of some practical isolators are shown.

tion) damping can also be supplied by external means. In leaf springs damping is caused by the rubbing action between individual leaves; in certain resilient mountings with steel mesh inserts it is caused by the rubbing of the wires on each other. Although it is unplanned in many cases, this form of damping occurs also in nonrigid mechanical joints where one part can slip relative to another.

For purposes of analysis, it is convenient to reduce coulomb and other nonviscous forms of damping to an equivalent viscous damping which uses energy dissipated per cycle as the basis of equivalence.[1]

Metal springs have the practical disadvantage of transmitting high frequencies very readily. For example, although the low natural frequency of an internal combustion engine (say, 15 Hz) can be isolated easily, the higher frequencies present are transmitted through the metal of the spring to the foundation. These higher frequencies in engines, for example, may range from two hundred to several thousand hertz and are due to detonation local resonances at the mounts and other sources. Transmission of these frequencies is minimized by ensuring that there be no direct contact between the spring and the supporting structure. In commercial mountings, this is accomplished by inserting rubber or felt pads between the ends of the spring and the surfaces to which it is fastened.

Helical Springs. For most applications, the design of helical compression or tension springs (e.g., core springs) has been reduced to nomographs or other handbook methods. Treatises have been written on the subject,[13] and there is extensive commercial and technical literature available. Some design considerations of importance in certain applications, such as buckling and lateral stiffness of helical springs, have also been investigated.

Timoshenko [14] reviewed the general subject of the buckling of helical springs. Haringx [15] has reported on several phases of an analytical and experimental study of the mechanical properties of helical springs and rubber rods in compression. In the first of these papers, he considers the buckling of helical springs with various end conditions.

If the unloaded length of a steel spring is denoted by l_0 and its diameter by D_0, for the case of a helical spring of circular wire cross section with ends hinged or constrained to remain parallel, the relative compression ($\xi = \Delta l/l_0$) at which buckling occurs is given by Haringx as

$$\xi = 0.8125 \left\{ 1 - \left[1 - 6.87 \left(\frac{D_0}{l_0} \right)^2 \right]^{1/2} \right\} \qquad (13.25)$$

where l_0/D_0 is the slenderness ratio.

This relationship is derived under the following simplifying assumption: the spring is idealized as an elastic prismatic rod made of a hypothetical material with stiffness in bending, compression, and shear which corresponds to that of the original spring. The experimental evidence appears to justify this assumption.

A plot of the relation given in Eq. (13.25) is shown in Fig. 13.11. The curve delineates a region wherein all combinations of spring slenderness and relative compressions lead to instability. It shows that there is a certain critical slenderness ratio ($l_0/D_0 = 2.62$ if the spring ends are hinged or constrained to remain parallel; or $l_0/D_0 = 5.24$ if the spring ends are clamped) below which the spring does not buckle, above which it does. Haringx has measured the relative compressions at which buckling occurs for the springs of several slenderness ratios. The results of these experiments are shown by the points superimposed on the curve in Fig. 13.11.

Belleville Springs. Coned disk springs, perhaps more widely known as "Belleville springs," consist essentially of circular disks dished to a conical shape as indicated in Fig. 13.12. A manual dealing with their

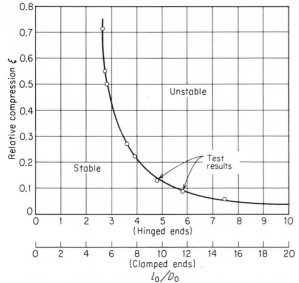

Fig. 13.11 Stable and unstable values of relative compression for coil springs.

design and use is available,[16] so that here we will only sketch their general characteristics.

Load-deflection characteristics of Belleville springs can be obtained in considerable variety by changing the ratio h/t (Fig. 13.12). In Fig. 13.13, the effect of changes in this ratio upon the load-deflection diagram is shown for three cases. For $h/t = 1.5$, we note that the spring has practically zero stiffness for a considerable range of deflection; that is, the load is almost constant over this range.

Belleville springs may be used singly or may be stacked in parallel (Fig. 13.14a) or in series (Fig. 13.14b). By stacking in series, the deflection for a given load is increased directly with the number of disks; in parallel, the load for a given deflection increases in proportion to the number of disks.

Advantages of Belleville springs include:

1. High energy storage for a relatively small space requirement in the direction of load application

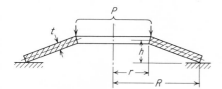

Fig. 13.12 Cross section of Belleville or disk spring.

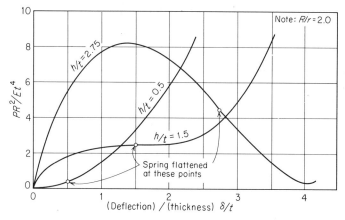

Fig. 13.13 Load-deflection characteristics of Belleville springs.

2. Controlled nonlinear stiffness characteristics (which can be obtained by varying the ratio h/t)

3. Coulomb damping (obtained from parallel stacking)

4. Variable stiffness (obtained by series or parallel stacking)

On the other hand,

1. The springs fatigue easily.

2. Guides must be provided to prevent sidewise buckling.

3. Clearance must be provided between the central tube or rod.

4. There is a possible "snapping" action as the springs pass through a flat shape.

In practice, the ratios of outside to inside diameter vary between 1.5

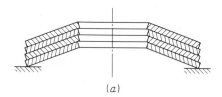

(a)

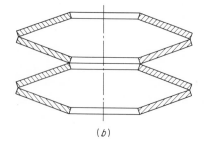

(b)

Fig. 13.14 Stacking of Belleville springs for increasing stiffness (parallel) or deflection (series): (a) stacked in parallel; (b) stacked in series.

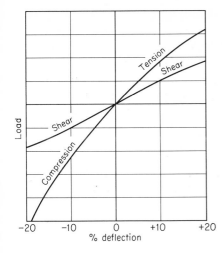

Fig. 13.15   Load-deflection curves for rubber in shear, tension, and compression.

and 3.5.   The optimum value of this ratio will depend on the type of loading as well as other design considerations.   Analyses indicate that for maximum energy storage, the ratio should be between 1.8 and 2.0. In springs of this type at stresses that are commonly used for static loading conditions, energy-storage capacities may be as high as 500 to 700 inch-pounds per pound of active material.

Rubber Mountings.   Rubber is used very effectively to isolate small machinery and mechanical devices — engines, motors, instruments, electronic gear, etc.   In Fig. 13.15 a typical set of static load-deflection curves for rubber is shown.   The exact values will vary with the composition but the qualitative behavior will be essentially the same.

It can be seen that the load-deflection curves for all three cases are linear over only limited ranges, although that for rubber in shear extends to the greatest deflections.   The soft characteristics of the latter are used widely in commercial resilient mountings to obtain relatively low natural frequencies at reasonable stress levels in the rubber.   In addition, the advantages of minimal effect of shape and a very great ability to withstand repeated variations in stress * outweigh the disadvantages of small energy-storage capacity and the more complex molding and bonding problems associated with fabricating an actual mounting.

Rubber in compression is used widely for applications that require high energy-storage capacity and an ability to support heavy loads without failure.   The stiffness of a compressive rubber pad depends on the end restraint against lateral bulging.   The *shape factor* of a compressive pad is defined as the ratio of the area of one loaded surface to the total area of free surface.   Roughly, the shape factor varies linearly with the

* Useful life of actual mountings in service at normal loads is 1.5×10^9 cycles or greater.

greatest dimension of the loaded surface and inversely with the unde-flected height of the pad.   For a given load, both the compressive and shear moduli increase with increasing shape factor.[17-19]   The maximum service life of rubber in compression is less than that for rubber in shear.

Rubber can seldom be used in tension to support the main load be-cause of its tendency to tear or separate at bonded edges under tensile loading.   There are no commercial resilient mountings which use rub-ber in this manner, although it may be used rarely in custom installa-tions.

In current design practice, it is common to find that mountings use rubber in shear as the primary elastic element and rubber in compres-sion as a secondary element which furnishes a snubbing action if the

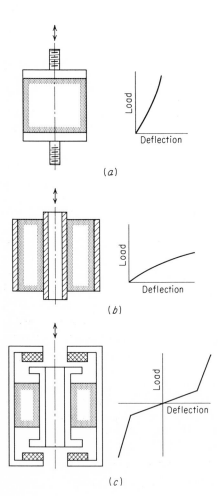

Fig. 13.16 Typical load-deflec-tion curves for three common types of rubber mounts: (*a*) com-pression; (*b*) shear (section); (*c*) shear compression (section).

TABLE 13.1   Various Physical Constants of Some Representative Rubber Stocks [17]
(Temperature of Rubber 70°F)

Durometer hardness, Shore A	Moduli (lb/in.²) * static			% volume compression at 1,000 lb/in.² hydrostatic pressure	Damping (log decrement)	Specific gravity	Specific heat	Velocity of sound in rubber rods	
	Compression	Tension	Shear					ft/sec	m/sec
30	185	175	50	0.34	0.041	1.01	0.47	115	35
40	270	230	70	0.31	0.055	1.06	0.43	165	50
50	375	305	95	0.29	0.14	1.11	0.40	210	64
60	550	450	140	0.27	0.23	1.18	0.38	345	105
70	750	610	195	0.24	0.35	1.25	0.35	750	228
80	1,200	1,025		0.22	0.47	1.31	0.33		

* N lb/in.² $= N \times 7,000$ N/m².

mounting is subjected to an overload.   Typical designs are shown in Fig. 13.16 with their appropriate load-deflection curves.

Most resilient mountings are designed to operate primarily in a preferred direction, usually an axis of symmetry.   The stiffness characteristics in other directions are usually not the same.   For example, in a shear mounting, the stiffness in the transverse directions is greater because displacements in these directions tend to compress the resilient element (Fig. 13.16b).   Mountings have been designed which are equally stiff in an axisymmetric direction and in the transverse directions as well.

In the design of compression mountings, account must be taken of the shape of the resilient element and, since the volume compressibility of solid rubber is extremely low,* it cannot be used successfully as a compression spring, unless space is provided for lateral expansion.   Ribbed construction will provide the space into which the rubber can bulge.

Detailed information on the properties of rubber and rubberlike materials is available in commercial and technical literature.   In Table 13.1, some representative engineering properties are given for various-hardness rubber stocks.†

* Almost all rubbers have less volume compressibility than water; e.g., at a hydrostatic pressure of 30,000 psi, the percentage decrease in original volume of water is 7.3%, of a 30-durometer (soft) rubber 6.5%, and of an 80-durometer (hard) rubber 4.7%.

† A Pearl pencil eraser has hardness roughly equivalent to 30-durometer rubber; a good Vulcanite pipe stem has that roughly equivalent to 80-durometer rubber.

Under dynamic conditions, the apparent stiffness of rubber is greater than under comparable static conditions. The work of many observers has shown that this phenomenon is a function of (1) the ambient temperature, (2) the frequency (forced or free) at which the observations are made, (3) the static strain in the specimen being tested, (4) the amplitude of vibration, and (5) the kind of rubberlike material of which the specimen is made.

It should be noted that when interpreting results in the literature obtained from dynamic tests it is important to distinguish between the behavior of a material in thin strips or rods or in bulk. For example, the velocity of a bulk compressional wave in a rubberlike material is greater than 3,000 ft/sec (900 m/sec) while that of a wave in a thin strip or rod is from 100 to 1,000 ft/sec (30 to 300 m/sec). Resilient mountings are almost always constructed in such a manner as to use rubberlike materials in bulk, so that in the discussion that follows, the dynamic properties in bulk only will be mentioned.

The dynamic modulus of the viscoelastic mathematical models which are used to characterize the properties of rubberlike materials is less dependent upon frequency as temperature is increased. This has been verified experimentally, although there is considerable variation in the degree of temperature dependence of dynamic modulus with the type of rubberlike material being tested. Usually, for a given frequency value, synthetic rubbers display a greater variation of dynamic modulus with temperature than does natural rubber. In general, dynamic modulus increases with increasing frequency.

Both static and dynamic moduli increase with increasing strain in the specimen. In addition, for any value of strain, the dynamic modulus is always greater than the corresponding static modulus.

With increasing amplitude of vibration, it has been shown that the dynamic modulus of most rubberlike materials decreases. Despite the many observations of this phenomenon, there is no agreement among workers in the field as to the reason for the dependence of modulus upon amplitude. Gehman [20] found that with decreasing modulus, there is an associated temperature increase, which might be attributed to higher hysteresis losses. However, the temperature rise is too small to account for the entire effect. Other tests indicate that if an initially stressed specimen is caused to vibrate, increasing amplitude of vibration causes a decrease in apparent dynamic modulus measured from the hysteresis loop. That is, at the original static strain, the slope of the principal axis of the hysteresis loop is less than that of the static stress-strain curve at that point. Whatever the mechanism, the overall effect is to reduce modulus with increasing amplitude of vibration.

There are other second-order dynamic effects on the properties of rubber which will not be mentioned here. The reader is referred to the extensive technical literature available.

Cork. One of the oldest materials used for vibration isolation is cork. It is used in compression or in a combination of compression and shear. Unlike a rubberlike material in compression, cork becomes less stiff at high loadings, displaying the same type of stress-strain curve as copper. The dynamic properties of cork are also very much dependent upon frequency.

Generally, in order to obtain sufficiently large deflections, the machine to be isolated is mounted on large concrete blocks which are separated from the surrounding foundation by several layers of cork slabs 1 to 6 in. (2 to 15 cm) thick. The recommended pressure to which the cork should be subjected for optimum performance is between 7 and 20 lb/in.^{2} (5,000 to 15,000 kg/m^2).[21]

Oil, water, and moderate temperature have little effect upon the operating characteristics of cork; but cork does tend to compress with age under an applied load. At room temperature its effective life may be measured in decades; at 200°F it is reduced to a fraction of a year. Figure 13.17 shows why cork must have sufficient load to give a reasonable resiliency. It is not a very effective isolator in the low-frequency range, since great thicknesses are needed to achieve the correspondingly large deflections which are required. Unless the slabs are properly spaced, this can lead to an unstable condition.

Felt. When felt is used as a vibration-isolation material, the greatest isolation efficiency is obtained by using the smallest possible area of the softest felt, in maximum thickness, under a static load that the felt will resist without excessive compression or loss of structural stability. It has a high damping factor and thus is particularly useful in reducing amplitude of vibration at resonance. The amplification factor at resonance is almost independent of amplitude and load and is about 4 for soft felt. For general purposes, felt mountings of $\frac{1}{2}$ to 1 in. (1 to 2.5 cm) thickness are recommended, with an area of 5% of the total area of the base if the machine has a flat bed. In installations where vibration is

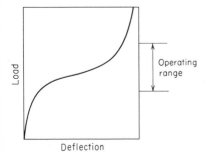

Fig. 13.17   Load-deflection curve for cork.[21]

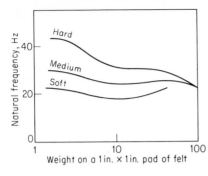

Fig. 13.18 Natural frequency as a function of weight (lb) loading on different grades of felt. (*After Tyzzer and Hardy.*[22])

not excessive, no bonding is necessary between the felt and the machine. Resonance curves of loaded felt pads in compressional vibration show the nonlinear character of their damping and stiffness properties. The compliance of a felt pad increases somewhat more slowly than the thickness of the pad and there is a decrease in stiffness with increasing amplitude of vibration. The increase of stiffness with pressure is so large that the ratio of elastic modulus to pressure (stiffness to mass) does not vary greatly between pressures of 3 to 100 lb/in.^{2} (2,000 to 70,000 kg/m^2). In this range, the natural frequency of a mass mounted on a felt pad is determined by the thickness of the pad rather than its area and static load (Fig. 13.18). In most cases, the effectiveness of felt in reducing vibration transmission is limited to frequencies above 40 Hz. Felt is particularly useful in reducing vibration transmission in the audio frequency range since it offers a large impedance mismatch to most engineering materials.[22]

REFERENCES

1. L. S. Jacobsen, Steady Forced Vibration as Influenced by Damping, *Trans. ASME*, vol. 52, p. APM169, 1930.
2. C. E. Crede, *Vibration and Shock Isolation*, Wiley, 1951.
3. W. T. Thomson, *Vibration Theory and Applications*, Prentice-Hall, 1965.
4. E. E. Ungar and C. W. Dietrich, High-frequency Vibration Isolation, *J. Sound Vibration*, vol. 4, pp. 224–241, 1966.
5. R. Plunkett, Interaction between a Vibrating Machine and Its Foundation, *Noise Control*, vol. 4, no. 1, pp. 18–22, January, 1958.
6. R. Plunkett, Experimental Measurement of Mechanical Impedance or Mobility, *J. Appl. Mech.*, vol. 21, pp. 250–256, 1954.
7. D. Muster, "Resilient Mountings for Reciprocating and Rotating Machinery," *Eng. Rept. 3*, ONR Contract N7 onr-32904, July, 1951; L. L. Beranek (ed.), *Noise Reduction*, fig. 18.17, p. 488, McGraw-Hill, 1960.
8. J. C. Snowdon, "Rubberlike Materials, Their Internal Damping and Role in Vibration Isolation," *J. Sound Vibration*, vol. 2, pp. 175–193, 1965.
9. Society of Automotive Engineers, Aerospace Shock and Vibration Committee G-5, "Design of Vibration Isolation Systems," 1962.

10. G. S. Mustin, "Theory and Practice of Cushion Design," The Shock and Vibration Information Center, DoD, 1968.
11. L. E. Smollen, "Generalized Matrix Method for the Design and Analysis of Vibration-isolation Systems," *J. Acoust. Soc. Am.*, vol. 40, pp. 195–204, 1966.
12. H. Himelblau and S. Rubin, "Vibration of a Resiliently Supported Rigid Body," in C. E. Crede and C. M. Harris (eds.), *Shock and Vibration Handbook*, vol. I, McGraw-Hill, 1961.
13. A. M. Wahl, *Mechanical Springs*, Penton, 1949.
14. S. Timoshenko, *Theory of Elastic Stability*, McGraw-Hill, 1936.
15. J. A. Haringx. "On Highly Compressible Helical Springs and Rubber Rods and Their Application for Vibration-free Mountings," *Philips Res. Repts.*, vol. 3, pp. 401–449, 1948; vol. 4, pp. 49–80, 206–220, 261–290, 375–400, 407–448, 1949.
16. Society of Automotive Engineers, *Manual on Design and Manufacture of Coned Disk Springs or Belleville Springs*, January, 1950. Reaffirmed without change, 1964.
17. "Some Physical Properties of Rubber," U.S. Rubber Company, 1941.
18. *Rubber in Engineering*, Chemical Publishing, 1946.
19. Goodyear Tire and Rubber Company, Inc., *Handbook of Molded and Extruded Rubber*, 1949.
20. S. D. Gehman, Rubber in Vibration, *J. Appl. Phys.*, vol. 13, p. 402, 1942.
21. S. L. Dart and E. Guth, Elastic Properties of Cork, *J. Appl. Phys.*, vol. 17, no. 5, pp. 314–318, 1946; vol. 18, no. 5, pp. 470–478, 1947.
22. F. G. Tyzzer and H. C. Hardy, Properties of Felt in the Reduction of Noise and Vibration, *J. Acoust. Soc. Am.*, vol. 19, pp. 872–878, 1947.

Chapter Fourteen
Damping of Panels

ERIC E. UNGAR

The dynamic responses and sound-transmission characteristics of structures are determined by essentially three structural properties: mass, stiffness, and damping. Of these, damping is the one which is least understood and most difficult to predict, and—unlike mass and stiffness—damping usually cannot be inferred from simple static measurements.

A vibrating structure, such as a panel, at any instant contains some kinetic energy and some strain (or potential) energy; the kinetic energy storage is associated with the mass, the strain energy storage with the stiffness. In addition, any real structure also dissipates some energy as it deforms. It is this energy dissipation—or, more precisely, this conversion of ordered mechanical energy into thermal energy—that is called *damping*.

Unlike mass and stiffness, damping does not refer to a unique physical phenomenon, and that is the reason damping is so much more difficult to predict in general. There are as many damping mechanisms as there are ways of converting mechanical into thermal energy. These mechanisms include interface friction, fluid viscosity, turbulence, acoustic

434

radiation, eddy currents, magnetic hysteresis, and mechanical hysteresis (also called internal friction or material damping). In order to analyze or predict the damping of a given structure, one should ideally take into account all possible damping mechanisms; however, in practical cases one or two mechanisms generally predominate, and one may then neglect the effects of the others.

The primary effects of increased panel damping are (1) reduction of vibration amplitudes at resonances, with attendant reductions in stresses, structural fatigue, and sound radiation; (2) more rapid decay of free vibrations, and corresponding reduction of noise generated by repetitive impacts on the panel; (3) attenuation of structureborne waves propagating along the panel, i.e., reduced conduction of vibratory energy along the panel; and (4) increased sound isolation (transmission loss) of the panel above its critical (coincidence) frequency. All these effects are generally beneficial from the standpoint of noise and vibration control, but, unfortunately, highly damped panels usually involve penalties of weight and complexity.

14.1   Measures and Measurement

14.1.1   Simple Oscillator with Viscous Damping   The primary quantities that characterize damping, and the principles which underlie their measurement, may be introduced best by referring to a classical mass-spring-dashpot system, like that shown in Fig. 14.1. If x represents the displacement of the mass m from its equilibrium position, then this displacement gives rise to an elastic restoring (spring) force of magnitude kx. Similarly, a velocity $\dot{x}$ produces a retarding force of magnitude $C\dot{x}$. Here k is known as the spring constant, and C as the *viscous damping coefficient*.

The equation of motion of the mass

$$m\ddot{x} + C\dot{x} + kx = F(t) \qquad (14.1)$$

may readily be deduced from a "free body" diagram of the mass, like that of Fig. 14.1b. For the case where the exciting force is sinusoidal with circular frequency ω and amplitude F_0, one finds from the equation of motion that in the steady state [1,2]

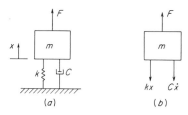

(a)   (b)

Fig. 14.1 (a) Mass-spring-dashpot system; (b) free-body diagram of mass.

$$\frac{Xk}{F_0} = [(1 - r^2)^2 + (2\zeta r)^2]^{-1/2} \tag{14.2}$$

where $X =$ peak displacement amplitude, m (in.)
$r = \omega/\omega_n = f/f_n =$ ratio of the exciting (circular) frequency to the undamped natural (circular) frequency, dimensionless
$f = \omega/2\pi =$ driving frequency, Hz
$f_n = \omega_n/2\pi =$ undamped natural frequency, Hz

$$\omega_n = \sqrt{k/m} \qquad \text{radians/sec} \tag{14.3}$$

$\zeta = C/C_c =$ *damping ratio* or *fraction of critical damping* (100ζ is usually called the *percent of critical damping*), dimensionless
$C_c =$ *critical viscous damping coefficient,* which is the smallest value of C for which the mass of Fig. 14.1 will execute no oscillations if it is displaced from equilibrium and released (in absence of other forces), N-sec/m (lb_f-sec/in.)

$$C_c = 2\sqrt{km} = 2m\omega_n \tag{14.4}$$

From Eq. (14.2) one finds that at low frequencies, where $r \ll 1$, the displacement amplitude obeys $X \approx F_0/k$; only the spring constant controls the displacement. At high frequencies, where $r \gg 1$, one finds $X \approx F_0/m\omega^2$, so that here the mass m is the only system parameter of importance. At the undamped natural frequency, which also is very nearly equal to the resonance frequency of the damped system for small damping ($\zeta^2 \ll 1$), that is, at $\omega \approx \omega_n$ and $r \approx 1$, one finds from Eq. (14.2) that at resonance

$$\frac{X_{res}k}{F_0} = Q = \frac{1}{2\zeta} \tag{14.5}$$

where $X_{res} =$ amplitude of the displacement at $r = 1$, m (in.)
$Q =$ *amplification at resonance,*[3,4] dimensionless
A plot of Eq. (14.2) exhibits a peak at $r \approx 1$ and steep slopes on both sides of the peak (see Fig. 14.2). From Eq. (14.2) one may readily calculate that X^2 takes on one-half of its maximum value X_{res}^2 at the so-called half-power points, or 3-dB-down points, which for small damping correspond to the frequencies $\omega_1 \approx \omega_n(1 - \zeta)$ and $\omega_2 \approx \omega_n(1 + \zeta)$. The

---

* The amplitude of the total force exerted on the ground by the spring and dashpot combination may be shown to be given by

$$F_G = [(kX)^2 + (C\omega X)^2]^{1/2} = kX[1 + (2\zeta r)^2]^{1/2}$$

At resonance ($r \approx 1$) and with small damping ($\zeta \ll 1$), $F_{G_{res}} \approx kX_{res}$. Thus, $kX_{res}/F_0 \approx F_{G_{res}}/F_0$ represents the "force amplification," i.e., the ratio of the amplitude of the force applied to the ground (at resonance) to that of the force acting on the mass.

frequency interval between these two half-power points is $\Delta\omega = \omega_2 - \omega_1$ $= 2\zeta\omega_n$, and the half-power *bandwidth* [4] is defined as

$$b = \frac{\Delta\omega}{\omega_n} = \frac{\Delta f}{f_n} \approx 2\zeta \tag{14.6}$$

Of the measures of damping described so far, the viscous damping coefficient C and the damping ratio ζ are based directly on a physical model, whereas the amplification at resonance Q and the bandwidth b are obtained from the steady-state response of the vibrating system. Since damping also affects the decay of free vibrations, it may also be measured in terms of this decay. If the mass of a lightly damped mass-spring-dashpot system is displaced from its equilibrium position by an amount X_0 and then released, one finds by solving the differential equation (14.1) that the displacement obeys

$$x(t) = X_0 e^{-\zeta\omega_n t} \cos(\omega_d t + \phi) \tag{14.7}$$

where $\omega_d = \omega_n \sqrt{1 - \zeta^2} \approx \omega_n$ is the "damped natural frequency," radians/sec

ϕ = phase angle (depends on the initial velocity), radians

Equation (14.7) has the well-known decaying sinusoid form typical of free vibrations of simple damped systems.

If the cosine of Eq. (14.7) is $+1$ at some instant t_0, then it will take on unity value again at $t_N \approx t_0 + N \times 2\pi/\omega_n$, where N is any positive integer. Then one finds from Eq. (14.7) that $x(t_0)/x(t_N) = e^{2\pi\zeta N}$. By taking the natural logarithm and dividing by the number of cycles N over which the decay is observed, one obtains the *logarithmic decrement* δ, which perhaps is one of the oldest measures of damping [4]

$$\delta = \frac{1}{N} \ln \frac{x(t_0)}{x(t_N)} = 2\pi\zeta \tag{14.8}$$

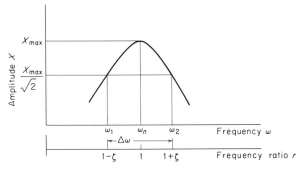

Fig. 14.2   Typical response curve in vicinity of resonance peak. Half-power points occur at ω_1 and ω_2; bandwidth is $\Delta\omega = \omega_2 - \omega_1$.

In dealing with logarithmic measures and instrumentation, one usually finds it useful to describe responses in terms of levels, and to define

$$L_x = 10 \log_{10} \left(\frac{x^2}{x_{\text{ref}}^2} \right) \tag{14.9}$$

where L_x = displacement level, dB
    x_{ref} = (constant) reference value of displacement,* m (in.)
The *decay rate* Δ_t is defined as the rate of decrease of L_x in a free oscillation [4]

$$\Delta_t = -\frac{dL_x}{dt} = -20 \ (\log_{10} e) \ \frac{dx/dt}{x} = 8.69\zeta\omega_n = 54.6\zeta f_n \quad \text{dB/sec} \tag{14.10}$$

The two rightmost expressions of Eq. (14.10) were obtained by considering only the envelope of the decaying function given by Eq. (14.7).

Another commonly used measure of the rate of decay of free oscillations is the *reverberation time* T_{60}, defined [5] as the time required for the vibration level to decrease by 60 dB. From Eq. (14.10)

$$T_{60} = \frac{60}{\Delta_t} = \frac{1.10}{\zeta f_n} \quad \text{sec} \tag{14.11}$$

It is instructive now to investigate briefly the various energy quantities that pertain to the simple mass-spring-dashpot system executing steady-state sinusoidal oscillations of the form $x(t) = X \sin (\omega t + \phi)$. The instantaneous kinetic and potential energies then obey

$$E_{\text{kin}} = \frac{m}{2} \dot{x}^2 = \frac{m}{2} \omega^2 X^2 \cos^2 (\omega t + \phi)$$

$$\tag{14.12}$$

$$E_{\text{pot}} = \frac{k}{2} x^2 = \frac{k}{2} X^2 \sin^2 (\omega t + \phi)$$

The force $-C\dot{x}$ acting through a differential displacement dx dissipates a differential amount of energy $C\dot{x} \, dx = C\dot{x}^2 \, dt$. The energy D dissipated in a whole cycle then is

$$D = CX^2\omega \int_0^{2\pi} \cos^2 (\omega t + \phi)d(\omega t) = \pi CX^2\omega \tag{14.13}$$

The *damping capacity* Ψ is defined [3,4] as the ratio of the energy dissipated per cycle to the maximum potential energy, or

$$\Psi = \frac{D}{E_{\text{pot, max}}} = \frac{2\pi\omega C}{k} = 4\pi\zeta r \tag{14.14}$$

and the *loss factor* or *dissipation factor* η is defined similarly, except that

* See Table 2.3.

the energy dissipated per radian is considered instead of that per cycle [3]

$$\eta = \frac{D/2\pi}{E_{\text{pot, max}}} = \frac{\omega C}{k} = 2\zeta r \qquad (14.15)$$

Usually, one is only concerned with damping at the resonance or of freely decaying vibrations, which also occur at the resonance frequency. For $\omega = \omega_n$ one finds [6] that

$$E_{\text{tot}} = E_{\text{kin}} + E_{\text{pot}} = 2E_{\text{kin, max}} = 2E_{\text{pot, max}} = kX^2$$

and the loss factor becomes

$$\eta = \frac{\Psi}{2\pi} = \frac{D}{2\pi E_{\text{tot}}} = \frac{D}{2\pi E_{\text{pot, max}}} = 2\zeta \qquad (14.16)$$

14.1.2 Interrelation among Measures of Damping

Because the loss factor η is defined in terms of energy quantities, it may be applied to any damping mechanism, not just to viscous damping. For small damping, the loss factor is related to the other previously discussed measures of damping at resonance as

$$\eta = \frac{\Psi}{2\pi} = 2\zeta = 2\frac{C}{C_c} = \frac{2.20}{f_n T_{60}} = \frac{\Delta_t}{27.3 f_n} = \frac{\delta}{\pi} = b = \frac{1}{Q} \qquad (14.17)$$

where η = loss factor, dimensionless; see Eq. (14.15)
  Ψ = damping capacity, dimensionless; see Eq. (14.14)
  ζ = fraction of critical damping, dimensionless; see after Eq. (14.3)
  C = viscous damping coefficient, N-sec/m (lb$_f$-sec/in.)
  C_c = critical damping coefficient, N-sec/m (lb$_f$-sec/in.); see Eq. (14.4)
  f_n = natural frequency, Hz; see after Eq. (14.2)
  T_{60} = reverberation time, sec; see Eq. (14.11)
  Δ_t = decay rate, dB/sec; see Eq. (14.10)
  δ = logarithmic decrement, dimensionless; see Eq. (14.8)
  b = half-power bandwidth, dimensionless; see Eq. (14.6)
  Q = amplification at resonance, dimensionless; see Eq. (14.5)
For large damping, the interrelations are more complex and less useful, because special analyses generally are required to deal with particular highly damped systems.

Equation (14.17) is often used to define an "equivalent viscous damping coefficient" for a system that is damped by other than a viscous mechanism. Use of this equivalent coefficient is justified as long as one is concerned only with the magnitudes of forced resonant responses or of freely decaying vibrations.

14.1.3 Complex Stiffness and Complex Moduli [3]   If one considers

purely sinusoidal excitation of the system of Fig. 14.1 and introduces standard exponential notation by setting

$$F(t) = F_0 e^{j\omega t} \qquad x = X e^{j\omega t}$$

then one finds that one may rewrite Eq. (14.1) as

$$(-\omega^2 m + j\omega C + k)X = F_0$$

This result is completely equivalent to

$$(-\omega^2 m + \bar{k})X = F_0$$

if one defines $\bar{k}$ as a complex quantity

$$\bar{k} = k + jk_i \qquad k_i = \omega C \tag{14.18}$$

By comparing the last two terms of Eq. (14.18) with Eq. (14.15) one finds that

$$\eta = \frac{\omega C}{k} = \frac{k_i}{k} \tag{14.19}$$

on the basis of which one may write

$$\bar{k} = k(1 + j\eta) \tag{14.20}$$

Thus, as far as sinusoidal vibrations are concerned, use of a spring with an appropriately defined complex stiffness is entirely equivalent to use of an elastic spring and a dashpot.

According to Eq. (14.19), the loss factor which corresponds to a constant viscous damping coefficient is proportional to frequency. Loss factors corresponding to other mechanisms have other frequency dependences (and may be modeled in terms of frequency-variable damping coefficients); and, in fact, loss factors or complex stiffnesses with appropriate frequency dependences are widely used to represent the effects of various damping mechanisms.

The loss factor (and the complex stiffness) of a system is independent of amplitude only if the system is mathematically linear, that is, if the system response is proportional to the excitation. Although amplitude dependences of loss factors and complex stiffnesses are occasionally reported, these quantities are primarily useful for linear systems.

If a uniform rod or pad of a real material is subjected to sinusoidal loading, then this material sample generally will dissipate energy, in addition to storing energy; i.e., it will exhibit a complex stiffness. The (elastic) stiffness of an axially loaded uniform bar, and thus its strain energy storage, is proportional to the modulus of elasticity E of the material. In order to account for energy dissipation, one may simply introduce a complex modulus $\bar{E} = E_r + jE_i$ so as to obtain the proper complex stiffness $\bar{k} = \bar{E}A/L$ of the bar, where A represents its cross-sectional area and L its length. In analogy to Eq. (14.19), which expresses the loss factor η of a *system* as the ratio of the imaginary to real stiffness component, one may express the extensional loss factor β_e of a *material* as the ratio of the imaginary to the real component of the elastic modulus

$$\beta_e = \frac{E_i}{E_r} \qquad (14.21)$$

A completely analogous definition applies to the shear loss factor β_s of a material, in terms of the complex shear modulus, $\bar{G} = G_r + jG_i$

$$\beta_s = \frac{G_i}{G_r} \qquad (14.22)$$

The real parts of the complex moduli often are called *storage* moduli, the imaginary parts, *loss* moduli.

14.1.4 Measurement
Most techniques for measuring complex moduli use a material sample as a spring. A variety of such techniques is described in Refs. 7 to 11. The remainder of this discussion is addressed to measurement of the damping of panels.

Because a panel is a multimodal system (whose response may be visualized as composed of the responses of many single-degree-of-freedom systems) great caution must be used in applying any of the single-degree-of-freedom concepts to the measurement of panel damping. At low frequencies (at and near the fundamental resonance), one may usually utilize single modes — say, by exciting with pure single-frequency sinusoids, fixing on response maxima, and measuring the associated bandwidths or decay rates. At higher frequencies, any excitation is likely to excite several modes simultaneously, and in a frequency sweep the response decrease near one peak may be affected by the response increase associated with another peak. Clearly, measurements of resonant amplification, bandwidth, or logarithmic decrement cannot be performed usefully at such frequencies.

Direct measurement [12] of power dissipation is difficult and requires relatively complex and carefully phase-matched equipment.

The most useful and convenient technique for characterizing the damping of panels consists of measurement of decay rates (or reverberation times) in convenient frequency bands. The panel is excited by a band of white noise; the excitation is then cut off, and the slope of the logarithm of the envelope of the decaying signal is measured to determine the decay rate Δ_t. The loss factor η is computed from this decay rate and the center frequency of the excitation band [see Eq. (14.17)].

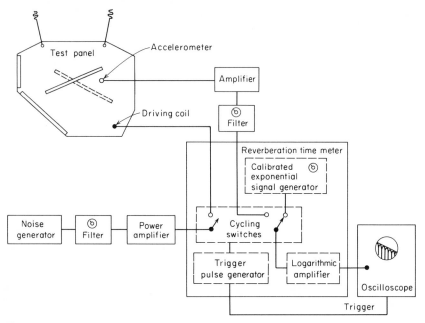

Fig. 14.3   System for measurement of reverberation time (decay rate) using a "reverberation time meter." [14]

Two types of instruments are particularly useful for such decay-rate measurements: a *graphic-level recorder* (see Chap. 4) or a *reverberation-time meter* (see Fig. 14.3).[13, 14]

In damping measurements, one must be extremely careful to prevent excessive extraneous energy dissipation. To minimize support-related damping, it is usually useful to suspend the test panel from two long strings (as indicated schematically in Fig. 14.3). To minimize damping by the exciter, a noncontacting drive system must be used. The accelerometer should be as light as possible, so as not to distort the responses, and the cables leading to it should be light and flexible. Finally, any electrical filter used must have a decay rate of its own that is shorter than that of the panel.

14.2   Consequences of Damping

14.2.1   Reduction of Resonant Responses and Stresses
The amplitude of the maximum stress σ_{max} in a uniform homogeneous panel at any resonance may be estimated conveniently from[15]

$$\frac{\sigma_{max}}{E} \approx \sqrt{3} \, \frac{V}{c_L} \qquad (14.23)$$

where E = modulus of elasticity of panel material, N/m² (lb$_f$/in.²)

$c_L = \sqrt{E/\rho_p}$ = longitudinal wavespeed in panel material, m/sec (in./sec)

ρ_p = material density, kg/m³ (lb$_m$/in.³)

V = amplitude of panel velocity, at location where this amplitude is greatest, m/sec (in./sec)

For the case [15] where a panel is exposed to a pressure $p_0(x,y,t) = p(x,y)$ cos ωt, whose distribution over the panel (i.e., over the x-y coordinate system) maintains the same shape but varies in magnitude sinusoidally with time, one may express the pressure distribution $p(x,y)$ in terms of the mode-shape functions $\psi_{mn}(x,y)$ of the panel

$$p(x,y) = \sum_{m,n=0}^{\infty} P_{mn}\psi_{mn}(x,y) \qquad (14.24)$$

where the P_{mn} are constants which may be computed from

$$P_{mn} = \frac{\displaystyle\int_A p(x,y)\psi_{mn}(x,y)\,dA}{\displaystyle\int_A \psi_{mn}^2(x,y)\,dA} \qquad (14.25)$$

where A denotes the area of the panel. The subscripts m and n identify the mode; $\psi_{mn}(x,y)$ represents the deflection shape corresponding to the m,n mode. The velocity distribution $V_0(x,y,t) = V(x,y)$ cos ωt over the panel is obtained by superposition of the modal contributions

$$v(x,y) = \sum_{m,n=0}^{\infty} V_{mn}\psi_{mn}(x,y) \qquad (14.26)$$

where the V_{mn} are constants.

At resonance of a particular mode, the response due to that mode predominates, and the above series may be approximated by a single term.

From the equation of flexural motion for a uniform panel, one finds that the modal velocity V_{mn} of a resonant mode depends on the modal pressure P_{mn} as

$$V_{mn} = \frac{P_{mn}}{\eta\mu\omega_{mn}} \qquad (14.27)$$

Here η = loss factor of the plate (corresponding to the m,n mode), dimensionless

$\mu = \rho_p h$ = plate mass per unit area, kg/m² (lb$_m$/in.²)

ρ_p = density of plate material, kg/m³ (lb$_m$/in.³)

h = plate thickness, m (in.)

ω_{mn} = resonance frequency of the m,n mode, radians/sec

Consider, for example, a rectangular panel on simple supports, exposed to a uniform pressure distribution $p(x,y) = p_c$. For such a panel with edge lengths a and b in the x- and y-coordinate directions ($0 < x < a$, $0 < y < b$), the mode shapes obey [2] $\psi_{mn} = \sin(m\pi x/a)\sin(n\pi y/b)$. Then, from Eq. (14.25) one finds that

$$P_{mn} = \begin{cases} 16p_c/\pi^2 mn & \text{if both } m \text{ and } n \text{ are odd} \\ 0 & \text{otherwise} \end{cases}$$

For a simply supported plate, the natural frequency ω_{mn} is given by

$$\omega_{mn} = \frac{\pi^2}{\sqrt{12}} c_L h \left(\frac{m^2}{a^2} + \frac{n^2}{b^2}\right)$$

Then, according to Eqs. (14.27) and (14.23), the maximum stress at the resonance of the m,n mode (where m,n both are odd) obeys

$$\sigma_{max} \approx \frac{\sqrt{3}E P_{mn}}{c_L \rho_p h \eta \omega_{mn}} = \frac{96}{\pi^4} \frac{p_c}{\eta h^2 mn \left(\dfrac{m^2}{a^2} + \dfrac{n^2}{b^2}\right)}$$

where η, ρ_p, h, $\omega_{m,n}$ and $P_{m,n}$ are given above
 E and c_L are given after Eq. (14.23)
 p_c = uniform rms sound pressure over the panel
Note that, as in Eq. (14.5), the response at resonance depends only on the damping and varies inversely as the loss factor.

For $p_c = 10^4$ N/m^2 (≈ 1.4 psi ≈ 0.1 atm) acting on a 0.25-cm (0.1-in.) thick plate with dimensions of 0.75 by 1.00 m (30 by 40 in.), the stress at the fundamental resonance ($m = n = 1$) turns out to obey $\eta \times \sigma_{max} \approx 5.7 \times 10^8$ N/m^2 ($\approx 8.2 \times 10^4$ psi). For a typical value of $\eta \approx 0.01$, one obtains the quite considerable value of $\sigma_{max} \approx 5.7 \times 10^{10}$ N/m^2 ($\approx 8.2 \times 10^6$ psi).

14.2.2 Reduction of Response to Random Excitation

Although no general simple relations are available for predicting the responses of panels to random excitation, one finds that generally, the root-mean-square displacement (or velocity) varies inversely as $\sqrt{\eta}$. Consider a panel exposed to a pressure $p_0(x,y,t) = p(x,y)g(t)$, where $g(t)$ is a random function of time.[15] Here the modal excitation $\bar{P}_{mn} = P_{mn}g(t)$. The mean-square modal excitation force is

$$\langle \bar{P}_{mn}^2 \rangle = P_{mn}^2 \langle h^2 \rangle = \int_0^\infty W_p(f)\, df$$

where the brackets $\langle\ \rangle$ indicate averaging with respect to time and $W_p(f)$ represents the spectrum density (see Chap. 2) corresponding to $h(t)$, expressed in terms of the frequency $f = \omega/2\pi$.

The response of a mode of vibration to random excitation is essentially a resonant response excited by all frequency components between the half-power points encompassing the resonance frequency. Thus, if $W_p(f)$ is a fairly smooth function near f_{mn}, then the mean-square effective resonant excitation acting on a mode is

$$\langle \bar{P}^2_{mn} \rangle \approx W_p(f_{mn})\Delta f = W_p(f_{mn})\eta f_{mn}$$

Substitution of the square root of this result into Eq. (14.27) then yields the root-mean-square modal velocity

$$(V_{mn})_{\text{rms}} \approx \frac{1}{2\pi\mu} \sqrt{\frac{W_p(f_{mn})}{\eta f_{mn}}}$$

Random (broadband) excitation is likely to excite several modes at a time. The total panel response may then be approximated by superposition of the various resonant modal responses. Modes that are highly damped and closely spaced in frequency will have their response curves overlap, and will respond coherently (essentially in phase with each other). Modes that are lightly damped and widely spaced will have no overlap and will respond incoherently. A convenient criterion for "modal overlap" may be obtained by comparing the half-power bandwidth $\Delta f = \eta f_n$ of Eq. (14.6) with the average frequency interval between modal resonances [16]

$$(\Delta f)_{\text{modes}} \approx \frac{hc_L}{A\sqrt{3}} \tag{14.28}$$

If ηf_n exceeds $(\Delta f)_{\text{modes}}$, then modal overlap and coherent responses will occur.

The number of modes that have their resonances in a given excitation frequency band $(\Delta f)_{\text{exc}}$ which is considerably wider than $(\Delta f)_{\text{modes}}$ is approximately $N_R = (\Delta f)_{\text{exc}}/(\Delta f)_{\text{modes}}$. If modal overlap occurs, then one may expect to obtain approximately $N_G = (\Delta f)_{\text{exc}}/\Delta f$ groups of modes, [16, 17] each group made up of $N_M = \Delta f/(\Delta f)_{\text{modes}}$ members, where all members of the group respond coherently with each other, but where the response of one group is incoherent with that of the others. The foregoing reasoning permits one to estimate the approximate rms-velocity distribution over the panel in particular cases, but the computation is tedious because one must take the mode-shape functions into account. However, one may obtain an upper bound $V_{\text{rms,ub}}$ on the rms velocity at any point on the panel by carrying out the calculation as if all mode shapes had their maxima at the same location. Because the velocities of coherent modes add directly, whereas for incoherent modes the total

mean-square velocity is the sum of the individual mean-square velocities, one finds that

$$V_{\text{rms,ub}} \approx N_M \sqrt{N_G} \ \overline{[(V_{mn})_{\text{rms}}]}$$

The bar over $(V_{mn})_{\text{rms}}$ indicates the average over all modes with resonances in the excitation band. Note that for the case of no modal overlap N_M reduces to unity and N_G to N_R in the above expression.

It is also of interest to note that narrowband excitation can typically
excite N_M modes at resonance, namely, those that may be expected to
have their response maxima (half-power bandwidth) encompass the
excitation frequency. Increased damping results in a greater number
N_M of excited modes responding coherently to the excitation. The
involvement of more modes tends to counterbalance the reduction of
the responses of the individual modes, so that increased damping may
not lead to a reduction in response.

14.2.3   Attenuation of Propagating Waves   In order to illustrate
the effect of damping on freely propagating waves, it is convenient to
consider a (nonspreading) straight-crested flexural wave. Such a wave
traveling in the x-coordinate direction along a panel, in the absence of
damping, may be described by [18, 19]

$$v(x,y,t) = v_0 \ \text{Re}\{e^{j(\omega t - k_b x)}\} \tag{14.29}$$

where $k_b = 2\pi/\lambda_b$ denotes the wavenumber and obeys

$$k_b = \left(\frac{\omega^2 \mu}{D}\right)^{1/4} = \left[12(1 - \nu^2)\frac{\rho_p \omega^2}{Eh^2}\right]^{1/4}$$

$$\approx \left[\frac{\sqrt{12}\omega}{hc_L}\right]^{1/2} \tag{14.30}$$

Here D represents the flexural rigidity of the plate, and the second expression shown applies for a homogeneous plate of uniform thickness
h, with Poisson's ratio ν.

If one now introduces the effect of damping by replacing E by $\bar{E} = E(1 + j\eta)$, one obtains a complex wavenumber $\bar{k}_b \approx k_b(1 - j\eta/4)$. Introduction of this wavenumber into Eq. (14.29) yields

$$v(x,y,t) = v_0 e^{-k_b \eta x/4} \ \cos(\omega t - k_b x)$$

which represents a propagating wave whose amplitude decreases
exponentially with distance.

As the propagation distance x increases by one wavelength $\lambda = 2\pi/k_b$,
the amplitude changes by a factor of $e^{-\eta\pi/2}$, and the decay in vibration
level per wavelength [25] obeys

$$\Delta_\lambda(\text{dB/wavelength}) = -20 \log_{10} (e^{-\eta\pi/2})$$

$$= 10 (\log_{10} e)\pi\eta = 13.6\eta \qquad (14.31)$$

where η is the loss factor at the frequency of interest.

14.2.4 Decay of Free Vibrations The decay rate Δ_t of a free vibration is directly proportional to the loss factor and to the center frequency of the band (or to the natural frequency of a mode) under consideration

$$\Delta_t(\text{dB/sec}) = 27.3\eta f_n \qquad (14.32)$$

Assume that a single impact produces a decaying velocity response with an envelope $V_1 e^{-\eta\omega_n t/2}$ [see Eq. (14.7)] and that successive impacts occur at intervals of T, in such a way that the responses due to the various impacts are uncorrelated. Then at any instant t after application of the most recent pulse ($t < T$) the panel will exhibit the effects of that pulse and all previous pulses. If energy addition (i.e., addition of squared variables) holds, then the mean-square velocity due to all pulses obeys

$$V^2 = V_1^2 e^{-\eta\omega_n t}(1 + e^{-\eta\omega_n T} + e^{-2\eta\omega_n T} + \cdots)$$

$$= \frac{V_1^2 e^{-\eta\omega_n t}}{1 - e^{-\eta\omega_n T}}$$

For large intervals T between impacts, that is, for $\eta\omega_n T \gg 1$, the foregoing relation reduces to

$$V^2 = V_1^2 e^{-\eta\omega_n t}$$

indicating that the effect of each impact is independent of the effect of the others. For intervals that are small, so that $\eta\omega_n T \ll 1$, one obtains

$$V^2 \approx \frac{V_1^2}{\eta\omega_n T}$$

which shows that the effects of the individual impacts are additive and contribute to build up a mean-square velocity which varies inversely as the loss factor and as the interval between impacts.

14.2.5 Increase of Transmission Loss From Chap. 11 of this book one notes that below about 0.6 times the critical frequency the transmission loss is independent of damping, and depends almost entirely on the panel mass per unit area. In the vicinity of the critical frequency, the transmission loss depends both on the damping and on the mass per unit area in a complicated manner. At frequencies above about 1.5 times the critical frequency, the transmission loss increases by 20 dB per tenfold increase in mass per unit area and by 10 dB per tenfold increase in loss factor.

14.3   Mechanisms and Magnitudes

14.3.1   Transport of Energy
All the usual damping measurement techniques, and particularly the decay-rate and the input power measurement methods, essentially determine the rate at which energy disappears from the measurement point (i.e., from the panel under study). These techniques do not distinguish between mechanical energy that is actually dissipated by conversion to heat and vibratory energy that is propagated away—either as mechanical waves to adjacent structures or as sound radiated into fluids in contact with the panel.

In fact, one of the greatest difficulties one encounters in attempting to measure "the damping" of a panel consists of supporting the panel in a manner which minimizes the transport of energy away from the panel via the supports. The best (and the only practical) way of avoiding the flow of energy from the test panel to adjacent structures is to disconnect the panel from these structures—or to use the softest possible connection if complete disconnection is impossible. Attempts at blocking energy flow by adding large amounts of stiffness or mass at connections to adjacent structures have proved fruitless;[17] there is available no practical way to measure only the energy-dissipation damping of a panel which is part of a larger structure, without removing the panel from the structure.

In the previous paragraph, the term "the damping" appeared in quotation marks, because in the previous sections damping has always implied energy dissipation only. If one carries out damping measurements on a panel which is part of a larger structure, then one measures what has been called the "effective damping," which accounts for both energy dissipation and transport.

It is clear that any steady-state input power measurement of damping will measure more than the dissipation damping, because it will also sense energy that is conducted away. Decay-rate measurements, however, may lead to damping values that are either larger or smaller than those associated with energy dissipation in the test panel. If after removal of the excitation there is a net energy flow from the test panel to the adjacent structures, one obtains a greater decay rate. If there is a net energy flow to the test panel from the adjacent structures, one obtains a slower decay.[20]

For practical panels vibrating in air, energy losses due to acoustic radiation generally are negligible, compared to those due to other damping mechanisms. The radiation resistance R_{rad} of a panel is defined so that the acoustic power P_A radiated from the panel is given by $P_A = R_{rad}$ $\langle v^2 \rangle$, where $\langle v^2 \rangle$ represents the spatial average of the mean-square velocity of the panel. Therefore the energy lost by the panel per cycle is $D = P_A/f$. Since the maximum kinetic energy of the panel is $M\langle v^2 \rangle$, where M denotes the total mass of the panel, one may use Eq. (14.16) to obtain a corresponding "radiation loss factor"

$$\eta_{rad} = \frac{R_{rad}}{\omega M} \tag{14.33}$$

For a panel for which the acoustic wavelength λ_c at coincidence is considerably greater than the longer edge dimension b, one may estimate η_{rad} from [21]

$$\frac{\eta_{rad} \times 2\pi f \mu}{\rho c} \approx \begin{cases} \left(1 - \dfrac{f_c}{f}\right)^{-1/2} & \text{for } f > f_c \quad (14.34a) \\[2ex] \dfrac{8}{\pi^2} \dfrac{l\lambda_c}{A} \sqrt{\dfrac{f}{f_c}} & \text{for } f \ll f_c \quad (14.34b) \end{cases}$$

where ρ = density of air, kg/m³ (lb$_m$/in.³)
    c = speed of sound in air, m/sec (in./sec)
    μ = panel mass per unit area, kg/m² (lb$_m$/in.²)
    A = panel area (one side), m² (in.²)
    f = vibration frequency, Hz
    f_c = critical frequency, Hz
    l = effective perimeter of panel, m (in.)
    λ_c = wavelength at critical frequency, m (in.)
The critical frequency f_c and the corresponding wavelength λ_c may be found from

$$\lambda_c = \frac{c}{f_c} = \frac{\pi h c_L}{\sqrt{3}\,c} \qquad \text{[from Eq. (11.4)]}$$

where c_L = longitudinal wavespeed in panel material, m/sec (in./sec)
    h = panel thickness, m (in.)
For a rectangular panel with edge lengths a and b, $l = 2(a + b)$; for a panel with ribs, the effective perimeter l is equal to twice the total length of the ribs plus the perimeter of the original (ribless) panel. Using the foregoing relations, one finds that the radiation loss factor at 500 Hz of an aluminum panel, which is 0.25 cm ($\approx$ 0.1 in.) thick and measures 1.00 by 1.25 m (40 by 50 in.), is $\eta_{rad} \approx 1.1 \times 10^{-3}$. This value corresponds to a perimeter $l = 2(1.00 + 1.25) = 4.50$ m.

Note that adding two 1.25-m-long ribs and three 1.0-m-long ribs would add $2(2 \times 1.25 + 3 \times 1.00) = 11.0$ m to the effective circumference, increasing l and η_{rad} by a factor of $(4.5 + 1.0)/4.5 \approx 3.5$.

14.3.2   Effect of Edges and Other Discontinuities   In order to account for the damping produced by linear reinforcing beams attached to a panel or by seams in the panel, and/or by panel edges, it is useful to introduce the concept of the absorption coefficient of a linear discontinuity, in direct analogy to the corresponding quantity used in the study of the acoustics of enclosed spaces. The absorption coefficient γ of a linear discontinuity is defined as the fraction of the panel bending-wave energy impinging on the discontinuity which is not returned to the panel; thus, γ may account for both dissipation and energy transport to adjacent structures.

By considering the flexural waves traveling on the panel to be uniformly distributed in direction (i.e., to constitute a "diffuse field"), and by accounting for the mean free path of such waves and for the energy carried by them, one may show that the loss factor η_{disc} due to discontinuities obeys [22]

$$\eta_{\text{disc}} = \frac{\lambda_p}{\pi^2 A} \sum_i \gamma_i L i \qquad (14.35)$$

where γ_i = absorption coefficient of ith discontinuity, dimensionless

L_i = effective length of ith discontinuity, m (in.)

A = panel area, m² (in.²)

λ_p = panel flexural wavelength, m (in.)

$\approx \sqrt{\pi h c_L / f \sqrt{3}}$

h = panel thickness, m (in.)

c_L = longitudinal wavespeed in panel material, m/sec (in./sec)

The "effective length" of a discontinuity is the length of the discontinuity on which panel waves can impinge. Since waves traveling on the panel can impinge on only one side of a plate edge, the effective length of any edge is equal to its actual length. However, waves can impinge on both sides of a seam or reinforcing beam located at several wavelengths from an edge; the effective length of such a discontinuity in the panel interior is twice its actual length.

Equation (14.35) permits one to find the loss factor due to discontinuities — and to add this to the loss factors due to all other significant damping mechanisms, so as to obtain the total loss factor — provided one knows the absorption coefficient associated with each discontinuity. The absorption coefficient γ_0 of a given type of discontinuity can only rarely be predicted from first principles, but it can be determined experimentally relatively easily by adding a discontinuity of length L_0 to a test panel, measuring the increase $\Delta\eta$ in loss factor it produces, and using

$$\gamma_0 = \frac{(\Delta\eta)\pi^2 A}{\lambda_p L_0}$$

which follows from Eq. (14.35).   If the absorption coefficient of a discontinuity depended on the length of the discontinuity and on its position on the panel, then one would require a very large catalog of absorption coefficient data before one could use Eq. (14.35) to predict the damping of a panel.   Fortunately, it has been found[22] that absorption coefficients practically are independent of discontinuity length and position, provided one restricts the problem to frequencies that are high enough so that the panel flexural wavelengths are considerably shorter than both a characteristic plate surface dimension and the discontinuity length.

The energy that panel seams or reinforcements attached to the panel can dissipate, and thus the absorption coefficient of such a discontinuity, depends very markedly on the fastening method used.   Reinforcements that themselves are not made of highly dissipative material (e.g., metal beams attached to metal panels and seams in metal panels) generally produce only very small absorption coefficients if they are continuously welded or joined by means of a rigid adhesive, but they result in significant absorption coefficients if they are multipoint-fastened, i.e., riveted, bolted, or spot-welded.[12]   It has been shown[12,14,23] that the dominant mechanism responsible for the damping of multipoint-fastened discontinuities is associated not with interface slip and friction, but with an "air-pumping" effect produced by motions of adjacent surfaces (between connection points) away from and toward each other.   The damping is due to the viscosity of the air (or other ambient gas, or any liquid that might be introduced between the surfaces).

The absorption coefficient pertaining to multipoint-fastened beams attached to a test panel in air has been studied extensively, both theoretically[23] and experimentally.[12,14]   This absorption coefficient has been found to be proportional to the width of the attached beam and to depend primarily on the ambient air pressure and on the ratio of bolt spacing (or distance between connection points) to the panel flexural wavelength.   Figure 14.4 summarizes the results of an extensive series of measurements, carried out on aluminum and steel panels of various thicknesses, with various beam widths and bolt spacings.   The data are presented in terms of a reduced absorption coefficient $\gamma_r = \gamma(w_0/w)$, plotted against a reduced frequency $f_r = f(d/d_0)^2(h_0/h)(c_{L0}/c_L)$, where w represents the beam width, d the bolt spacing, h the panel thickness, and c_L the longitudinal wave velocity in the panel material; the subscript "0" indicates reference values of the various quantities, which are listed in the lower right part of Fig. 14.4.   The reduced absorption coefficient accounts for the proportionality of γ to beam width, and the reduced frequency accounts for the dependence of γ on bolt spacing/wavelength ratio.   In order to find, for example, the damping increase

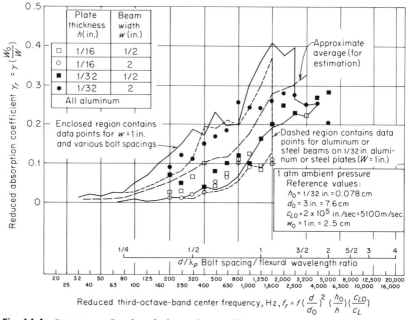

Fig. 14.4 Summary of reduced absorption coefficient data for beams of several widths on aluminum and steel plates of several thicknesses.[14] λ_p is defined after Eq. (14.35).

at 400 Hz produced by fastening to a 0.10-cm-thick copper panel ($c_L \approx$ 3,000 m/sec) of 1 m² area, a 5.0-cm-wide beam of 0.6 m length, using rivets spaced 6.1 cm apart, one first finds the reduced frequency (using values from Fig. 14.4) to be $f_r = 340$ Hz. From Fig. 14.4, one obtains $\gamma_r \approx 0.08$; thus $\gamma = \gamma_r w/w_0 \approx 0.16$. From Eq. (14.35), then, $\eta_{\text{disc}} / \lambda_p \gamma L/\pi^2 A \approx 1.1 \times 10^{-3}$. Observe that if the rivet spacing were doubled, f_r would be increased by a factor of 4, to 1,360 Hz, for which $\gamma_r \approx 0.25$ and $\eta_{\text{disc}} \approx 1.75 \times 10^{-3}$.

14.3.3   Internal Friction (Material Damping) of Solids

Fortunately, the loss factors of most metals and common architectural structural materials are relatively independent of amplitude (as long as the stress amplitude remains well below the fatigue limit), temperature (as long as the material temperature remains well below the melting point), and frequency (within the audio range). Typical loss factor values are listed in Table 14.1.

Ferromagnetic materials often exhibit much greater damping than common structural metals, and also have loss factors which vary markedly with stress amplitude. For example, Nivco 10 (an alloy containing 72% Co and 23% Ni) has a loss factor which is nearly proportional to the stress amplitude and which reaches a value of 0.05 for a stress amplitude of 10^8 N/m^2 ($\approx$ 15,000 psi).   Certain alloys of magnesium and cobalt have loss factors approaching 0.1 at low to medium stress levels.[3] Judging from the fact that the stiffness of titanium alloys decreases considerably with increasing temperature, one would expect that the loss factor of titanium is also quite temperature sensitive; however, no data appear to be available.

In contrast to metals and "hard" structural materials, plastics and rubbers [3, 25, 26] have stiffnesses (moduli of elasticity) and loss factors which vary strongly with frequency and temperature.   Figure 14.5 shows how the shear modulus G (which for all plastics and rubbers is very nearly equal to one-third of the modulus of elasticity) and loss factor β in shear (which for plastics and rubbers is very nearly equal to the loss factor in extension) of a typical plastic material vary with frequency and temperature.

At low frequencies and/or high temperatures, the material is soft and sufficiently "mobile" for the material strain to follow an applied stress

TABLE 14.1   Typical Loss Factors of Structural Materials at Small Amplitudes, at Room Temperature, and in the Audio Frequency Range [19, 24]

Material	Loss factor η
Aluminum	10^{-4}
Brass, bronze	$<10^{-3}$
Brick	1 to 2×10^{-2}
Concrete	
light	1.5×10^{-2}
porous	1.5×10^{-2}
dense	1 to 5×10^{-2}
Copper	2×10^{-3}
Cork	0.13 to 0.17
Glass	0.6 to 2×10^{-3}
Gypsum board	0.6 to 3×10^{-2}
Lead	0.5 to 2×10^{-3}
Magnesium	10^{-4}
Masonry blocks	5 to 7×10^{-3}
Oak, fir	0.8 to 1×10^{-2}
Plaster	5×10^{-3}
Plexiglass, Lucite	2 to 4×10^{-2}
Plywood	1 to 1.3×10^{-2}
Sand, dry	0.6 to 0.12
Steel, iron	1 to 6×10^{-4}
Tin	2×10^{-3}
Wood fiberboard	1 to 3×10^{-2}
Zinc	3×10^{-4}

without appreciable phase shift, so that the energy losses and β are relatively small. The material is then said to be in its "rubbery" state. At high frequencies and/or low temperatures, the material is relatively stiff and lossless, and its behavior resembles that of an elastic material; the material then is said to be in the "glassy" region. At intermediate frequencies and temperatures, the damping is greatest and the stiffness takes on intermediate values.

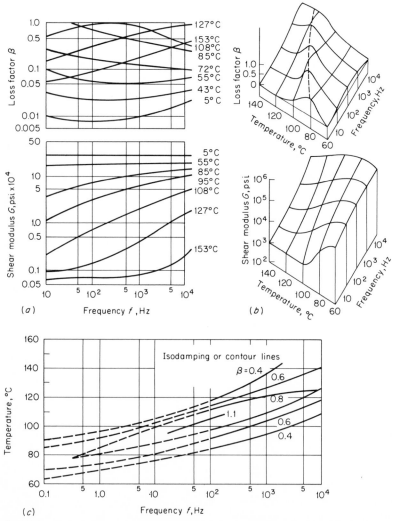

Fig. 14.5 Dependence of shear modulus and loss factor of polyester plastic on frequency and temperature. (*After Lazan*[3] *and Ungar and Hatch.*[25]) Note: 1 psi $\approx$ 7,000 N/m². (*a*) Functions of frequency at constant temperatures; (*b*) isometric plots; (*c*) contours of constant loss factor.

As may be seen especially clearly from the upper isometric plot of Fig. 14.5b, the transition region (around the maximum β ridge indicated by the dotted curve) shifts to higher frequencies as the temperature is increased. The β curves of Fig. 14.5a also show this trend. If one begins with the material at a given temperature and frequency and increases the temperature a certain amount, then one may return to the same values of the properties by increasing the frequency appropriately. Thus, an increase in temperature is "equivalent to" (i.e., leads to the same changes in material properties as) a decrease in frequency. From the isodamping contour plot of Fig. 14.5c, one sees that for this particular plastic, a 10°C ($\approx$ 18°F) increase in temperature is equivalent to slightly more than a decade decrease in frequency.[27]

The behavior indicated in Fig. 14.5 is typical of most polymeric and elastomeric materials, in that all exhibit rubbery, transition, and glassy regions. Data for a number of viscoelastic materials are given in Table 14.2. However, the modulus and loss factor magnitudes and the temperature-frequency equivalences vary widely and depend strongly on material composition. Plastics composed of several polymeric components may even exhibit multiple damping peaks. It is well known that the addition of plasticizers softens plastics (and shifts the transition region toward higher frequencies and/or lower temperatures), whereas the addition of rigid fillers produces stiffening.[26,28] Because the dynamic properties of plastics depend strongly on their composition and processing, it is likely that two nominally identical samples of a given plastic bought from different sources, or coming from different manufacturing batches, will exhibit widely different stiffness and loss factor characteristics.

14.3.4 Other Mechanisms

The most widely studied and best understood means for increasing the damping of a panel by a predictable amount consists of attaching to the panel one or more layers of a viscoelastic material which itself has high damping. The damping of panels by means of attached viscoelastic layers is discussed in the next section.

Dry friction damping, although of possible importance for the low-frequency (low-modal) motions of built-up structures,[29,30] is generally of little importance for panels.

Energy can also be removed from a vibrating panel by permitting it to "rattle" against another object, or by arranging other objects to rattle against it. Damping here occurs by energy transport, due to energy losses in materials at contact points where large stresses may be obtained, and due to a transfer of energy from low-frequency to higher-frequency plate modes. Analyses of particular "impact dampers" are available, but are of little general utility.[31,32]

Granular materials, notably sand, have been used to provide structural

TABLE 14.2 Dynamic Elastic Properties of Some Viscoelastic Materials at or near Room Temperature

Material	Max. loss factor, β_{max}	Frequency for β_{max}, Hz
Polysulfide rubber (Thiokol H-5)	5.00	1,000
Butyl rubber (Enjay 9-262-4)	4.02	3,100
Urethane rubber (Disogrin IDSA 9250)	2.59	3,000
Butyl rubber (Enjay 9-262-1)	2.56	3,000
(Enjay 9-262-3)	2.20	3,000
Polyvinyl butyral	2.0	2
3M tape adhesive, No. 466	1.82	1,000
Butyl gum (Type U-50), vulcanized	1.80	10,000
Buna N (Type B-5), carbon-filled, vulcanized	1.60	10,000
GR-S rubber	1.60	10,000
Urethane rubber (82% Pb powder filler)	1.40	3,000
Fluoro rubber (3M No. IF4)	1.30	4,000
Neoprene GRT	1.18	1,775
Polyvinyl chloride acetate	1.14	100
Neoprene (Type GN-50), vulcanized	1.10	10,000
Buna N (Type B-O), unvulcanized	1.10	1,000
Hycar 1014	0.76	1,850
3M damping tape adhesive, No. 435	0.73	4,000
Neoprene (Type CG-1), vulcanized	0.60	10,000
Fluoro silicone rubber (Silastic LS53)	0.56	4,000
Polysulfide rubber (Thiokol ST)	0.53	1,300
Urethane rubber (Disogrin IDSA 7560)	0.51	4,250
GR-S rubber (23.5% styrene)	0.47	1,700
(3.0% styrene)	0.38	1,400
GR-S (Type S-50) carbon-filled, vulcanized	0.35	10,000
Silicone gum (Linde No. Y-1032)	0.33	400
Natural gum rubber (Type N-1)	0.30	1,000
GR-S rubber (Krylene)	0.30	10,000
Hevea rubber (filled)	0.25	10,000
(vulcanized)	0.20	10,000
Natural rubber (tire tread stock)	0.20	500

damping. These materials typically are placed in voids in the primary structures, for example in honeycomb cells or in specially designed cavities in the structures.[33,34] Although layers of granular materials are known to exhibit high loss factors (the loss factor of dry sand is of the order of 0.1),[35,36] the mechanisms responsible for this damping or for the damping which granular materials provide to structures are not fully understood. The damping produced by granular materials gen-

Loss factor at five frequencies, Hz					Test temperature, °C	Shear modulus G_1 at β_{max},* lb/in.²	Reference no.
1	10	100	1,000	10,000			
0.17	0.50	1.20	5.00	....	25 ± 50	1,500	47
......	....	0.40	0.92	>2.0	21	1,500	48
......	....	0.10	0.13	~0.50	25	2,900	48
......	....	0.55	0.84	>1.80	21	2,900	48
......	....	0.45	0.68	>1.40	24	2,900	48
......	0.14	0.06	0.06	....	30 ± 70	29,000	8
......	~1.00	1.17	1.82	~0.80	25	360	49
......	....	....	1.20	1.80	20	1,000	50
......	0.40	0.75	1.20	1.60	20	1,500 to 4,400	50
......	0.20	0.50	1.00	1.60	20	1,500	51
......	....	0.40	0.49	....	25	2,500	48
......	~0.30	0.46	0.79	....	45 ± 55	17	49
......	....	0.60	1.01	....	20	220	52
......	....	1.14	0.73	0.90	40 ± 20	580	49
......	....	....	0.70	1.10	20	290	50
......	0.50	0.65	1.00	....	15	1,200	50
......	....	0.40	0.63	....	20	200	52
......	~0.50	0.58	0.67	~0.70	10 ± 50	37	49
......	....	....	0.02	0.60	20	4,350	50
......	~0.10	0.16	0.46	....	30 ± 70	87	49
......	....	0.40	0.51	....	20	160	52
......	....	0.10	0.22	....	25	1,500	48
......	....	~0.30	0.39	....	20	200	52
......	....	~0.25	0.37	....	20	13	52
......	....	....	0.20	0.35	20	435	50
......	~0.15	0.20	0.40	....	60 ± 40	2.5	49
0.04	0.08	0.14	0.30	....	20	150	50
0.11	0.12	0.13	0.15	0.3	20	100	53
0.11	0.11	0.11	0.14	0.25	20	870	53
0.03	0.03	0.03	0.05	0.20	20	870	54
0.16	0.18	0.20	......	....	25 ± 50	150	47

* To convert to N/m², multiply tabulated values of shear modulus by 7,000.

erally increases markedly with amplitude, especially as the structural acceleration approaches the acceleration of gravity.

14.4 Damping of Panels by Viscoelastic Layers

If one adds one or more layers of a material with high inherent damping to a relatively lightly damped panel, then the resulting composite panel will have higher damping than the original one. Bending of a panel, which is made up of a number of layers of different materials, generally causes each layer to bend, extend, and deform in shear. With each type of deformation in each layer there is associated some storage of strain energy, and with each component of energy storage there is associated some energy dissipation.

The damping action of a composite panel may be understood by considering the panel action like that of a series-parallel array of springs (e.g., Fig. 14.6), where each spring represents an energy-storage mechanism. A given bending deformation of the composite panel then corresponds to some displacement x of the top of the array with respect to its base. This displacement causes each spring to deflect some amount; if the ith spring, whose stiffness is K_i, deflects an amount x_i, the energy it stores is $W_i = K_i x_i / 2$. If the displacements are cyclic, then, in view of the definition of the loss factor, the energy dissipated per cycle by the spring is $D_i = 2\pi\eta_i W_i$, where η_i is the loss factor of the ith spring. The loss factor η of the array of springs may then be found by applying the definition of loss factor

$$\eta = \frac{\Sigma D_i / 2\pi}{\Sigma W_i} = \frac{\Sigma \eta_i W_i}{\Sigma W_i} = \frac{\Sigma \eta_i K_i x_i^2}{\Sigma K_i x_i^2} \qquad (14.36)$$

The summations are taken over all springs.

If the ith spring is to make a significant contribution to the loss factor of the array (i.e., to the loss factor of the composite panel), then the value of $\eta_i K_i x_i^2$ must be a significant fraction of the sum which appears in the numerator of Eq. (14.36). This means that the ith spring must have a loss factor of considerable magnitude, must have a reasonably large stiffness, and must also be so arranged in the array that it experi-

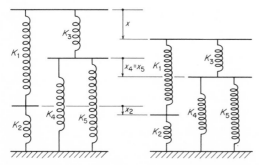

Fig. 14.6 Series-parallel array of springs (representing energy-storage mechanisms) in undeflected (left) and deflected (right) positions. (Deflections $x_1 = x - x_2$, $x_3 = x - x_4$ are not indicated.)

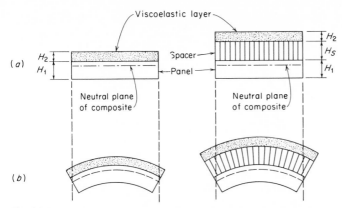

Fig. 14.7 Section of panel with free viscoelastic layer, without and with spacing layer: (*a*) flat (underformed) and (*b*) deformed.

ences a considerable deformation x_i when the top of the array is displaced.*

Equation (14.36) also shows that the composite loss factor is the weighted average of the loss factors of the individual mechanisms, with energy-storage magnitudes acting as weighting factors. If only one of the mechanisms, say the first, has a nonzero loss factor associated with it, then Eq. (14.36) reduces to

$$\eta = \frac{\eta_1 W_1}{\Sigma W_i} = \frac{\eta_1}{1 + \dfrac{W_2 + W_3 + \cdots}{W_1}} \qquad (14.37)$$

Here η is always smaller than η_1, but η may be made to approach η_1 by arranging the array and stiffnesses so that the energy W_1 stored by mechanism 1 becomes a larger fraction of the total energy storage ΣW_i.

14.4.1 Free Viscoelastic Layers; Extensional Damping [37,38] †

If a free (i.e., uncovered) viscoelastic layer is attached to a panel which otherwise has a very small loss factor (Fig. 14.7), then bending typically produces both flexure and extension of the two layers. Shear here generally has little effect on energy storage. By considering both flexural and extensional energy storage, one finds that the loss factor of the composite panel is

---

* Since each spring dissipates the fraction η_i of the energy it stores, a spring (or energy-storage mechanism) can dissipate a significant amount of energy only if the energy storage is considerable *and* if the dissipated fraction thereof is significantly large.

† All the theoretical results discussed in this chapter pertaining to damping by means of viscoelastic layers have been derived on the basis of one-dimensional (beamlike) flexure, and they have been found satisfactory for all practical purposes.

$$\eta = \frac{\beta_2}{1 + \dfrac{k^2(1 + \beta_2^2) + (r_1/H_{12})^2[(1 + k)^2 + (\beta_2 k)^2]}{k\{1 + (r_2/H_{12})^2[(1 + k)^2 + (\beta_2 k)^2]\}}} \qquad (14.38)$$

where $\beta_2 =$ loss factor of the viscoelastic layer, dimensionless
$k = K_2/K_1 = E_2 H_2/E_1 H_1 =$ ratio of extensional stiffnesses, dimensionless
$r_1 = H_1/\sqrt{12} =$ radius of gyration of elastic layer, m (in.)
$r_2 = H_2/\sqrt{12} =$ radius of gyration of viscoelastic layer, m (in.)
$H_{12} = (H_1 + H_2)/2 =$ distance between neutral planes of the two layers, see Fig. 14.8, m (in.)
$E_i =$ modulus of elasticity of ith layer, N/m² (psi)
$H_i =$ thickness of ith layer, m (in.)

Figure 14.8 is a plot based on Eq. (14.38), under the added restriction that $\beta_2^2 \ll 1$.

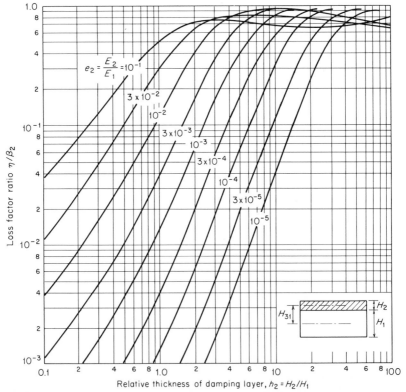

Fig. 14.8   Dependence of loss factor η of panel with free viscoelastic layer on relative thickness and relative elastic modulus of viscoelastic layer [37] (β_2^2, which equals the loss factor of viscoelastic material, is assumed to be small compared to unity).   From Eq. (14.38).

In most practical situations, the added layer has less extensional stiffness than the basic panel, and $k \ll 1$. Since β_2 for useful materials never
exceeds unity by very much, then $k\beta_2 \ll 1$ also. For this case Eq. (14.38)
may be rewritten as

$$\eta = \frac{\beta_2}{1 + \dfrac{E_1}{E_2} \dfrac{H_1^3/H_2}{H_2^2 + 12H_{12}^2}} = \frac{\beta_2}{1 + \dfrac{1}{e_2 h_2 (3 + 6h_2 + 4h_2^2)}} \qquad (14.39)$$

where $e_2 = E_2/E_1 =$ ratio of moduli of elasticity
$\quad\;\; h_2 = H_2/H_1 =$ ratio of thicknesses
Equation (14.39) shows that η depends both on the material properties
(via β_2 and the ratio e_2) and on the geometry [via the factor $h_2(3 + 6h_2$
$+ 4h_2^2)$]. The equation shows that the loss factor η of the composite is
proportional to the loss factor β_2 of the viscoelastic material and increases
with increasing e_2 and with increasing h_2. As h_2 and/or e_2 becomes very
large, η approaches β_2.
If $k\beta_2 \ll 1$, so that Eq. (14.39) holds, η increases monotonically with
increasing relative viscoelastic layer thickness h_2. For larger k, the value
of η/β_2 tends toward unity with increasing h_2.
For the case where $e_2 h_2(3 + 6h_2 + 4h_2^2) \ll 1$, Eq. (14.39) reduces to

$$\eta \approx \beta_2 e_2 h_2 (3 + 6h_2 + 4h_2^2) = \frac{E_2''}{E_1} \frac{H_2}{H_1} \left[3 + 6 \left(\frac{H_2}{H_1}\right) + 4 \left(\frac{H_2}{H_1}\right)^2 \right] \qquad (14.40)$$

where $E_2'' = \beta_2 E_2$ is the imaginary part of the complex modulus of elasticity [see Eq. (14.21)], which is also known as the *loss modulus,* of the viscoelastic material. Viscoelastic materials typically have much smaller
moduli than primary structural materials, so that generally $e_2 < 10^{-2}$.
For this value of e_2, Eq. (14.40) applies for thickness ratios h_2 up to about
1.5; for smaller values of e_2, the equation is applicable up to larger values
of h_2.
If two viscoelastic layers are applied to a basic structural panel, one
layer on each side, and if each layer satisfies the usually practically applicable condition $k\beta_2 \ll 1$, then the loss factor of the composite may be
computed as the sum of the contributions of the two layers. In other
words, one may simply apply Eq. (14.39) to each viscoelastic layer and
then add the results.
What happens if one divides a layer of thickness H_2 of a given viscoelastic material on one side of a panel in half and applies half to the other
side of the panel? From Fig. 14.8 it may be seen that η varies as h_2^2 for
small values of h_2, so that halving h_2 reduces η by a factor of 4, and the
two half-thickness layers may be expected to result in about one-half as
much damping as one full-thickness layer. At intermediate values of
h_2, where η varies as the first power of h_2, two half-thickness layers have

about the same damping effect as a single full-thickness layer. For the region of Fig. 14.8 where η is nearly independent of h_2, the previously described procedure for calculating the combined effect of two layers does not apply; for such thick layers of viscoelastic material, one obtains practically the same damping, no matter how the viscoelastic material is divided between the two sides of the primary panel.

14.4.2 Spaced Free Viscoelastic Layers [37]   One may increase the extensional strain energy storage in a given free viscoelastic layer by use of a "spacing layer" (see Fig. 14.7). The spacing layer ideally should store no strain energy itself, should be extremely rigid in shear, and should be constructed so that elements normal to the undeflected panel surface remain normal to the deflected surface.

A good spacing layer may be obtained, for example, by placing a layer of cubes on the panel, bonding the top faces of the cubes rigidly to the viscoelastic layer, and the bottom faces to the basic panel, and making no connection between one cube and another. The honeycomb portions of practical honeycomb sandwich panels often behave nearly like ideal spacers.

The damping of panels with (ideally) spaced free viscoelastic layers may be obtained directly from Eq. (14.38) or from the first form of Eq. (14.39), except that one must now take the distance between the neutral planes as $H_{12} = H_s + (H_1 + H_2)/2$, where H_s represents the thickness of the spacing layer. If the spacing layer is not ideal, the damping obtained from the spaced viscoelastic layer will be less than that obtained with an ideal spacing layer. Expressions for the loss factor of a panel with a non-ideally spaced viscoelastic layer are cumbersome [37] and of practical interest only for special applications.

14.4.3 Constrained Viscoelastic Layers; Shear Damping [37,38]   If a viscoelastic layer is attached to a basic structural panel, and if an elastic "constraining" layer is attached atop the viscoelastic layer (Fig. 14.9), then bending of the composite produces not only bending and extensional strains in all three layers, but also shear, primarily of the middle (viscoelastic) layer. The shear-strain energy storage tends to dominate the damping action of constrained viscoelastic layers.

If the extensional stiffness of the viscoelastic layer is small compared to the stiffnesses of the other two layers, as is often the case in practice, the loss factor η of the composite panel is given by

$$\eta = \frac{\beta_2 YX}{1 + (2 + Y)X + (1 + Y)(1 + \beta_2^2)X^2} \qquad (14.41)$$

Here Y is a *stiffness parameter,* defined by

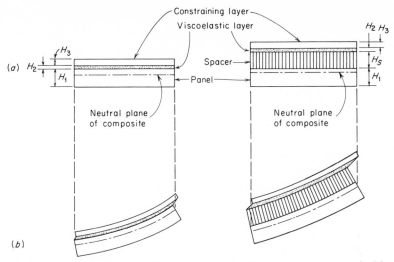

Fig. 14.9 Section of panel with constrained viscoelastic layer, without and with spacing layer: (a) flat and (b) deformed.

$$\frac{1}{Y} = \frac{E_1 H_1^3 + E_3 H_3^3}{12 H_{31}^2} \left(\frac{1}{E_1 H_1} + \frac{1}{E_3 H_3} \right) = \frac{1 + e_3 h_3^3}{12 h_{31}^2} \left(1 + \frac{1}{e_3 h_3} \right) \quad (14.42)$$

where E_1, E_3 = moduli of elasticity of the two elastic layers, N/m² (psi)

H_1, H_3 = thicknesses of the two elastic layers, m (in.)

$H_{31} = H_2 + (H_1 + H_3)/2$ = distance between neutral planes of elastic layers, m (in.)

In the second form of the above equation, the ratios $e_3 = E_3/E_1$, $h_3 = H_3/H_1$, and $h_{31} = H_{31}/H_1$ have been introduced. If the two elastic layers have the same modulus of elasticity, then $e_3 = 1$ and Y becomes purely a *geometric parameter*.

The variable X has been called the *shear parameter* and is defined as

$$X = \frac{G_2}{p^2 H_2} \left(\frac{1}{E_1 H_1} + \frac{1}{E_3 H_3} \right) \quad (14.43)$$

where G_2 = real part of complex shear modulus, N/m² (psi)

$p = 2\pi/\lambda$ = wavenumber, m⁻¹ (in.⁻¹)

λ = wavelength of flexural vibration of composite panel, m (in.)

Figure 14.10 shows some curves of η vs. X, for several values of β_2 and Y, based on Eq. (14.41).

The square root of the shear parameter X is proportional to the ratio of the bending wavelength λ to the "decay distance" for a shear disturbance (the distance within which a localized shear disturbance decays to $1/e$ of its initial value, e being the base of natural logarithms). Thus, X is a measure of how well the viscoelastic layer couples the flexural motions of the two elastic layers. In fact, the complex flexural rigidity $\bar{B}$ (N-m or in.-lb$_f$) of the composite panel obeys

$$\bar{B} = (B_1 + B_3)\left(1 + \frac{\bar{X}Y}{1 + \bar{X}}\right) \qquad (14.44)$$

where $B_1 = E_1 H_1^3/12$ and $B_3 = E_3 H_3^3/12$ denote the flexural rigidities of the individual elastic layers by themselves, and where $\bar{X} = X(1 + j\beta_2)$. If the shear stiffness of the viscoelastic layer is very small, so that $X \approx 0$, then $\bar{B} \to B_0 = B_1 + B_3$; that is, the flexural rigidity of the composite is merely equal to the sum of the rigidities of the separate elastic layers, which act as if they were flexurally uncoupled. For very large values of X, on the other hand, $\bar{B} \to B_\infty = (B_1 + B_3)(1 + Y)$; the flexural rigidity one obtains here is found to correspond to presence of a fully shear-rigid coupling between the elastic layers.

In the light of the foregoing discussion it is also of interest to note that one may express the stiffness parameter or geometric parameter Y as

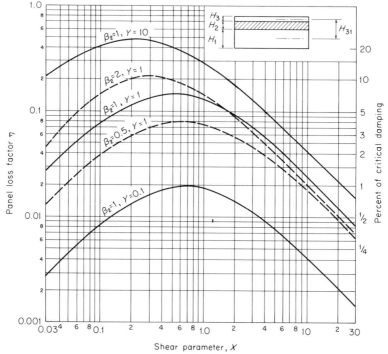

Fig. 14.10   Loss factor η of a panel with a constrained viscoelastic layer vs. shear parameter X, for several values of viscoelastic material loss factor β_2 and stiffness parameter Y. From Eq. (14.41).[37]

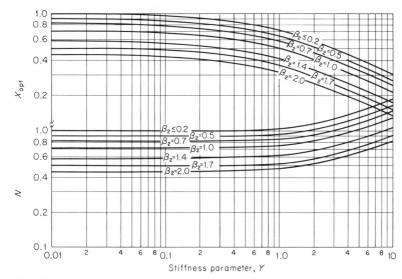

Fig. 14.11   Optimum shear parameter X_{opt} and function N [Eqs. (14.45) and (14.48)] vs. stiffness parameter Y for several values of loss factor β_2 of viscoelastic material.[37]

$$Y = \frac{B_\infty - B_0}{B_0} = \frac{B_{\text{coupled}} - B_{\text{uncoupled}}}{B_{\text{uncoupled}}}$$

and that Y represents the increase in rigidity obtained by proceeding from uncoupled to fully coupled elastic layers, as a fraction of the uncoupled rigidity.

From Eq. (14.41) and Fig. 14.10, one observes that with a given value of β_2 and Y, η/β_2 is small if X is either very small or very large. For small X, η varies as X; for large X, η varies as $1/X$. At some intermediate value $X = X_{\text{opt}}$, where

$$X_{\text{opt}} = \frac{1}{\sqrt{(1 + Y)(1 + \beta_2^2)}} \qquad (14.45)$$

η takes on its greatest magnitude,

$$\eta_{\max} = \frac{\beta_2 Y}{2 + Y + \dfrac{2}{X_{\text{opt}}}} \qquad (14.46)$$

The dependence of X_{opt} on Y and β_2 is indicated in Fig. 14.11. Figure 14.12 shows $\eta_{\max}$ as a function of Y with β_2 as a parameter.

With the aid of the two foregoing equations, one may rewrite Eq. (14.41) in the reduced form

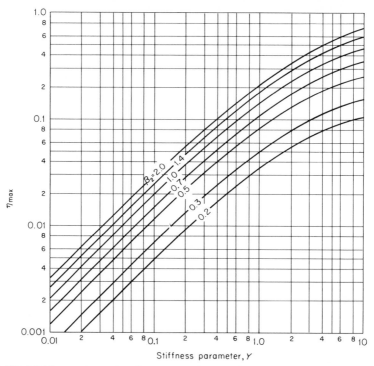

Stiffness parameter, Y

Fig. 14.12   Maximum loss factor η_{max} of panel with constrained viscoelastic layer, as function of stiffness parameter Y and loss factor β_2 of viscoelastic material.   From Eq. (14.46).[37]

$$\frac{\eta}{\eta_{max}} = \frac{2(1 + N)X/X_{opt}}{1 + 2NX/X_{opt} + (X/X_{opt})^2} \qquad (14.47)$$

where N is defined as

$$N = \frac{(2 + Y)X_{opt}}{2} = \frac{2 + Y}{2\sqrt{(1 + Y)(1 + \beta_2^2)}} \qquad (14.48)$$

N is independent of the shear parameter X.   The lower part of Fig. 14.11 shows a plot of Eq. (14.48), indicating how N varies with Y for given values of β_2.   The variation of η/η_{max} with X/X_{opt}, computed from Eq. (14.47), is shown in Fig. 14.13.

Calculation of Loss Factor of Given Panel.[37,39]   The foregoing section summarizes all the relations one needs in order to calculate the damping of any constrained viscoelastic layer configuration.   However, one still needs a relation between the wavelength λ_p and the frequency of flexural vibration of the composite.

The usual bending relation

$$\frac{1}{p^2} = \left(\frac{\lambda_p}{2\pi}\right)^2 = \frac{1}{\omega}\sqrt{\frac{B}{\mu}} \tag{14.49}$$

applies, with

B = flexural rigidity of composite panel, N-m (lb_f-in.)

μ = mass per unit area of panel, kg/m^2 ($lb_m/in.^2$)

Unfortunately, however, B depends on X according to Eq. (14.44), and one must know p^2 to find X.

One may combine Eqs. (14.49), (14.44), and (14.43) in order to obtain an expression for X from which the variables p^2 and B have been eliminated. However, this expression is a cubic in X and cannot readily be solved. For practical purposes, one usually does better by using an iteration procedure, which consists of the following steps:

1. Calculate X_{opt} from Eq. (14.45) or Fig. 14.11.
2. Using $\bar{X} \approx X_{opt}$ in Eq. (14.44), find B.
3. Compute $1/p^2$ from Eq. (14.49).

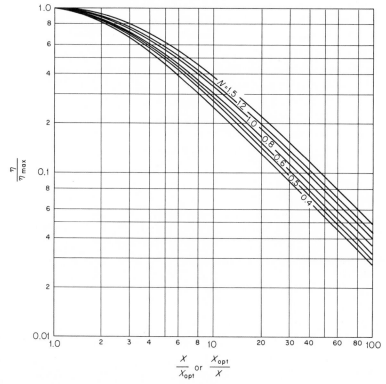

Fig. 14.13 Effect of deviations of X from X_{opt} on loss factor η of panel with constrained viscoelastic layer. From Eq. (14.47).[37]

4. Calculate X from Eq. (14.43).

5. If this X does not differ much from the previously used value, stop. Otherwise, use this new value in Eq. (14.44) to find B.

6. Repeat steps 3 to 5 as often as necessary, then use final value of X for further calculations.

For example, consider the loss factor at 500 Hz of a composite panel made from two aluminum sheets, one 0.50 cm ($\approx$ 0.2 in.), the other 0.25 cm ($\approx$ 0.1 in.) thick, with 0.075 cm ($\approx$ 0.03 in.) of a viscoelastic material between them; the viscoelastic material has $G_2 \approx 7 \times 10^7$ N/m² ($\approx 10^4$ psi) and $\beta_2 = 0.8$ at 500 Hz and at the temperature of interest, and a density $\rho_2 = 7,000$ kg/m³ ($\approx$ 0.25 lb/in.³). Here one finds from Eq. (14.42) that $Y \approx 2.88$, for which from Eq. (14.45) or Fig. 14.11 $X_{opt} = 0.40$.

The total plate mass per unit area is found to be $\mu = 38.5$ kg/m², and $B_1 + B_3 = 820$ N-m. In the first iteration, then, using $X = X_{opt}$ one obtains $B = 1,490$ N-m. This leads to $1/p^2 = 2.0 \times 10^{-3}$ m², from which one calculates a new value $X = 1.57$.

Because the final value $X_{opt} = 1.57$ is considerably different from the value $X = 0.40$, which was initially assumed, one repeats the foregoing calculations, beginning with this latest value of X, and finds a new X. The process is repeated as often as necessary until no further significant change in X is obtained. In this example, the sequence of X values found from successive iterations (to slide-rule accuracy) is 1.73, 1.96, 1.99, 2.00. Then $X/X_{opt} \approx 2.00/0.40 = 5.0$.

From Fig. 14.11 one obtains $N = 0.93$, and from Fig. 14.12, $\eta_{max} = 0.22$. Finally, from Fig. 14.13, $\eta/\eta_{max} = 0.53$. Hence, $\eta \approx 0.12$.

Panel Design.[37,39] The procedure for designing a panel with a constrained viscoelastic layer of a prescribed material to provide maximum damping at a given frequency is relatively straightforward. Because η_{max} increases with increasing Y (see Fig. 14.12), it is clear that one should strive for the greatest possible value of Y.

For the case where space or weight limitations require the constraining layer and the viscoelastic layer to be much thinner than the base panel, so that $e_3h_3 \ll 1$, Eq. (14.42) reduces to

$$Y \approx 3e_3h_3 = \frac{3E_3H_3}{E_1H_1}$$

In this case one should therefore choose E_3 and H_3 as large as possible.

Where the constraining layer and the basic plate have the same modulus of elasticity, one obtains the greatest value of Y, namely, $Y \approx 3$, by choosing $H_3 = H_1$. Values of H_3 near H_1 result in values of Y near 3; greater deviations of H_3 from H_1 result in values of Y that are progressively smaller than 3; a value of $H_3 = nH_1$ results in the same value of Y as a value $H_3 = H_1/n$, where n is any constant.

For the case where the constraining layer is to be made of a material with a different modulus of elasticity than the basic panel, there exists no simple general rule of thumb for selecting the optimum constraining layer thickness. One generally does best to use Eq. (14.42) to calculate Y as a function of h_3 for the applicable value of e_3, and then to select that h_3 which yields the maximum Y.

Once one has chosen the maximum attainable value of Y, one may proceed as follows:

1. Find X_{opt} from Eq. (14.45) or Fig. 14.11.
2. Set $\bar{X} = X_{opt}$ in Eq. (14.44) and calculate B.
3. Compute $1/p^2$ from Eq. (14.49), neglecting the mass of the visco-elastic layer.
4. Choose H_2 so that $X = X_{opt}$, where X obeys Eq. (14.43).
5. Calculate the mass of the viscoelastic layer. If it is small compared to the masses of the other two layers, continue with step 6. Otherwise, return to step 3, but include the mass of the viscoelastic layer. Repeat until mass change is negligible.
6. The loss factor then will be $\eta = \eta_{max}$, as given by Eq. (14.46).

For example, consider the design of a constrained layer damping treatment for a 0.5-cm ($\approx$ 0.2-in.) thick aluminum panel, using a constraining layer of a decorative plastic having $E_3 = 7 \times 10^9$ N/m^2 ($\approx 10^6$ psi) and $\rho_3 = 4,200$ kg/m^2 ($\approx$ 0.15 lb/in.^{3}), to give optimum damping at 400 Hz. The viscoelastic material to be considered here is an adhesive which has $\rho_2 = 5.5$ grams/cm^3 ($\approx$ 0.2 lb/in.^{3}), $G_2 = 3.5 \times 10^7$ N/m^2 ($\approx$ 5,000 psi), and $\beta_2 = 0.5$ at 400 Hz, at the temperature of interest. Here, because $E_1 = 7 \times 10^{10}$ N/m^2 for aluminum, $e_3 \approx 10^{-1}$. From Eq. (14.42), or from its simplified form $Y = 3e_3h_3(1 + h_3)^2/(1 + e_3h_3)^2(1 + e_3h_3^3)$, which applies for $h_2 \ll 1$, one may calculate values of Y that correspond to assumed values of h_3, and proceed by trial and error until one finds the greatest value of Y. Here, that value is $Y = 2.70$, corresponding to $h_3 = 4.5$. Then one finds $H_3 \approx 2.25$ cm, $X_{opt} = 0.46$, $B = 13,600$ N-m, $\mu = 108.5$ kg/m^2, $1/p^2 = 4.55 \times 10^{-3}$ m^2, $H_2 = 0.32 \times 10^{-2}$ m, and $\eta = \eta_{max} = 0.15$.

14.4.4 Spaced Constrained Layers [37,40]

As in the case of free (unconstrained) viscoelastic layers, introduction of a spacing layer (see Fig. 14.9) can serve to increase the strain energy storage—and thus the damping—of a constrained viscoelastic layer. Again, an ideal spacing layer should be infinitely rigid in shear and should have zero extensional stiffness.

Equations (14.41) to (14.49) apply also for constrained viscoelastic layers with ideal spacing layers, except that one must here include the thickness H_s of the spacing layer in calculating H_{31}; thus, here $H_{31} = H_s + H_2 + (H_1 + H_3)/2$. With spacing layers whose properties deviate from those of the ideal, one obtains less damping than with ideal spacing layers.

It should be noted that it does not matter on which side of the visco-elastic layer one places a given spacing layer. The spacer's effect in increasing the strain energy storage in the viscoelastic layer is the same in both cases; the equations reflect this independence of spacer location.

14.4.5 Other Design Considerations

Configuration Design and Material Selection.   In practice, one usually needs to design damping configurations that perform well over extended ranges of frequency and temperature. Viscoelastic materials and viscoelastically damped structures which have high loss factors exhibit these high losses only in limited frequency and temperature regions; usually, the lower the peak loss factor, the narrower are the regions over which the material maintains reasonably high damping.[25, 28]

In designing constrained layer damping configurations for broad frequency ranges, one desires to keep the loss factor as near as possible to the value $\eta_{\max}$ over the frequency (and temperature) range of interest. One thus needs to keep X near X_{opt}, which requires a viscoelastic material whose shear modulus G_2 is proportional to frequency ω, while its loss factor β_2 remains at a high value over the ranges of interest. Unfortunately, there seem to be no viscoelastic materials with high loss factors that have shear moduli that vary with higher powers of frequency than 0.5.[25]

Partial Coverage.[41]   In many cases, particularly where weight savings are important, one may do well to cover only part of the panel surface with (unconstrained or constrained) viscoelastic layers. Where one is concerned with damping of only particular modes, one may readily select and eliminate those damping layer locations at which relatively little strain energy is stored (and dissipated).

In free-layer configurations, strain energy storage in the viscoelastic material occurs primarily by extension, which is greatest at the antinodes and least at the nodes; therefore one may expect to affect the loss factor of the composite plate relatively little if one removes some of the damping material near the nodes. However, one must keep the uninterrupted stretches of damping material long enough, so that they will be made to extend significantly as the composite is flexed. As a rule of thumb, the dimensions of the uninterrupted stretches should be kept greater than 40% of the flexural wavelength; one may then expect to obtain a loss factor which is not less than 80% of that obtained with a fully covered panel.

In constrained layer configurations, as has been discussed, the visco-elastic material dissipates energy primarily by virtue of the energy it stores in shear. Shear in the viscoelastic layer is produced by relative shearing displacements of the elastic layers, and these are greatest at the nodes. Thus, one may expect to affect the loss factor of a constrained layer damped panel relatively little by removing the damping treatment (i.e., the viscoelastic layer, the constraining layer, or both) near the anti-nodes. However, interrupting the constraining layer may result in increased shear at the points of interruption, thus increasing damping, and the total situation becomes more complicated. To a first approxi-mation, the location of the interruption is found to be relatively unim-portant, as long as one keeps the uninterrupted lengths of the constrain-ing layer greater than 60% of a flexural wavelength of the composite; then one obtains a loss factor which is no less than 50% of that obtained with a fully covered panel. Continuity of the viscoelastic layer under a continuous constraining layer is unimportant.

Thick Viscoelastic Layers.[42]   All the foregoing discussions were based on the assumption that the viscoelastic layers are thin (i.e., that the thick-ness of the layer is much smaller than any elastic half wavelength in it) and experience only negligible thickness deformations. These assump-tions hold in most practical cases for the frequency ranges of interest.

However, a thick free viscoelastic layer may have standing-wave reso-nances associated with thickness deformations in the frequency range of concern. These resonances occur in a viscoelastic layer of thickness H_2 at frequencies

$$f_n = \frac{2n - 1}{4H_2} \sqrt{\frac{\kappa}{\rho_2}} \qquad n = 1, 2, 3, \ldots$$

where ρ_2 denotes the density of the viscoelastic material and κ is an ap-propriate thickness modulus. For viscoelastic layers whose other dimen-sions greatly exceed the thickness,

$$\kappa = \frac{1 - \nu}{(1 + \nu)(1 - 2\nu)} E_2 = \frac{2(1 - \nu)}{1 - 2\nu} G_2$$

where E_2 = real part of modulus of elasticity of viscoelastic material, N/m² (psi)

     G_2 = real part of shear modulus of viscoelastic material, N/m² (psi)

     ν = Poisson's ratio of viscoelastic material, dimensionless

For viscoelastic layers whose other dimensions are much smaller than the thickness (e.g., for viscoelastic layers consisting of small squares or buttons, which usually are not useful for panel damping), $\kappa \approx E_2$. Unfortunately, data on Poisson's ratio for most viscoelastic damping

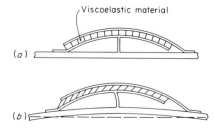

(a)

(b)

Fig. 14.14 Lazan's "corrugated" damping configuration[45] for amplified shearing motion of viscoelastic layer: (a) undeflected and (b) deflected.

materials are not available, and most such materials have Poisson's ratios near 1/2, for which κ depends very strongly on the specific value of ν.

The greatest loss factor contribution due to a thickness resonance in a free viscoelastic layer occurs at the lowest thickness resonance frequency f_1; if the loss factor β_2 of the viscoelastic material is less than unity, then the loss factor η_1 of the composite at this frequency is given by

$$\eta_1 \approx \frac{\beta_2}{1 + 1.23\beta_2^2\mu_1/\rho_2 H_2}$$

where μ_1 denotes the mass per unit area of the basic panel and $\rho_2 H_2$ that of the viscoelastic layer.

14.4.6 Special-purpose Configurations   The use of viscoelastic layers at panel edges, between the panels and their supports has been studied extensively,[43] but generally is not a practical means for obtaining useful amounts of damping. Use of viscoelastic layers inserted between mating surfaces at riveted joints has also been studied,[44] but also appears to be of limited practical value because the joint must be designed so that the rivets can withstand the static loads and thus the rivets limit the energy that can be stored in the viscoelastic layer.

Configurations that amplify the energy storage in a given amount of viscoelastic material for a limited range of panel modes include those making use of some leverage idea, like the "corrugated" treatment suggested by Lazan [45] (see Fig. 14.14), and those using dynamic amplification, e.g., by means of various point-attached vibration absorbers.[46]

In general, in designing special-purpose configurations one must consider (1) tuning the configuration to the desired frequency, (2) adjustment of the geometry and proper positioning of localized attached configurations, and (3) matching of the impedances of attached devices to the panel impedances.[41]

REFERENCES

1. D. Muster and R. Plunkett, Chap. 13 of this book.
2. E. E. Ungar, "Mechanical Vibrations," in H. A. Rothbart (ed.), *Mechanical Design and Systems Handbook*, Chap. 6, McGraw-Hill, 1964.

3. B. J. Lazan, *Damping of Materials and Members in Structural Mechanics*, Pergamon, 1968.
4. R. Plunkett, "Measurement of Damping," in J. F. Ruzicka (ed.), *Structural Damping*, Sec. 5, The American Society of Mechanical Engineers, New York, 1959.
5. R. D. Bruce, Chap. 4 of this book.
6. E. E. Ungar and E. M. Kerwin, Jr., "Loss Factors of Viscoelastic Systems in Terms of Energy Concepts," *J. Acoust. Soc. Am.*, vol. 34, pp. 954–957, July, 1962.
7. A. W. Nolle, "Methods for Measuring Dynamic Mechanical Properties of Rubberlike Materials," *J. Appl. Phys.*, vol. 19, pp. 753–774, August, 1948.
8. H. C. Rorden and A. Grieco, "Measurement of Dynamic Internal Dissipation and Elasticity of Soft Plastics," *J. Appl. Phys.*, vol. 22, pp. 842–845, June, 1951.
9. W. Fletcher and A. Gent, "Apparatus for the Measurement of Dynamic Shear Modulus and Hysteresis of Rubber at Low Frequencies," *J. Sci. Instr.*, vol. 29, p. 186, 1952.
10. E. Fitzgerald and J. Ferry, "Method for Determining the Dynamic Mechanical Properties of Gels and Solids at Audio Frequencies; Comparison of Mechanical and Electrical Properties," *J. Colloid Sci.*, vol. 8, pp. 1–34, 1953.
11. A. Schlägel, "Measurements of Modulus of Elasticity and Loss Factor for Solid Materials," *Brüel and Kjaer Tech. Rev.* 1 (January, 1958), pp. 1–11, and 2 (April, 1958), pp. 1–15.
12. E. E. Ungar, "Energy Dissipation at Structural Joints; Mechanisms and Magnitudes," *U.S. Air Force Rept.* FDL-TDR 64-98, July, 1964.
13. Spencer-Kennedy Laboratories, Inc., "Model 507 Reverberation Time Meter," Boston, May 10, 1963.
14. E. E. Ungar and J. R. Carbonell, "On Panel Vibration Damping due to Structural Joints," *AIAA J.*, vol. 4, pp. 1385–1390, August, 1966.
15. E. E. Ungar, "Maximum Stresses in Beams and Plates Vibrating at Resonance," *Trans. ASME*, ser. B, J, "Eng. for Ind.," vol. 84, pp. 149–155, February, 1962.
16. P. W. Smith, Jr., and R. H. Lyon, "Sound and Structural Vibration," *NASA Contr. Rept.* CR-160, March, 1965.
17. E. E. Ungar and K. S. Lee, "Considerations in the Design of Supports for Panels in Fatigue Tests," *U.S. Air Force Rept.* AFFDL-TR-67-86, September, 1967.
18. R. B. Lindsay, *Mechanical Radiation*, McGraw-Hill, 1960.
19. L. Cremer and M. Heckl, *Körperschall*, Springer-Verlag, 1967.
20. E. E. Ungar and J. E. Manning, "Analysis of Vibratory Energy Distributions in Composite Structures," in G. Herrmann (ed.), *Dynamics of Structured Solids,* The American Society of Mechanical Engineers, 1968.
21. G. Maidanik, "Response of Ribbed Panels to Reverberant Acoustic Fields," *J. Acoust. Soc. Am.*, vol. 34, pp. 809–826, 1962.
22. M. A. Heckl, "Measurements of Absorption Coefficients on Plates," *J. Acoust. Soc. Am.*, vol. 34, pp. 803–808, 1962.
23. G. Maidanik, "Energy Dissipation Associated with Gas-pumping at Structural Joints," *J. Acoust. Soc. Am.*, vol. 40, pp. 1064–1072, 1966.
24. I. Vér and C. Holmer, Chap. 11 of this book.
25. E. E. Ungar and D. K. Hatch, "Your Selection Guide to High-damping Materials," *Prod. Eng.*, vol. 32, no. 16, pp. 44–56, Apr. 17, 1961.
26. G. W. Becker and H. Oberst, "Über das dynamisch-elastische Verhalten linearer, vernetzter und gefüllter Kunststoffe," *Kolloid Z.*, vol. 148, pp. 6–16, 1956.
27. M. L. Williams, R. F. Landel, and J. D. Ferry, "The Temperature Dependence of Relaxation Mechanisms in Amorphous Polymers and Other Glass-forming Liquids," *J. Am. Chem. Soc.*, vol. 77, pp. 3701–3707, 1955.
28. H. Oberst, "Werkstoffe mit extrem hoher innerer Dämpfung," *Acustica,* vol. 6, Beiheft 1, pp. 144–153, 1956.
29. L. E. Goodman, "A Review of Progress in Analysis of Interfacial Slip Damping," in

J. E. Ruzicka (ed.), *Structural Damping*, Sec. 2, The American Society of Mechanical Engineers, 1959.

30. T. H. H. Pian, "Structural Damping of Simple Built-up Beam with Riveted Joints in Bending," *J. Appl. Mech.*, vol. 24, pp. 35–38, 1957.

31. T. S. Lundgren, C. C. Chang, and Y. C. Whang, "Damping of Rectangular Plate Vibrations," *U.S. Air Force WADC Tech. Rept.* 59-544, March, 1960.

32. W. H. Reed, III, "Hanging-chain Impact Dampers: A Simple Method for Damping Tall Flexible Structures," International Research Seminar: Wind Effects on Buildings and Structures, Ottawa, Canada, September, 1967.

33. N. D. Wolf, "Results of Loss Factor Measurements on Steel and Concrete Beams Using a Viscoelastic or Sand Damping System," *U.S. Air Force Tech. Doc. Rept.* ASD-TDR-62-717, Sept., 1962.

34. W. Kuhl and H. Kaiser, "Absorption of Structure-borne Sound in Building Materials without and with Sand-filled Cavities," *Acustica*, vol. 2, pp. 179–188, 1952.

35. H. Schmidt, "Die Schallausbreitung in Körnigen Substanzen," *Acustica*, vol. 4, pp. 639–652, 1954.

36. G. Kurtze, "Körperschalldämpfung durch Körnige Medien," *Acustica*, vol. 6, Beiheft 1, pp. 154–159, 1956.

37. D. Ross, E. E. Ungar, and E. M. Kerwin, Jr., "Damping of Plate Flexural Vibrations by Means of Viscoelastic Laminae," in J. E. Ruzicka (ed.), *Structural Damping*, Sec. 3, The American Society of Mechanical Engineers, 1959.

38. E. E. Ungar, "Loss Factors of Viscoelastically Damped Beam Structures," *J. Acoust. Soc. Am.*, vol. 34, pp. 1082–1089, 1962.

39. J. E. Ruzicka, T. F. Derby, D. W. Schubert, and J. S. Pepi, "Damping of Structural Composites with Viscoelastic Shear Damping Mechanisms," *NASA Contr. Rept.* CR-742, March, 1967.

40. E. M. Kerwin, Jr., "Damping of Flexural Waves in Plates by Spaced Damping Treatments Having Spacers of Finite Stiffness," *Proc. 3d Intern. Congr. Acoust., 1959*, Elsevier, 1961, pp. 412–415.

41. E. M. Kerwin, Jr., "Macromechanisms of Damping in Composite Structures," *Internal Friction, Damping, and Cyclic Plasticity*, spec. publ. 378, American Society for Testing and Materials, 1965, pp. 125–147.

42. E. E. Ungar and E. M. Kerwin, Jr., "Plate Damping due to Thickness Deformations in Attached Viscoelastic Layers," *J. Acoust. Soc. Am.*, vol. 36, pp. 386–392, 1964.

43. T. J. Mentel, "Vibrational Energy Dissipation at Structural Support Junctions," in J. E. Ruzicka (ed.), *Structural Damping*, Sec. 4, The American Society of Mechanical Engineers, 1959.

44. D. J. Mead and D. C. G. Eaton, "Interface Damping at Riveted Joints," part I: "Theoretical Analysis," *U.S. Air Force Rept.* ASD TR 61-467, September, 1961.

45. B. J. Lazan, A. F. Metherell, and O. Sokol, "Multiple-band Surface Treatments for High Damping," *U.S. Air Force Rept.* AFML-TR-65-269, September, 1965.

46. D. I. G. Jones, "Effect of Isolated Tuned Dampers on Response of Multispan Structures," *J. Aircraft*, vol. 4, pp. 343–346, 1967.

47. W. Philipoff, "Mechanical Investigations of Elastomers in a Wide Range of Frequencies," *J. Appl. Phys.*, vol. 24, no. 6, pp. 685–689, June, 1953.

48. E. J. Cook, J. A. Lee, E. R. Fitzgerald, and J. W. Fitzgerald, "Dynamic Mechanical Properties of Materials for Noise and Vibration Control," *Chesapeake Instrument Co., Shadyside, Md., Rept.* 101, Jan. 1, 1960, Contract NOnr-2678 (00).

49. B. J. Lazan and L. E. Goodman, "Material and Interface Damping," in C. M. Harris and C. E. Crede (eds.), *Shock and Vibration Handbook*, Chap. 36, McGraw-Hill, 1961.

50. A. W. Nolle, "Dynamic Mechanical Properties of Rubberlike Materials," *J. Polymer Sci.*, vol. 5, 1–54, 1950.

51. J. C. Snowdon, "Choice of Resilient Materials for Anti-vibration Mountings," *Brit. J. Appl. Phys.*, vol. 9, no. 12, pp. 461–469, December, 1958.
52. R. E. Morris, R. R. James, and C. W. Guyton, "A New Method for Determining the Dynamic Mechanical Properties of Rubber," *Rubber Age*, vol. 78, pp. 725–730, February, 1956.
53. J. C. Snowden, "Reduction of the Response to Vibration of Structures Possessing Finite Mechanical Impedance, Part I," *Willow Run Lab. Univ. of Mich., Ann Arbor, Tech. Rept.* 2892-4-T, November, 1959.
54. E. R. Fitzgerald, J. W. Fitzgerald, and A. E. Woodward, "Dynamic Mechanical Properties of Plastic Materials," *Chesapeake Instrument Co., Shadyside, Md., Final Rept.*, June 1, 1959, Contract NObs-72100.

Chapter Fifteen

Wrappings, Enclosures, and Duct Linings

THEODORE J. SCHULTZ

Wrappings and acoustical enclosures are ways of controlling the noise from machines, pipes, ducts, and other small sound sources. Ordinarily, these acoustic barriers have no function to fulfill except to contain the sound, in contrast to rooms where the boundaries must usually carry structural loads and resist damage from occupancy by people. (A major exception is the soundproofing treatment applied to an aircraft fuselage.)

This chapter builds on previous chapters, which have covered the acoustical properties of spaces, porous materials, solid structures, vibration isolators, and panel damping. No attention is paid here to visual attractiveness or to the economics of installation, both of which may strongly influence the engineering design for a particular application.

15.1 Wrappings

Wrappings are usually made either from a layer of porous material alone, or from one or more layers of porous material covered and separated by impermeable layers. Wrappings are typically applied to noisy

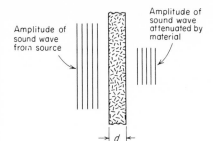

Amplitude of sound wave from source

Amplitude of sound wave attenuated by material

Fig. 15.1 Attenuation of a plane sound wave as it passes at normal incidence through a porous acoustic blanket of thickness d.

pipes, valves, or mixing chambers in hydraulic or ventilating systems, and to the interior of aircraft fuselages to attenuate noise generated outside the passenger spaces. Most reported studies of wrappings have dealt with aerospace equipment, not architectural noise control structures.[1,3,4]

15.1.1 Transmission Loss by a Single Porous Layer
A porous material, when used as a sound-attenuating layer, attenuates a sound wave partly by acting as a reflecting surface—as does a solid wall—and partly by conversion of the acoustic energy of the sound that penetrates the material to heat by viscous losses in the interstices. So effective at high frequencies are some materials, such as fine-fiber glass or mineral-wool blankets, that a sound wave at 1,000 Hz is attenuated 60 dB in traveling 30 cm (1 ft)! The mass surface density of a "wall" of this material and thickness is only 3 kg/m² (0.6 lb/ft²). The weight of a solid partition necessary to provide this much attenuation at 1,000 Hz would be over 50 times as great! At low frequencies the story is much different. For example, at 250 Hz the attenuation for a 30-cm fine-fiber layer would be only 10 dB, while that for a 150-kg/m² (30-lb/ft²) solid wall would be about 42 dB (see Fig. 11.5).

Let us consider a sound wave impinging at normal incidence on the left side of a porous blanket of thickness d and emerging with reduced amplitude from the right side, as shown in Figure 15.1.

The parameters are as follows *

d = thickness of blanket, m
R_f = flow resistance of blanket of thickness d, mks rayls
R_1 = flow resistivity of blanket material = R_f/d, mks rayls/m (1 cgs rayl/in. = 394 mks rayls/m)
ρ_m = bulk density of blanket material, kg/m³
M_s = mass surface density of blanket, kg/m²
w_s = surface weight of blanket, lb/ft²
b = complex propagation constant = $\alpha + j\beta$ (see Table 10.4)

* See Appendix C for conversion factors.

α = attenuation of sound wave while traveling unit distance inside the blanket material, nepers/m; to convert from nepers to decibels, multiply nepers by 8.69; 1 neper/m = 0.22 dB/in.

β = phase shift of sound wave while traveling unit distance inside the blanket material, radians/m; to convert from radians to degrees, multiply radians by 57.3

$c_m = \omega/\beta$ = speed of sound inside the blanket material, m/sec

$\lambda_m = c_m/f = 2\pi/\beta$ = wavelength of sound wave inside the blanket material, m

c = speed of sound in air, m/sec (about 345 m/sec, or 1,128 ft/sec, at normal temperature and pressure)

f = frequency of sound, Hz

ω = angular frequency of sound = $2\pi f$, radians/sec

ρc = characteristic resistance of air = 410 mks rayls for normal temperature and pressure (10 mks rayls = 1 cgs rayl)

$Z_s = |Z_s|\angle\theta$ = specific normal acoustic impedance of the blanket material [see Eq. (10.17)]; $|Z_s|$ and θ are, respectively, the magnitude and phase angle of Z_s, which in general is a complex number equal to the product of the effective complex density ρ' and the complex speed of sound, c_m, in the material. In the case of a plane progressive wave, Z_s is the "characteristic resistance" of the medium $\rho_m c_m$. Often the specific normal acoustic impedance is normalized to the characteristic resistance of air ρc, yielding the specific acoustic impedance ratio $Z_s/\rho c$

In the case of Fig. 15.1, we shall distinguish among three regions of frequency in which the mechanism of attenuation of sound waves differs and for which different models of the acoustical behavior of the blanket are required.

Region A. At low frequencies the blanket possesses insufficient inertia to remain motionless under the excitation of the sound wave; it moves as a whole because of the action of the air particles pumped back and forth by the sound pressure through the pores of the blanket. This region includes all frequencies for which the thickness of the blanket is less than one-tenth the wavelength of sound within the blanket. Under these circumstances, the blanket can be treated in terms of lumped constants, with no consideration of sound-wave propagation within the blanket.

Region C. At high frequencies the blanket as a whole has sufficient inertia to remain still while the air particles move through it. A portion of the incident sound wave is reflected back from the left-hand surface (see Fig. 15.1) of the material, and a second portion of the wave propagates on through the blanket, so that at different distances within the blanket different amplitudes and phases occur. Some of this second portion is reflected back into the blanket from the right-hand blanket-air interface, but, for the purposes of the present treatment, it is assumed to be sufficiently attenuated in its return pass through the blanket that it may be neglected. The remainder of the second portion exits from the blanket and comprises the attenuated wave shown at the right. Region C occurs at frequencies for which the total propagation loss within the blanket exceeds unity, $\alpha d > 1$ neper or, in practical units, $\alpha d > 9$ dB.

Region B. In the frequency range of transition between regions A and C the blanket behavior exhibits some aspects of both the models described above. The attenuation is marked by a smooth transition between the kinds of behavior derived from the models for regions A and C.

Attenuation in Region A. The sound attenuation in region A can be calculated by regarding the blanket in terms of lumped constants, analogous to electrical circuit theory, as shown in Fig. 15.2. The attenuation is defined in terms of the pressure existing in the absence of the blanket, p_0, and the attenuated pressure p_2 passing on from the blanket into perfectly absorbing space with characteristic resistance ρc. This analysis yields the transmission loss

$$\mathscr{R} = 10 \log \left\{ 1 + \frac{\dfrac{R_f}{\rho c} \left(\dfrac{\omega M_s}{\rho c}\right)^2 \left(4 + \dfrac{R_f}{\rho c}\right)}{4 \left[\left(\dfrac{R_f}{\rho c}\right)^2 + \left(\dfrac{\omega M_s}{\rho c}\right)^2\right]} \right\} \quad \text{dB} \tag{15.1}$$

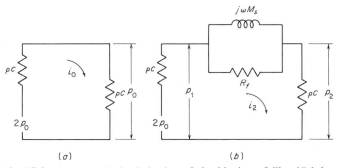

(a)   (b)

Fig. 15.2 Analysis of the behavior of the blanket of Fig. 15.1 in terms of its lumped mass M_s and flow resistance R_f per unit area proceeds by analogy to electric circuit theory. The incident wave from the left is regarded as coming from a generator whose internal impedance is the characteristic resistance ρc of the medium. (a) Rms sound pressure p_0 existing in the air before addition of porous blanket; (b) rms sound pressure p_1 and p_2 on the two sides of the porous blanket for the same source of sound at the same frequency as for (a).

This equation is plotted in Fig 15.3; the lower abscissa is in dimension-less units, while the upper abscissa gives the values in practical units, frequency (f, Hz) times surface weight (w_s, lb/ft²), assuming standard temperature and barometric pressure. The parameter $R_f/\rho c$ is also dimensionless.

Attenuation in Region C. In region C, three different components $\mathscr{R}_1$, $\mathscr{R}_2$, and $\mathscr{R}_3$ contribute to the total transmission loss $\mathscr{R}$ provided by the blanket. The first contribution $\mathscr{R}_1$, usually the largest of the three, affects the portion of the sound wave that enters and travels through the blanket, and comprises the loss suffered by the wave in a single passage through the blanket:

$$\mathscr{R}_1 = \alpha d \text{ nepers}$$
$$= 8.69\ \alpha d \text{ dB} \tag{15.2}$$

That component of the transmission loss caused by a part of the inci-dent sound being reflected back toward the source from the left-hand air-blanket interface is called the reflection loss $\mathscr{R}_2$. A similar reflection loss $\mathscr{R}_3$ occurs as the wave exits at the right-hand blanket-air interface $\mathscr{R}_3$, which is equal in magnitude to $\mathscr{R}_2$.

Both $\mathscr{R}_2$ and $\mathscr{R}_3$ are determined by the ratio of the characteristic im-pedance of the blanket material to the characteristic resistance of the air. Curves are given in Fig. 15.4 for determining either $\mathscr{R}_2$ or $\mathscr{R}_3$ from the

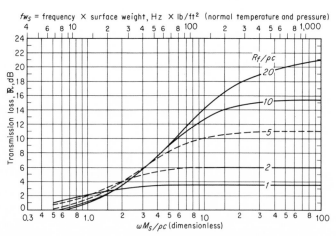

Fig. 15.3 Transmission loss $\mathscr{R}$ in region A through a porous acous-tic blanket, under the assumption that wavelength of sound in blanket is greater than 10 times the blanket thickness. Surface density M_s is measured in kg/m² and $\rho c = 410$ mks rayls for normal pressure and temperature. The upper abscissa is the product of frequency and surface weight in Hz-lb/ft².

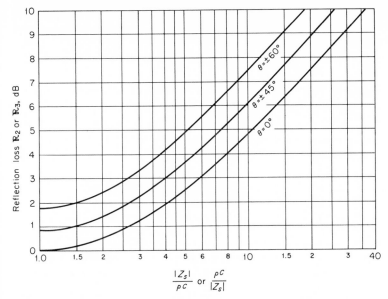

Fig. 15.4 Reflection loss at air-blanket or blanket-air interface in region C. Use either $|Z_s|/\rho c$ or its reciprocal, whichever is greater than unity.

magnitude $|Z_s|$ and the phase angle θ of the characteristic impedance of the blanket.

The total transmission loss $\mathscr{R}$ for the blanket in region C is given by the sum of $\mathscr{R}_1$, $\mathscr{R}_2$, and $\mathscr{R}_3$.

Attenuation in Region B. Because the behavior of the blanket in region B conforms to neither of the clear-cut models appropriate to regions A and C, no simple calculation procedure is offered here. It is generally adequate to calculate the transmission loss in regions A and C, to plot the results, and to estimate the transmission loss in region B by fairing a curve smoothly between the curve segments for regions A and C.

Example 15.1 Determine the transmission loss of a 5-cm (2-in.) layer of glass-fiber blanket (e.g., OCF type PF 105). The fiber of this material has a mean diameter of about 10^{-6} m (4×10^{-5} in.).

SOLUTION:

Step 1: The relevant acoustical properties of this material were calculated in Chap. 10 and are given in Table 10.5. The frequency-dependent parameters needed are plotted in Fig. 15.5.

The flow resistivity R_1 of this blanket is 4.1×10^4 mks rayls/m (104 cgs rayls/in.) and the bulk density $\rho_m = 9.6$ kg/m^3 (0.6 lb/ft^3). Thus, a 5-cm layer would have a flow resistance R_f of $R_1 d = 4.1 \times 10^4 \times 0.05 = 2,050$ mks rayls, so that $R_f/\rho c = 5$; the surface mass M_s is 0.48 kg/m^2 (0.1 lb/ft^2).

Step 2:   Region A (low frequencies) is the frequency range for which $d < 0.1$ λ_m, or $\lambda_m > 0.5$ m.   From Fig. 15.5, this is seen to include all frequencies below about 250 Hz.

Step 3:   Determine the transmission loss $\mathcal{R}$ for region A from Fig. 15.3, using

$$R_f/\rho c = 5 \quad \text{and} \quad M_s = 0.48 \text{ kg/m}^2 \ (w_s = 0.10 \text{ lb/ft}^2)$$

Frequency, Hz	fM_s, Hz-kg/m²	fw_s, Hz-lb/ft²	$\mathcal{R}$ (region A), dB
125	60	12.5	1.0
250	120	25.0	4.0

Step 4:   Region C (high frequencies) is the frequency range for which $\alpha d > 1$ neper, or 9 dB.   With $d = 5$ cm this corresponds to $\alpha > 177$ dB/m (4.5 dB/in.).   From Fig. 15.5, it is seen that the attenuation α exceeds this value for all frequencies above about 1,000 Hz.

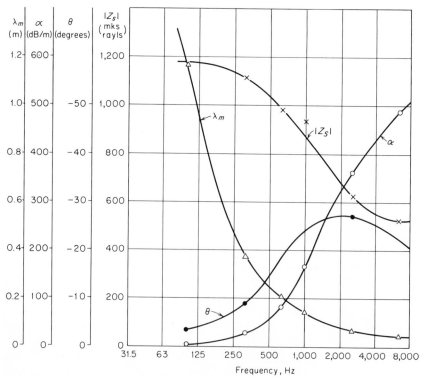

Fig. 15.5 Acoustical parameters for Owens-Corning Fiberglas Type PF 105, whose average fiber diameter is 1 micron and whose bulk density is 9.6 kg/m³ (0.6 lb/ft²).

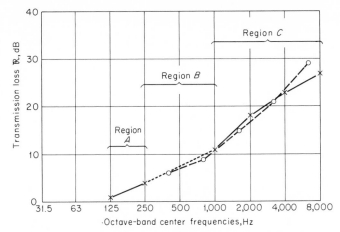

Fig. 15.6 Transmission loss for blanket of Example 15.1. Calculated values (x's) are shown for regions A and C. The dotted curve smoothly connects these segments to give the values for intermediate region B. The dashed curve (o's) shows measured data for a 5-cm blanket of similar material. (*Measurements courtesy of Douglas Aircraft Co., Long Beach, Calif.*)

Step 5: Determine transmission-loss components $\mathscr{R}_1 = \alpha d$ (from Fig. 15.5), $\mathscr{R}_2$ and $\mathscr{R}_3$ (from Fig. 15.4), and their sum $\mathscr{R}$ for region C.

Fre-quency, Hz	α, dB/m	dB/in.	$\mathscr{R}_1 = \alpha d$, dB	$\lvert Z_s \rvert$, mks rayls	$\lvert Z_s \rvert / \rho c$, dimen-sionless	θ, degrees	$\mathscr{R}_2 + \mathscr{R}_3$, dB	$\mathscr{R}$ (region C), dB
1,000	170	4.3	8.6	870	2.12	−24	2.0	11
2,000	320	8.1	16.2	680	1.66	−27	1.6	18
4,000	430	10.9	21.8	550	1.34	−25	1.0	23
8,000	510	13.0	26.0	520	1.27	−20	1.0	27

Step 6: The transmission-loss values for regions A and C are plotted for this example in Fig. 15.6; the values for region B are found by fairing a curve to connect smoothly the results for regions A and C. For comparison, measured data are shown for a 5-cm blanket of similar material. The agreement is good.

Oblique Incidence. If the sound wave impinges on a blanket at some angle other than normal incidence, the wave will travel a greater distance inside the blanket and thus will suffer a greater propagation loss than for normal incidence. However, the distance traveled inside the blanket is not as great as might be expected. Moreover, there are com-

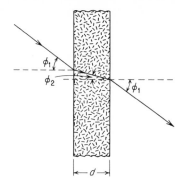

Fig. 15.7 Sound transmission through a blanket at oblique angles, showing incidence and emergence angle ϕ_1 in air and refraction angle ϕ_2 in blanket.

pensating effects for $\mathcal{R}_2$ and $\mathcal{R}_3$ that tend to stabilize the total blanket attenuation for various angles of incidence.

The value of $\mathcal{R}_1$ for sound at oblique incidence (see Fig. 15.7) is governed by Snell's law:

$$\sin \phi_m = \frac{c_m}{c} \sin \phi \qquad (15.3)$$

where ϕ = angle of incidence of sound on blanket

ϕ_m = angle of travel inside blanket; $\phi = \phi_m = 0$ for normal incidence

The speed of wave propagation within the blanket c_m is less than the speed of sound in air c. For example, at about 2,000 Hz in the PF 105 blanket of Example 15.1 above, $c_m = 176$ m/sec. If we assume a sound wave impinging on the blanket at an angle $\phi = 45°$, then, from Eq. (15.3),

$$\sin \phi_m = \frac{176}{344} (\sin 45°) = 0.36$$

$$\phi_m = 21°$$

The angle of travel for the wave inside the blanket is much closer to the normal than the wave outside the blanket, and the distance traveled by the wave in the blanket is, therefore, not much greater than the blanket thickness. In this case, the travel distance d_2 inside the blanket is

$$d_2 = \frac{d}{\cos \phi_m} = \frac{d}{\cos 21°} = 1.06 \, d$$

and the value of $\mathcal{R}_1$, in decibels, is increased by 6% over the value for normal incidence.

Compensating changes occur in $\mathcal{R}_2$ and $\mathcal{R}_3$ for oblique incidence because the loss associated with reflections at the boundaries due to impedance mismatch is of the form

$$\mathcal{R} = 10 \log \left| 1 + \frac{Z_s \cos \phi}{2\rho c} \right|^2 \qquad (15.4)$$

which decreases for increasingly oblique angles of incidence ϕ, depending on the magnitude of $Z_s/\rho c$. [The cosine term was omitted in Eq. (15.1) since for normal incidence $\cos \phi = 1$.] Whether this decrease in $\mathcal{R}_2$ and $\mathcal{R}_3$ completely compensates the increase in $\mathcal{R}_1$ depends on $|Z_s|/\rho c$, but the tendency is to prevent the transmission loss of the blanket from changing greatly with incidence angle, and, therefore, the transmission loss for random incidence is not much different than for normal incidence.

15.1.2   Attenuation of Sound by a Porous Blanket Added to a Plate or a Pipe
A porous blanket may be added to a plate, as shown in Fig. 15.8a, or used to wrap a pipe or ventilation duct. We shall show that this type of treatment is practically useless for increasing the attenuation of sound at low frequencies but can be very effective at high frequencies.

The same three frequency regions A, B, and C are distinguished as for the previously discussed case. In this case, the appropriate measure of performance of the added blanket is insertion loss.

Insertion Loss in Region A $(d < \lambda_m/10)$. The motion of a plate or pipe of Fig. 15.8a will be unaffected by mounting a lightweight blanket against it. Thus, as there is no sound wave arriving on the blanket from the left, there will be no attenuation due to reflection at that blanket surface as expressed in Eq. (15.1) and Fig. 15.3 for the blanket alone. Instead, for the frequency range below which $d < \lambda_m/10$, the same amount of air is pumped in and out of the right-hand side of the material (Fig. 15.8a) as is moved by the plate alone. The combination of plate and blanket behaves just like the plate itself; the insertion loss due to the addition of the blanket is thus zero throughout region A.

Attenuation in Region C $(\alpha d > 9 \text{ dB})$. In region C the blanket is thick

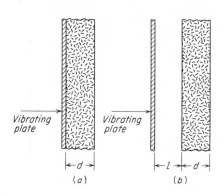

Vibrating plate

Vibrating plate

$\leftarrow d \rightarrow$

$\leftarrow l \rightarrow \leftarrow d \rightarrow$

(a)

(b)

Fig. 15.8 Examples of blanket (a) adjacent to, and (b) spaced away from a vibrating plate.

enough that the sound wave radiated by the plate is attenuated by propagation loss $\mathscr{R}_1$ within the material.

Since the blanket lies in contact with the plate on the left, the attenuation due to impedance mismatch at this interface does not occur, so that $\mathscr{R}_2 = 0$. On the exit side of the blanket, however, $\mathscr{R}_3$ is determined as for the blanket alone in region C. Thus, the total insertion loss L_{IL} is given by the sum of $\mathscr{R}_1 = \alpha d$ and $\mathscr{R}_3$ from Fig. 15.4.

Attenuation in Region B. Starting at the frequency for which the wavelength in the blanket material equals 10 times the blanket thickness, that is, the upper-frequency boundary of region A, gradually increase the attenuation from zero (dB) to meet the attenuation curve in region C.

If there is an airspace between the plate and the blanket, as in Fig. 15.8b, start increasing the attenuation from zero at the frequency for which the *average* wavelength in the air and the blanket equals 10 times the total thickness of the airspace and the blanket.

Example 15.2   Determine the insertion loss of the blanket of Example 15.1 added to an airplane fuselage structure having a 0.127-cm (0.050-in.) aluminum skin, with mass equal to 3.55 kg/m² (0.727 lb/ft²).

SOLUTION:

Step 1: Regions A, B, and C are defined by the same frequencies as in Example 15.1.

Step 2: The insertion loss due to the blanket is negligible in region A since the blanket is in contact with the skin and does not add substantially either to the impedance mismatch or to the mass of the skin.

Step 3: In region C, $\mathscr{R}_1$ and $\mathscr{R}_3$ are the same as in Example 15.1, but $\mathscr{R}_2 = 0$, because the blanket is in contact with the skin and no reflection loss can occur. These attenuations are increments to be added to the transmission loss of the aircraft fuselage structure.

Frequency, Hz	$\mathscr{R}_1$, dB	$\mathscr{R}_2$, dB	$\mathscr{R}_3$, dB	L_{IL} (region C), dB
1,000	8.6	0	1.0	10
2,000	16.2	0	0.8	17
4,000	21.8	0	0.5	22
8,000	26.0	0	0.5	27

Step 4: Plot the insertion loss values for region C and determine the values for region B by connecting a smooth curve from zero attenuation at the upper frequency boundary of region A (in this case, 250 Hz) to meet the curve for region C.

In Fig. 15.9 the measured insertion loss is plotted for a bare fuselage structure with 0.127-cm aluminum skin (solid line, full dots). To the values of this

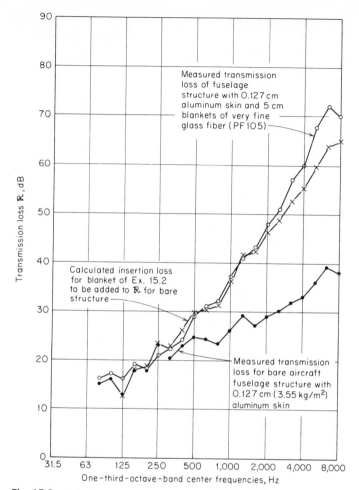

Fig. 15.9 Calculated and measured transmission loss for an aircraft fuselage structure with 0.127-cm (0.05-in.) aluminum skin and a 5-cm blanket of AA glass fiber. (*Measurements courtesy of Douglas Aircraft Co., Long Beach, Calif.*)

curve have been added the incremental attenuation values calculated in this example for the blanket mounted against a skin (x's). Measured data are given for such a skin/blanket combination (dashed line, open circles). The agreement is good below 2,000 Hz.

15.1.3 Attenuation through a Porous Blanket Wrapped with an Impermeable Membrane, Added to a Plate or a Pipe   While the blanket wrapping described in the previous example achieves useful attenuation at high frequencies, it does little good below about 500 Hz. This is true in general for blanket wrappings.[3,4]   A significant improvement

can be attained at only a modest increase in weight by covering the blanket wrapping with a limp, impermeable membrane. The membrane might be asphalted paper, thin sheet metal, linoleum, neoprene sheeting, lead-loaded vinyl plastic, etc., the heavier and limper the better. This impermeable membrane will not only protect the porous blanket from weather and mechanical damage, but also considerably increase the attenuation of the wrapping.

It is assumed that the membrane is supported only by the blanket wrapping, for if it comes into mechanical contact with the vibrating plate or pipe, the porous blanket will be flanked by the vibration through the supporting members, which will excite the outer sheet and cause it to radiate sound much as the original plate did. Here also, the measure of performance of the added structure is insertion loss.

We again identify three frequency regions of interest:

Region A. This is the frequency range below 1.5 f_0, where f_0 is the frequency of the resonance introduced by the addition of the impermeable wrapping, that is, the resonance caused by the mass of the membrane and the stiffness of the layer of air trapped in the blanket between the membrane and the plate.

Region B. This is the frequency range of transition between regions *A* and *C*.

Region C. This is the frequency range for which the propagation loss in the blanket $\alpha d > 9$ dB, as before.

Insertion Loss in Region A. The basic frequency of resonance for two limp plates coupled by an air layer between them, in response to a sound wave incident at an angle ϕ, is given by

$$f_0 = \frac{60}{\sqrt{Md \cos^2 \phi}} \quad \text{Hz} \quad (15.5)$$

where $d =$ thickness of airspace between plates, assumed less than 0.1 the wavelength in air, m

$M =$ equivalent surface mass $M_1 M_2/(M_1 + M_2)$, kg/m²

$M_1, M_2 =$ surface masses, respectively, of the two limp plates, kg/m²,

or

$$f_0 = \frac{170}{\sqrt{wd \cos^2 \phi}} \quad \text{Hz} \quad (15.6)$$

where $d =$ thickness of the airspace between the plates, assumed less than 0.1 the wavelength in air, in.

$w =$ equivalent surface weight $= w_1 w_2/(w_1 + w_2)$, lb/ft²

$w_1, w_2 =$ surface weights respectively of the two limp plates, lb/ft²

If there is a blanket in the airspace, set $\cos^2 \phi \approx 1$, because of the refraction at low frequencies, as discussed above. Also, the constant in the numerator is diminished by $1/\sqrt{2}$, because of isothermal conditions in the blanket. If $M_1 = M_2$ ($w_1 = w_2$) then $M = M_1/2 = M_2/2$ ($w = w_1/2 = w_2/2$). If $M_1 \gg M_2$ ($w_1 \gg w_2$) then $M = M_2$ ($w = w_2$). This is usually the case for an impermeable membrane enclosing a blanket-wrapped plate or pipe.

Adapting Eqs. (15.5) and (15.6) to the wrapping context, region A is defined to lie below 1.5 f_0, with

$$f_0 = \frac{42}{\sqrt{Md}} \qquad \text{Hz} \qquad (15.7)$$

where M = impermeable-membrane surface mass, kg/m^2, or

$$f_0 = \frac{120}{\sqrt{wd}} \qquad \text{Hz} \qquad (15.8)$$

where w = impermeable membrane surface weight, lb/ft^2

In region A, the insertion loss of the added blanket and the impermeable wrapping is essentially zero; in practice, when this treatment is applied to a thin plate or skin, it may even be a few decibels negative: that is, the insertion loss for the plate plus blanket plus wrapping may be a few decibels less than for the blanket alone.

Insertion Loss in Region C *($\alpha d > 9$ dB).*   There are two contributions to incremental attenuation in region C: the first is $\mathcal{R}_1$, due to propagation loss in the blanket; this is calculated from αd, as in the previous cases. There are no losses associated with impedance mismatch at the blanket faces, and so $\mathcal{R}_2$ and $\mathcal{R}_3$ of the previous examples are both zero. But there is now the transmission loss, $\mathcal{R}_4$, of the impermeable membrane, which, since the membrane is assumed to be limp, can be calculated on a mass-law basis. (See Fig. 11.5.)

In case the blanket and membrane are used to wrap a noisy source, the transmission loss $\mathcal{R}_4$ of the membrane should be calculated for *normal incidence*. If the blanket and membrane are applied as "soundproofing" treatment to a lightweight partition, such as the side walls and roof of a house trailer, a railway car, or a subway car, where the sound on the exterior surface may come from various angles, the *field-incidence* mass law should be used to determine $\mathcal{R}_4$.* In addition, the contribution $\mathcal{R}_1$ due to propagation loss in the blanket may be increased by 5 to 10%.

Insertion Loss in Region B.   The insertion loss of the blanket and membrane in region B is determined graphically by fairing a smooth curve from zero at a frequency of 1.5 f_0 to join the curve plotted for region C.

---

* According to Ref. 2 when impermeable skins are applied to the two faces of a blanket, it is the propagation angle on the side of the skin *opposite* the source (i.e., the "exit side" of each skin) that enters into the calculation of transmission loss in Eq. (15.4).

Example 15.3 Determine the insertion loss achieved by adding to an airplane fuselage structure having a 0.127-cm (0.050-in.) aluminum skin, a 7.5-cm (3-in.) blanket of the same material used in Examples 15.1 and 15.2, and an impermeable membrane having a mass of 2.5 kg/m² (0.5 lb/ft²).

SOLUTION:

Step 1: Define region A: For an impermeable layer with surface mass of 2.5 kg/m² over a 7.5-cm layer of blanket, the basic resonance frequency is found from Eq. (15.7)

$$f_0 = \frac{42}{\sqrt{2.5 \times 0.075}} = 97 \text{ Hz}$$

Note that we have used the mass of the impermeable membrane alone, even though its surface density (2.5 kg/m²) is comparable to that of the aluminum skin (3.55 kg/m²), to which the additional treatment is applied [a condition

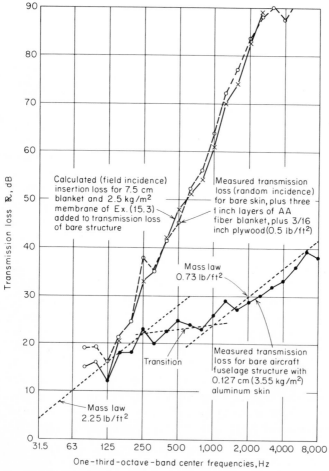

Fig. 15.10 Calculated and measured transmission loss for structure of Example 15.3 (random incidence).

that would ordinarily suggest the use of Eq. (15.5) or (15.6)]. Our procedure is justified by the peculiar transmission-loss behavior typical of bare aircraft structure (see the dotted curve of Fig. 15.10). At low frequencies (large wavelengths or correlation areas), such a structure exhibits mass-law behavior corresponding to the weight of the total structure, including frames, longerons, and skin; at high frequencies (short wavelengths or small correlation areas) the skin moves independently of the heavier structure, i.e., the heavier structure remains relatively still, so that the whole panel shows mass-law behavior corresponding to the weight of the skin alone. Between these two frequency ranges there is a smooth transition. Therefore, for the calculation of the double-wall resonance frequency, which is expected to be low, the appropriate surface density for the basic plate is that of the entire fuselage structure, not just the skin alone. This density is enough greater than that of the added membrane to justify the use of the modified Eq. (15.7).

The upper limit of region A is $1.5 f_0 \approx 148$ Hz; below this frequency, we estimate zero insertion loss for the blanket and membrane.

Step 2: With the blanket thickness increased to 7.5 cm region C is defined as the frequency range in which $\alpha > 120$ dB/m (3 dB/in.); thus region C extends upward from 800 Hz. Moreover, the values of $\mathcal{R}_1$ are increased by 50% as the blanket thickness is increased from 5 to 7.5 cm; $\mathcal{R}_2$ and $\mathcal{R}_3$ are both zero.

Step 3: Determine for region C the transmission loss $\mathcal{R}_4$ of the 2.5-kg/m^2 membrane; find values for both normal incidence of the sound, as for a wrapping or an aircraft fuselage, and field incidence, as for a trailer wall.

Step 4: Determine the incremental attenuation values for region B as before, by drawing a smooth curve from zero at 148 Hz, to meet the curve(s) for region C.

The following table gives the incremental values of attenuations $\mathcal{R}_1$ and

Frequency, Hz	Propagation loss inside blanket		Transmission loss of 2.5-kg/m^2 membrane		Total insertion loss	
	$\mathcal{R}_{1(\phi=0)}$ αd, dB	$\mathcal{R}_{1(\text{field inc.})}$ $\alpha d + 10\%$, dB	$\mathcal{R}_{4(\phi=0)}$, dB	$\mathcal{R}_{4(\text{field})}$, dB	$L_{\text{IL}(\phi=0)}$, dB	$L_{\text{IL}(\text{field inc.})}$, dB
Region A $\begin{cases}125 \\ 148\end{cases}$	......	......	0	0	0	0
	......	......	0	0	0	0
Region B $\begin{cases}250 \\ 500\end{cases}$	......	......	*	*	*	*
	......	......	*	*	*	*
Region C $\begin{cases}1,000 \\ 2,000 \\ 4,000 \\ 8,000\end{cases}$	12.9	14.2	26	21	39	35
	24.3	26.8	32	27	56	54
	32.7	36.0	38	33	71	69
	39.0	43.0	44	39	83	82

* To be determined graphically, as described in step 4.

$\mathscr{R}_4$ that represent either the reduction of sound from a noisy source wrapped with this treatment, or the incremental insertion loss to be added to that of a skin or plate to which it is applied. Both normal-incidence and field-incidence transmission losses for the overall transmission loss are given. The results are plotted in Fig. 15.10, along with measured data for an equivalent structure.

Detailed calculations for a variety of such structures have been given by Beranek and Work [3],* and by Mangiarotty.[4]

15.1.4 Attenuation of Aircraft Structures   Mangiarotty has developed a computer program based on the equations of Beranek and Work and with it has calculated the transmission losses of a number of lightweight structures suitable for use as aircraft fuselages.[4] He sought to optimize the distribution of a given weight of materials, and arrived at the following conclusions:

1. A number of impervious elements (i.e., impermeable layers separated by porous layers) provide a greater transmission loss than a single element of equivalent weight in the frequency region of mass-law behavior.

2. The optimum transmission loss of a multiple-element configuration is achieved in the mass-law frequency region when the masses of the impervious elements are all approximately equal. However, nonuniform masses are more desirable in the frequency region where the individual elements resonate, to prevent the resonances of the layers from coinciding. A good compromise is to make the individual masses different from each other but of the same order of magnitude.

3. Judicious distribution of the mass in a typical soundproofing configuration can, according to calculation, increase the transmission loss (without added weight) by approximately 3 to 6 dB at the lower frequencies, and 8 to 14 dB in the speech-interference frequency range.

4. Increasing the airspaces between the elements to about 7.5 cm (3 in.) results in a theoretical improvement of about 8 to 10 dB over most of the frequency range of interest, but this construction is not very practical in the neighborhood of the fuselage frames, and it is also wasteful of space. This method of improving the transmission loss is attractive only in situations where no addition of mass is permissible.

5. Replacing airspaces with acoustical blankets does not yield sufficient improvement (only 2 to 8 dB) to overcome the risk of structural vibration transmission between contacting blankets.

* These calculations have been criticized [5] as giving the Noise Reduction rather than the Transmission Loss of the structure, and thus yielding misleadingly high values. It has been shown by Mulholland, Price, and Parbrook,[2] however, that it is a simple matter to correct this apparent discrepancy so that the Beranek-Work procedure is applicable to all their panel constructions.

6. Terminating a soundproofing treatment with an acoustically absorbing treatment in lieu of a hard impervious surface does not result in a very great improvement in transmission loss, i.e., only about 2 to 6 dB over the frequency range of interest (300–4,800 Hz).

7. Reduction of the ambient pressure in an aircraft cabin to an equivalent pressure of 8,000 ft altitude increases the transmission loss of a typical configuration by 3 to 6 dB.

Note that it is proper to assume normal incidence of sound on the fuselage for high-speed aircraft because boundary-layer excitation of the panel is dominant during cruise conditions. Normal-incidence calculations are also valid for those portions of the fuselage near propellers; they would not be valid for attenuation of the noise of jet engines, for which the sound source extends over a considerable region, but one does not ordinarily design the treatment to control noise during the brief takeoff period when jet engine noise is most prominent.

15.2 Acoustic Enclosures

Enclosures are rigid, usually airtight, boxes, closely surrounding noise sources. Such configurations differ strikingly, both in their acoustical behavior and in the corresponding design concepts, from architectural structures in which the isolation is governed primarily by mass-law behavior. Small enclosures are essentially stiffness controlled; that is to say, they operate mostly in a frequency range below the fundamental resonance of the constituent elements. Such enclosures exhibit characteristics that seem contrary to our intuition based on experience with architectural walls and floors. The account presented here is adapted from studies reported by Jackson.[6,7]

Acoustic enclosures often provide satisfactory noise isolation for machinery and electrical equipment, such as large electrical transformers. Such a configuration may be idealized as shown in Fig. 15.11a, where both the transformer (the primary noise source) and the attenuating hood are regarded as pulsating boxes whose sides, considered as rigid planes, radiate essentially plane waves to point P. We are interested in finding the reduction in the sound wave at point P due to the attenuating box, i.e., the insertion loss of the box. Figure 15.11b represents a further idealization of the arrangement: two flat, infinite, parallel plates are separated by a distance l and immersed in air characterized by density ρ and sound velocity c. The plate at the left is regarded as the driving plate whose motion is given; this plate would, if it stood alone, account for the original pressure appearing at point P. With the addition of the attenuating plate at the right, which is driven into motion by the pressure from the driving plate, the pressure at P is usually reduced, although under certain conditions the pressure at P may actually be increased (see below). The driven plate is characterized by stiffness S, resistance R, and mass M.

The analysis proceeds under the assumptions that the enclosure is not in mechanical contact with any part of the source, that the presence of the enclosure does not affect the vibration of the enclosed source, and that no direct transmission of vibration can occur through the supports of the sound source.

15.2.1 One-dimensional Model of an Enclosure

The sound insertion loss of the enclosure is taken to be $20 \log v_0/v_1$, that is, the logarithmic ratio of the rms velocities of the driving and the driven plates, which yields

$$L_{IL} = 20 \log \frac{v_0}{v_1} = 10 \log \left[1 - \frac{2 \sin \theta \, (X \cos \theta - R \sin \theta)}{\rho c} \right.$$

$$\left. + \frac{\sin^2 \theta \, (X^2 + R^2)}{\rho^2 c^2} \right] \quad \text{dB} \qquad (15.9)$$

where $\theta = \omega l/c$

$\omega = 2\pi f =$ angular frequency, radians/sec

$f =$ frequency of excitation, Hz

$X = (\omega M - S/\omega)$, N-sec/m³

$M =$ mass surface density (mass per unit area) of plate B, the driven or attenuating plate, kg/m²

$S =$ elastic restraint per unit area of plate B, N/m³

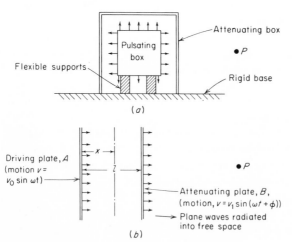

Fig. 15.11 Models for analysis of hood performance: (*a*) ideal sealed enclosure; (*b*) plane-wave equivalent of (*a*).

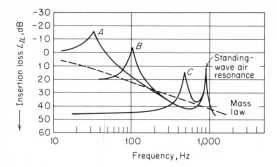

Fig. 15.12 Theoretical curves for insertion loss derived from Eq. (15.9) for $M = 16$ kg/m^2: $l = 0.19$ m ($7\frac{1}{2}$ in.). Curve A: $f_0 = 0$ Hz (see text), $Q = 30$ at 33 Hz; B: $f_0 = 100$ Hz, $Q = 30$ at 105 Hz; C: $f_0 = 475$ Hz, $Q = 30$ at 475 Hz.

$Q = \omega M/R$, damping coefficient associated with the motion of plate B, dimensionless

$R =$ mechanical resistance per unit area of plate B, N-sec/m^3

$\rho c =$ characteristic resistance of air, normally about 410 N-sec/m^3

The ratio of Eq. (15.9) depends upon the mass, the damping, and the stiffness of the attenuating plate and also on the distance between the two plates. The insertion loss is illustrated, for various values of plate stiffness, by the family of curves shown in Fig. 15.12. For comparison, the dashed line shows the normal-incidence mass-law transmission loss.

The three curves A, B, and C of Fig. 15.12 indicate solutions to Eq. (15.9) for three conditions of elastic restraint. The first, corresponding to $f_0 = 0$ Hz,* is for the case of a plate with no stiffness, which may be regarded as a pure mass. Even in this case a resonance appears at a low frequency (approximately 33 Hz) where the stiffness reactance of the air trapped between the plates balances the mass reactance of the attenuating plate. This resonance represents the lower limiting frequency for the family of resonances that would be obtained by allowing the elastic constraint of the plate to take on successively higher values, as illustrated by curves B and C.

Note from Fig. 15.12 that the driving plate, characterized by a constant rms velocity v_0, is not a constant-sound-power generator; the peaks associated with curves A and B represent negative attenuation and imply that more sound is radiated from the attenuating plate at these frequencies than from the unhooded source. As the attenuating plate stiffness is increased, the resonance frequency increases correspondingly, as shown by the peaks in curves B and C. The unlabeled peak appearing

---

*f_0 is the lowest resonance of the plate alone, at which the stiffness reactance of the plate balances its mass reactance; the total effective stiffness of the enclosure is greater by the addition of the stiffness of the trapped air.

farthest to the right in all three curves corresponds to a standing-wave air resonance between the two plates.

The meaning of the "elastic restraint" of an infinite plate is as follows. Any flat plate simply supported at its edges and subjected to an oscillatory excitation normal to its face will exhibit preferred modes of vibration. A rectangular plate, for example, has modal frequencies given by [see Eq. (11.29)]

$$f_{(m,n)} = \frac{\pi}{2} \left[\frac{Et^3}{12M(1 - \sigma^2)} \right]^{1/2} \left(\frac{m^2}{a^2} + \frac{n^2}{b^2} \right) \tag{15.10}$$

where $f_{(m,n)}$ = set of modal frequencies of the simple edge-supported
        rectangular plate
M = mass surface density of plate, kg/m²
$\rho_p = M/t$ = density of plate, kg/m³
m, n = positive integers, 1, 2, 3, . . .
E = Young's modulus of the plate, N/m²
σ = Poisson's ratio, usually about 0.3
t = plate thickness, m
a = length of plate, m
b = width of plate, m

The lowest modal frequency occurs when $m = n = 1$; for excitation frequencies below this value the diaphragm may be replaced by an equivalent piston having a uniform stiffness per unit area. It is this stiffness that is intended by the term S in Eq. (15.9).

It is clear from the curves of Fig. 15.12 that in order to attain good attenuation in the frequency range below the lowest resonance of the plate, the lowest resonance frequency must be as *high* as possible. From Eq. (15.10) it can be seen that, for a given size of plate, the lowest modal frequency is

$$f_{(1,1)} = 0.48 \sqrt{\frac{Et^3}{M}} \left(\frac{1}{a^2} + \frac{1}{b^2} \right)$$

$$= 0.48 c_L t \left(\frac{1}{a^2} + \frac{1}{b^2} \right) \tag{15.11}$$

where $c_L = \sqrt{E/\rho_p}$, speed of longitudinal waves in the plate, m/sec

Thus, for a plate of a given thickness, the material with the highest longitudinal wave velocity is most desirable. On an equal weight basis ($M = \rho_p t$), the material having the highest value of E/ρ_p^3 should be selected. Table 15.1 compares values of E/ρ_p^3 for various materials that might be used for the attenuating plate. The best material for an acoustical hood is one having high stiffness and low weight; this is just the opposite of the cases usually encountered in architectural acoustics where mass-law behavior [i.e., with a low fundamental (drumhead)

TABLE 15.1   Values of E/ρ_p^3 for Common Materials

Material	Young's modulus E, N/m²	Density, lb/ft³	Density ρ_p, kg/m³	E/ρ_p^3, mks units	E/ρ_p^3 normalized to steel
Steel	20×10^{10}	486	7.8×10^3	0.42	1.0
Aluminum	7.0×10^{10}	168	2.7×10^3	3.58	8.5
Crown glass	7.1×10^{10}	156	2.5×10^3	4.54	10.8
Flint glass	5.5×10^{10}	224	3.6×10^3	1.18	2.8
Bakelited paper	1.0×10^{10}	81	1.3×10^3	4.64	11.1
Rubber (Shore 50)	0.001×10^{10}	81	1.3×10^3	0.0045	0.011
Teakwood	1.7×10^{10}	56	0.9×10^3	22.8	54
Poplar wood	1.0×10^{10}	29	0.46×10^3	102	243

resonance frequency and a high critical frequency] is significant and we seek heavy, limp materials.

15.2.2 Steel, Rubber, and Glass Enclosures

Measurements of insertion loss for two small enclosures (with dimensions about 0.3 × 0.15 × 0.14 m) were made, using a constant-velocity loudspeaker inside the enclosure as the acoustic source. The results are shown in Figs. 15.13 and 15.14, and should be compared with Fig. 15.12.

Because the frequency range of interest is the region below the lowest resonance frequency, all resonances should be made as high in frequency as possible. For the steel box (Fig. 15.13), the plate resonance shows up at about 500 Hz. The peaks at high frequencies are air resonances. Thus, the box is successful as an acoustic enclosure below about 300 Hz where the average attenuation is slightly over 40 dB. For the soft rubber box (Fig. 15.14), by contrast, the plate wall resonance occurs at 80 Hz owing to the limpness of the rubber. No useful attenuation is achieved at any part of the low-frequency range. For frequencies above 400 Hz, the mean insertion loss is 30 dB, interrupted by a number

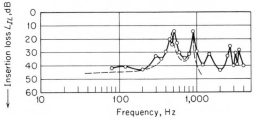

Fig. 15.13 Insertion loss produced by steel box 0.305 × 0.152 × 0.14 m. Box properties: wall thickness $t = 0.203$ cm; $M = 15$ kg/m²; unoccupied volume of air enclosed = 0.00475 m³. Loudspeaker source inside box. Dotted curve is theoretical curve C from Fig. 15.12.

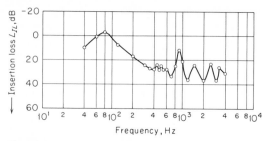

Fig. 15.14 Insertion loss produced by soft rubber box $0.305 \times 0.152 \times 0.14$ m. Wall thickness $t = 1.27$ cm; $M = 17$ kg/m^2; unoccupied volume of air enclosed $= 0.00475$ m^3.

of air resonances. This behavior resembles the mass-law behavior of architectural structures because the soft rubber amounts to a limp mass.

15.2.3 Effect of Enclosed Volume

At low frequencies Eq. (15.9) reduces to

$$L_{\mathrm{IL}} = 10 \log \left(1 + \frac{2lS}{\rho c^2} \right) \quad \mathrm{dB} \qquad (15.12)$$

which implies that if the volume of air, proportional to l, inside a given enclosure is increased, a corresponding increase will occur in the insertion loss. If the second term is large compared to 1 (the stiffness S is large), increasing the volume by a factor 2 increases L_{IL} by 3 dB.

To illustrate this effect, a cube with steel walls was used to enclose a small loudspeaker unit. Then an incompressible material was added to reduce the volume to produce curve A of Fig. 15.15. Both above and below the plate resonance (approximately 170 Hz) the enclosure containing the greater volume shows the higher insertion loss, by an amount $10 \log (0.0311/0.0197) \approx 2$ dB.

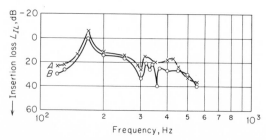

Fig. 15.15 Effect of varying enclosed volume of air in an acoustic enclosure. Box with 0.355-m sides; $M = 25$ kg/m^2. Curve A: volume of air enclosed $= 0.0197$ m^3; curve B: volume of air enclosed $= 0.0311$ m^3.

15.2.4 Damping of Enclosure Resonances   Two distinct resonance systems exist for an enclosure made up of finite plates: the plate resonances and the air resonances. Different methods of damping them are required.

A detailed picture of the plate resonance behavior is revealed in Fig. 15.16. (Note that the finite plate size introduces antiresonances due to higher-order vibrational modes in the plate, ignored above.) At the fundamental resonance frequency A the plate bends as shown at the left; at any instant of time, all points on the plate move in the same direction. Thus, the plate curvature is comparatively small and plate damping treatments are least effective for this mode of vibration. Above this fundamental resonance the plate no longer moves unidirectionally; the outer regions now move in a direction opposite to that of the center region. The net volume displacement of the plate decreases with increasing frequency until a condition is reached where the net volume flow is zero, corresponding to a peak in attenuation B. At a slightly higher frequency C the next resonance mode occurs, but in this case the plate flexes more sharply than for the fundamental mode and so is more amenable to mechanical damping.

With respect to air resonances, a plate spacing of 0.19 m (7.5 in.) leads to standing waves at approximately 900 Hz and its integral multiples, as shown in Figs. 15.12 to 15.14. Larger spacing would lead to a correspondingly lower standing-wave frequency. At these frequencies of air resonance, the trapped air behaves as a very high impedance coupling the two plates together so that the attenuating plate must move with nearly the same motion as the driving plate. This form of resonance can be suppressed only by applying sound-absorptive treatment in the trapped air itself. In Fig. 15.17, curve A shows the buildup of sound pressure that occurs owing to the presence of the enclosure. The inclusion of sound-absorptive material (see curve B) reduces the sound pres-

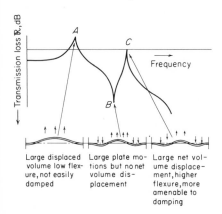

Fig. 15.16   Motions of a clamped plate subjected to uniform sound pressure.

sure near the standing-wave frequencies. The curves also show that either with or without the absorptive treatment, there is a buildup of sound pressure inside the hood below 500 Hz, reaching a value of 15 dB at 50 Hz.

Jackson [6,7] gives experimental results for a number of large panels of various constructions, suitable for building practical enclosures.

15.2.5 Double Windows (Double Glazing)   A common means of efficiently utilizing the allowable mass is to use a double-skinned construction, that is, two plates separated by an airspace, as in the double glazing of windows.   While significant advantages at high frequencies can often be realized with this type of construction in architectural design, there may be difficulties at low frequencies for which acoustical enclosures are usually intended.   An example of the insertion loss for a double glass window is illustrated in Fig. 15.18.   The two curves show, respectively, the attenuation for a single sheet of 4-mm glass (the full curve) and for two sheets of the same material spaced 5.7 cm ($2\frac{1}{4}$ in.) apart.   For frequencies below 160 Hz, no benefit is derived from the extra sheet of glass.   In fact, at 90 Hz the insertion loss is diminished with double glazing, because resonance occurs owing to the mass of the two plates and the stiffness of the trapped air [see Eqs. (15.7) and (15.8)]. This resonance cannot easily be damped without affecting transparency, and so resonances of this kind must be moved to a frequency outside the working range, either by increasing the spacing or increasing the mass of the individual panels.

15.2.6 Unsealed Enclosures   Sometimes the sound source cannot be sealed off completely because some gaps are required to provide exits for pipes, shafts, ventilation, etc.   In these cases, no matter how ef-

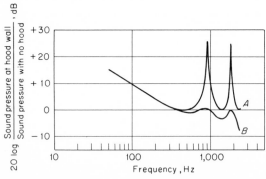

Fig. 15.17 Buildup of sound pressure owing to presence of enclosure [plane-wave model with $l = 7\frac{1}{2}$ in. (19 cm)].   Curve A: no acoustic absorption between plates; curve B: 2-in. superfine glass-fiber blanket at hood surface.

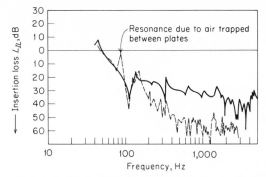

Fig. 15.18 Insertion loss for double windows (glazing): ———, one sheet of 4-mm glass, 0.76 m (30 in.) square; – – –, two sheets of same glass spaced 5.7 cm ($2\frac{1}{4}$ in.) apart. The sheets are so close together that little benefit is derived from the second sheet below 160 Hz.

ficient the hood itself, transmission of sound will always occur through the openings and it is essential to know the extent to which the overall attenuation of the system may be thus affected. The theoretical problem resembles that of the familiar Helmholtz resonator, which has an insertion loss at low frequencies of the form

$$L_{\mathrm{IL}} = 10 \log \left[\left(1 - \frac{V_0 P \omega^2}{c^2 A} \right)^2 + \omega^2 r^2 \right] \quad \mathrm{dB} \quad (15.13)$$

where c = speed of sound in air, m/sec
$\quad V_0$ = enclosed air volume, m³
$\quad A$ = cross-sectional area of the leak, m²
$\quad P$ = equivalent length of the leak, m
$\quad r$ = damping term due to externally radiated energy and to losses in the leak outlet

The performance typical of a "leaky" small enclosure is shown in Fig. 15.19 where magnification is seen to occur at the resonance frequency indicated by the arrow and the formula. The area and length of the

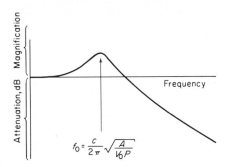

Fig. 15.19 Attenuation in the neighborhood of resonance for a Helmholtz resonator.

leak outlet and the volume of the enclosed air should be chosen so that this frequency lies as far as possible below the working range. Thus, if the frequency f_0 is to be low, we must either *decrease* the cross-sectional area of the leak A, *increase* the length of the leak opening, or *increase* the basic volume of the enclosure, V_0.

15.2.7 Ventilation for Enclosures (Ducts and Mufflers)   Frequently when an enclosure must be provided for a noisy machine, some means must be found to provide a continuous flow of ventilating air. This requires the use of some type of muffler, if the transmission loss of the enclosure is not to be compromised.

The various types of dissipative mufflers and their expected performances are the subject of Chap. 12. Alternate possibilities are

1. Lined ducts (Sec. 15.3 below, Sec. 12.3, and Ref. 11)
2. Lined bends (Sec. 12.4)
3. Plenum chambers (Sec. 12.5)
4. Prefabricated dissipative mufflers (Sec. 12.8)

15.3   Design Procedure for Rectangular, Lined Ducts *

The following procedure permits calculation of the attenuation of sound propagating along a duct with absorbing walls; the design of a muffler to meet given requirements proceeds by trial and error, with the guidance of a few general principles. The calculation [9, 10] is applicable to the principal (plane-wave) mode of sound propagation in the duct and is based on the normal impedance Z of the duct lining. It is valid both for the case of a low rectangular channel of height h with one absorbing and three rigid walls (Fig. 15.20a and b) and, on account of the mirror principle, for a duct of height $2h$ with two absorbing walls facing one another (Fig. 15.20c and d).

The attenuation, which typically exhibits a more-or-less broad maximum at a certain frequency, depends upon the duct height h, the depth d of the absorptive lining, the frequency of the sound, expressed here in terms of a parameter that depends on the ratio of the duct height to the wavelength, $\eta = 2h/\lambda$, and the flow resistance R_f of the absorptive lining material. This material may be in the form of a thin porous layer spaced out a distance d from the duct wall (Fig. 15.20b and d),† or it may be a continuous blanket of thickness d (Fig. 15.20a and c).

---

* Less accurate but simpler methods for estimating the attenuation of sound in lined ducts are given in Sec. 12.3. The reader is also referred to Ref. 8.

† In this case, in order to prevent the propagation of sound behind the treatment the airspace behind the porous layer must be subdivided into cells no larger than approximately $\lambda_0/4$, where λ_0 is the wavelength corresponding to the frequency of maximum attenuation.

The design procedure begins with the choice of the relevant parameters for the duct and its lining and goes on to a calculation of the attenuation, using the curves of Fig. 15.21 and the design charts of Fig. 15.22, according to the following steps. If the problem is not one of design but rather of calculating the attenuation of a given duct, proceed directly to step 6 below.

The choice of the duct proportions (i.e., the value of h, in step 1 below) governs the bandwidth of the curve of attenuation vs. frequency (high and narrow vs. low and broad). The design of the duct lining consists in selecting other parameters to yield the desired values for the real and imaginary parts of the normal acoustic impedance of the lining. The optimum value for this impedance (leading to the highest attenuation per unit length of duct) is[9] always

$$\frac{Z}{\rho c} = \frac{R}{\rho c} + \frac{jX}{\rho c} = (0.92 - j\,0.77)\eta$$

This occurs for $\phi = 0.7$ rad (40°) and yields, for values of $\eta \leqslant 0.3$, a maximum attenuation as much as 19 dB in a length of duct equal to h.

Note that the optimum impedance for the lining is not ρc, as it would be to achieve maximum absorption for sound at normal incidence; nor is it a value greater than ρc, as it would be to achieve the greatest absorption for sound at oblique incidence. It is not even a pure resistance: maximum attenuation does not occur for $X = 0$, that is, for $d = \lambda/4$; some stiffness reactance is required, depending on both the frequency and the duct height.

To achieve this optimum impedance over a wide frequency range would require the effective value of stiffness to vary with frequency in such a way as to maintain $X = 0.77\eta \approx 1.5h/\lambda$, a condition that imposes practical difficulties. In practice one designs to achieve this optimum impedance at the frequency where greatest attenuation is required, by proper choice of the lining depth d.

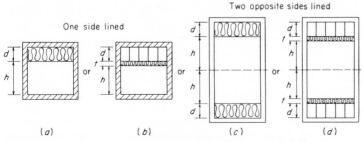

Fig. 15.20 Cross section of rectangular lined duct: (a) and (b) lined on one side only; (c) and (d) lined on two opposite sides.

The flow resistivity of the lining material is then chosen so that, for the dimensions required by the optimum choice of reactance, the optimum value is achieved for the real part of the lining impedance.

Step 1: Choose the duct height h. Since the attenuation increases rapidly with decreasing duct height, one ordinarily chooses h as small as is consistent with the permissible pressure drop in the lined duct for the expected gas flow. Note that in choosing h the related frequency parameter η (see definition in step 2) must also be considered according to whether one wants great attenuation in a narrow band of frequencies (η small) or less attenuation over a broader frequency band (η large); this choice is also affected by the choice of the frequency of maximum attenuation (see next step).

Step 2: For the design frequency f_0 at which maximum attenuation is desired, and for the value of duct height h chosen in step 1, find the frequency parameter $\eta_0 = 2hf_0/c$, where f_0 is in hertz and h and c are in compatible units. At normal room temperature

$$\eta_0 = hf_0/172 \qquad (h \text{ in meters})$$
$$= hf_0/6{,}780 \qquad (h \text{ in inches})$$

Step 3: Find the depth d of absorptive treatment that will yield the optimum reactance $X = 0.77\eta_0$ for the duct lining at frequency f_0, as follows: *

1. For duct treatment consisting of a thin porous layer over a subdivided airspace, the depth of the airspace should be

$$d = 74/f_0\eta_0 \qquad\qquad \text{m}$$
$$= 2.9 \times 10^3/f_0\eta_0 \qquad \text{in.}$$

2. For a solid blanket of porous material, the duct lining thickness should be

$$d = 60/f_0\eta_0 \qquad\qquad \text{m}$$
$$= 2.4 \times 10^3/f_0\eta_0 \qquad \text{in.}$$

Step 4: Choose a dissipative lining material to yield the optimum normalized flow resistance

$$(R_f/\rho c)_{\text{opt}} = 0.92\eta_0$$

1. If the dissipative material is in the form of a thin sheet of thickness $t(< \lambda_0/10)$ spaced out a distance d from the rigid duct wall, choose a porous material with flow resistivity

* The specific reactance due to an air layer of depth d is $X/\rho c = -j \cot (2\pi d/\lambda)$, which should be equated to $0.77\eta_0$ to optimize the value of d. In the equations of this step, the cotangent function is approximated by the first term of a series expansion: $X/\rho c \approx -\lambda/2\pi d$.

$$R_1 = 377\eta_0/t \qquad \text{(mks rayls/m, } t \text{ in meters)}$$
$$= 1.5 \times 10^4 \eta_0/t \qquad (t \text{ in inches})$$

2. If the dissipative material is in the form of a blanket of thickness d, the material should have a flow resistivity

$$R_1 = 1,130\eta_0/d \qquad \text{(mks rayls/m, } d \text{ in meters)}$$
$$= 4.5 \times 10^4 \eta_0/d \qquad (d \text{ in inches})$$

Step 5: From the available materials that satisfy all other design requirements (temperature, corrosion or moisture resistance, cost, mechanical properties, etc.) select one with a value of nominal flow resistance as close as possible to the optimum value found in step 4. It may be necessary to choose a material with nominal flow resistance less than optimum and compress it in assembling the lining to give the required value. (See Figs. 10.2 and 10.3.) This flow resistance is assumed to be the same for all frequencies in the following calculations.

Calculate the actual value of normalized flow resistance for the chosen material, either $R_f/\rho c = R_1 t/\rho c$ for the thin layer spaced away from the duct wall, or $R_f/\rho c = R_1 d/3\rho c$ for the continuous blanket.

This completes the choice of design parameters for the duct and lining; its attenuation is calculated as follows:

Step 6: Set up a computation sheet with the following values of the frequency parameter η across the top line (these values will always be

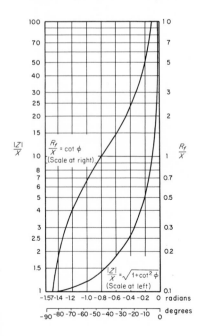

Fig. 15.21 Design curve relating the magnitude $|Z|$, the phase angle ϕ, and the real R and imaginary X parts of the complex normal acoustic impedance of the duct lining. (This chart should be reproduced on a transparent sheet and used as a moveable overlay on Fig. 15.22.)

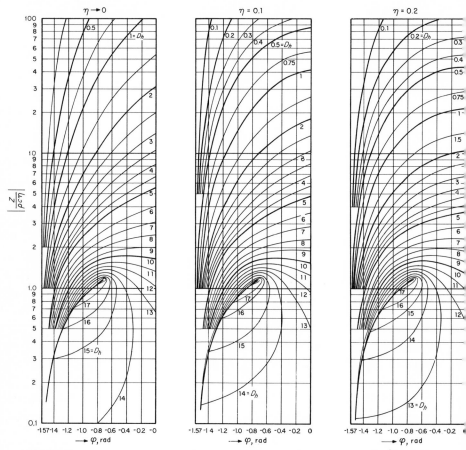

Fig. 15.22   Series of six design charts for various values of the frequency parameter η; these charts show contours of constant D_h as functions of the magnitude $|Z|$ and phase

the same, since they are the values of η for which design curves of Fig. 15.22 have been prepared): $\eta = 0.05$, 0.1, 0.2, 0.4, 0.6, and 1.0.

Step 7: On the next line, enter the values of frequency (Hz) corresponding to these values of η and the duct height h

$$f = 172\eta/h \qquad (h \text{ in meters})$$
$$f = 6{,}280\eta/h \qquad (h \text{ in inches})$$

Step 8: Using the duct lining thickness d (found in step 3 or given) calculate for each frequency the normalized reactance of the lining:

   1. For thin porous layer over airspace:

$$X/\rho c = 57/fd \qquad (d \text{ in meters})$$
$$= 2.23 \times 10^3/fd \qquad (d \text{ in inches})$$

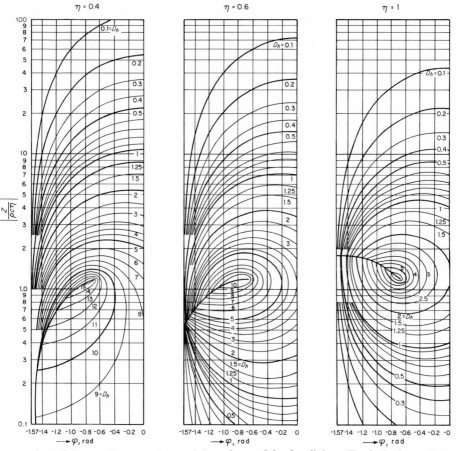

$\eta = 0.4$ $\eta = 0.6$ $\eta = 1$

angle ϕ of the complex normal acoustic impedance of the duct lining. To obtain the total attenuation of a treated duct of length L, multiply D_h by L/h (see Fig. 15.20).

2. For continuous blanket lining:

$$X/\rho c = 46/fd \qquad (d \text{ in meters})$$
$$= 1.8 \times 10^3/fd \qquad (d \text{ in inches})$$

Enter these values on line 3.

Step 9: From the values of η in line 1 and $X/\rho c$ in line 3, calculate $X/\rho c\eta$ for each frequency, and enter these values on line 4.

Step 10: From the value of $R_f/\rho c$ found in step 5 and $X/\rho c$ in line 3, calculate for each frequency the value of $R_f/X = \dfrac{R_f/\rho c}{X/\rho c}$, and enter these values on line 5.

Step 11: Now refer to Fig. 15.21. This figure shows the relationships between the real R and imaginary X parts of the lining impedance in terms of the magnitude $|Z|$ and phase angle ϕ of this impedance. The point at the lower left-hand corner, corresponding to $|Z|/X = 1$ and $\phi = -1.57$ radians ($-90°$), represents the case where the entire impedance is made up of the pure reactance of the lining, a value determined by the choice of the lining depth d and the frequency, the flow resistance of the lining assumed to be zero. As the lining flow resistance increases from zero, the ratio $|Z|/X = \sqrt{1 + \cot^2 \phi}$ increases according to the lower curve (read on the scale at the left), and the ratio $R_f/X = \cot \phi$ increases according to the upper curve (read on the scale at the right). For each frequency, on line 2, enter the right-hand vertical scale of Fig. 15.21 with the value of R_f/X (from line 5; count all $R_f/X \leqslant 0.1$ as 0.1) and project a horizontal line to the left to meet the curve marked $R_f/X = \cot \phi$; drop a vertical line from this intersection point to meet the curve marked $|Z|/X$; now for this intersection point, read the corresponding values of $|Z|/X$ from the left-hand vertical scale and of ϕ from the horizontal scale (ϕ will always be negative for mufflers of the type discussed here) and enter these values on lines 6 and 7, respectively, of the computation sheet.

Step 12: For each frequency multiply the value of $|Z|/X$ on line 6 by the value of $X/\rho c\eta$ on line 4 to get $|Z|/\rho c\eta$; enter these values on line 8.

Step 13: For each frequency on line 2, select the design curve of Fig. 15.22 with the corresponding value of η; for $\eta = 0.05$, use the design curve for $\eta = 0$. (This calculation procedure may be extended, with some loss of accuracy, to higher frequencies corresponding to values of η equal to 2 or 3; use the design curve for $\eta = 1$. At these higher frequencies, high-order modes of propagation come into play and this method of calculation probably underestimates the attenuation that will actually be achieved.)

Step 14: Determine for each frequency the attenuation D_h in decibels per length of treated duct equal to the duct height h, as follows:

Enter the design chart selected in step 13 with the coordinates $|Z|/\rho c\eta$ (from line 8) and ϕ (from line 7) to locate a point that determines the corresponding value of D_h, by interpolation among the contours of equal D_h on the design chart.

Step 15: To get the total attenuation of a treated duct of length L, multiply D_h by L/h. This step completes the calculation of attenuation for a section of treated duct, neglecting reflections at the ends (see Sec. 12.3.3).

Alternate Procedure to Steps 11 to 14. Steps 11 to 14 can be simplified and speeded up by the use of a transparent version of Fig. 15.21. This transparency may be prepared by tracing the two curves of that figure onto a piece of thin tissue or tracing paper. It is important that this figure be the exact size of the individual charts of Fig. 15.22 and have the markings on both horizontal and vertical scales. This transparency is used in connection with the design curves of Fig. 15.22 to determine the attenuation D_h in decibels per length of treated duct equal to the duct height h. Once prepared, this transparency should be kept with the book for future use.

Step 16: Superimpose the transparency of Fig. 15.21 over the proper design chart for the frequency in question (selected as in step 13), taking care that the vertical (left-hand) axes of the two graphs are aligned and that the lower left-hand corner of the grid on the transparency — corresponding to $|Z|/X = 1$ and $\phi = -1.57$ radians — lies directly over the value $X/\rho c\eta$ (from line 4) on the vertical scale of the selected design chart. With the two figures thus superimposed, one can see the interplay of different choices for the real and imaginary parts of the lining impedance with the contours of the design chart, particularly the "maximum crest," and can also appreciate the consequences of varying R.

Step 17: Keeping the transparency and the design chart carefully aligned as in step 16, enter the right-hand vertical scale of the transparency with the value of R_f/X (from line 5) (count all $R_f/X \leq 0.1$ as 0.1), and project a horizontal line to the left to meet the curve of $R_f/X = \cot \phi$; drop a vertical line from this intersection point to meet the curve of $|Z|/X$. Mark the point *on the design chart* that lies directly under this last intersection, and read the corresponding value of D_h by interpolation among the contours of equal D_h on the design chart. (Return to step 15 or go on to step 18.)

Step 18: By an extension of steps 16 and 17, one can modify or optimize a lining design as follows: aligning the vertical (left-hand) axes of the transparency and the design charts for the frequency in question, slide the transparency up or down until the $|Z|/X$ curve on the transparency is tangent to a contour of desired (usually maximum) D_h on the design chart. Then the combination of duct and lining parameters required to achieve this attenuation is found as follows: First, project a vertical line upward from the point of tangency to meet the R_f/X curve, and, from that intersection, draw a horizontal line to meet the right-hand vertical axis of the transparency, to find the desired value of R_f/X. Also note the number on the left-hand vertical axis of the design chart that lies directly beneath the lower left-hand corner of the grid on the transparency: this number gives the desired value of $X/\rho c\eta$. Knowing η (from the design curve being used), one can get the relation of the frequency f to the duct height h (either of which may be chosen arbitrarily, see step 2), as well as the value of $X/\rho c$ and hence also of $R_f/\rho cf$ (from the value read for R_f/X); knowing $X/\rho c$ and the frequency f one can find the required depth of duct lining d (from the equations in step 8). It will not be possible to adjust the parameters to achieve maximum attenuation at all frequencies; the procedure given in steps 1 to 5 optimizes the attenuation for the frequency originally chosen. (Return to step 15.)

Kurze [11] has given a solution of the duct attenuation problem in closed form that is useful, when many such attenuation calculations must be made, if the services of a computer are available. The form is not particularly convenient for design purposes, for example, finding the proper value of duct lining impedance to yield a desired amount of attenuation:

$$\gamma = jk \sqrt{1 - \left(\frac{2}{kh}\right)^2 \left[1 + \frac{1}{1 + \frac{4Z}{jkh\rho c}} \pm \sqrt{1 + \frac{1}{\left(1 + \frac{4Z}{jkh\rho c}\right)^2}}\right]} \qquad (15.14)$$

where γ = complex propagation constant

$\mathrm{Re}(\gamma)$ = attenuation of the fundamental mode, in nepers per unit length of duct; multiply by 8.69 to get the attenuation per unit length in decibels: 8.69 $\mathrm{Re}(\gamma) = D/h$; the duct length is measured in meters if k and h are in m^{-1} and m, respectively; or in inches if k and h are in in.$^{-1}$ and in., etc.

$\mathrm{Im}(\gamma)$ = phase shift per unit length of duct, radians

$k = \omega/c = 2\pi/\lambda = 2\pi f/c$, the wavenumber, in m^{-1}

The other quantities are defined as before.

The significance of the two signs under the radical is as follows: in the vicinity of optimum attenuation there is some ambiguity in identifying the "fundamental mode" of wave propagation in a duct. In practice, one can define the fundamental mode as the *least rapidly attenuated mode*. Accordingly, in using Eq. (15.14), check whether the positive or the negative sign yields lower value of $\mathrm{Re}(\gamma)$ and use that lower value for the duct attenuation. Except in the region of maximum attenuation (near the crests in the design charts of Fig. 15.22), the negative sign will usually give the lesser attenuation.

REFERENCES

1. Leo L. Beranek (ed.), *Noise Reduction,* p. 360, McGraw-Hill, 1960.
2. K. A. Mulholland, A. J. Price, and H. D. Parbrook, "Transmission Loss of Multiple Panels in a Random Incidence Field," *J. Acoust. Soc. Am.,* vol. 43, pp. 1432–1435, 1968.
3. Leo L. Beranek and George A. Work, "Sound Transmission through Multiple Structures Containing Flexible Blankets," *J. Acoust. Soc. Am.,* vol. 21, pp. 419–428, 1949.
4. R. A. Mangiarotty, "Optimization of the Mass Distribution and the Air Spaces in Multiple-element Soundproofing Structure," *J. Acoust. Soc. Am.,* vol. 35, pp. 1023–1029, 1963.
5. Pritchard H. White and Alan Powell, "Transmission of Random Sound and Vibration through a Rectangular Double Wall," *J. Acoust. Soc. Am.,* vol. 40, pp. 821–832, 1966.
6. R. S. Jackson, "The Performance of Acoustic Hoods at Low Frequencies," *Acustica,* vol. 12, pp. 139–152, 1962.

7. R. S. Jackson, "Some Aspects of the Performance of Acoustic Hoods," *J. Sound Vibration*, vol. 3, pp. 82–94, 1966.

8. American Society of Heating, Refrigerating and Air-Conditioning Engineers (ASHRAE), *Guide and Data Book*, (revised frequently). 345 East 47th St., New York, N.Y. 10017.

9. L. Cremer, "Theorie der Luftschall-Dämpfung in Rechteckkanal mit Schluckender Wand und das sich dabei ergebende höchste Dämpfungsmass," *Acustica*, vol. 3, Beiheft 2, p. 249, 1953.

10. U. Kurze, "Untersuchungen an Kammerdämpfern," *Acustica*, vol. 15, no. 3, pp. 139–150, 1965.

11. U. Kurze, "Schallausbreitung im Kanal mit periodischer Wandstruktur," *Acustica*, vol. 21, no. 2, pp. 74–85, 1969.

Chapter Sixteen

Noise of Gas Flows

HANNO H. HELLER and PETER A. FRANKEN

16.1   Introduction

Noise associated with high-speed or unsteady gas flows and with interaction of gas flows and solid objects is usually called *aerodynamic noise.*
Such noise is associated with numerous industrial processes and has
gained considerable importance over the past few years.

A working knowledge of aerodynamic noise and its control is based on
an understanding of the source types involved.   At subsonic flow speeds,
such noise is created by one or more of three basic types of aerodynamic
sources, namely, monopole, dipole, and quadrupole sources.

In addition to describing the characteristics of these three basic types
of aerodynamic sources and to deriving scaling laws for aerodynamic
noise, this chapter treats a number of practical aerodynamic noise generators; in particular, jets, flow spoilers, diffusers, flow-through grids,
and air valves.

16.2   Aero-acoustic Source Types

Aerodynamic Monopole.   Aerodynamic monopole radiation occurs
when mass or heat is introduced into a fluid at a non-steady rate.   Typi-

cal monopole sources are pulse jets (where high-speed air is periodically ejected through a nozzle), sirens (where a steady airflow is periodically chopped), or propellers at zero pitch (where air is periodically displaced every time a blade passes a given point).

A monopole source is like a pulsating sphere (Fig. 16.1a). The acoustic pressure-pulse fronts are always in phase, thereby producing a spherical directivity pattern.

The dimensional relation between radiated sound power and the important parameters that produce it is

$$W_{A_\text{monopole}} \propto \rho L^2 \frac{u^4}{c} = \rho L^2 u^3 M \qquad (16.1)$$

where W_A = radiated sound power, watts
ρ = density of gas, kg/m³
c = speed of sound in the gas, m/sec
u = flow speed, m/sec
L = pertinent length dimension, m
M = Mach number equal to u/c, dimensionless

Aerodynamic Dipole. Aerodynamic dipole radiation occurs when a flow of gas interacts with a body to produce unsteady forces. This source type is found in compressors where turbulent flow interacts with stator or rotor blades and on propellers with nonzero pitch. The "sing-

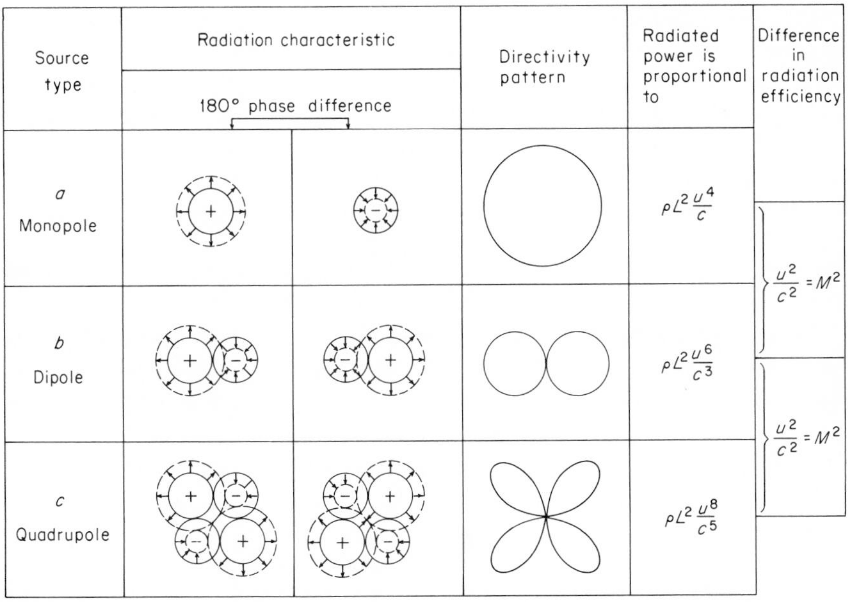

Source type	Radiation characteristic 180° phase difference		Directivity pattern	Radiated power is proportional to	Difference in radiation efficiency
a Monopole				$\rho L^2 \dfrac{u^4}{c}$	
					$\dfrac{u^2}{c^2} = M^2$
b Dipole				$\rho L^2 \dfrac{u^6}{c^3}$	
					$\dfrac{u^2}{c^2} = M^2$
c Quadrupole				$\rho L^2 \dfrac{u^8}{c^5}$	

Fig. 16.1 Aero-acoustic sources and their dimensional properties.

ing" telegraph wire, and edge-tone phenomena, where flow is shed from solid objects in a periodic manner, can also be related to aerodynamic dipole radiation. Other examples are gas flow through grids and valves.

An oscillating force, i.e., an aerodynamic dipole, is analogous to a pair of monopole sources separated a distance small compared to a wavelength and oscillating out of phase (Fig. 16.1b). While one monopole is expelling gas, the other is sucking it in; subsequently, the process reverses.

The dimensional dependence of radiated dipole sound power is

$$W_{A_{\text{dipole}}} \propto \rho L^2 \frac{u^6}{c^3} = \rho L^2 u^3 M^3 \qquad (16.2)$$

Dipole radiation is a factor of M^2 different from that of a monopole. For subsonic speeds (where $M < 1$) the dipole is less efficient than the monopole. Because of the pressure cancellation that occurs in the plane normal to the dipole axis (see Fig. 16.1b, column 4), the contributions from the positive and negative sources of the dipole cancel on this plane, yielding a directivity pattern with a figure-of-eight shape, strongest in the direction of the dipole axis and weakest at right angles to the axis.

Aerodynamic Quadrupole. Aerodynamic quadrupole radiation results from the viscous stresses within a turbulent gas flow in the absence of obstacles. The resulting forces can occur only in opposing pairs, since no counterforce from an obstacle is available. Such fluctuating force pairs, called quadrupoles, can be associated with corresponding dipole pairs that radiate 180° out of phase (see Fig. 16.1c).

Aerodynamic quadrupole sources constitute the dominant source type in high-speed, subsonic, turbulent-air jets. Quadrupole-source strength is large in regions where both turbulence and mean-velocity gradient are high, such as in the highly turbulent mixing layer of a jet.

The dimensional dependence of radiated quadrupole sound power is

$$W_{A_{\text{quadrupole}}} \propto \rho L^2 \frac{u^8}{c^5} = \rho L^2 u^3 M^5 \qquad (16.3)$$

Quadrupole efficiency differs by a factor M^2 compared to that of the dipole. At subsonic speeds ($M < 1$), quadrupole radiation efficiency is lower than that of a dipole because of the double cancellation effect illustrated in the fourth column of Fig. 16.1c.

Although monopole, dipole, and quadrupole sources decrease in their respective radiation efficiency for subsonic flows, the dependences of radiated sound powers on flow speeds for them show the opposite trend; that is to say, the total radiated sound power varies as the fourth, sixth, and eighth power of flow speed u for monopole, dipole, and quadrupole sources, respectively. Thus, in spite of inherently low efficiency, radiation from a quadrupole source may dominate over that from the other types if flow speed is high.   Quadrupole radiation for a jet engine with high exhaust speed usually predominates, although other internal sources, such as rough burning (predominantly a monopole source) or compressor noise (predominantly a dipole source), are present and contribute to the total noise.

The constant of proportionality for each type of source may take on different values, depending on the particular sound-generating process. Thus, the constant for a singing-wire phenomenon is different from that for an edge-tone phenomenon, although both result from aerodynamic dipole radiation.   However, the proportionality relations of Eqs. (16.1) to (16.3) are useful in estimating the effect that changing one or more parameters will have on the sound power radiated.   Increasing the exhaust speed u of a jet (quadrupole-type source) by a factor of 2 increases the sound-power level by 24 dB, whereas an increase of the exhaust nozzle area S (proportional to L^2) by the same factor increases the sound-power level by only 3 dB.   Because the thrust of a jet engine is proportional to Su^2, it would be better to achieve a doubling of thrust by increasing the nozzle area by a factor of 2.0 rather than the exhaust speed by a factor of 1.4 because the inherent increase in noise will be much less.

16.3   Noise from Gas Jets

The simplest free jet is an airstream emanating from a large reservoir through a rounded convergent nozzle (Fig. 16.2).   The gas accelerates from near zero velocity in the reservoir to peak velocity in the narrowest cross section of the nozzle.   Sonic flow speed occurs at a pressure ratio $p_0/p_s = 1.89$, where p_0 is the steady pressure upstream of a nozzle and

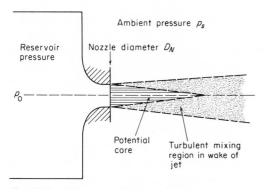

Fig. 16.2   Subsonic turbulence- free jet.

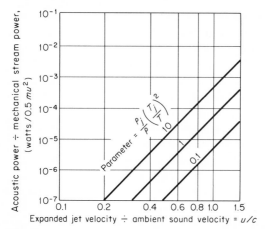

Fig. 16.3 Ratio of acoustic power of jet noise to mechanical stream power. The abscissa is the ratio of the expanded jet velocity to the ambient sound velocity. The parameter is the ratio $(\rho_j/\rho)\,(T_j/T)^2$ of jet and ambient densities and absolute temperatures. It is implicit in this chart that the acoustic power increases as the eighth power of the jet velocity.

p_s is ambient pressure downstream of the nozzle. An increase above this critical pressure ratio does not further increase the flow speed for this particular nozzle configuration,* but rather introduces a phenomenon called *choked flow*. (See Sec. 16.6, on air-valve noise.)

In this idealized jet flow, no interaction is assumed between the gas flow and adjacent solid surfaces. The noise is solely a result of the turbulent mixing processes in the shear layer. The sources of noise are distributed over a considerable length downstream from the nozzle. The very high frequencies of such jet noise are predominantly radiated from the flow region close to the nozzle where eddy sizes are small. Lower frequencies are radiated further downstream where eddy sizes are much larger. The noise of such a jet flow can be attributed to quadrupole-type sources.[1]

The total sound power radiated by a gas jet flow may be expressed in terms of the mechanical stream power in the jet, equal to

$$W_{\text{mech strm}} = \tfrac{1}{2}\,mu^2 \tag{16.4}$$

where $m =$ mass flow of gas, kg/sec
$\qquad u =$ speed of flow at exit nozzle, m/sec

The ratio of the sound power radiated to the mechanical stream power is plotted in Fig. 16.3 as a function of the ratio of the jet speed to

* Supersonic speeds can be obtained for an increased pressure ratio only if the convergent part of the nozzle is followed by a divergent section.

the ambient sound speed with the parameter equal to $(\rho_j/\rho)(T_j/T)^2$, where ρ_j and T_j are jet density and absolute temperature, respectively, and ρ and T are ambient density and absolute temperature.

The sound field from a gas jet is directive, with maximum noise being radiated in the annular region of 30° to 45° from the jet axis. A typical directivity pattern for the total sound level of an air jet at ambient temperature, emerging at sonic speed, is shown in Fig. 16.4. The details of the pattern depend upon the jet velocity, but the typical directivity pattern may be used as an estimate for all jets.

In the far field the frequency spectrum of a free jet exhibits broadband character with a shallow peak (see Fig. 16.5) occurring at a frequency designated here as f_p. The location of this peak in the frequency spectrum can be determined from the nondimensional relation

$$S = \frac{f_p D_N}{u} = 0.2 \tag{16.5}$$

where S = Strouhal number, dimensionless
$\quad D_N$ = nozzle diameter, m
$\quad u$ = exit velocity, m/sec

Figures 16.3 to 16.5 together with Eq. (16.5) comprise a general procedure for estimating the total sound power, the directivity, and the frequency spectrum of free-jet noise.

Example 16.1 Find the sound-pressure-level spectrum at 45° from the jet axis and 3 m (9.8 ft) from the nozzle of a 0.03-m (1.2-in.) diameter jet. The jet exhausts into ambient air (21°C, 0.75 m Hg) at sonic ($M = 1$) exit velocity u, that is, 344 m/sec (1,128 ft/sec). The mass flow m is 0.2 kg/sec (0.44 lb/sec).

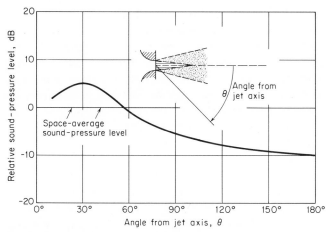

Fig. 16.4 Directivity pattern of subsonic air jet.

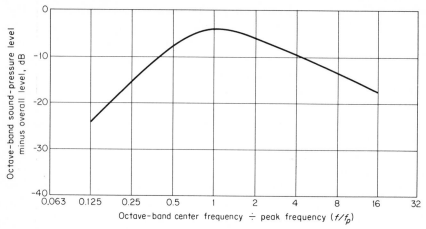

Fig. 16.5   Octave-band sound-pressure-level spectrum of a turbulent jet, where 0 dB is the overall sound-pressure level at the angle θ (see also Fig. 16.4).

SOLUTION: The mechanical stream power $\frac{1}{2} m u^2$ is 1.18×10^4 watts. The ratios $(\rho_j/\rho)(T_j/T)^2$ and u/c are unity; therefore, from Fig. 16.3, the ratio of sound power to mechanical stream power is 6×10^{-5}. The resulting overall sound power is 0.71 watt and the sound-power level is 118.5 dB re 10^{-12} watt.

From Fig. 16.4 the directivity correction at 45° is $DI_{45°} = 3$ dB. Taking inverse-square-law $(-10 \log 4\pi r^2)$ spreading of the jet noise into account, we obtain an overall sound-pressure level at 45° and a distance of 3 m [see Eq. (6.29)] equal to $118.5 + 3 - 10$ log $[4\pi(3)^2] = 101$ dB re 2×10^{-5} N/m². With the peak Strouhal number equal to 0.2, we obtain from Eq. (16.5) a spectrum peak of 2,300 Hz. The sound-pressure-level spectrum is then found from Fig. 16.5, by adding 101 dB to the values along the ordinate.

Realistic jets which emanate from pipes or engine nozzles do not provide smooth, low-turbulence entrainment of flow, but rather flow that has been disturbed or spoiled before leaving the nozzles. In this case, the above procedure is not valid unless the jet-exhaust speed exceeds approximately 100 m/sec. Below this speed major portions of the aerodynamically generated noise usually must be attributed to interaction between flow and obstacle, i.e., to dipole-source-type noise. In this lower speed range, the eighth-power-of-jet-velocity law that is implicit in Fig. 16.3 does not any longer apply.

16.4   Spoiler Noise

In the context of this chapter we will define *spoiler-generated noise* as noise resulting from pipe or duct-flow interactions with rigid objects or obstructions within a "semi-infinite" pipe or duct. In Fig. 16.6 a flow-spoiler noise-generating system is shown schematically.   Due to the impingement of the gas flow, forces of unsteady drag and lift are set up on the spoiler body which create noise within the hard-walled pipe. Such an obstruction in a pipe may be a strut, stringer, or guide vanes. The special case of a valve generally involves a pressure ratio exceeding the critical pressure ratio of 1.89, so that choked-flow phenomena occur, and this case will be considered separately later in the chapter.

Some simplifying restrictions and idealizations must be introduced at this point:

1. The flow spoiler (although it can be of arbitrary shape) is small compared to the pipe cross-section dimensions.   This restriction ensures that the flow speeds at the constriction do not rise excessively above the pipe mean-flow speeds.   The sound generation should then be due to flow/rigid-body interaction (a dipole-source mechanism), rather than to a turbulent mixing process (a quadrupole-source mechanism).

2. The pipe wall is acoustically hard.   Radiation owing to fluctuating lift forces (which are directed orthogonally to the mean pipe-flow direction) are canceled by the image reflection effect of the nearby pipe walls. Only radiation from the fluctuating drag-force component, pointing along the direction of the pipe axis, can propagate along the pipe.   The obvious implication for a low-noise spoiler system is to use bodies of low drag, such as airfoils, rather than grids or other oddly shaped bodies.

With these restrictions, a prediction scheme for the overall sound-power radiation from an enclosed-spoiler flow system is presented. Assuming that the noise generated arises only from the fluctuating

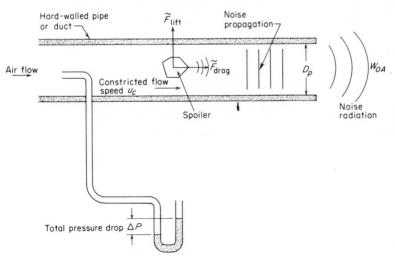

Fig. 16.6 Flow-spoiler system. The sound power is generated by the spoiler and radiated from the open end of the pipe.

forces on the body, the sound power generated per bandwidth is, for sea-level standard conditions,[2]

$$W_A(f) = 2.2 \times 10^{-8} f^2 [\tilde{F}(f)]^2 \qquad \text{watts/Hz} \qquad (16.6)$$

where $W_A(f)$ = sound power per unit bandwidth, watts/Hz

$F(f)$ = fluctuating force per unit bandwidth, N/Hz

Recent substantiations [3,4] indicate that Eq. (16.6) can be used directly for prediction. Unfortunately, the experimental assessment of the magnitude and direction of $\tilde{F}$ on bodies is very tedious. It turns out, however, that for several elementary flow-spoiler configurations, such as the one shown in Fig. 16.6, *the overall fluctuating forces are directly proportional to the steady-state drag force that the body experiences.*[5,6] This drag force is a power function of the pressure drop ΔP across the spoiler (as measured by an upstream and a downstream total-pressure probe). We can therefore write a semiempirical equation for the broadband noise generated by the spoiler and radiated into the free space outside the pipe (see Fig. 16.6) *

$$W_{OA} = \frac{k \, \Delta P^3 D_p^2}{\rho^2 c^3} \qquad \text{watts} \qquad (16.7)$$

where W_{OA} = overall sound power generated, watts

k = constant of proportionality, dimensionless

ΔP = total pressure drop across the spoiler, N/m²

D_p = pipe diameter, m

ρ = atmospheric density, kg/m³

c = speed of sound in the ambient medium, m/sec

Specific information on spoiler geometry is missing in this equation, but it is implicitly present in the pressure drop. Also, because $\Delta P \propto u^2$, Eq. (16.7) indicates that spoiler noise can be represented by a dipole model. From a variety of experimental spoiler configurations the constant of proportionality k is found to be about 2.5×10^{-4} for air.

With the restrictions enumerated earlier, and with $c = 344$ m/sec (1,128 ft/sec) and $\rho = 1.18$ kg/m³ (0.0735 lb/ft³), Eq. (16.7) is presented in parametric form in Fig. 16.7.

Equation (16.7) predicts the *broadband* sound power generated by the spoiler configuration. Discrete-frequency noise components may occur under certain conditions (e.g., excitation of edge tones) and could exceed the broadband levels. Once such an excitation mechanism is identified, the generation of discrete tones can usually be controlled by detuning or by rounding off sharp corners or edges, or cutting feedback paths by

---

* The overall sound power radiated from the pipe has been shown to equal three times the overall sound power that would be generated by the same flow-spoiler system located in free space.[3]

treating reflecting surfaces with a sound-absorbent layer. Thus, the broadband sound-power level, as calculated by Eq. (16.7), should be considered as a lower bound that occurs in the absence of discrete-frequency sound.

Another important restriction in the use of Eq. (16.7) is that the frequency at which the noise spectrum peaks must be lower than the "cut-off" frequency of the pipe, which is defined as

$$f_{co} = \frac{0.293 \ c}{r} \qquad \text{Hz} \qquad (16.8)$$

where c = speed of sound, m/sec
$\quad$ r = radius of pipe, m
or

$$f_{co} = \frac{0.5 \ c}{w} \qquad \text{Hz} \qquad (16.9)$$

where w = largest transverse dimension of a rectangular pipe, m
If this restriction is not met, the propagated sound power near and above the peak frequency (which determines most of the overall sound power) will be affected by the transverse resonances of the pipe.

The frequency spectrum measured outside the pipe for the noise

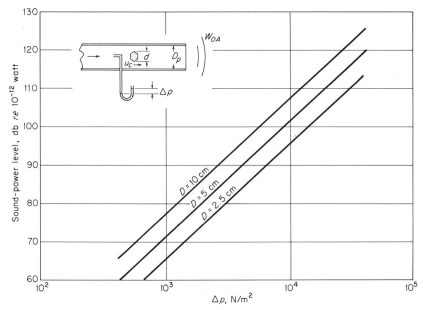

Fig. 16.7 Nomogram for determining noise generation from in-pipe flow spoiler. Sound power W_{OA} is radiated from the end of the pipe.

generated by the spoiler in the pipe exhibits a haystack shape with a peak at a frequency given by

$$f_p \approx b\, \frac{u_c}{d} \qquad \text{Hz} \qquad\qquad (16.10)$$

where u_c = constricted flow speed, m/sec
   d = projected width of the spoiler (see Fig. 16.7)
   b = constant which equals 0.2 for pressure differences on the order of 4,000 N/m^2 and 0.5 for pressure differences on the order of 40,000 N/m^2

The constricted flow speed u_c for cold air can be determined from Table 16.1 as a function of pressure difference ΔP. Values in between those in the table can be interpolated.

Example 16.2   Find the sound-power-level spectrum of the noise produced by an inpipe-spoiler configuration where the pipe diameter D_p is 5 cm and the flow spoiler stretching across the pipe is a flat plate of 2-cm width. A pressure drop across the spoiler equal to $\Delta P = 10,000$ N/m^2 was measured.

SOLUTION: From Fig. 16.7 we obtain the total sound power $L_w = 102$ dB *re* 10^{-12} watt. The peak frequency f_p of the spectrum can be expected to be at about

$$f_p = \frac{0.35\, u_c}{d}$$

for a pressure drop of 10,000 N/m^2.

From Table 16.1, we find $u_c = 124$ m/sec; hence $f_p \approx 2,170$ Hz. From Fig. 16.8 the sound-power-level spectrum in octave bands can be plotted. Finally, add 102 dB to the ordinate.

16.5   Grid or Grille Noise

The acoustical phenomena associated with flow through grids, grilles, diffusers, or porous plates, which often terminate air-conditioning ducts, are similar to those of spoiler configurations. The main distinctions are that for these types of elements, (1) the grid is located at the end of a duct, (2) the duct has a sizable cross-sectional area (say, 0.2- to 1-m square), and (3) the speed of air flow in a typical duct is low. However, with each passing year, flow speeds in newly installed air-conditioning

TABLE 16.1   Constricted Flow Speed u_c for Cold Air as a Function of the Pressure Drop

ΔP	(N/m^2) = 2,500	5,000	10,000	20,000	30,000	40,000
u_c	(m/sec) = 63	90	124	173	209	238

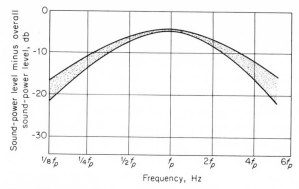

Fig. 16.8 Generalized octave-band spectrum for in-pipe generated spoiler noise, referenced to the total power and the peak frequency.

ducts increase, although even these "high" values, say, on the order of 30 m/sec, are low compared to sonic speed (344 m/sec). Even though the local speed through the air passages is higher than the average flow speed by a factor equal to the ratio of the area of the duct to the open area of the diffuser, the speed of the "air jets" emerging from the individual air passages (or orifices) in a diffuser is generally so low that the jets do not represent the dominant source of noise. As a simple rule, consider a jet speed of approximately 100 m/sec as the boundary below which jet-related noise is not important.

Grid noise is a result of an interaction between flow and rigid body. Thus, duct airflow interacts with diffuser guide vanes, and constitutes a dipole sound-generating mechanism.

When an element made from a grid of rods (Fig. 16.9) or a perforated sheet terminates a duct, additional, though similar, excitation mechanisms come into play. A series of circular rods may shed periodic vortices, which give rise to lift-force fluctuations on each rod, which in turn radiate noise. In this case, sound is produced at a discrete frequency given by

$$f = 0.2 \frac{u}{D_R} \qquad (16.11)$$

where u = mean-flow speed, m/sec

D_R = diameter of the rods, m

The emerging airflow from a duct terminated by a perforated plate containing sharp-edged circular orifices of the shape shown in Fig. 16.9 may become periodic, controlled in frequency and amplitude by a feedback mechanism. A feedback effect occurs only for orifices whose diameters are one to two times their lengths. Flow through typical air-conditioner grids, however, almost always gives rise to dipole-type power radiation, and noise prediction methods for grids and grilles involve the typical dipole, sixth-power-of-velocity dependence for sound power.

Similar to spoiler noise, grid noise can be related to the cube of pressure drop, which is the characteristic quantity for the overall radiated sound power. Hence, there is no need to consider the specific geometry of a grid, as long as the pressure drop across it is known.

Pressure drop ΔP across a variety of different air-conditioning diffusers has been measured and a simple prediction scheme devised by Hubert.[7] He defines a normalized pressure drop coefficient

$$\xi = \frac{\Delta P}{0.5 \rho u^2} \tag{16.12}$$

where ρ = density of air, kg/m³

u = mean-flow speed in the duct prior to the grid, m/sec

In Fig. 16.10a to c three typical diffuser configurations are presented together with their ξ values. Values of ξ for similar diffuser configurations may be estimated from this figure.

A relation for the overall sound-power level from air-conditioning diffusers has been derived[7]

$$L_w = 10 + 10 \log S + 30 \log \xi + 60 \log u \qquad \text{dB } re \ 10^{-12} \text{ watt}$$

$$\tag{16.13}$$

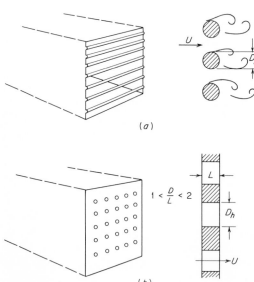

Fig. 16.9 Special cases of duct termination: (a) circular rods; (b) sharp-edged holes.

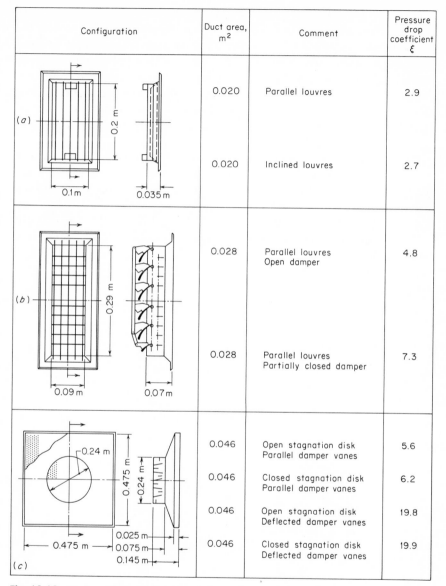

Configuration	Duct area, m²	Comment	Pressure drop coefficient ξ
(a) 0.2 m / 0.1 m / 0.035 m	0.020	Parallel louvres	2.9
	0.020	Inclined louvres	2.7
(b) 0.29 m / 0.09 m / 0.07 m	0.028	Parallel louvres Open damper	4.8
	0.028	Parallel louvres Partially closed damper	7.3
(c) 0.24 m / 0.475 m / 0.24 m / 0.475 m / 0.025 m / 0.075 m / 0.145 m	0.046	Open stagnation disk Parallel damper vanes	5.6
	0.046	Closed stagnation disk Parallel damper vanes	6.2
	0.046	Open stagnation disk Deflected damper vanes	19.8
	0.046	Closed stagnation disk Deflected damper vanes	19.9

Fig. 16.10   Various duct termination configurations and their pressure drop coefficients ξ equal to $\Delta P/q$ [see Eq. (16.12)].   The duct area S is in m².   (*After Hubert.*[7])

where S = area of the duct cross section prior to diffuser, m^2

ΔP = pressure drop through diffuser, N/m^2

Noise spectra of various diffuser configurations do not exhibit identical shapes even when normalized to similar flow speeds and exhaust areas. Minor construction differences may emphasize different frequency regimes. Some diffusers will, if improperly designed, radiate discrete frequency noise. In practical noise control problems, however, a general spectrum shape can be used that fits most diffuser noise spectra, as shown in Fig. 16.11.

To determine the sound-power spectrum radiated by a certain diffuser configuration, first determine the relevant spectrum for the particular flow speed from Fig. 16.11. To account for the duct area and the pressure drop coefficient, the spectrum must then be shifted in level (i.e., vertically only) according to Fig. 16.12 and Fig. 16.13, respectively.

The following example illustrates the procedure of Eq. (16.13). (See Fig. 16.14.)

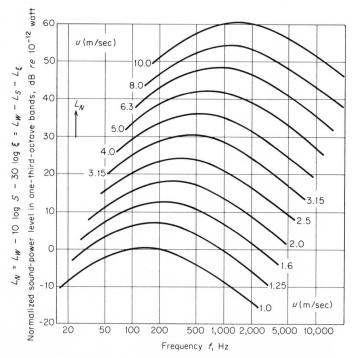

Fig. 16.11 Normalized one-third-octave sound-power-level spectra for noise radiated from diffusers for various flow velocities. (*After Hubert.*[7])

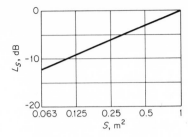

Fig. 16.12 Nomogram to determine effect of duct area. $L_s = 10 \log S$, where S is in m².

Example 16.3 Find the sound-power-level spectrum that results from the airflow through a diffuser that terminates an air-conditioning duct. Mean-flow air speed through the duct upstream of the diffuser is 4.5 m/sec (15 ft/sec) and the duct cross-sectional area is 0.125 m² (195 in.²). A pressure drop ΔP of 75 N/m² (0.3 in. water) across the diffuser was measured at this flow speed. SOLUTION: Determine: $q = \frac{1}{2}\rho u^2 = 12$ N/m² for an atmospheric density $\rho = 1.2$ kg/m³. Thus $\xi = \Delta P/q = 6$. From Fig. 16.11 obtain a spectrum for $u = 4.5$ m/sec; this spectrum must be shifted vertically by -9 dB to account for the area of the ducts (from Fig. 16.12) and, in turn, the new curve shifted by $+22$ dB, to account for the pressure drop coefficient ξ (from Fig. 16.13). Figure 16.14 illustrates the procedure. The resultant curve peaks to $L_w = 52$ dB in the 630-Hz one-third-octave band. A margin of ± 5 dB should be allowed to account for the uncertainty of the procedure.

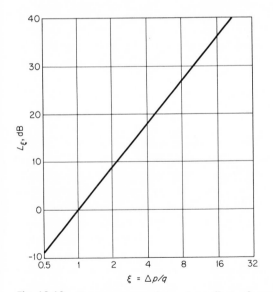

Fig. 16.13 Nomogram to determine effect of pressure drop coefficient $L_\xi = 30 \log (\Delta p/q)$, where $q = 0.5\ \rho u^2$, and Δp is the pressure drop across the diffuser.

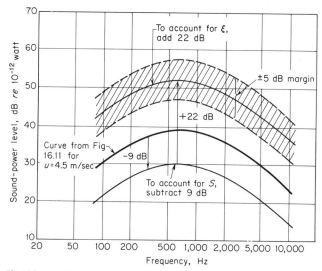

Fig. 16.14 Example of grid-noise spectrum calculation (see Example 16.3). Note that the heavy solid line is for $u = 4.5$ m/sec; $\xi = 1$; and $S = 1$ m². To this are added the shifts for the area of duct, 10 log S, and for the pressure drop coefficient 30 log ξ.

16.6   Air-valve Noise

A valve generally controls the flow of a gas through a pipe. In the case of airflow, the control is achieved by introducing a constriction into the pipe. Valves are usually operated with a pressure ratio (ratio of upstream to downstream pressures) sufficient to make the Mach number of the flow at the exit equal to unity. In symbols, $u = c$. Under this condition the flow is said to be *choked,* and the mass flow reaches a maximum value.

Two noise producing mechanisms are associated with choked flow. The first mechanism is the *turbulent mixing process* downstream from the valve (see Fig. 16.2). This mechanism is associated with quadrupole radiation. The second mechanism results from *interaction between the turbulence and the complex flow field* that forms downstream from the valve. This mechanism is often designated as *shock noise* and is not associated with the elemental source types. For valves with pressure ratios less than about 3, it is necessary to consider the noise associated with both mechanisms. For valves with pressure ratios greater than 3, shock noise predominates, and the noise from the turbulent mixing mechanism may be neglected.

Calculation Procedure. To determine the noise of air valves, first calculate the mechanical stream power 0.5 mc^2, where m is the mass flow of the air, in kg/sec, passing through the valve, and c is the speed of sound in air at the valve. This quantity has the units of power (watts). Then determine the power levels for the turbulent mixing and shock noise mechanisms as in the next two paragraphs. This calculation procedure is demonstrated in Ex. 16.5 in the next section.

Turbulent Mixing Mechanism. For the first (turbulent mixing) mechanism, assuming ambient air conditions, the ratio of the overall sound power radiated (watts) to the quantity 0.5 mc^2 (watts) is approximately 6×10^{-5} (see Fig. 16.3). The sound-power spectrum has a "haystack" shape, peaking at frequency f_p. This peak frequency is

$$f_p = \frac{S_0 c}{D} \quad \text{Hz} \tag{16.14}$$

where D = narrowest cross dimension of the valve opening, m
$\quad S_0$ = peak Strouhal number, dimensionless

Figure 16.15 shows a method for selecting the cross dimension D for a partially open simple valve.

The peak Strouhal number for turbulent mixing is $S_0 = 0.2$. The octave-band spectrum typical of a turbulent mixing process is identical to that of a free-jet process (see Fig. 16.5).

Shock Noise Mechanism. For the shock noise mechanism, the ratio of the overall sound power radiated to the mechanical stream power 0.5 mc^2 depends on the pressure ratio across the valve, as shown in Fig. 16.16. The peak Strouhal number S_0 is dependent on the pressure ratio, as shown in Fig. 16.17, and determines f_p, as before, from Eq. (16.14). The sound-pressure-level octave-band spectrum has a "haystack" shape that is slightly more peaked than that for the turbulent mixing noise, as shown in Fig. 16.18.

Top view through valve

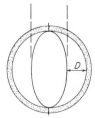

Cross-sectional view

Fig. 16.15 Selection of the cross-dimension D associated with the opening of a simple valve.

16.7 Transmission of Airflow Noise through Cylindrical Pipes

The engineering procedures presented so far in this chapter apply to ideal airflow noise sources. If the flow exhausts directly into the environment of concern, say, a test space, then the procedures are directly useful for estimating the noise levels in that medium. In many cases, however, the flow and the noise generating configuration, say, the valve, are contained within a cylindrical pipe. If we are interested in the noise power and spectrum radiated into

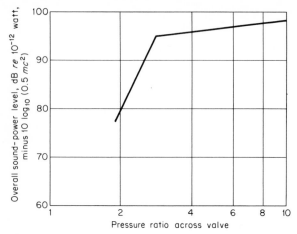

Fig. 16.16 Overall sound-power level of choked air valves for shock noise mechanism.

the space surrounding that pipe, we must know the noise reduction properties of the pipe.

The acoustic behavior of a pipe can be considered in two frequency regions. In the upper region, the noise reduction (or transmission-loss) properties of the pipe correspond to those of a flat panel of the same construction. Methods for estimating transmission loss of flat plates are given in Secs. 11.2 to 11.4, particularly in Sec. 11.2. In the lower frequency region the cylindrical stiffness of the pipe becomes important, changing the acoustic behavior (see Sec. 11.4.9). The frequency at which these two regions meet is called the *ring frequency*, which is the frequency at which the pipe circumference is equal to the wavelength of sound traveling through a sheet as a compressional wave. The ring frequency is given by

$$f_r = \frac{c_L}{D_p \pi} \quad \text{Hz} \tag{16.15}$$

where D_p = nominal diameter of the pipe, m

c_L = longitudinal speed of sound in metal = 5,100 m/sec for steel

In the frequency region below f_r, the pipe transmission loss depends on the ratio t/D_p, where t is the thickness of the pipe in meters. The transmission loss below f_r is found in two steps. First determine the "limiting transmission loss" from Fig. 16.19. Second, modifications are made to the limiting transmission loss in the frequency region below

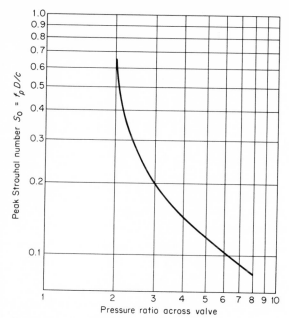

Fig. 16.17 Peak Strouhal number of choked air valves for shock noise mechanism.

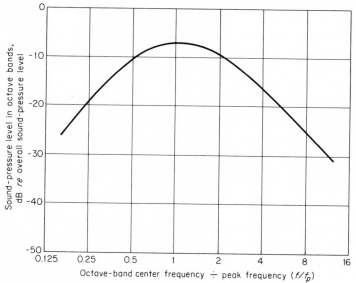

Fig. 16.18 Design octave-band-sound-level spectrum for choked air valves for shock noise mechanism.

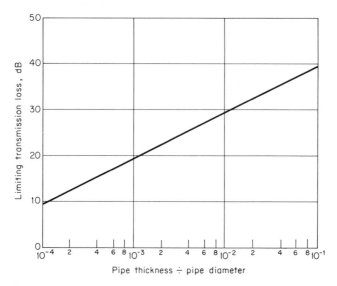

Fig. 16.19   Limiting transmission loss of cylindrical steel pipe below the ring frequency. The abscissa is t/D_p.

f_r as given in Table 16.2. The modified values, added algebraically to the limiting transmission loss, form the curve of transmission loss vs. frequency below f_r.

Example 16.4   An airflow source of noise is contained in a cylindrical steel pipe with a nominal diameter of $D_p = 0.3$ m (1 ft) and a wall thickness of $t = 6 \times 10^{-3}$ m (0.25 in.). Find the transmission loss as a function of frequency. SOLUTION: From Eq. (16.15) we determine the ring frequency as 5,400 Hz. Because the density of steel is 7.8×10^3 kg/m³ (490 lb/ft³), the surface mass of the pipe is 47 kg/m² (9.6 lb/ft²). *Above the ring frequency,* the transmission loss is determined by the product of surface mass times frequency, according to Fig. 11.5. For example, at a frequency of 12,000 Hz, this product is 5.6×10^5 Hz-kg/m², yielding a field-incidence transmission-loss value from Fig. 11.5 of 67 dB at 12,000 Hz. *Below the ring frequency,* the transmission loss is determined by the ratio of pipe thickness to pipe diameter. For the dimensions here, $t/D_p = 2 \times 10^{-2}$. From Fig. 16.19 the value of the limiting transmission

TABLE 16.2   Corrections for Cylindrical Pipe Wall Transmission Loss below the Ring Frequency

$\dfrac{f}{f_R}$	0.025	0.05	0.1	0.2	0.3	0.4	0.5	0.6	0.7	0.8
Correction, dB	−6	−5	−4	−3	−2	−2	−2	−2	−2	−3

loss is 32 dB. Using the values in Table 16.2 to modify the limiting transmission loss, we obtain the estimated transmission loss below the ring frequency shown in Fig. 16.20.

Because of the multiple reflections that occur when a noise source is located inside a large pipe, we assume, in making engineering estimates, that the resulting sound field is at "field incidence" (see Chap. 11). Let us now combine, by example, the material on valve noise with the material on pipe transmission to estimate the noise produced by an air valve in a cylindrical pipe.

Example 16.5 A choked air valve operates at a pressure ratio of 3, with a mass flow of $m = 8$ kg/sec (17.6 lb/sec) and a characteristic open cross dimension of $D = 0.1$ m (4 in.). The valve is contained in the cylindrical steel pipe of the previous example [$D_p = 0.3$-m (1-ft) diameter and wall thickness $t = 6 \times 10^{-3}$ m (0.25 in.)]. Estimate the sound-power levels radiated outside the pipe.
SOLUTION: Calculate the mechanical stream power $= 0.5\ mc^2 = 4.7 \times 10^5$ watts. Multiply this quantity by 6×10^{-5} (see Fig. 16.3 for $u = c$ and air at normal condition) to obtain the estimated *power of turbulent mixing noise* $= 28.2$ watts, which corresponds to an overall sound-power level $= 134.5$ dB *re* 10^{-12} watt. With the peak Strouhal number equal to 0.2 (Fig. 16.17), we obtain a spectrum peak of 680 Hz [see Eq. (16.14)]. The spectrum shape for the sound-power level is given by Fig. 16.5 (see Col. 2 of the table below).
For a pressure ratio of 3, Fig. 16.16 indicates that the overall shock noise sound-power level $= 10 \log (0.5\ mc^2) + 95$ dB $= 152$ dB. Figure 16.17 gives the peak Strouhal number as 0.2, so that f_p is the same as for the turbulent mixing noise. The spectrum shape for the shock noise sound-power level is given by Fig. 16.18. The two spectra and the total levels are tabulated here.

Octave-band mid-frequency, Hz	Turbulent mixing band levels L_W, dB	Shock noise band levels L_W, dB	Total L_W, dB
58	112	122	122
115	120	133	133
330	127	142	142
660	130	145	145
1,320	128	143	143
2,640	125	136	136
5,280	122	127	127.5
10,560	117	119	121

For this example, the shock noise predominates in all frequency bands, and there is a significant contribution from the mixing noise only in the two highest bands. Because there is no standard filter set with these mid-fre-

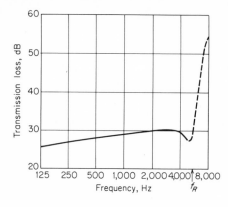

Fig. 16.20 Example of transmission loss through a cylindrical steel pipe with diameter of 0.3 m (12 in.) and wall thickness of 0.006 m (0.25 in.) (see Example 16.4).

quencies, the points are displaced when plotted on standard frequency paper. The resultant sound-power-level spectrum inside the pipe is shown by the upper line in Fig. 16.21. Using the transmission loss for the pipe from Fig. 16.20, we obtain the estimated sound-power level of the radiated sound outside the pipe (lower line of Fig. 16.21).

16.8 Scaling Procedures Based on Dynamic Similarity

The estimation procedures that have been presented in this chapter have utilized several nondimensional quantities, such as ratio of sound power to mechanical power and peak Strouhal number. Inherent in these procedures is a general scaling principle that is useful in handling dynamically similar systems.

Two systems are dynamically similar if they are geometrically similar and if the Mach number (ratio of flow speed to sound speed) and the Reynolds number (ratio of flow speed times a length dimension to kinematic viscosity, i.e., $\text{Re} = ul/\nu$) are the same. The condition of Mach-number similarity ensures that effects of compressibility are accounted for, and the condition of Reynolds-number similarity ensures that the flow is dynamically similar everywhere at corresponding locations. The radiated noise can then be measured on the model and scaled directly to the full-scale situation. The procedure requires only that data be taken on the scale model at the scaled point of interest, and that the data be analyzed in constant-percentage frequency bands (such as octaves or one-third octaves). The sound-pressure-level estimate for the full-scale configuration will be the same as the sound-pressure level measured on the model, with the frequency scale for the model tests divided by the scale factor of the model.

While it is relatively easy to maintain geometric and Mach-number

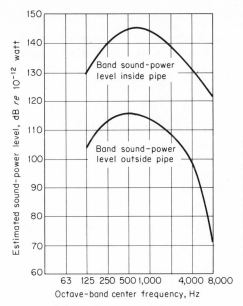

Fig. 16.21 Example of noise calculation for choked air flow through a valve in a cylindrical steel pipe. The upper curve is the calculated sound-power level inside the pipe. After subtracting the transmission loss from Fig. 16.19, the sound-power level outside the pipe is given by the lower curve.

similarity, in many cases Reynolds-number similarity cannot be achieved. Suppose we have a one-fourth-scaled model. Obtaining an identical Reynolds number (for the same kinematic viscosity) requires an increase of flow speed by a factor of 4. This change in flow speed could change the Mach number to an extent that would grossly alter the acoustical phenomena. Such a change is readily conceivable if the flow speed had to be changed from a subsonic to a supersonic Mach number. To maintain the Reynolds number one might use a different test gas, or change the density, either of which would change the kinematic viscosity. For some situations, such as a subsonic jet, the noise generated is relatively insensitive to Reynolds number, so that the same gas may be used for the jet or for the full-scale device. Stated simply, one must be confident that the essential flow features are roughly the same for the model and full-scale situations.

REFERENCES

1. M. J. Lighthill, "On Sound Generated Aerodynamically," *Proc. Roy. Soc. (London)*, vol. A211, p. 564, 1952.
2. N. Curle, "The Influence of Solid Boundaries upon Aerodynamic Sound," Proc. Roy. Soc. (London), vol. A231, pp. 501–514, 1955.
3. H. H. Heller and S. E. Widnall, "Sound Radiation from Rigid Flow Spoilers Correlated with Fluctuating Forces," *J. Acoust. Soc. Am.*, vol. 47, pp. 924–936, 1970.
4. P. J. F. Clark and H. S. Ribner, "Direct Correlation of Fluctuating Lift with Radiated Sound for an Airfoil in Turbulent Flow," *J. Acoust. Soc. Am.*, vol. 46, pp. 802–805, 1969.

5. E. Y. Yudin, "The Acoustic Power of the Noise Created by Airduct Elements," *Sov. Phys.-Acoust.*, vol. 1, pp. 383–398, 1955 (English).
6. C. Gordon, "Spoiler-generated flow noise," Part I, "The Experiment," and Part II, "Results," *J. Acoust. Soc. Am.*, vol. 43, pp. 1041–1048, 1968, and vol. 45, pp. 214–223, 1969.
7. M. Hubert, "Untersuchungen über Geräusche durchströmter Gitter," doctoral dissertation, Berlin Tech. Univ., 1970.

Chapter Seventeen

Damage-risk Criteria for Hearing

ARAM GLORIG, M.D. *

There is no straightforward, simple statement that will suffice as a damage-risk criterion for any biological hazard. Certain arbitrary assumptions and compromises must be agreed upon and accepted before proposed criteria can be established. Damage-risk criteria for hearing are no exception. Many disciplines outside of biology are involved. Many value judgments must be made. Furthermore, since the hazard affects so many and the cost could be so great, these judgments must be made by the entire interested community on the basis of information arising from medical, legal, sociological, and economic factors.

17.1 Definition of Terms

Audiometer, discrete-frequency — an instrument for determining the auditory threshold for discrete pure tones, usually 250, 500, 1,000, 2,000, 3,000, 4,000, 6,000, and 8,000 Hz. Measurements are sometimes also made at 125 and 1,500 Hz. At each test frequency, an audiometer

* Director, Callier Hearing and Speech Center, Dallas, Tex.

zero has been established by international agreement, which roughly corresponds to the average pure-tone auditory threshold for a population of young persons, with little exposure to high noise levels.

Auditory threshold — the sound-pressure level of the minimal acoustic signal that evokes an auditory sensation for a specified fraction of the number of times the signal is presented to the ear.   The auditory threshold for spondaic words, for example, is defined as the minimal sound-pressure level at which a subject can repeat correctly 50% of the words presented to him.   Auditory thresholds are different for different stimuli, and the particular stimulus used must be identified. Typical stimuli are pure tones, warbled tones, bands of noise, and words.

Auditory threshold shift — a measured change in the auditory threshold, usually of an individual, from some threshold determined at a prior time.   It may be temporary or permanent.   A *temporary threshold shift* recovers in a few seconds up to a few days.   A *permanent threshold shift* usually shows no recovery after 2 to 3 weeks.

Damage-risk criteria — specify limitation on personnel noise exposures which, if not exceeded, ensure that the risk of hearing damage to a group is acceptably small.

Hazardous noise — is any sound capable of causing permanent damage to the sense of hearing of human beings.

Hearing level — the number of decibels by which some quantity related to sound or hearing is above (+) or below (−) the standardized audiometric zero at a specified frequency.   For example, one can quantify speech, or noise, or a threshold of hearing at a particular frequency in terms of its "hearing level."

Hearing loss — a hearing level in decibels that relates specifically to a raised (+) permanent auditory threshold for an individual.   For example, if a particular person's auditory threshold at 1,000 Hz is 40 dB above audiometric zero at that frequency, he has a "hearing loss" of 40 dB.

Hearing handicap (hearing impairment) — exists whenever, for an individual, the average of his hearing losses at 500, 1,000, and 2,000 Hz exceeds 25 dB (as measured with an audiometer conforming to International Standards Organization, ISO/R-389-1964).*   This is also called "hearing impairment for communication by conversational speech."

Degree of hearing handicap — (percentage impairment of hearing) begins from an average hearing loss of 25 dB (zero hearing handicap) and increases at the rate of 1.5 percentage points for each decibel of average hearing loss until a loss of 92 dB (or 100% impairment) is achieved.

* If the ASA-1951 Audiometer Reference Level is used, a hearing handicap of 25 dB (ISO) is about 15 dB (ASA).

Percentage risk—defined as the difference between the percentage of people with a hearing handicap in a noise exposed group and the percentage of people with a hearing handicap in a non-noise exposed (but otherwise equivalent) group.

Noise level, A-scale weighting (unit dBA)—the reading obtained when measuring noise to determine hearing hazard by a standard sound-level meter, with the A-weighting network switched into the electrical circuit. The meter response is set to "slow." In general, the meter is observed for about 5 sec, and the value observed (averaging any fluctuations) is recorded as the A-scale noise level in dBA. If the fluctuations exceed a range of 6 dBA, use as the average a value that is only 3 dBA below the peak of the fluctuations. A measurement is usually made at the position of an exposed ear, without the person present.

17.2 Need

Various studies reported by Gallo and Glorig,[1] Baughn,[2] and Robinson[3] show that large percentages of groups exposed to loud noise over long periods of time sustain permanent auditory threshold shifts (see Fig. 17.1). The shifts first occur mainly in the auditory frequencies above 2,000 Hz, but in time affect the thresholds at 500, 1,000, and 2,000 Hz, particularly at 2,000 Hz. If all existing cases of permanent threshold shifts in this country were to be converted into compensation costs to

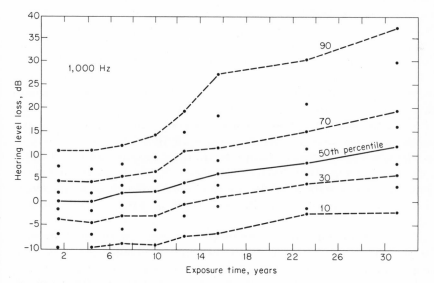

Fig. 17.1 Distribution of hearing levels at 1,000 Hz as a function of exposure years in the indicated percentiles. (*From Gallo and Glorig.*[1])

industry, the total amount of compensation involved would probably be staggering. Likewise, the deleterious socioeconomic effects on people with these hearing deficiencies, involving such large numbers, are also probably very large.

17.3   What Is Meant by Damage Risk?

To define "damage risk" it is necessary to have prior judgments on "damage" and "risk." For example, if we were to assume that everyone's hearing must be protected when at work against any change in threshold at any audible frequency, the task of protection would literally be impossible because non-occupational-noise causes of hearing loss, such as age, disease, ambient noise, and individual differences, cannot be determined separately. Consequently, it is essential that workable definitions of "damage" and "risk" should be agreed upon by responsible groups (medicine, science, regulatory agencies).

If one looks at those daily functions related to hearing that are necessary to economic and social survival in our complex civilization, it seems reasonable to conclude that communication by speech is by far the most important. Thus, present-day damage-risk criteria have been chosen to protect exposed groups of workers against loss of hearing for everyday speech, which loss is, by agreement, related specifically to the arithmetic average of the hearing losses for an individual at 500, 1,000, and 2,000 Hz.[4]

The dynamic range of everyday speech varies between a whisper and a loud shout. For a typical person, the sound-pressure levels L_p of everyday speech in sentence form vary between roughly 55 and 65 dB when the talker and listener are comfortably conversing at 3 to 6 ft apart. Particular syllables will range in level above and below this average by about 15 dB. This range of average L_p's corresponds to about 35 to 45 dB hearing level, averaged for the frequencies 500, 1,000, and 2,000 Hz. For practical purposes, a hearing loss (averaged at the same frequencies) not in excess of 25 dB will permit speech to be understood satisfactorily. Above an average hearing loss of 25 dB, a person's hearing is said to be *handicapped or damaged.* As the loss of hearing increases above 25 dB, the *degree of handicap* increases.[5] Hearing handicap and degree of hearing handicap are defined in the first section.

Risk. For the purpose of hearing-conversation criteria, risk is defined as the difference between the percentage of people with a hearing handicap in a noise exposed group and the percentage of people with a hearing handicap in a non-noise exposed (but otherwise equivalent) group. Table 17.1 shows expected percentage risk as a function of continuous sound level at work in dBA (measured on the A-scale of an American National Standards Institute standardized sound-level meter) and years of exposure (the age of a person is assumed to be equal to years of exposure plus 20 years). For each noise exposure level, two rows of percentages are given in Table 17.1. The upper row, labeled "total," gives the percentage of people who have hearing impairment due to the combination of noise *and age,* as a function of years of exposure. The lower row, labeled "due to noise," takes out the effects of aging (presbycusis), as discussed below.

It is obvious from Table 17.1 that if we are to assume no risk of hearing handicap whatsoever, the noise exposure levels must be 80 dBA or lower. It is also known that attaining such levels in industry with machines as presently designed is economically impractical. Thus, the total community must accept a compromise and assume a certain risk that is consistent with that compromise.[6]

Although risk is based on hearing handicap, it must be emphasized

TABLE 17.1 Percentage Risk of Developing a Hearing Handicap

Age, years		20	25	30	35	40	45	50	55	60	65
Exposure, years (*re* age 20)		0	5	10	15	20	25	30	35	40	45
80	Total	0.7	1.0	1.3	2.0	3.1	4.9	7.7	13.5	24.0	40.0
	Due to noise				No increase in risk at this level of exposure						
85	Total	0.7	2.0	3.9	6.0	8.1	11.0	14.2	21.5	32.0	46.5
	Due to noise	0.0	1.0	2.6	4.0	5.0	6.1	6.5	8.0	8.0	6.5
90	Total	0.7	4.0	7.9	12.0	15.0	18.3	23.3	31.0	42.0	54.5
	Due to noise	0.0	3.0	6.6	10.0	11.9	13.4	15.6	17.5	18.0	14.5
95	Total	0.7	6.7	13.6	20.2	24.5	29.0	34.4	41.8	52.0	64.0
	Due to noise	0.0	5.7	12.3	18.2	21.4	24.1	26.7	28.3	28.0	24.0
100	Total	0.7	10.0	22.0	32.0	39.0	43.0	48.5	55.0	64.0	75.0
	Due to noise	0.0	9.0	20.7	30.0	35.9	38.1	40.8	41.5	40.0	35.0
105	Total	0.7	14.2	33.0	46.0	53.0	59.0	65.5	71.0	78.0	84.5
	Due to noise	0.0	13.2	31.7	44.0	49.9	54.1	57.8	57.5	54.0	44.5
110	Total	0.7	20.0	47.5	63.0	71.5	78.0	81.5	85.0	88.0	91.5
	Due to noise	0.0	19.0	46.2	61.0	68.4	73.1	73.8	71.5	64.0	51.5
115	Total	0.7	27.0	62.5	81.0	87.0	91.0	92.0	93.0	94.0	95.0
	Due to noise	0.0	26.0	61.2	79.0	83.9	86.1	84.3	89.5	70.0	55.0

Exposure level in dBA

that there are degrees of hearing handicap extending from very little at 25 dB average hearing loss to total hearing handicap at 92 dB average hearing loss. Therefore, even though by definition a specified number of people will sustain a hearing handicap, the *degree* of handicap is dependent on the number of decibels average hearing loss above 25 dB. If, in industry, risk were to be based on the percentage of people with *any* hearing handicap, corresponding noise levels that would affect only a small percentage of people should be chosen as criteria for permissible exposure. On the other hand, if risk is to be related to compensation costs, two factors are involved, the percentage of people handicapped and the degree of handicap expected from the exposure (see Table 17.1).

A different criterion for avoiding hearing handicap may be desired for exposures to non-occupational-noise conditions, such as encountered in recreation equipment, home power equipment, and the like. Cohen, Anticaglia, and Jones [7] believe that acceptable noise levels under these conditions should be 15 dBA lower than those acceptable in industry for the same exposure times. They argue that a worker in industry receives a paycheck for assuming certain job risks and can be compensated should an occupation-related hearing handicap occur. By contrast, in off-job situations in which persons may be subjected to hearing-damage risk no direct benefits are received from the noise-maker.

17.4   Measurement of Noise Exposure

Table 17.1 shows that the length of exposure of a person to noise is at least as important as the level itself in producing hearing handicap. Thus, it is imperative that we keep in mind the concept of *noise exposure*. Also, the noise levels to which a noise exposed population is subjected usually fluctuate with time while the resulting noise induced hearing loss is a gradual process accumulating over years.

Several methods are available for measuring noise levels. These include (1) measurement of sound-pressure levels in octave bands, (2) determination of sound level A, and (3) determination of total sound-pressure level (sound level C).

When broad interest in noise induced hearing loss first began, noise level measurements were usually made in octave bands. With more research, it became apparent that the average of the sound-pressure levels in the higher bands only correlated well with noise induced hearing loss. The conclusion was reached that a single number taken with a meter that discriminates against energy content at low frequencies is adequate. Such a number is the A-weighted noise level in dBA.

Robinson[3] concludes that the magnitude of error in hearing-loss calculations is on the order of ±2 dB when sound level A is used to specify the energy even when rising and falling spectra are compared. Similar findings have been reported by Baughn[8] in a study where over 600 spectra have been analyzed. Consequently, sound level A has been accepted for measurements concerning conservation criteria.

17.5 Hearing-damage Theories

The most reliable data relating hearing damage to noise exposure were obtained in situations (weaving mills, paper mills, etc.) where workers were exposed to the same noise levels every day for years of time. On the other hand, the relation between hearing handicap and part-day or intermittent exposures is much harder to quantify. As a result, theories are relied on to set limits for shorter daily exposures. Two theories have been advanced. They do not yield the same results either for continuous or for intermittent noises. However, one seems better for continuous noise and the other for interrupted noise.

The *equal-energy theory*[3,9] reasons that hearing-damage risk is determined by the total amount of noise energy to which the ear is exposed each day (8 hr). Thus, for half-day exposure (4 hr), regardless of how the exposure is distributed throughout the 8-hr period, 3-dB-greater noise levels are permissible for the same risk. Similarly, each halving of energy permits adding 3 dB to the noise levels permissible. The equal-energy theory has been verified for continuous noise, i.e., a single long exposure each day.

The *equal temporary effect theory*[10] says that hearing-damage risk is related to the temporary loss of hearing (temporary threshold shift, called TTS) in young ears that results from noise exposure.

This theory is based on the observation that those noise exposures that ultimately produce permanent hearing loss also produce temporary hearing loss in young ears. The converse is also true.

The TTS studies indicate that intermittent noise is much less harmful than steady noise. As proof, Sataloff, Vassallo, and Mendulse[11] found that iron ore miners exposed intermittently to intense rock drill noise underground showed no greater incidence of impaired hearing than do workers exposed to steady noise 20 dB lower. Other investigations also agree with this conclusion.[12]

17.6 Damage-risk Criteria

The most widely accepted damage-risk criteria for avoiding noise induced hearing impairment were prepared by the National Academy of

TABLE 17.2 Permissible Average Noise Levels, in dBA, for Steady and Interrupted Noise Exposures

To use this table, select the column headed by the number of times the noise occurs per 8-hr workday, read down to the average A-weighted sound level of the noise, and locate directly to the left in the first column the total duration of dangerous noise for any 24-hr period.

Cumulative exposure	Number of noise interval exposures per 8-hr workday						
	1	3	7	15	35	75	150 or more
8 hr	90						
6 hr	91	92	93	94	94	94	94
4 hr	93	94	95	96	98	99	100
2 hr	96	98	100	103	106	109	112
1 hr	99	102	105	109	114	(115)	
30 min	102	106	110	114	(115)		
15 min	105	110	115				
8 min	108	115					
4 min	111						

NOTE: *Noise intervals* are periods of time during which the noise remains above 80 dBA. The *average noise level* is the mean A-weighted SPL during this time. An *interruption* of exposure occurs when the noise drops below 80 dBA for either (1) more than 5 min, or (2) a length of time equal to one-fifth of the average noise interval duration.

Sciences — National Research Council, Committee on Hearing, Bioacoustics and Biomechanics, generally referred to as CHABA.[13] The CHABA method is valid for all-day exposures to continuous or intermittent noise that are repeated almost every day over a period of 10 years or more. Exposures to noise are considered acceptable if over 90% of any group so exposed suffers no hearing handicap,* i.e., reduction in ability to understand speech, as defined above.

For intermittent noises, it is assumed that the spacing and the duration of noise bursts both remain constant throughout the day. The method is not intended to apply to (1) noises of impacts, (2) brief noises lasting less than a second, (3) noises composed of several pure tones, or (4) noises too variable to be measured accurately with a standard sound-level meter.[14]

The CHABA conclusions state that significant hearing-damage risk for workers begins when continuous sound levels throughout a working day are 90 dBA or more. When the noise is intermittent throughout the day, the number of times per day that it occurs must be known. With

---

* Apparently, CHABA used a table different from Table 17.1, as the latter would indicate protection for approximately 85% of any group so exposed to 90 dBA. Table 17.1 has been adopted in American National Standards Institute Document ANSI S3.40.

this information, the total duration for each noise level (in dBA) that is permissible is found from Table 17.2.

The values in Table 17.2 for a *single* exposure are in conformance with the equal-energy relation (3 dB per halving of exposure time). The permitted levels with about 7 noise intervals are about 5 dB per halving of exposure time; moreover, considerably higher noise levels are permitted when, as for miners, there are 35 or more intervals of noise per day, as in coal mines.

The exposure limits of Table 17.1 apply to noises where the C and A levels differ by less than about 5 dB. If the value of the C level minus the A level is 5 dB or more, the numbers in Table 17.2 are too strict. This bias can be removed by reducing the levels in Table 17.2 by the number of decibels shown in Table 17.3, before determining the exposure limit from Table 17.2.[14] The C minus A deduction shown in Table 17.3 is small for most common noises and usually can be ignored.

17.7 Noise Exposure Rating [15]

The severity of a noise exposure is indicated by a *noise exposure rating,* defined as the ratio of the duration of a hazardous noise to that allowed by Table 17.2. A noise exposure is considered acceptable for all values of exposure rating that do not exceed unity. The damage risk to hearing increases as the noise exposure rating becomes progressively greater than unity.

If a noise exposure consists of two or more successive noise conditions

TABLE 17.3 Deductions for Types of Noise Where the C-scale Reading of Noise Level Exceeds the A-scale Reading by 5 or More Decibels

To use the table, find the actual A-scale average noise level in the first column and read horizontally in the column headed by the C−A differences the number of decibels to be subtracted from the actual A-scale level before entering Table 17.2 to find the cumulative exposure time limit.

A-scale noise level, dBA	C − A				
	5	10	15	20	25
90	0	1	1	1	1
100	2	4	4	5	6
110	4	6	8	9	10
120	6	9	12		

having levels above 89 dBA, the *noise exposure rating for the whole exposure is the sum of the noise exposure ratings for the individual noise conditions;* i.e.,

$$\frac{C_1}{T_1} + \frac{C_2}{T_2} + \frac{C_3}{T_3} + \cdots = \text{Noise exposure rating}$$

where C_n = total time of exposure at an actual particular level

T_n = total time of exposure permitted by Table 17.2 (including the corrections of Table 17.3) at that particular level

For example, consider the hazard of an exposure to 91 dBA all morning and 100 dBA of 10-sec duration repeated twice a minute all afternoon. From Table 17.2, we find that 91 dBA is allowed for 6 hr, so that only 4 hr at 91 dBA gives an exposure rating of 0.67. The 100 dBA occurs 480 times for a total of 80 min. From Table 17.2, we see that 240 min is allowed, so that the exposure rating is 0.33. The exposure rating for the whole exposure totals 1.0, which is just acceptable. Noise levels below 90 dBA are not considered in these calculations.

17.8   Walsh-Healey Public Contracts Act (The ACGIH Recommendations)

The American Conference of Governmental Industrial Hygienists (ACGIH) adopted (1969) a simplified set of permissible noise levels (in dBA) vs. hours of exposure during each 8-hr day as shown in Table 17.4. This table corresponds closely to that column of Table 17.2 headed 7 noise interval exposures per 8-hr day. It embodies the implicit presumption that practically all noise exposures are interrupted at least a few times a day by meal or rest periods, machinery stoppages, and so on.

TABLE 17.4 Permissible Noise Exposures

Duration per day, hr	Noise level dBA slow response
8	90
6	92
4	95
3	97
2	100
$1\frac{1}{2}$	102
1	105
$\frac{1}{2}$	110
$\frac{1}{4}$ or less	115 max.

The ACGIH also adopted the "noise exposure rating" described in Sec. 17.7.

The ACGIH recommendations were the basis for the American government's Walsh-Healey regulation.

Walsh-Healey Public Contracts Act.[16] The Bureau of Labor Standards, U.S. Department of Labor, in May, 1969, added a safety regulation on industrial noise exposure to the Walsh-Healey Act. This regulation states that any manufacturer or fabricator whose contract with the federal government exceeds $10,000 must initiate feasible engineering or administrative controls to reduce noise exposure levels to within those shown in Table 17.4.

According to the regulation, when the daily noise exposure for workers is composed of two or more periods of noise exposure of different levels, their combined effect should be considered, rather than the individual effect of each. If the sum of the fractions $C_1/T_1 + C_2/T_2 + \cdots + C_n/T_n$ (i.e., the noise exposure rating) exceeds unity, then the mixed exposure should be considered to exceed the limit value. C_n indicates the total time of exposure at a specified noise level, and T_n indicates the total time of exposure permitted at that level.

The Bureau of Labor Standards regulation also states that when the equivalent noise level exceeds 90 dBA, i.e., the noise exposure rating exceeds unity, if administrative or engineering controls fail to reduce the equivalent noise level exposure to that shown in Table 17.4, personnel (ear) protective devices must be provided and worn to reduce noise levels within the levels of the table.

When noise levels are available that were determined by octave-band analyses, the equivalent A-weighted sound level may be determined from Fig. 17.2 as follows: The octave-band sound-pressure levels are converted to the equivalent A-weighted sound level by plotting them on Fig. 17.2 and noting the A-weighted sound level corresponding to the point of highest penetration into the sound-level contours. This equivalent A-weighted sound level, which may differ from the actual A-weighted sound level of noise, may then be used to determine noise exposure limits from Table 17.4. Figure 17.2 also incorporates the corrections of Table 17.3 for those noises where the C-scale levels exceed the A-scale levels significantly.

The Bureau of Labor Standards emphasizes that the *purpose of the Walsh-Healey regulations is to ensure that a worker's ears are protected.* Thus, even though the noise levels in the regulation are complied with, if the hearing of an appreciable number of workers is being damaged, the company will be held in violation of the act. On the other hand, if the noise levels in a plant are above those shown as permissible, and a well-

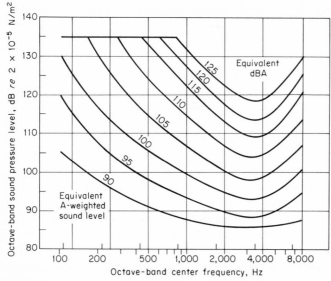

Fig. 17.2 Bureau of Labor Standards contours for determining equivalent A-weighted sound level.

executed audiometric program shows that no damage occurs, the company will not be held in violation.

The meaning of *administrative controls* is that the noise exposure rating (see Sec. 17.7) will be reduced to unity or less by controlling the lengths of exposure of a worker to hazardous noise. *Engineering controls* means control of the noise levels at the workers' operating positions to achieve a noise exposure rating of unity or less. The Bureau of Labor Standards advises that these two means of control must be fully exploited before personnel protective devices are utilized to satisfy the regulation.

The Intersociety Committee on Guidelines for Noise Exposure Control, composed of representatives from the American Industrial Hygiene Association, the Industrial Medical Association, the American Conference of Governmental Industrial Hygienists, the American Academy of Occupational Medicine, and the American Academy of Opthalmology and Otolaryngology, has also proposed a guideline document quite similar to the recommendations of CHABA (i.e., to Tables 17.1 and 17.2 and Fig. 17.2).

17.9 Measurement of Intermittent Noise

A major difficulty in relating hearing-damage risk to a given noise situation is that allowances for intermittency can only be approximated since it has not before now been possible to account for the innumerable variables of time and level in an actual industrial environment. A noise hazard meter [17, 18] has been proposed which measures an intermittent noise in such a way as to yield results consistent with laboratory studies of the effects of noise on temporary threshold shift.[10, 19] This meter, when commercially available, will greatly assist those responsible for prevention of hearing handicap in determining whether a noise hazard exists.

17.10   Criterion for Impulse Noise

Impulse noise is a sudden impulse of pressure like that created by gunfire, a balloon bursting, or a hammer striking an anvil. It may be a pure impulse or a reverberant impulse (see Fig. 17.3).

Criteria for impulse noise have been studied but no definitive standardization has been completed.[20] The Walsh-Healey regulation simply says that exposure to any *single* impulse noise should not exceed 140 dB, peak sound-pressure level. Two equipments for measuring impulse noise in accordance with this regulation are commonly accepted: (1) a calibrated storage cathode-ray oscilloscope, and (2) a true peak-reading meter (see Sec. 4.5 and Fig. 4.11). Both methods permit storage of the impulse for subsequent reading.

Most of the data for impulse noise is derived from studies of gunfire, an example that is probably not equivalent to all types of impulse noise. Many variables make hearing-conservation criteria for impulse noise difficult to formulate. Some of the variables are the rise and fall times of peak, repetition rate, environment, and amount of "ring." If we assume that the amount of energy that reaches the neural elements of the inner ear is the controlling factor producing hearing loss, it is quite apparent that the effects of impulse noise would not be the same as steady noise. Even though the peak levels are high, the exposure time is so short that the energy per peak is very small. Consequently, any criterion must account for peak duration and number of peaks per unit of time. Furthermore, if one looks only at peak duration without accounting for "ring" duration, the criterion will be too high. Therefore, any criterion must consider both duration of peak and duration of "ring," if any is present.[20, 21]

A tentative proposal from CHABA, Working Group 57,[22] suggests a criterion for impulse noise based on Figs. 17.3 and 17.4. These figures provide for duration with and without ring and repetition rates. The features of this damage-risk criterion are as follows: (1) The maximum peak pressure permitted is 164 dB (flat scale) (without ear protection) for the shortest pulses (25 microseconds); (2) as the duration increases, the permitted peak pressure level decreases at a rate of 2 dB for each doubling of duration, dropping to a terminal level of 138 dB for AD durations of 200 to 1,000 milliseconds; and (3) a similar decrease takes place for AC durations, except that a terminal level of 152 dB is reached at about 1.5 milliseconds.

This damage-risk criterion of Fig. 17.3 represents the tolerance limits for 100 impulses distributed over a period of 4 min to several hours on any single day. It is assumed that the pulses reach the ear from a source to one side of the head. If the pulses arrive from a source in front or in back of the head, the curves can be shifted upward by 5 dB (5 dB

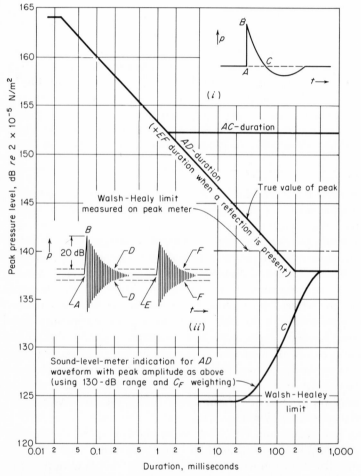

Fig. 17.3 Damage-risk criteria for impulse noise. (i) Pure impulse such as is created by gunfire in free space. (ii) Impulse consisting of a series of damped oscillations which may be followed by a reflected wave at a lower level. *(From Coles, Garinther, Hodge, and Rice,*[21] *and recent recommendations of the Intersociety Committee on Guidelines for Noise Exposure Control.)*

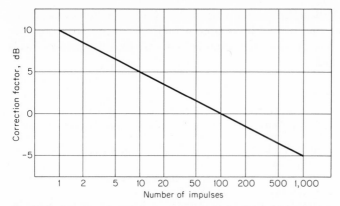

Fig. 17.4 Correction factor for rating impulse noise as a function of number of impulses per day. The correction is applied to the criterion obtained from Fig. 17.3. It requires a reduction in acceptable peak level of 5 dB for each factor of 10 increase in number of impulses.

more noise is permitted). A 5-dB change in the criterion is to be made for each tenfold change in number of impulses. Thus, for example, if 1,000 impulses occur within an 8-hr period, 5 dB less noise is permitted (see Fig. 17.4).

It appears from limited studies of hearing loss from impulse-type noise that this criterion is reasonable, and should serve until more data are available.[23]

At this writing, the Bureau of Labor Standards, Department of Labor, accepts, for Walsh-Healey purposes, a sound-level-meter reading in lieu of a bona fide impulse noise measurement. This alternative is based on recommendations of the Intersociety Committee on Guidelines for Noise Exposure Control. The measurement procedure is limited to impulses of greater than 5 millisecond duration. The standard sound-level meter is set on its 130 dB range and on "C_F" weighting. If the reading is 125 dBC or less, the impulse noise peak may be considered to be less than the 140 dB peak-sound-pressure level stated in the regulation. However, this technique is only useful for the evaluation of impulse noises whose durations lie in the range of 5 to about 30 milliseconds. As shown in Fig. 17.3, the C_F reading will produce a below-125 dB reading, as called for in the recommendation, for durations in the 5 to 30 millisecond range. Longer duration impulse noises, though at the 140 dB peak level established in the Walsh-Healey regulation, would produce C_F readings in excess of 125 dBC, and, hence, the sound-level meter would rate the impulse noise as unacceptable. For example, if a 200-millisecond, 140-dB-peak-level impulse noise were measured with the sound-level meter, the reading would be 134 dBC, or 9 dBC higher than the regulation allows. Thus, for durations above 30 milliseconds, the sound level meter might indicate unacceptable impulse noise levels, even though their actual peak sound-pressure levels were below 140 dB.

REFERENCES

1. R. Gallo and A. Glorig, "Permanent Threshold Shift Changes Produced by Noise Exposure and Aging," *Am. Ind. Hyg. Assoc. J.*, vol. 25, pp. 237–245, May–June, 1964.
2. William Baughn, "Noise Control—Percent of Population Protected," *Intern. Audiol.*, vol. 5, pp. 331–338, September, 1966.
3. D. W. Robinson, "The Relationships between Hearing Loss and Noise Exposure," *Nat. Phys. Lab., London, AERO Rept.* ac 32, July, 1968.
4. "Guide for the Evaluation of Hearing Impairment, A Report of the Committee on Conservation of Hearing," *Trans. Am. Acad. Ophthalmol. Otolaryngol.*, March–April, 1959.
5. H. Davis and S. Silverman, *Hearing and Deafness*, p. 176, Holt, 1962.
6. A. Glorig, W. Dixon Ward, and J. Nixon, "Damage Risk Criteria and Noise Induced Hearing Loss," *Arch. Otolaryngol.*, vol. 74, pp. 413–423, October, 1961.
7. A. Cohen, J. Anticaglia, and H. H. Jones, *Sound Vibration (USA)*, vol. 4, p. 20, November, 1970.
8. William Baughn, personal communication.
9. *Guide for Conservation of Hearing*, suppl. to the trans. of the American Academy of Ophthalmology and Otolaryngology, Rochester, Minnesota, 1969.
10. W. D. Ward, A. Glorig, and D. L. Sklor, "Temporary Threshold Shift from Octave Band Noise; Application to Damage Risk Criteria," *J. Acoust. Soc. Am.*, vol. 31, pp. 522–528, 1959.
11. J. Sataloff, L. Vassallo, and H. Mendulse, "Hearing Loss from Exposure to Interrupted Noise," *AMA Arch. Environ. Health*, vol. 18, pp. 972–981, 1969.
12. Other reports on hearing damage to miners agree with Sataloff et al.[11] Bláho and Slepicko (Czechoslovakia), *Czech. Hyg.*, vol. 12, pp. 521–527, 1967; Jönsson (Germany), *Deutsche Gesundh.-Wes.*, vol. 22, pp. 2286–2289, 1967; report by Derzay and Goodwin, U.S. Department of the Interior, Dec. 19, 1969.
13. CHABA, "Hazardous Exposure to Intermittent and Steady-State Noise," *J. Acoust. Soc. Am.*, vol. 39, pp. 451–464, 1966; and special report by CHABA to Bureau of Occupational Health, early 1970.
14. J. H. Botsford, "Current Trends in Hearing Damage Risk Criteria," *Sound Vibration (USA)*, vol. 4, pp. 16–19, April, 1970.
15. This method of rating exposure is based on a proposal of the Physical Agents Committees of the American Conference of Governmental Hygienists, 1968.
16. *Federal Register*, vol. 34, no. 96, May 20, 1969, Rule 50-204.10. Errata published in *Federal Register*, vol. 34, no. 96, July 15, 1969.
17. J. H. Botsford, "Noise Hazard Meter," *J. Acoust. Soc. Am.*, vol. 47, p. 90A, 1970.
18. L. M. Moser, "Noise Hazard Meter Based on Temporary Threshold Shift," *J. Acoust. Soc. Am.*, vol. 48, p. 105A, 1970.
19. W. D. Ward, "The Use of TTS in the Derivation of Damage Risk Criteria for Noise Exposure," *Intern. Audiol.*, vol. 5, pp. 309–316, 1966.
20. Aram Glorig et al., "Damage Risk Criteria and Noise Induced Hearing Loss," *Arch. Otolaryngol.*, vol. 74, pp. 413–423, October, 1961.
21. R. R. A. Coles, G. R. Garinther, D. C. Hodge, and C. G. Rice, "Hazardous Exposure to Impulse Noise," *J. Acoust. Soc. Am.*, vol. 43, pp. 336–343, 1968.

22. W. D. Ward (ed.), *Proposed Damage Risk Criterion for Impulse Noise (Gunfire) (U)*, Report of Working Group 57, NAS-NRC, Committee on Hearing, Bioacoustics, and Biomechanics, 2101 Constitution Ave., Washington, D.C., July, 1968.
23. D. C. Hodge and G. R. Garinther, "Validation of the Single-impulse Correction Factor of the CHABA Impulse-Noise DRC," *J. Acoust. Soc. Am.*, vol. 48, p. 97A, 1970.

Chapter Eighteen

Criteria for Noise and Vibration in Communities, Buildings, and Vehicles

LEO L. BERANEK

The cost of any noise control project is related directly to the criterion selected as the goal for the end result. In the past, the criteria for design have generally been under the control of the owners or the manufacturers of machinery, vehicles, or buildings. In a relatively few instances cities, states, or airports have adopted codes or regulations that specify maximum external factory noise levels or require adequate muffling of vehicles, or impose limits on the noise levels of aircraft as heard in communities in the immediate vicinity of airports.

Federal regulation of permissible noise levels in communities is likely to increase. The Federal Aviation Administration now requires that new designs of aircraft meet federal noise standards under prescribed conditions of operation. Existing aircraft are expected soon to have to be retrofitted to produce lower community noise levels. Maximum noise levels for other modes of transportation, including recreational equipment, are under consideration by the U.S. Department of Transportation.

The U.S. Department of Housing and Urban Development is planning to issue a policy statement and guidelines relating to noise associated with dwellings, appliances, and neighborhood noise sources. Individual states and cities in increasing numbers are adopting building codes, zoning restrictions, and noise limit regulations designed to control noise levels in residential living. The newly formed U.S. Environmental Protection Agency is expected to take an active part in regulating noise in this country. Similar activities are found in many nations of the world.

This chapter briefly discusses the present state of two subjects: (1) scales and procedures for rating noise with respect to its effects on people, and (2) workable criteria for the degree of noise and vibration control required in engineering design and city planning.[1-40]

18.1   Methods for Rating Noise and Vibration [3, 4]

Noise and vibration may affect people's lives in a number of ways. Noise or vibration may, if sufficiently loud or intense,
1. Damage hearing or health
2. Interfere with work tasks
3. Interfere with speech communication
4. Affect interroom privacy
5. Interfere with sleep
6. Cause annoyance

Physical measurements of noise and vibration can produce data that may be used to determine ratings specific to each of the six effects listed above. For noise, sound-pressure levels are generally measured in contiguous octave bands or one-third-octave bands within the audible frequency range. In addition, A-weighted and C-weighted levels are often measured. Equipment and techniques of measurement are discussed in Chap. 4. For vibration, acceleration levels are generally measured either at the principal frequencies of mechanical resonance of the structure in question, at frequencies of excitation, or in octave or one-third-octave bands covering a frequency range that extends from about 1 to 1,000 Hz.

Single-number ratings, that correlate with subjective judgments and that are suitable for use in particular situations (such as design specifications, codes, regulations, or evaluations), can be derived from measured sound and vibration band levels according to several accepted procedures.

The choice of a criterion for vibration or noise control design is not, however, without difficulty. Each human activity is influenced differently by noise. For example, loud noise as an expression of emotion or excitement is a desirable factor among those at a football game, though perhaps not for the neighbors. Sleep may be affected by only slight sounds if they carry information, such as the implication of danger. By contrast, the noise requirements for interroom privacy and intraroom speech communication can be stated more precisely.

Underlying the whole subject of the response of human beings to noise are the two important facts of substantial differences among people and the fluctuating nature of most noises. For example, the range in normal voice levels of a random selection of people is likely to be of the order of 5 to 10 dB. As we shall show later, some people are hardly bothered by noise at all, while others react to almost any noise that they do not make themselves. Some noises, such as ventilation noise, are quite steady, while others, such as nearby road traffic or aircraft flyovers, may fluctuate over a wide range of intensities. Thus, any criterion for design must represent the result of a statistical evaluation of how a sample of people, involved in a particular mode of behavior, react to noise or vibration (or both) in a given situation.

Because people's response to noise and vibration has received widespread attention in the United States only relatively recently, it is not surprising that the many different constructs that have evolved for estimating the several aspects of human response from physical data have not yet been reduced to a few that have wide acceptance. Analyses of many of these alternatives are given in Refs. 3 to 5. In this chapter, the author presents those rating procedures that he has found most useful for gauging people's response to noise and vibration. First, these procedures are named and briefly described. Immediately thereafter, they are presented in detail.

1. *Speech interference level.*[1-5] This procedure rates steady noise according to its ability to interfere with conversation between two people in an environment free of nearby reflecting surfaces that might strengthen the talkers' voices. It is well founded on the science of speech communication.

2. *Loudness and loudness level.*[6-8] Originally loudness (or its logarithm, loudness level) was a means for rating a noise based on the judgment of a person or a listening jury. Nowadays it is almost always determined by calculation. Thus, the procedure for calculating loudness rates the relative loudness of a noise, as it would be judged by a population of listeners. The method described here has been standardized both in the United States and internationally. An alternate calculation procedure called *perceived noise level,*[5] now widely used in the aerospace industry, is, for steady sounds, a variant of loudness level.

3. *NC or PNC rating curves.*[2,5,9,10] The NC, or noise criteria, curves have found extensive use in the rating of noises in buildings, including the rating of air-conditioning noises. The PNC, or preferred noise

criteria, curves are a recent modification that corrects several deficiencies in the older NC curves. The two ratings differ from the two procedures listed above, in that a single rating number (e.g., PNC-35) is assigned to a *curve* of octave-band sound-pressure levels that should not be exceeded in an architectural space for a given use. The NC rating method has been validated extensively in practice since its introduction in 1957. The NC or PNC method combines the results of the speech-interference and loudness procedures above because it specifies simultaneously the maximum permissible speech-interference level and the maximum permissible loudness level of a noise for a given space. There is a further important advantage to the set of numbers that comprise a noise criterion curve; namely, it requires measurement of the noise in octave bands of frequency. Thus, the set of measured octave-band levels required to determine whether a given noise satisfies a specified NC or PNC curve can also be used by an engineer to identify the mechanisms in a source that are responsible for the noise rating.

4. *A-weighted sound level.*[4, 11–15] This quantity is measured directly by a sound-level meter equipped with an electronically modified frequency response known as A-weighting. For reasons not completely understood, there is an impressive degree of correlation between A-weighted level, speech-interference level, loudness level, and NC or PNC level. A considerable advantage of the A-weighted level is that very small, hand-held, battery-operated instruments, for which all relevant acoustical and electrical characteristics are standardized, are available to measure steady A-weighted levels. If the noise to be rated fluctuates, the running A-weighted level provides a single-channel signal that can be used to yield statistical data as a function of time. Because human reactions to noise vary widely, and because the other noise ratings are so highly correlated among themselves, it has been argued that A-weighted sound levels are as good as other methods for rating noise.

5. *Noise pollution level.*[4, 16, 17] This rating procedure is one of the first (1969) successful methods for dealing with fluctuating noises. Each of the four procedures above assumes a single type of noise (e.g., air-conditioning noise) that is reasonably steady with time. Those procedures offer no solution to a mixture of several noises, some steady and some intermittent, or to one type of noise that varies over a wide range of levels (e.g., traffic noise). The noise pollution level L_{NP} uses as its data base instantaneous A-weighted level (or some other weighted level such as loudness level), recorded over a sample period of time sufficiently long (e.g., 30 min) to establish the statistics of the total noise exposure for some longer period of time (e.g., 7 A.M. to 7 P.M.). The NPL concept teaches that annoyance owing to noise is equal to the sum

of two terms.   The first is determined by the mean energy of the noise, the second by the range of variation of the noise.* Thus, one encounters the unexpected finding that if the mean energy is increased by adding more noise events, such as the number of airplanes per day, this usually means that the variability term is decreased because the noise is more continuous.   Consequently, the L_{NP} can go through a maximum and eventually decrease with an increasing number of events.   The L_{NP} prediction of how annoyance varies with increasing frequency of events may indeed be correct for the limited number of examples already investigated, but it is too early to conclude that it is generally applicable to all complex noises.

18.1.1   Speech-interference Level, PSIL [1-5]   The speech-interference level of a noise has had two definitions, one related to octave-band filter sets with the "old" cutoff frequencies, the other with the new (preferred) band-center frequencies.[18]

The SIL (old) is the arithmetic average of the sound-pressure levels in the three octave bands: 600 to 1,200, 1,200 to 2,400, and 2,400 to 4,800 Hz.

The PSIL (new, preferred) is the arithmetic average of the sound-pressure levels in the three octave bands with the preferred, geometric-mean, center frequencies at 500, 1,000, and 2,000 Hz.

The speech-interference level of a steady, continuous noise is a good guide to the interfering effect of that noise on speech communication. In Table 18.1 we see the PSILs for which about 60% reliable communication of individual words and numbers out of context is possible outdoors at various levels of voice effort and at various separation distances between a listener and a talker.   For example, we see from Table 18.1 that if the PSIL of the intruding noise were 70 dB, the average person would need to shout to make himself understood outdoors at a distance of 2 m (6 ft).   If the noise fluctuates in level, speech sounds will be heard better in quieter than in noisier periods.   To determine the effective PSIL for a slowly fluctuating noise, one must determine the average PSIL over the time period of interest.   For fairly rapid fluctuations, the answer is more complicated.[21]

Lacking an octave-band filter set, the PSIL can be estimated from A-weighted sound levels (see Chap. 4 for equipment to measure L_A) by the relation PSIL $\approx L_A - 7$ dB.   Obviously, this approximation is most accurate for cases where the noise spectrum has most of its energy in the frequency range between 500 and 3,000 Hz, or where the band

---

* In Germany, the energy level alone is used in rating traffic noise (DIN Norms 18005 and 45641).

TABLE 18.1   Relations among PSIL, Voice Effort, and Background Noise

Speech interference levels (PSIL) of steady continuous noises in decibels at which reliable speech communication is barely possible * between persons at the distances and voice efforts shown.   The interference levels are for average male voices (reduce the levels 5 dB for female voices) with speaker and listener facing each other, using unexpected word material.   It is assumed that there are no nearby reflecting surfaces that aid the speech sounds.

Distance between talker and listener, ft (m)	PSIL, dB †			
	Talker's voice effort			
	Normal	Raised	Very loud	Shouting
0.5 (0.15)	74	80	86	92
1 (0.3)	68	74	80	86
2 (0.6)	62	68	74	80
4 (1.2)	56	62	68	74
6 (1.8)	52	58	64	70
12 (3.7)	46	52	58	64

* Corresponding to an articulation index of about 0.40.[19,20]
† SIL (calculated from old octave bands) ≈ PSIL −3 dB.

TABLE 18.2   Rough Estimation of PSIL or L_A by the "Walkaway Test" [4]

The numbers give the average distance at which a male talker with an average voice at conversational level can just *not* be understood by an average listener on a relatively wind-free day.   Ranges in the columns are given because the average number in one column is associated with the spread shown in other columns; for example, 17.5 ft would occur for a PSIL of 57 ± 4 dB, or conversely, a PSIL of 57 dB would yield separation distances in the range of 14 to 21 ft depending on the persons involved.

Distance apart, ft (m)	PSIL, dB	L_A, dBA
6 to 9 (2 to 3)	64 to 70	71 to 77
9 to 14 (3 to 4.5)	59 to 66	66 to 73
14 to 21 (4.5 to 7)	53 to 61	60 to 68
21 to 32 (7 to 10)	47 to 55	54 to 62
32 to 50 (10 to 15)	41 to 49	48 to 56
50 to 75 (15 to 23)	35 to 43	42 to 50

levels below 500 Hz slope upward with decreasing frequency less rapidly than the downward slope of the A-weighting network and where at the same time the band levels slope off above 3,000 Hz.

If no instrumentation at all is available, the PSIL (or L_A) may be estimated approximately by a "walkaway test." This crude procedure is based on using two male talkers with average voices and determining the separation distance required so that they just do *not* understand each other.[4] The test should be performed on a reasonably calm day. One person reads in a conversational tone a passage of text unfamiliar to the listener, who then moves progressively farther away from the speaker to the point where he can *just barely understand a word or two in a 10-sec period* of listening. The distance at which this occurs is measured. Then the test is repeated several times, alternating the speakers. Of course, each speaker must be careful to maintain normal conversational voice level rather than trying to maintain communication with the listener. The averaged distances so obtained are approximately related to the PSIL or the L_A by Table 18.2.*

18.1.2   Loudness S and Loudness Level L_L [4,6-8]   The calculated loudness and loudness level of a noise that is steady and free of pure tones or narrow bands of energy (which would make the frequency spectrum highly irregular) are determined by a calculation procedure that converts the one-third-octave-band spectrum of the noise into a single number having the units of sones or phons, respectively. The procedure is designed to approximate the mean of the responses of a large number of individuals who were asked to make subjective judgments of loudness level or loudness for that noise.

Loudness Level.   The loudness level of a given noise is defined as the sound-pressure level of a narrow band of noise (or a pure tone) at 1,000 Hz, frontally presented to a listener, that sounds to him equal in loudness to the noise being rated. Thus, in judging loudness level experimentally, a noise whose loudness level is to be determined would be frontally presented, say, by a loudspeaker, to a person. While switching back and forth between that noise and the narrowband reference noise the listener would adjust the level of the latter until the two noises sound equally loud. The sound-pressure level (*re* 2×10^{-5} N/m²) of the reference noise would then be measured at the position of the listener's head, with the listener removed. This measured level would be called the loudness level of the given noise in phons, as

---

* Note by comparison with Table 18.1 that an increase in noise level of 17 dBA is sufficient to reduce the level of intelligibility from that of "60% reliable recognition of *unfamiliar isolated words*" to "recognition of an occasional word or two in *connected speech* in a 10-sec period."

judged by that particular listener.   When the loudness level is computed from the procedure given below, the result is called *calculated loudness level*.

Loudness.   Loudness differs from loudness level, although they are related logarithmically.   Loudness is a numerical designation that, for a given sound, is proportional to the subjective magnitude of the "loudness" of that sound as judged by a normal listener.   The unit of loudness is the sone.   By definition, one sone is the loudness of a sound whose loudness level is 40 phons.

In judging loudness experimentally, a person might, for example, be presented alternately with a sound of unknown loudness and a 1,000-Hz narrow band of (reference) noise presented frontally with a sound-pressure level of 40 dB.   Thus, the reference noise would have a loudness level of 40 phons or a loudness of $S_{40} = 1$ sone.   He would be asked, "How many times as loud is the test sound as the reference sound?"   If, for example, the listener responded "10," the loudness of the test sound would be 10 sones by his judgment.

Loudness vs. Loudness Level.   A doubling in loudness in sones is commonly taken to be equal to an increase in loudness level of 10 phons.[8]   Recent experimental data indicate that the majority of listeners would probably judge a doubling in loudness to correspond to an increase of about 9 phons in loudness level.*   To be consistent with current American and international standards we shall assume here that a change of 10 phons corresponds to a doubling or halving of loudness.

Procedure for Calculating Loudness Level.†   The most widely used procedure for calculating loudness level first involves a table that provides a means for converting the sound-pressure level in each one-third-octave band to a loudness index for that band.   A supplementary formula provides a rule for converting the set of loudness indices for all the one-third-octave bands into the total calculated loudness in sones.   Another

* Privately communicated by Professor S. S. Stevens of Harvard University, Cambridge, Mass.   Stevens has in preparation a complete revision of Ref. 6 titled "Assessment of Noise: Calculation Procedure Mark VII."   He says that either Mark VI or Mark VII will calculate loudness level with adequate precision for practice, but Mark VII provides a closer model of perception that would accord with the performance of the human hearing mechanism.

† An alternate procedure for calculating loudness level advanced by E. Zwicker is given in Ref. 8.   (See also *Frequenz*, vol. 13, p. 234, 1959.)   It yields numbers about 5 phons higher than the method of S. S. Stevens.   The evidence in the literature indicates that Stevens' values are more correct (see *Sound and Vibration*, vol. 16, pp. 16–28, October, 1969; *Acustica*, vol. 21, p. 134, 1969; Urbanek, 1969; and Ref. 6).   Stevens' method is the only one standardized in the United States.   He recommends that his Mark VI procedure be used with one-third-octave-band data for best accuracy.   Both methods are standardized in ISO R-532 (December, 1966).   Only the Zwicker method is standardized in Germany (DIN Norm 45631).

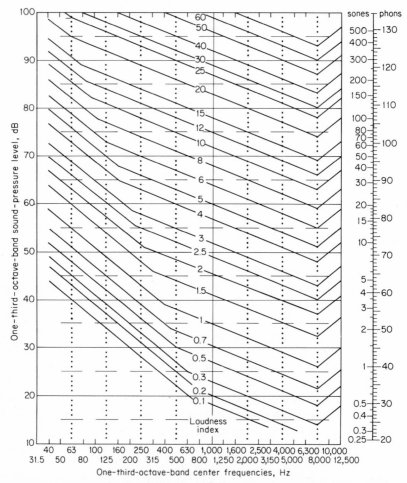

Fig. 18.1 Contours of equal-loudness index.[6-8] The abscissa is the geometric-mean frequency of a one-third-octave band of noise. The ordinate is the sound-pressure level in the band. The parameter gives the loudness index for that band. The bar graph at the right gives the conversion from total loudness in sones (see text) to loudness level in phons.

formula converts from calculated loudness to calculated loudness level in phons. The procedure is as follows: [6,7]

1. Enter the geometric-mean frequency of a one-third-octave band along the abscissa of Fig. 18.1. Then from the sound-pressure level for that band determine the loudness index for that band. Repeat for each band.

2. Find the total calculated loudness S_t by

$$S_t = I_m + 0.15[(\Sigma I) - I_m] \qquad \text{sones} \qquad (18.1)$$

where I_m = greatest of the loudness indices
ΣI = sum of loudness indices of all one-third-octave bands
3. The total calculated loudness in sones is converted into calculated loudness level in phons by

$$L_L = 40 + 10 \log_2 S_t \quad \text{phons} * \tag{18.2}$$

A nomogram for this relation is given along the right-hand edge of Fig. 18.1.

Example 18.1 Determine the loudness and loudness level of a noise with the one-third-octave-band spectrum given in the second column below. Also, determine the speech interference and the A-weighted sound levels for the noise.

Frequency, Hz	One-third-octave-band level (measured)	Loudness index (from Fig. 18.1)
31.5	75	3.0
40	74	3.4
50	73	3.6
63	72	3.8
80	71	4.0
100	70	4.5
125	69	4.7
160	68	5.0
200	67	5.0
250	66	5.0
315	65	5.0
400	64	5.0
500	63	5.0
630	62	5.0
800	61	5.0
1,000	60	5.0
1,250	59	5.0
1,600	57	4.8
2,000	56	4.6
2,500	55	4.7
3,150	55	5.0
4,000	54	5.0
5,000	54	5.2
6,300	54	5.5
8,000	54	5.7
10,000	54	5.2

* Because the argument of a logarithm must be dimensionless, S_t is actually the ratio S_t/S_{40}, where $S_{40} = 1$ sone.

SOLUTION: The loudness indices, as determined from Fig. 18.1, are tabulated in the third column above. The sum $\Sigma I = 122.7$ and $I_m = 5.7$ (8,000 Hz). From Eq. (18.1), $S_t = 5.7 + 0.15(122.7 - 5.7) = 5.7 + 17.5 = 23.2$ sones. From the nomograph on Fig. 18.1, a loudness of 23.2 sones corresponds to 85.5 phons. The speech-interference level, PSIL, equals the average of the levels for the 500-, 1,000-, and 2,000-Hz octave bands. Combining one-third-octave-band levels, using Fig. 2.7, gives the equivalent octave-band levels of 68, 65, and 61 dB, respectively. Thus the PSIL = 65 dB. The A-weighted sound level is found by first adding (paying due regard to the signs) the values in Table 4.1 to the 26 one-third-octave-band levels. Then combine the new numbers, again using Fig. 2.7. This procedure yields the A-weighted sound level $L_A = 71$ dBA. We note that here L_A is 6 dB greater than PSIL.

18.1.3   NC or PNC Noise Criteria Curves [2, 9, 10]

1957 NC Curves.   One of the most widely used sets of values as a basis for noise criteria indoors is the noise criteria (NC) curves of Fig. 18.2 and the equivalent numbers of Table 18.3.   The NC values apply to steady noises and specify the maximum noise levels permitted in each octave band for a specified NC curve.   For example, if an architectural specification calls for noise levels not to exceed the NC-20 criterion curve, the sound-pressure levels in all eight octave bands must be less than those of the NC-20 curves of Fig. 18.2 or those along the second line of Table 18.3.   Alternatively, the "NC rating" of a given noise equals the highest penetration of any band of that noise into the curves. For example, if a noise spectrum had octave-band levels, starting with the 63-Hz band, of 50, 55, 58, 60, 55, 50, 45, and 39 dB, respectively, the noise would be given a rating of NC-57, because the fourth (500-Hz) band penetrates to a level that is 2 dB above the highest NC curve approached, i.e., the NC-55 curve.   In practice, in determining whether a noise conforms to a specified NC (or PNC level described below) the curve is passed through the average of the levels in the three contiguous bands falling nearest the curve.   However, in locating a NC (or PNC) curve relative to the datum points the level in the highest of these three bands must not lie more than 2 dB above the average for the three.   For example, if the levels were 1 dB *below* the curve in the 500-Hz band, 2 dB *above* the curve in the 1,000-Hz band, and 1 dB *below* the curve in the 2,000-Hz band, the rating curve could correctly be passed through the average of the levels for these three bands.   Otherwise the tangent method first described above must be used.

The 1957 NC curves have been criticized recently by some users.   It has been demonstrated that if a noise spectrum is deliberately generated with octave-band levels equal to those of a particular NC curve, the sound heard is not a pleasant, or "neutral" sound, but is both "hissy" and "rumbly."   This is particularly true of noise from air-conditioning sys-

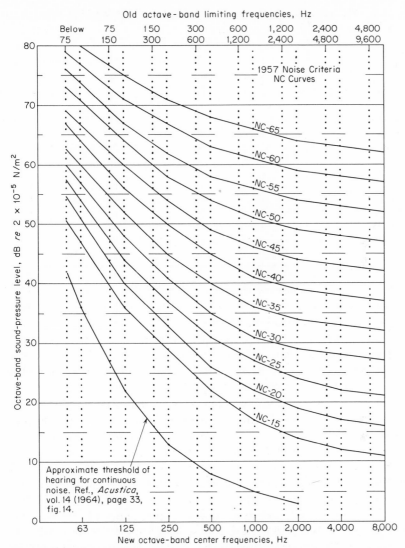

Old octave-band limiting frequencies, Hz

Fig. 18.2 Indoor noise criteria (NC) curves (1957). Refer to Table 18.3 for the numerical values of sound-pressure levels of these curves.

tems, where duct or fan turbulence adds to the "rumble" of the random-noise sound. To achieve a more acceptable noise "quality" it sometimes has been necessary to lower the levels by about 5 dB in both the very low and the very high frequency bands, as compared to those permitted by the 1957 NC curves.

It has also been pointed out that the original 1957 NC curves were determined for the old octave-frequency bands given along the top of

TABLE 18.3   Octave-band Sound-pressure Levels Associated with the 1957 Noise Criterion (NC) Curves of Fig. 18.2

1957 noise criterion curves	63 Hz	125 Hz	250 Hz	500 Hz	1,000 Hz	2,000 Hz	4,000 Hz	8,000 Hz
NC-15	47	36	29	22	17	14	12	11
NC-20	51	40	33	26	22	19	17	16
NC-25	54	44	37	31	27	24	22	21
NC-30	57	48	41	35	31	29	28	27
NC-35	60	52	45	40	36	34	33	32
NC-40	64	56	50	45	41	39	38	37
NC-45	67	60	54	49	46	44	43	42
NC-50	71	64	58	54	51	49	48	47
NC-55	74	67	62	58	56	54	53	52
NC-60	77	71	67	63	61	59	58	57
NC-65	80	75	71	68	66	64	63	62

Fig. 18.2. Thus, the use of those curves with the old "below-75-Hz" band converted into *two* new "31.5-Hz and 63-Hz" bands is incorrect. Also, it has been pointed out that the lower end of the curves does not conform to a more recently (1964) determined threshold-of-hearing curve.[23]

1971 PNC Curves. To overcome the objections listed above to the 1957 NC curves, a number of suggestions have been made for modifying them, which has led to the 1971 *preferred noise criteria (PNC) curves* shown in Fig. 18.3.[51] These PNC curves have values that are about 1 dB lower than the NC curves in the four octave bands at 125, 250, 500, and 1,000 Hz for the same curve rating numbers. In the 63-Hz band, the permissible levels are 4 or 5 dB lower; in the highest three bands, they are 4 or 5 dB lower. Because these curves have not received national or international overview at this writing, they must be regarded as provisional. Otherwise, they are used exactly in the same way as the NC curves. Parenthetically, we note that in open-plan spaces (see Sec. 18.3.3) people prefer a background noise with a spectrum more like the PNC than the NC curves, although they like the spectrum of Fig. 18.15 the best. The numerical values of the curves are given in Table 18.4.

18.1.4 A-weighted Sound Level L_A [4, 11–15]

The A-weighted sound level L_A may be measured with any standard sound-level meter. This particular sound level is identified by an "A." To avoid confusion, we also tag the unit decibel with an A, i.e., dBA, although many writers prefer dB(A), or simply dB. The A-weighting network of a sound-level meter discriminates against sound pressures at frequencies below 500 Hz and above 10,000 Hz (see Table 4.1).

The A-weighted sound levels are immediately appealing, both because they are easy to obtain and because they correlate reasonably well with the best single-number ratings used for all six of those effects of noise on people that are listed in Sec. 18.1. For this reason, most regulatory or law enforcement groups or agencies may be expected to use A-weighted sound levels read from simple, hand-held, battery-operated instruments, unless the new "ear-weighting" levels, in dBE, described in the appendix to this chapter should be adopted.

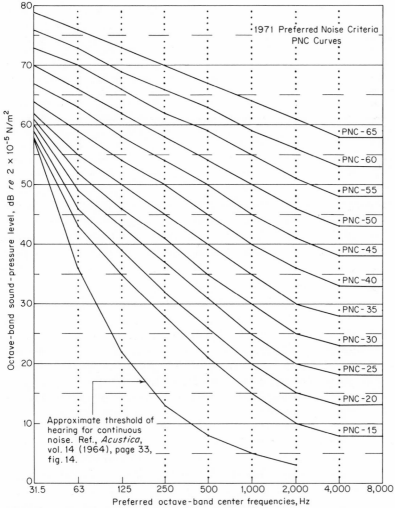

Fig. 18.3 Preferred noise criteria (PNC) curves (1971). Refer to Table 18.4 for the numerical values of sound-pressure levels of these curves. These curves have not yet been acted on by any standardization group.

TABLE 18.4   Octave-band Sound-pressure-level Values Associated with the 1971 Preferred Noise Criterion (PNC) Curves of Fig. 18.3 *

Preferred noise criterion curves	31.5 Hz	63 Hz	125 Hz	250 Hz	500 Hz	1,000 Hz	2,000 Hz	4,000 Hz	8,000 Hz
PNC-15	58	43	35	28	21	15	10	8	8
PNC-20	59	46	39	32	26	20	15	13	13
PNC-25	60	49	43	37	31	25	20	18	18
PNC-30	61	52	46	41	35	30	25	23	23
PNC-35	62	55	50	45	40	35	30	28	28
PNC-40	64	59	54	50	45	40	36	33	33
PNC-45	67	63	58	54	50	45	41	38	38
PNC-50	70	66	62	58	54	50	46	43	43
PNC-55	73	70	66	62	59	55	51	48	48
PNC-60	76	73	69	66	63	59	56	53	53
PNC-65	79	76	73	70	67	64	61	58	58

* These curves have not yet been acted on by any standardization group.

18.1.5   Noise Pollution Level L_{NP} [4, 16, 17]

Many noises, of which community noise is an example, fluctuate between relatively quieter periods and noisier periods, such as before and while trucks pass by or aircraft fly over. The previous rating methods took no account of such variations. To handle fluctuating signals, the *noise pollution level* L_{NP} has been devised by Robinson.[16] It takes into account two general observations:

1. Annoyance is related to the intensity or perhaps even to the total energy of the noise measured throughout a period of time, such as a 24-hr day, or during sleeping or waking hours.

2. Annoyance increases with variability of the noise, given the same total energy of the noise for the time period.

The procedure for determining L_{NP} has been applied to existing data involving the intrusion of aircraft and traffic noise into neighborhoods. It has not yet been correlated with subjective reactions of people to a wide variety of fluctuating noises. To date, only one research program has been designed especially to assess the merit of noise pollution levels for rating noise in communities to the east of Heathrow Airport, London.[24] The results confirm that "L_{NP} is a good predictor of median, general dissatisfaction for existing communities and existing noises. The results might be very different if either new noises or new populations or both were to be considered." *Thus, L_{NP} must be viewed as a tentative rating quantity.* Nevertheless, the concept of average A-weighted sound level and of augmentation of this level to account for fluctuations accords well with intuition and with general observations. Also, as we said before, the A-weighted sound level generally correlates well with other measures of human response to steady noise.

The definition of noise pollution level is

$$L_{NP} = L_{eq} + 2.56\sigma \qquad \text{dB(NP)} \qquad (18.3)$$

where L_{eq} = mean-square, A-weighted sound level of a sufficiently long sample of the noise, dBA

σ = standard deviation (see Chap. 5) for the same sample of A-weighted sound level, dBA

To determine L_{eq} and σ, some type of apparatus suitable for determining the statistical (probability) distribution of the A-weighted levels of the noise must be used. Such an analysis will yield, as shown in Fig. 18.4, a histogram (curve A) and a cumulative distribution (curve B) of the noise levels during the period of observation.

If the probability distribution of the instantaneous A levels is reasonably Gaussian, the $L_{10\%}$, $L_{50\%}$, and $L_{90\%}$ levels, such as found from curve B in Fig. 18.4, may be employed as follows to obtain L_{NP}. Otherwise the true standard deviation σ must be determined.

The approximate formulas for L_{NP} are

$$L_{NP} = L_{eq} + (L_{10\%} - L_{90\%}) \qquad \text{dB(NP)} \qquad (18.4)$$

or

$$L_{NP} = L_{50\%} + d + \frac{d^2}{60} \qquad \text{dB(NP)} \qquad (18.5)$$

where $d = (L_{10\%} - L_{90\%})$, dBA

$L_{eq} \approx L_{50\%} + \dfrac{d^2}{60}$

$L_{10\%}$ = decile noise level exceeded 10% of the time during the observation period, dBA

$L_{90\%}$ = decile noise level exceeded 90% of the time during the observation period, dBA

$L_{50\%}$ = median decile noise level, dBA

Example 18.2 Assume an urban traffic noise with the standard deviation and the three decile levels, measured over the time period from 8 A.M. to 6 P.M., shown on Fig. 18.4. Determine L_{NP}.

SOLUTION: First for Eq. (18.5) find d and d^2. From Fig. 18.4, $d = 11.5$ and $d^2 = 132$. So, from Eq. (18.5) $L_{NP} = 80.5 + 11.5 + (132/60) \approx 94$ dB(NP). As a check, from Eq. (18.3), $L_{NP} = 82.5 + 2.56 \times 4.5 = 94$ dB(NP). We must observe, however, that many traffic noises are not Gaussian. Because truck noise averages about 12 dB higher than passenger car noise, a bimodal probability distribution curve may result. In this case a true value for σ should be calculated for use in Eq. (18.3).

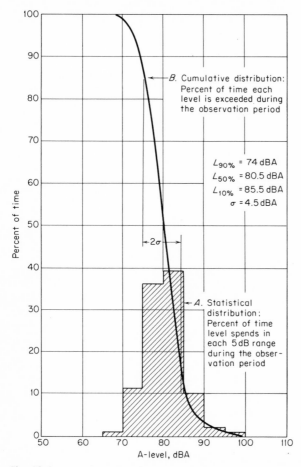

Fig. 18.4 (*A*) Histogram showing statistical distribution and (*B*) curve showing cumulative distribution, both presenting A-weighted noise levels at a site near a roadway.

18.2   Criteria for Noise in Communities [37]

Urban noises are generally not steady. To estimate the reaction of people to such noises accurately, some sort of statistical approach is required, which also has to take into account people's greater annoyance with noise at night when they are attempting to sleep than during waking hours. Thus, we need to speak of "noise exposure" at a site, meaning the whole description of time-varying sound level, throughout a 24-hr period, rather than some simple average level. Generally, we also need to differentiate between neighborhoods where loud noises have already existed for many years and those neighborhoods where the onset of increased noise levels has occurred recently.

18.2.1 Differences in Human Response to Noise [40]   The subjective responses of people to community noise are affected by many factors. As diagrammed in Fig. 18.5, the physical noise stimulus is obviously a prime factor. The characteristics of a noise, its level, statistical properties, frequency distribution, variation with time, and so on, can be determined by *engineering measurements*. The degree of interruption of some human activity, such as speech communication or sleep, or the relative annoyance of different noises can be studied in the laboratory by psychologists who evoke *subjective responses* from test listeners. Subjective responses can also be studied in communities directly, though much less conveniently. Subjective responses of people are the result not only of the stimulus and the activity affected, but also of individual differences, previous exposure to noise, and attitudes toward noise or toward the parties producing the noise. *Acceptability* (judged in terms of complaints) is the result of all these factors, plus other psychosocial factors including the degree of organization of affected parties in the community and the existence of a known place to register complaints.

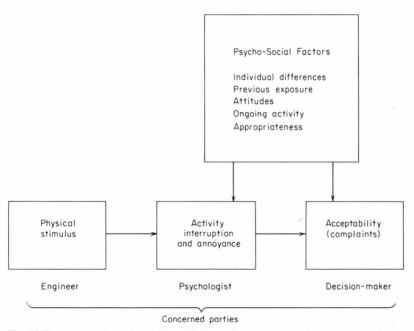

Fig. 18.5 Simplified model of the subjective response of people to noise as related to three concerned parties.

In confirmation of the statement that not all people are affected equally by the same community noise exposures, refer to Figs. 18.6 and 18.7. Apparently, about 10% of a typical population are so sensitive to noise that they object to any noise not of their own making. Another sizable portion (about 25%) seems to be almost imperturbable. We may expect that in neighborhoods where people have been subjected to loud noise levels for years, many of those who have not moved away are in the "imperturbable" category. For any new noise situation, a variety of re-

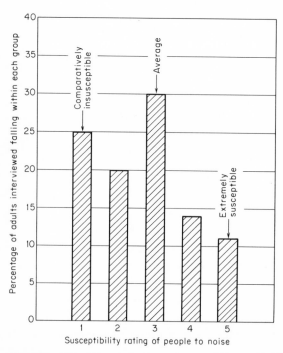

Fig. 18.6 This bar graph shows the percentage of 1,377 adult residents interviewed in depth in a 1961 Central London community survey for each of five categories of noise susceptibility rating. The susceptibility rating was derived from the answers to six questions on a 40-item questionnaire that evoked statements from the interviewees about their sensitivity to noise as follows: (1) Does noise ever bother, annoy, or disturb you in any way? (2) When you hear the noise most annoying to you in your home do you feel very annoyed, moderately annoyed, a little annoyed, or not at all annoyed? (3) Would you say you were more sensitive or less sensitive than other people to noise? (4) Is there too much or too little fuss made about noise nowadays? (5) How far would you agree that noise is one of the biggest nuisances of modern times? (6) Could you sum up your opinion by saying whether you find noise in general: very disturbing, disturbing, a little disturbing, or not at all disturbing?

OUTDOOR NOISES HEARD INDOORS
1. Road traffic noise
2. Aircraft noise
3. Noises from neighbors' dwellings
   (children and adults' voices, radio/TV,
   bells, footsteps, banging, etc.)
4. Noise of pets

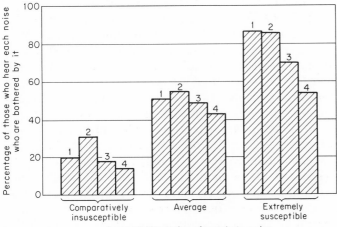

Fig. 18.7 This bar graph shows the percentage of the 1,377 adults interviewed in Central London who said they are bothered by particular outdoor noises that they hear when they are in their homes as related to their susceptibility rating to noise. The survey concluded that the extent of annoyance both with noise in general, and with particular noises, is very strongly related to the susceptibility rating (see Fig. 18.6).

actions should be expected from the people exposed. Evidently noise control in urban situations must be directed toward the two-thirds of the population who lie between these two extremes and who would be expected to express appreciation of the results, although all persons would surely benefit.

There are also indications of different tolerances to noise levels from one climate to another. Comparisons between noise tolerances in colder areas, England and Scandinavia, and in warmer countries, France and Italy, reveal significant differences.[4] Those countries that live with open windows almost the year around seem to be more tolerant of noise by 5 to 10 dBA than those that enjoy heavy, well-closed windows half of the year or more. This difference has been noticed in the United States by Bolt Beranek and Newman Inc., which finds, from an unpublished traffic-noise survey, that people in Los Angeles appear to be less sensitive to traffic noise than those who live in Detroit or Boston, though whether this is due to open-window/closed-window conditions or some more subtle factors is not known.

18.2.2 A-level Rating Method for Community Noise   For predicting general public acceptability to most community noises, it appears adequate to measure the A-weighted sound level of a noise.   For situations where the economic factors are of major importance, such as for noises produced by jet aircraft, more complex measurements and evaluations are justified.   For example, for jet-aircraft noise, the "effective perceived noise level in EPNdB" has been adopted for use in governmental certification of individual aircraft.[41, 42]   But in communities, where noises arise from many sources and fluctuate widely, statistically measured A-weighted sound level and the noise pollution level derived from it appear to be satisfactory for rating noise exposure.

Approximate mean A-weighted noise levels and the relative loudnesses of some noises typically found in indoor and outdoor environments are given in Table 18.5.   There are wide differences among different manufacturers' products in the noise that they produce, and so the levels in the table must be treated as only representative.

A complete description of the noise exposure at a site would distinguish between weekdays and weekends and among daytime, late evening, and sleeping-time noise level distributions.   Where this is not possible, sample measurements can be made at convenient times and the details at other times estimated by reference to available noise surveys.[4, 38, 39]

18.2.3 Criteria for Non-aircraft Community Noise[2, 4]   Most nonaircraft neighborhood noise arises from industry, surface transportation, and construction equipment.

Criteria for External Industrial Noise.   Generally, existing statutes governing industrial noise, as measured at the boundaries of surrounding residential areas, are expressed in terms of octave-band sound-pressure levels, similar in shape to the NC curves.   They also usually apply to steady noises.   Nighttime (2200 to 0700) limits in American cities lie[4] in the range between NC-35 and NC-50, with the median at about NC-43.   This median corresponds to a median A-weighted level of about $L_{50\%} = 50$ dBA.   Where daytime limits are given in American statutes, they are usually about 5 dBA, and sometimes 10 dBA, higher.

If these criteria are extrapolated to fluctuating noises, and if we assume an $L_{10\%} - L_{90\%}$ difference of 12 dBA, we obtain a median noise pollution level of $L_{NP} \approx 65$ dB(NP).

The city of Chicago specifies the following permissible maximum sound-pressure levels at points on the boundary of a residence district adjacent to each particular type of manufacturing district.   For Districts B and C the levels specified must be met either as for district A or at 125 ft from the nearest property line of a plant, whichever is greater.

Octave-band center frequency	A Restricted manufacturing zoning district	B General manufacturing zoning district	C Heavy manufacturing zoning district
31.5	72 dB	72 dB	75 dB
63	71	71	74
125	65	66	69
250	57	60	64
500	51	54	58
1,000	45	49	52
2,000	39	44	47
4,000	34	40	43
8,000	32	37	40
A-scale level (for monitoring purposes)	55 dBA	58 dBA	61 dBA

Criteria for External Truck, Bus, Motorcycle, and Automobile Noise.[45] Regulations in the United States governing noise made by automobiles and trucks operated on highways are based on measurements made at a position 50 ft from the center line of the traffic lane; the permissible limits for speeds above 35 mph (56 kph) generally fall between 82 and 92 dBA, with the median at 87 dBA. California has separate categories of maximum noise levels measured at 50 ft for (1) automobiles, (2) heavy trucks and buses, and (3) motorcycles, when "operated at anytime or under any condition of [road] grade, load, acceleration or deceleration."[43,44] They specify, for speeds greater than 35 mph, 82 dBA for (1), 90 dBA for (2), and 86 dBA for (3). For speeds under 35 mph, after 1972, the values become 76 dBA for (1), 86 dBA for (2), and 82 dBA for (3).

The California vehicle code also specifies that *new* motor vehicles sold in California after 1972 must not produce noise in excess of 84 dBA for (1) and 86 dBA for (2) and (3) under a specified set of max-noise tests.

In a recent noise ordinance (July 1971) the City of Chicago requires that "no person shall sell, or offer for sale, a new motor vehicle that produces a maximum noise exceeding the following noise limit at a distance of 50 ft from the center line of travel, under test conditions specified in SAE Standards J331 and J986; and SAE Recommended Practices J366 and J184: (1) Motorcycles manufactured before January 1970, 92 dBA; after January 1970, 88 dBA; after January 1973, 86 dBA; after January 1975, 84 dBA; after January 1980, 75 dBA. (2) Passenger cars, motor-driven cycles or any other motor vehicle with a gross weight

TABLE 18.5 Noise Level and Relative Loudness of Typical Noises in Indoor and Outdoor Environments

L_A (dBA)	Subjective impression	Community * (outdoor)	Home or industry * (indoor)	Relative loudne (human judgme of different sound levels)
130		Military jet aircraft takeoff with afterburner from aircraft carrier at 50 ft (130 dBA)		32 times as loud
120	Uncomfortably loud		Oxygen torch (121 dBA)	16 times as loud
		Turbofan aircraft at takeoff power under flight path at 200 ft (118 dBA)	Riveting machine (110 dBA). Rock-n-roll band (108–114 dBA)	
110		Same jet flyover at 1,000 ft (103 dBA). Boeing 707, DC-8 at 6,080 ft before landing (106 dBA). Bell J-2A helicopter at 100 ft (100 dBA)		8 times as loud
100	Very loud	Boeing 737, DC-9 at 6,080 ft before landing (97 dBA).	Newspaper press (97 dBA)	4 times as loud
90		Motorcycle at 25 ft (90 dBA)		2 times as loud
		Car wash at 20 ft (89 dBA). Prop. plane flyover at 1,000 ft (88 dBA). Diesel truck, 40 mph at 50 ft (84 dBA). Diesel train, 45 mph at 100 ft (83 dBA). Power mower at 25 ft (85 dBA)	Food blender (88 dBA) Milling machine (85 dBA) Garbage disposal (80 dBA)	
80	Moderately loud	High urban ambient sound (80 dBA). Passenger car, 65 mph at 25 ft (77 dBA). Freeway at 50 ft from pavement edge 10 A.M. (76 ± 6 dBA)	Living room music (76 dBA) TV-audio, vacuum cleaner (70 dBA)	Reference loudne
70			Cash register at 10 ft (65–70 dBA). Electric typewriter at 10 ft (64 dBA). Dishwasher, rinse at 110 ft (60 dBA). Conversation (60 dBA)	$\frac{1}{2}$ as loud
60		Air-conditioning condensing unit at 15 ft (55 dBA). Large transformers at 100 ft (50 to 60 dBA)		$\frac{1}{4}$ as loud
50	Quiet	Bird calls (44 dBA). Lower-limit urban daytime ambient noise (40 dBA)		$\frac{1}{8}$ as loud
40		[Scale interrupted]		$\frac{1}{16}$ as loud
10	Just audible	Threshold of hearing		
0				

* Numbers in parentheses are A-weighted levels.

of under 8,000 lbs, manufactured before January 1973, 86 dBA; after January 1973, 84 dBA; after January 1975, 80 dBA; after January 1980, 75 dBA. (3) Any motor vehicle with a gross weight of 8,000 lbs or more, after January 1968, 88 dBA; after January 1973, 86 dBA; after January 1975, 84 dBA; after January 1980, 75 dBA."

The Chicago ordinance also specifies that "no person shall operate within the speed limits specified in this section either a motor vehicle or combination of vehicles . . . at any time or under any condition of grade, load, acceleration or deceleration in such manner as to exceed the following noise limits (see table below) for the category of motor vehicle, based on a distance of not less than 50 ft from the center line of travel under the test procedures [listed above]."

Type of vehicle	Noise limit in relation to posted speed limit	
	35 mph or less	Over 35 mph
(1) Any motor vehicle with a maximum gross weight of 8,000 lbs or more		
—before January 1973	88 dBA	90 dBA
—after January 1973	86 dBA	90 dBA
(2) Any motorcycle other than a motor-driven cycle		
—before January 1978	82 dBA	86 dBA
—after January 1978	78 dBA	82 dBA
(3) Any other motor vehicle		
—after 1 January 1970	76 dBA	82 dBA
—after 1 January 1978	70 dBA	79 dBA

The Society of Automotive Engineers has adopted a standard (SAE J986a) that suggests as limits the maximum A-weighted noise level of new passenger cars or light trucks (below 6,000 lb gross weight) to 86 dBA measured at 50 ft. For new trucks exceeding 6,000 lb gross weight, it suggests a maximum noise level of about 90 dBA, measured at 50 ft when measured under steady operating conditions that produce maximum sound, usually maximum engine load.

With good mufflers, modern automobiles traveling on smooth road surfaces at speeds of 65 mph on level grade can easily meet a 78-dBA requirement at 50 ft from the center line of the traffic lane.[45] Trucks, with good mufflers, can easily meet an 86-dBA requirement at 50 mph up a 4% grade. Trucks with straight stacks (no muffler) produce a maximum A-weighted noise level of 95 dBA at 50 mph up a 4% grade, measured at 50 ft.

Criteria for Construction Machinery, Powered Equipment, and Hand Tools. In addition, the Society of Automotive Engineers recommends

(SAE Recommended Practice J952b) that other outdoor engine-powered equipment (construction and industrial machinery, such as crawler tractors, dozers, loaders, power shovels and cranes, motor graders, paving machines, off-highway trucks, ditchers, trenchers, compacters, scrapers, and wagons) not produce levels in excess of 88 dBA, measured at 50 ft. The Chicago noise ordinance specifies, for (1) these same powered equipments or powered hand tools, the following noise limits at a distance of 50 ft, under test conditions specified in SAE Standard J952 and SAE Recommended Practice J184, "manufactured: after January 1972, 94 dBA; after January 1973, 88 dBA; after January 1975, 86 dBA; after January 1980, 80 dBA.   (2) Agricultural tractors and equipment manufactured: after January 1972, 88 dBA; after January 1975, 86 dBA; after January 1980, 80 dBA.   (3) Powered commercial equipment of 20 HP or less intended for infrequent use in a residential area, such as chain saws, pavement breakers, log chippers, powered hand tools, etc., manufactured: after January 1972, 88 dBA; after January 1973, 84 dBA; after January 1980, 80 dBA.   (4) Power equipment intended for repetitive use in residential areas, including lawn mowers, small lawn and garden tools, riding tractors, snow removal equipment, etc., manufactured: after January 1972, 74 dBA; after January 1975, 70 dBA; after January 1978, 65 dBA."

Criteria for Engine-powered Water Craft and Motor-driven Recreational or Off-highway Vehicles.   The Chicago noise ordinance specifies that "no person shall operate any engine-powered pleasure vessel, craft, or motorboat . . . in such a manner as to exceed the following noise limit, as measured at a distance of not less than 50 ft: before January 1975, 85 dBA; after January 1975, 76 dBA.   No person shall sell or offer for sale a new motor-driven recreational or off-highway vehicle, including dunebuggies, snowmobiles, all-terrain vehicles, go-carts and mini-bikes that produces a maximum noise exceeding the following limit at a distance of 50 ft from the center line of travel, under test procedures in substantial conformity with SAE Standards J331 and J986 and SAE Recommended Practice J184: Snowmobiles manufactured after January 1971, 86 dBA; after June 1972, 82 dBA; after June 1974, 73 dBA.   The other vehicles manufactured after January 1971, 86 dBA; after January 1973, 82 dBA; after January 1975, 73 dBA." The ordinance also specifies that "no person may operate a motor-driven vehicle not subject to registration for road use, at any time or under any condition of load, acceleration, or deceleration, in such a manner as to exceed the following noise limit at any point on property zoned for business or residential use at a distance of not less than 50 ft from the path of travel: before January 1973, 86 dBA; after January 1973, 82 dBA."

Criteria for General Zoning Considerations.[4] Many countries, including the United States through its Department of Housing and Urban Development, have sponsored studies of acceptable noise levels for dwellings. These noise levels may be used either as a guide to zoning or as a guide to siting and construction of dwellings where the noise situation and the zoning are already established.

In general, these studies indicate that *indoor* noise levels should preferably not exceed those listed below for

1. *Listening to radio and television, indoor A-weighted levels:*
   $L_{50\%} = 35$ to 45 dBA
   $L_{10\%} = 41$ to 51 dBA
   $L_{NP} = 50$ to 60 dB(NP)
2. *Sleeping, indoor A-weighted levels:*
   $L_{50\%} = 25$ to 50 dBA, depending on the nature of the surroundings, e.g., country vs. urban areas
   $L_{10\%} = 31$ to 56 dBA, same comment
   $L_{NP} = 40$ to 65 dB(NP), same comment

The above-listed indoor levels may be converted to *outdoor* levels by the following approximate values

Open windows:
   Add 10 dB to the above numbers.
Closed windows with openable sashes:
   Add 20 dB to the above numbers.

A summary of approximate outdoor noise levels that exist today in different areas (parts) of American cities, by day and by night, are given in Fig. 18.8.

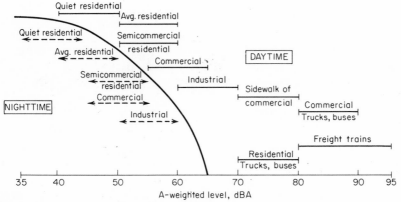

Fig. 18.8 Summary of existing A-weighted noise levels by day and by night in different U.S. city areas.

Based on social surveys in the United States and the international "consensus" given above, the Department of Housing and Urban Development has adopted a set of "guideline criteria" for noise exposure at residential construction sites.[4]   In the policy circular that sets out the HUD noise abatement policy, these criteria are expressed in terms of outdoor A levels not to be exceeded for more than so-many minutes per 8- or 24-hr period.   However, they can also be approximated in standard statistical format (cumulative A-level distributions) as shown in Fig. 18.9.   Converted to L_{NP} (noise pollution levels), these *outside noise level* criteria become

Clearly acceptable: L_{NP} = less than 62 dB(NP)

Normally acceptable: L_{NP} = between 62 and 74 dB(NP)

Normally unacceptable: L_{NP} = between 74 and 88 dB(NP)

Clearly unacceptable: L_{NP} = more than 88 dB(NP)

By way of comparison, a limit of L_{NP} = 74 dB(NP) corresponds to an outdoor level at the upper limit of the international consensus on permissible noise levels for "sleeping" with open windows.   It also agrees well

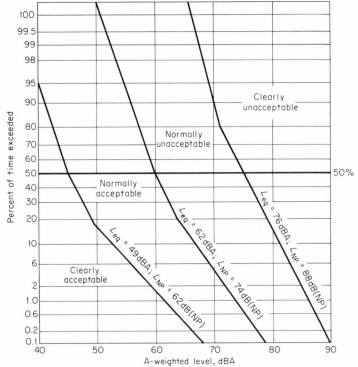

Fig. 18.9   Criteria adopted by U.S. Department of Housing and Urban Development for non-aircraft noise measured outdoors in residential areas. L_{eq} is the mean-square A-weighted sound level. L_{NP} is the noise pollution level [see Eq. (18.3)].

with Robinson's recommendation of 73 dB(NP) as a limitation on noise in urban residential areas where satisfactory speech intelligibility is desired indoors with open windows.   Robinson indicates that for neighborhoods adjacent to transportation, the limit can probably be in the neighborhood of 88 dB(PN), the dividing line in Fig. 18.9 between normally and clearly unacceptable.

18.2.4   Criteria for Aircraft-community Noise [4-6, 14, 17, 28-32, 35]   For those people living in the vicinity of airports, aircraft noise far surpasses all other noises in its effect on their daily living.   Because of its overriding importance and because of huge costs that are associated with its control, rating schemes for aircraft-neighborhood noise have reached a high state of refinement worldwide.

Effective Perceived Noise Level. [4, 5, 30, 31, 35, 41, 42]   The basic number for rating the event of an individual aircraft flying overhead is the effective perceived noise level L_{epn} in the unit EPNdB.   It embodies the results of extensive psychological measurements of the relative "noisiness" or "annoyance" of aircraft-flyover sounds.   In principle, it is to the first order the calculated *loudness level* of the noise at the instant during flight when the noise on the ground reaches its maximum *plus* an adjustment for the *duration* of the sound, which effectively yields a kind of time-integrated loudness level.   It also embodies a correction for possible *pure-tone content*, resulting from the whine of turbine machinery in the engine. The calculations are complicated and are covered in detail in Refs. 41 and 42 and any amendments that have been issued since.

Noise Exposure Forecast, NEF. [4, 35]   The total noise exposure at a point in a community in the neighborhood of an airport, where other noise can be neglected, is viewed as the (daytime-nighttime)-weighted sound-pressure-squared level of the noise produced by all the different aircraft flying their different flight paths.   For a specific *class of aircraft i* on a *flight path j,* the $\text{NEF}_{(ij)}$ is expressed

$$\text{NEF}_{(ij)} = \text{EPNL}_{(ij)} + 10 \log \left[\frac{n_{D(ij)}}{K_D} + \frac{n_{N(ij)}}{K_N} \right] - C \qquad (18.6)$$

where $\text{EPNL}_{(ij)}$ = effective perceived noise level produced at a point in the neighborhood by a particular class of aircraft following a particular flight path

$n_{D(ij)}$ and $n_{N(ij)}$ = numbers of operations, for daytime (0700–2200) and nighttime (2200–0700) respectively of aircraft class i on flight path j (note that each landing is counted as one operation, as is each takeoff)

$K_D = 20$ ⎫
$K_N = 1.2$ ⎬   The choice of these constants signifies that a single nighttime flight contributes as much to the NEF as 17
$C = 75$ ⎭   daytime flights.   The value of C is arbitrarily chosen

so that NEF numbers typically lie in a range where they are not likely to be confused with other composite noise ratings. (We note that 5 nighttime flights *per hour* is equivalent to 50 daytime flights *per hour*, or a factor of 10 difference. The number 17 arises from the fact that there are 15 daytime hours and 9 nighttime hours yielding $10 \times 15/9 \approx 17$.)

The total NEF at a given ground position is determined by summation of all the individual $NEF_{(ij)}$ values on an energy (pressure-squared) basis

$$NEF = 10 \log \sum_i \sum_j \text{antilog} \left[\frac{NEF_{(ij)}}{10} \right] \qquad (18.7)$$

For a single part of the day, assuming all aircraft of one class, the NEF increases on an energy basis; i.e., the term in the summation is $10 \log N$, where N is the number of aircraft.

Aided by a computer program that takes into account the noise from each class of aircraft, the flight profile, the distances and aspect angles between the point in the neighborhood and the aircraft at each point along its flight path, and, further, knowing the number of flights of each class of aircraft from each runway of an airport during daytime and nighttime, contours of NEF can be drawn around an airport.

It has been accepted generally that noise exposure forecast levels greater than NEF-40 are generally unacceptable to people, while levels less than NEF-30 are acceptable, considering that an air transportation system is necessary. It is also considered to be the responsibility of legislative bodies to determine the permissible NEF noise level that represents the best compromise between the rights of individuals and the need for a viable air transportation system.

A rough comparison of NEF to EPNdB can be made. For example, for 200 operations (100 takeoffs and 100 landings) between 0700 and 2200 hours, the NEF-40 contour is the same line as the 108 EPNdB contour. If 6 percent of the flights occur at night, the two are the same for 100 operations.

As a *very rough guideline* to gauge the areas involved inside the NEF-30 and NEF-40 contours,[4] let us define an *airport traffic parameter X*

$$X = N_d + 17 N_n \qquad (18.8)$$

where N_d, N_n = respectively the number of jet operations (takeoffs plus landings) during the daytime (0700–2200) and nighttime (2200–0700) at the airport

The parameter X assumes the types of aircraft and the style of operations projected for the year 1975 at American airports. No retrofitting of pre-1970 aircraft for noise abatement purposes is assumed.

Using 30 American airports as a guide, for which detailed computer calculations of NEF-30 and NEF-40 contours have been made, to represent predicted 1975 operations, airports are divided into four groups and for these the approximate NEF contour sizes are shown in Table 18.6.

With this table we can construct, *around each runway,* two "lozenge-shaped" contours approximating the NEF-30 and NEF-40 contours (see the example of Fig. 18.10). The dimensions of each "lozenge" are given by a pair of numbers (such as "6.0 mi/1.5 mi"). The length L of the "lozenge" is the distance along the ground under the flight path *from either end* of the runway. The width W of the "lozenge" is the distance along the ground to *either side* of the runway.

The final NEF-30 (or NEF-40) contour for the whole airport (all runways) is the curve that encloses *all* the areas lying within *any* of the NEF-30 (or NEF-40) zones for the various runways and flight paths. The accurately calculated NEF contours for the example of Fig. 18.10 are taken from Ref. 46.

We observe from Fig. 18.10 that such a simplified procedure is conservative; that is, it often results in NEF contours that enclose areas that are too large compared to the computer determination of NEF contours for each runway (including the exact mix of aircraft, night and day). In some cases, particularly where a runway is not used very often, the difference can be large. However, the simplified procedure is useful for rough planning. For more detailed planning, where land acquisi-

TABLE 18.6   Approximate Procedure for NEF Contours

Length L and width W dimensions for determining the
shape of a "lozenge" that *roughly approximates* the NEF-30
and NEF-40 noise exposure forecast contours
around runways

Airport-traffic parameter [see Eq. (18.8)]	Contour size	
	NEF-30 L/W	NEF-40 L/W
X less than 50	1 mi/$\frac{1}{4}$ mi	None
$X = 50$ to 500	3 mi/$\frac{1}{2}$ mi	1 mi/1,000 ft
$X = 500$ to 1,300	6 mi/$1\frac{1}{2}$ mi	$2\frac{1}{2}$ mi/2,000 ft
X more than 1,300........	10 mi/2 mi	4 mi/3,000 ft

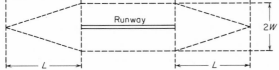

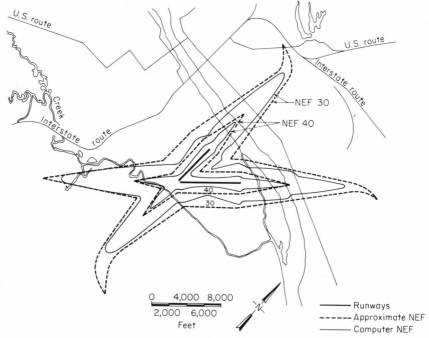

Fig. 18.10 Approximate (dashed) and exact (solid) noise exposure forecast (NEF) contours for the year 1975 at an international airport in the United States with $X \approx 300$.

tion or other large expenditures are contemplated, an accurate determination of the NEF contours must be made in order to avoid serious errors.

18.3 Criteria for Noise in Buildings [1-4]

18.3.1 Noise Criteria for Mechanical (Ventilation) Equipment

Noise criteria in buildings have been developed both scientifically and empirically.[1,2,9,10] The results of experience in applying these criteria to indoor spaces are presented as NC or PNC ratings and approximate A-weighted sound levels in Table 18.7. In nearly every type of space, a range of ratings is indicated. The lower edge of the range should be used in buildings where high-quality environments are desired. The upper edge should only be used for situations where economy or the physical situation imposes conditions that make it impractical to achieve lower levels.

In the specification of an NC or PNC curve, it is intended that the sound-pressure level in all the octave frequency bands not exceed the levels given by the appropriate curve. It is common practice, however, to permit the noise level in one octave band to exceed the corresponding value on the specified criterion curve by as much as 2 dB, *provided* the levels in the two adjacent (one above and one below) octave bands in the same frequency region are only 1 dB below the criterion curve.

The consultant or architect must sometimes use his judgment in selecting a curve for a particular specification because of unusual

TABLE 18.7 Recommended Category Classification and Suggested Noise Criteria Range for Steady Background Noise as Heard in Various Indoor Functional Activity Areas

Type of space (and acoustical requirements)	NC or PNC curve (see Table 18.3 or Table 18.4)	Approximate L_A, dBA
Concert halls, opera houses, and recital halls (for listening to faint musical sounds)	10 to 20	21 to 30
Broadcast and recording studios (distant microphone pickup used)	10 to 20	21 to 30
Large auditoriums, large drama theaters, and large churches (for very good listening conditions)	Not to exceed 20	30
Broadcast, television, and recording studios (close microphone pickup used only)	Not to exceed 25	34
Small auditoriums, small theaters, small churches, music rehearsal rooms, large meeting and conference rooms (for very good listening), or executive offices and conference rooms for 50 people (no amplification)	Not to exceed 35	42
Bedrooms, sleeping quarters, hospitals, residences, apartments, hotels, motels, etc. (for sleeping, resting, relaxing)	25 to 40	34 to 47
Private or semiprivate offices, small conference rooms, classrooms, libraries, etc. (for good listening conditions)	30 to 40	38 to 47
Living rooms and drawing rooms in dwellings (for conversing or listening to radio and television)	30 to 40	38 to 47
Large offices, reception areas, retail shops and stores, cafeterias, restaurants, etc. (for moderately good listening conditions)	35 to 45	42 to 52
Lobbies, laboratory work spaces, drafting and engineering rooms, general secretarial areas (for fair listening conditions)	40 to 50	47 to 56
Light maintenance shops, office and computer equipment rooms, kitchens and laundries (for moderately fair listening conditions)	45 to 55	52 to 61
Shops, garages, power-plant control rooms, etc. (for just acceptable speech and telephone communication). Levels above NC- or PNC-60 are not recommended for any office or communication situation	50 to 60	56 to 66
For work spaces where speech or telephone communication is not required, but where there must be *no risk* of hearing damage	60 to 75	66 to 80

situations, such as people's attitudes toward noise, local expectations, or the need for extreme economy. In all cases, careful analysis of each design problem is required to ensure that the proper specifications for allowable noise levels are achieved.

18.3.2 Noise Criteria for Speech-privacy Design   A prime acoustical problem in many architectural spaces, such as multifamily dwellings, offices, and banks, is that of isolating speech sounds produced in one space from possible listeners in other spaces. In quiet spaces, speech privacy may be achieved only by heavy solid partitions, most likely multilayered, between adjacent rooms of concern. The noise isolation of the intervening walls is of prime importance but so are the background noise levels in the rooms. After passing through a wall, the reduced speech levels are understandable only if the noise is not intense enough to mask them. Thus the architect has two means available to ensure speech privacy between two adjacent rooms:

1. Noise reduction by the intervening wall
2. Level of the background noise in the receiving room

Watters first conceived of an analytical means for determining the degree of speech privacy between two adjacent rooms.[2] Cavanaugh, Farrell, Hirtle, and Watters developed a comprehensive scheme for achieving speech privacy in buildings.[47] This procedure takes into detailed account the characteristics of speech and hearing, wall transmission loss as a function of frequency, and background noise spectrum, as well as the source room characteristics and the degree of privacy required. Their procedure is recommended for precise design. They also point out that special consideration must be given to rooms used for special purposes or where the occupants may have special requirements, e.g., hospitals, embassies, and so on. They did not consider possible differences among people's desires in different nations.

A simplified procedure, based on the fundamental work of Cavanaugh, Farrell, Hirtle, and Watters, has been devised by Young,[15] which uses the standard single-number rating for walls, namely, sound-transmission class (STC) and the A-weighted background noise level in the listening space. Young's modification, presented here, comprises nine steps involving three graphs.

Step 1: Select the voice level. This level depends, of course, on the person speaking and on his voice level. For male voices use 60 dBA for conversational voice, 66 dBA for raised voice, and 72 dBA for loud voice. Call this quantity S.

Step 2: Determine the source room function. This number, when added to S, gives the typical voice level throughout the source room. Determine this quantity from Fig. 18.11 and call it F.

Step 3: Select the degree of speech privacy desired by the occupants of the building.   Use 15 for "confidential," 9 for "normal," or other values selected according to need.   Use larger numbers for absolute privacy at one extreme, or lower numbers for extreme economy at the other. Call this quantity P.

Step 4: Determine or select the sound-transmission class (STC) (see appendix to Chap. 11 for definition and references to sources of STC values) *of the partition (common wall) between the two rooms.*[48]   This number is also equal to the ISO index (International Standards Organization, Document 43-346, August, 1964).   Call this quantity I.

Step 5: Determine the effects of the size of the partition (common wall), compared to the size of the receiving room, and the effect of the amount of sound absorption in the receiving room.   This determination assumes that in many furnished rooms the reverberation time is near 0.5 sec and that the room height does not exceed 10 ft (3 m).   It also assumes that the reverberation time of a bare room is about 1.5 sec at mid-frequencies.   Obtain this quantity from Fig. 18.12 and call it K.

Step 6: Sum I and K.   Call this quantity D.

Step 7: Determine or select the A-weighted background noise level in the receiving room.   Call this quantity N.

Step 8: Compute the sound excess

$$X = (S + F) + P - D - N \qquad (18.9)$$

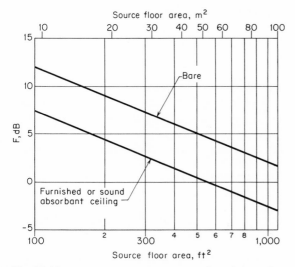

Fig. 18.11   Estimate of source room function F from room dimensions, for stud-and-plaster construction and room heights of about 10 ft.

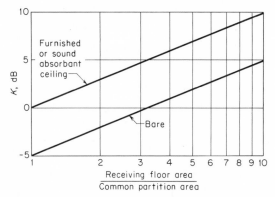

Fig. 18.12 Estimate of K from room dimensions, for stud-and-plaster construction and room heights of about 10 ft (3 m). The abscissa is the ratio of the floor area in the receiving room to the area of the common partition.

Step 9: Read the expected degree of satisfaction with the acoustical privacy from Fig. 18.13. If the degree of satisfaction indicated by Fig. 18.13 is too low or too high, either the STC of the wall or the background noise level, or both, may be adjusted accordingly. For maximum satisfaction, the background noise should conform to the lower limits of Table 18.7.

Example 18.3 Determine the degree of satisfaction with the speech privacy between two offices where voice levels are often raised, the office has a sound-absorbant ceiling, "confidential" privacy is desired, a 0.5-in.-gypsum-board-on-metal-stud partition is used, and the background noise is 40 dBA. Also, the floor area equals 500 ft² and is four times that of the common partition.
SOLUTION:
Step 1. $S = 66$ dBA
Step 2. $F = 1$ dB
Step 3. $P = 15$ dB
Step 4. $I = 39$ dB
Step 5. $K = 6$ dB
Step 6. $D = I + K = 45$ dB
Step 7. $N = 40$ dBA
Step 8. $X = 67 + 15 - 45 - 40 = -3$ dB
Step 9. Satisfaction

18.3.3 Background Noise for Open-plan Architectural Spaces

Open-plan architectural spaces differ from conventional ones in that the only allowable barriers for separating personnel acoustically are partial-height free-standing screens. In such spaces speech privacy is one of the most important design considerations, both to permit persons conversing to do so confidentially (banks, financial offices, and the like), and to allow personnel not conversing to concentrate (offices and schools). The problem to be solved is similar to that of the previous section except that the design must be more exact.

Speech privacy in an ordinary partitioned space requires that the background noise *exceed* some A-weighted level. This required level ordinarily does not hinder speech intelligibility (face-to-face) because the noise level need not be very high. In fact, with a good common wall, a range of background noise may be permitted, bounded at the lower limit by the speech-privacy requirement [see Eq. (18.9) and Fig. 18.13] and at the upper limit by the need to preserve speech intelligibility between people who may wish to converse (see Table 18.7).

In an open-plan space by contrast, often neither the speech-privacy requirement nor the speech-intelligibility requirement is ideally met. That is to say, the necessary background noise required for privacy is higher than that required for face-to-face intelligibility. Thus, a compromise must sometimes be made to arrive at the background noise that best satisfies the privacy and the speech-intelligibility requirements taken together.

In an open-plan space, a procedure for determining the compromise background noise may be utilized which resembles that of the preceding section. This procedure is based on several important assumptions: that the length and width of the open-plan space under consideration are large compared to its height; that the space be relatively dead acoustically with no large sound-reflecting areas on the ceiling (large light fixtures can be a problem). Also, open spaces with subareas largely enclosed by partial-height partitions do not qualify, because they will have acoustical characteristics like those of a small room. The computational steps are as follows:

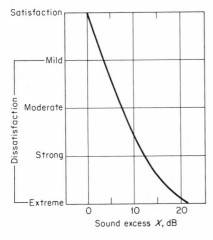

Fig. 18.13 Expected satisfaction with acoustical privacy vs. sound excess X.

Step 1: Select the expected voice level of the occupants. Use 54 dBA for low voice, 60 dBA for conversational voice, or 66 dBA for raised voice. These numbers are the voice levels produced by an average male talker at 3 ft (1 m). Generally, persons in an open-plan office must learn to use conversational voice, so that they will not disturb their neighbors. Thus, in design, a voice level of 60 dBA is chosen. Call this quantity S.

Step 2: Select the degree of speech privacy desired by the occupants of the space. Use 15 for "confidential" (financial institutions) or 9 for "normal" (concentration). Generally, one selects 9 for "normal" on the assumption that for "confidential" privacy, the persons involved must learn to use a "low" voice or will retire to enclosed conference rooms. Comparison of steps 1 and 2 shows that if voices are kept lower for "confidential" the sums of voice level and degree of privacy are alike for both normal and confidential occasions (that is, $54 + 15 = 60 + 9$). Thus the design for required separation between people and for partial-height partitions, if used, is the same. Call this quantity P.

Step 3: Determine from the table below the reduction in the voice level of the talker at the position of the nearest listener, a distance D away. At this point, we are interested only in the smallest value of D measured between people who want privacy relative to each other. It is assumed that people who wish to converse will be closer together. Call this quantity G.

	Quantity G			
	Distance D			
Room absorption	3 ft	6 ft	12 ft	24 ft
Acoustical ceiling only (NRC ≥ 0.75)	0	5	10	15
Acoustical ceiling and carpet (both with NRC ≥ 0.75)	0	6	12	18

Step 4: Determine the "equivalent STC" for the combination of source-receiver distance and barrier height. Use Fig. 18.14 for this purpose. [Note also that the barrier *length* must exceed the sum of the spread of positions (parallel to the baffle) of the talker or talkers and $4H$.] Call the "equivalent STC" I_{eq}.

Step 5: Determine the necessary A-weighted sound level L_A of the desired background noise from the formula

$$L_A = S + P - G - I_{eq} \qquad (18.10)$$

This noise level should yield reasonable satisfaction in regard to speech privacy for the person who "does not wish to be disturbed by intruding normal speech of others," rather than for "confidential privacy." As we said above, for greater privacy the person speaking must lower his voice. To convert L_A to approximate NC or PNC ratings, subtract 5 to 8 dB from L_A, using the difference between the two numerical columns of Table 18.7 as a guide.

Step 6: To determine whether the space will be acceptable with the background noise level of step 5 for two people who wish to converse, calculate the maximum distance they may be separated and still maintain satisfactory intelligibility. This distance is determined from the first column of Table 18.1 by inserting $L_A - 7$ dB $\approx$ PSIL into the column for "normal" voice effort. Also, the NC or PNC values (which also approximately equal $L_A - 7$ dB) can be entered into Table 18.7 to determine the adequacy of the space by comparison with partitioned spaces where the STC of the intervening wall is high. We generally find for good open-plan design that $L_A - 7$ dB $\approx$ PSIL is in the range of 43 to 48 dB. A level

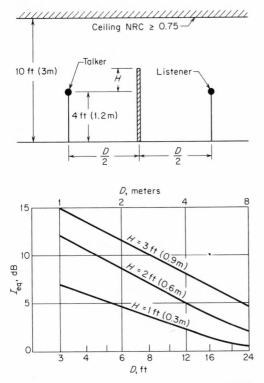

Fig. 18.14   Determination of the "equivalent STC" I_{eq} for a combination of source-listener distance D and excess barrier height H. The quantity H is the excess in barrier height above 4 ft (1.2 m). The room height is assumed to be 10 ft (3 m), and the ceiling is covered with a sound-absorbing material with NRC $\geqslant$ 0.75.

in this range permits men with average voices at normal speaking effort to converse up to about 12 ft, and women up to about 6 ft. These levels are high for high-quality office spaces but the aesthetic features of the open-plan concept are said to outweigh the disadvantages of the higher noise levels.

Design Considerations. If the distance between two people who wish to converse, as determined in step 6, or the NC or PNC rating number as entered into Table 18.7 is not satisfactory, either the occupants of the space who desire privacy from each other must be located farther apart so that G increases, or the partial-height barriers must be made taller (and also wider) so that the I_{eq} increases, or the degree of speech privacy must be compromised or lectures to large groups (more than about 12) must be held in a lecture hall with full-height partitions.

A crucial factor in open-plan design is *uniformity of background noise over the entire space,* as measured at the ear height of seated persons. This uniformity is necessary at all frequencies between about 200 and 5,000 Hz. One method of producing the background noise is to locate loudspeakers (in box baffles) above the acoustical tiles in a hung acoustical ceiling. Acoustical tiles generally transmit speech frequencies well. The loudspeakers must be chosen to have a conically shaped pattern of sound radiation at these frequencies such that the entire array, when properly spaced, gives the necessary uniformity. Low-grade loudspeakers usually have very narrow conical radiation patterns at the higher frequencies and thus "gaps" in the background noise distribution occur at locations between any two loudspeakers. A more efficient arrangement is to face the cones of the loudspeakers *upward* toward the hard slab above the hung ceiling. If not blocked by ductwork, this arrangement has the same effect as would increasing the height of the loudspeakers above the listeners by twice the distance between the slab and hung ceiling.

Imagine the result of a situation where a person talking sits beneath a loudspeaker that directs most of the noise downward while a nearby person desiring privacy sits between the loudspeakers and receives much less noise. The person talking would probably raise his voice to accommodate the louder noise, so that the one wishing to concentrate and who is sitting in the quieter place would hear the disturbing voice very clearly.

Spectrum Shape. The remaining question relates to the preferred shape of the background noise spectrum. If the background noise is produced electronically from loudspeakers located at frequent intervals, say above the acoustical tiles in a hung ceiling, both the level and the shape of the noise spectrum, as measured in the room at seated head height, can be adjusted. Lacking more specific information, it would seem logical to choose a noise spectrum that would have a shape such that in each frequency band the level would be related to the importance of that band in reducing the intelligibility of speech.[20] Assuming measurement in one-third-octave bands, a noise spectrum meeting this requirement would have its peak at about 2,000 Hz and would drop off at lower and higher frequencies. However, this spectrum, if used for background noise, would have a rather unpleasant sound, sounding shrill and unnatural in comparison with other background noises such as those from air-handling systems and from the occupants themselves.

The experience of the staff at Bolt Beranek and Newman Inc. suggests that noise spectra lying within the shaded region of Fig. 18.15 provide the desired quality of sound and can be produced by reasonably priced audio equipment, except for frequencies below 250 Hz. Background noise below 250 Hz usually is provided by the air-handling or air-conditioning equipment. The suggested spectrum of Fig. 18.15 is shaped similar to the spectrum of speech, which tends to make the background noise less noticeable in occupied offices. The spectrum

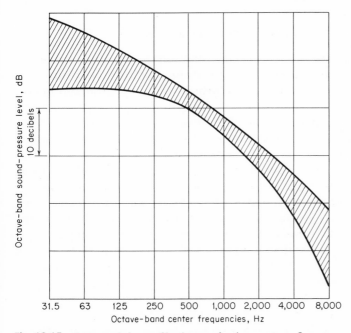

Fig. 18.15 Suggested shape of background noise spectrum for open-plan architectural spaces. If the noise spectrum is exactly that of the upper edge of the shaded area, add 7 dB to the level for the 1,000-Hz band to obtain L_A, the A-weighted level for the noise, in dBA. For a noise with the shape of the lower edge of the shaded area, $L_A = L_{1,000 \text{ Hz}} + 6$, dBA.

should, of course, be smooth, that is to say, no band should significantly protrude above or fall below the average line drawn through the individual band levels.

Noise may be produced electronically, as has been suggested above, or by air-driven sources. One such type of air-driven source is the air-handling diffuser that is usually located in the ceiling of each space. Two difficulties arise in the case of air diffusers. First, usually they are not located near enough to each other to provide a spatially uniform noise in the space. Second, the noise often differs considerably from one diffuser to another because of nonuniformity of air supply from the duct system to which they are connected. Another type of noise source could be a set of aerodynamically driven noise generators that are connected by flexible tubing to some form of constant-pressure air supply. Such a system would essentially be maintenance free and could be installed by the same trade union that installs the air-handling equipment. On the other hand, the level and spectrum of such a noise source are less easily changed to suit local requirements than those of noise from an electronic source.

18.4   Criteria for Vibration and Noise Inside Vehicles

Vibration and noise are controlled inside vehicles for one or more of four reasons, comfort, ease of voice communication, freedom from hearing-damage risk, and ability to hear warning signals that originate outside.

In the previous chapter, criteria for damage risk to hearing were discussed. Those criteria will be compared here to the practices followed by manufacturers for truck cabs and aircraft flight decks.

Vibration may cause discomfort, decreased ability to perform tasks, or motion sickness. The greatest research has been done on reduced comfort in the frequency range of 1 to 300 Hz, above the range for motion sickness. In this range, resonances in the body cause tissues to deform which results in discomfort or, for large amplitudes, in pain. Research dealing with the degradation of task performance includes such items as ability to respond to stimuli, to track a moving dot on a screen by hand movement of a control stick, or to read numbers or dials. Intense vibration may cause nervovascular disturbances, or damage to bones and joints or numbness and pain in muscles.

Face-to-face speech communication has been discussed in Sec. 18.1.1. With noise-canceling microphones and noise-excluding earphones, speech communication can take place over telephone equipment in exceedingly noisy environments,[20] such as in military tanks, on airport ramps, in ships' engine rooms, and in metal processing factories.

18.4.1 Criteria for Human Response to Vibration   Criteria for human response to vibration are useful for designing new vehicles or for evaluating and improving existing vehicles.   The literature on this subject has been compiled and reviewed [49] and compared to a proposed ISO standard.[50]

Motion sickness is very difficult to relate to magnitudes or frequencies of motion because visual cues, odors, ambient temperatures, fatigue, and exposure history are also factors of prime importance.   In brief, peak accelerations in excess of $0.15g$ at frequencies between 0.1 and 1.0 Hz can cause motion sickness.

Bender and Collins [49] conclude that the ISO standard fits well the available data on human comfort, decreased proficiency, fatigue, or damage to body, except at frequencies below 2 Hz where measurements are difficult to perform.   In addition, they recommend extension of the ISO standard up to the frequency of 1,000 Hz.

The ISO tentative criteria are in the form of one set of curves, related to three different scales.   As seen in Fig. 18.16, the three scales are for (1) the limit of daily exposure as determined by extreme discomfort or bodily damage, (2) decreased proficiency at manual or visual tasks owing to fatigue, and (3) reduced comfort.   The numbers on the ordinate scales are either the amplitudes (peak values) of sinusoidal acceler-

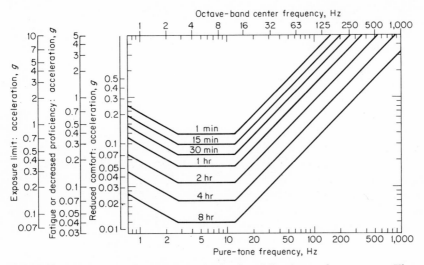

Fig. 18.16 Recommended vibration criteria for daily times of exposure. The numbers on ordinate scales are the amplitudes (peak values) of sinusoidal accelerations with the frequencies indicated on the lower scale or $\sqrt{2}$ times the rms values of random vibrations measured in octave bands with the center frequencies indicated on the upper scale. The daily time of exposure is the parameter on each curve. Three degrees of limits to human response are shown: (1) exposure limit; (2) fatigue or decreased efficiency; (3) discomfort.

ations or $\sqrt{2}$ times the rms values of random vibrations, measured in octave bands. The unit is $g = 9.8$ m/sec^2 (32.2 ft/sec^2). The abscissa is the sinusoidal frequency or the octave-band center frequency in hertz. Each curve applies to an exposure per day in minutes or hours.

The criteria in Fig. 18.16 are drawn for vertical vibration with the person either standing or sitting on a vibrating surface or chair. The curves are to be lowered by 3 dB (a factor of $\sqrt{2}$) on the graph for horizontal vibration (chest-to-back or side-to-side). That is to say, the body is more sensitive to horizontal than to vertical vibration.

For discrete-frequency motion, compare the amplitude of the sine wave of acceleration with the criteria. For example, a vertical excitation of $0.1g$ at 10 Hz would meet the reduced-comfort criterion if the duration were less than about 15 min, would not lead to fatigue and decreased proficiency if the exposure were under about 2 hr, and would not exceed the exposure-limit boundary if the excitation were less than about 6 hr.

When vertical excitation consists of several discrete frequencies, or of bands of frequencies, the component values are weighted and summed by the curves of Fig. 18.16. For example $0.1g$ at 16 Hz plus $0.4g$ at 32 Hz equals $0.1g + 0.2g = 0.3g$ at 16 Hz, or $0.17g + 0.4g = 0.57g$ at 32 Hz (note that $0.3g$ at 16 Hz lies on the same contour as $0.57g$ at 32 Hz).

18.4.2 Criteria for Noise Levels inside Ground Vehicles

Truck Cabs. The primary requirement for noise control in truck cabs is that the driver not suffer damage to his hearing. The Society for Automotive Engineers has adopted a recommended practice (SAE J336) that specifies that the noise level in a cab, measured 6 in. to the right of the driver's ear, should not exceed about 88 dBA. The test is conducted on a smooth dry road and a gear ratio is specified that produces a truck speed of approximately 50 mph at rated engine speed. The noise is measured during acceleration at full throttle from a beginning engine speed of one-half rated engine speed up to the rated speed. This noise level of 88 dBA, if not exceeded by air-conditioning levels, meets the Walsh-Healey requirements of the previous chapter.

Passenger Automobiles. Criteria for allowable noise inside passenger automobiles are not standardized because "quiet" is an important factor in competition among manufacturers. The most important considerations in noise control are to provide "quality" sound free from rapidly varying changes as a function of speed, particularly during acceleration and deceleration. Interference with speech communication among passengers is also an important consideration.

Typical noise spectra in compact and medium-priced 1970-model American automobiles are shown in Fig. 18.17. A survey of A-weighted

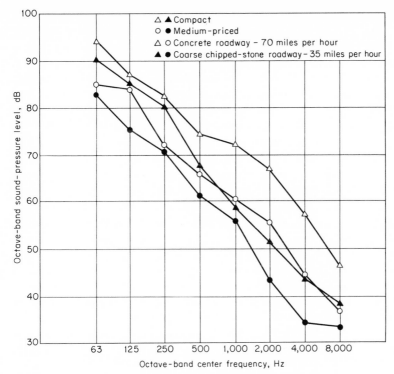

Fig. 18.17 Octave-band sound-pressure levels in two 1970 American automobiles at front-seat passenger-ear location. Windows were closed.

noise levels measured in 39 different 1970-model automobiles from the five principal American manufacturers is given in Fig. 18.18, for three conditions of operation.   In the better cars, the A level at the front-seat passenger-ear location is less than 70 dBA (about PSIL < 63 dB, which corresponds to "raised voice" at 4 to 5 ft).

Below 40 mph, road-tire excited sounds predominate in the passenger compartment.   At higher steady speeds, noise that is generated by wind flow through cracks (i.e., airflow resulting from aerodynamically induced pressure differences) is a significant component in the total noise. During high acceleration the engine and exhaust are major sources of noise.   The sound-transmission loss (STL) of fire wall and floor areas is very important in control of the engine and exhaust noise that enters the passenger compartment.

18.4.3 Noise Levels inside Jet Transports

Passenger Cabin.   Noise levels inside an airplane are a factor in a passenger's choice of the airline he chooses to fly.   Elements that influence his

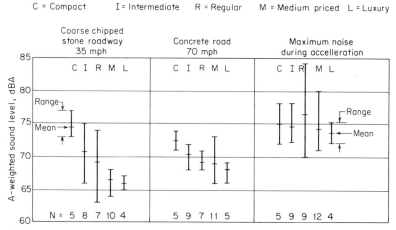

Fig. 18.18 Sound levels in 39 different 1970-model American passenger cars at front-seat passenger-ear location. *N* = sample size in each category. Windows were closed.

decision are his ability to converse with passengers in adjacent seats and the smoothness and steadiness of the noise spectrum. The interior noise levels, especially during cruise flight conditions, should not be too loud and should not have discernible discrete tones, or beats and pulsations.

The amount of noise reduction achievable in an airplane at frequencies below 800 Hz is determined primarily by the stiffness and weight of the basic structure of the fuselage. At frequencies above 800 Hz, some additional noise reduction (beyond that provided by the fuselage structure) can be achieved by the installation of absorptive acoustical materials. The amount of total noise reduction achievable is set by allowable added weight, space available for treatment, and cost, by the amount of flanking sound transmission through windows, floors and structural attachment points for interior-trim and airline-operator items, and by noise generated in the ventilation system or other onboard equipment.

The range of noise levels in the passenger cabin of four-engine commercial jet-transport aircraft is indicated in Fig. 18.19. These aircraft were powered by turbojet or by low-bypass-ratio turbofan engines. The noise levels pertain to normal long-range cruise conditions and were measured at head height at aisle and window-seats, with cabin air conditioning in normal operation and with individual passenger "eyeball" ventilators closed. The levels at the upper boundary of the range are found near the rear of the cabin where the boundary layer over the fuselage is relatively thick. The levels at the lower boundary are found near the front of the cabin where the boundary layer is relatively thin.   The noise levels at the window seat locations are usually similar to those at corresponding aisle seat locations except that the levels in the low frequencies tend to be somewhat higher at the window seats than at the aisle seats.

Cockpit.   The range of levels in the cockpit, measured at head positions of the pilot and copilot under normal long-range cruise conditions, is shown in Fig. 18.20.   Comparing these results to those shown in Fig. 18.19, the sound pressure levels in the cockpit tend to be somewhat higher at frequencies above 500 Hz and somewhat lower at frequencies below 500 Hz than in the passenger cabin.   As a result, the speech interference levels in the cockpit tend to be higher than in the passenger cabin.

The reason for the higher levels in the cockpit at frequencies above 500 Hz is considered to be primarily the result of higher-level boundary-layer noise outside the cockpit in conjunction with less fuselage noise reduction than achieved with the structure and the insulation that exists in the cabin.   The boundary-layer noise is higher because of flow separation and high local Mach numbers that occur in regions of rapid contour changes around the nose of the aircraft and the cockpit windows.   The air conditioning system and the equipment located near the cockpit may also contribute to the high-frequency interior noise levels.   The

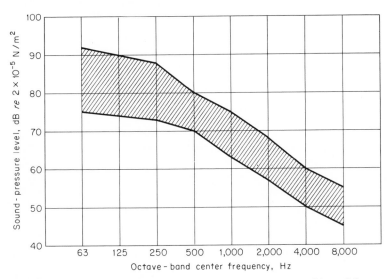

Fig. 18.19 Typical noise levels measured inside passenger cabins of first-generation long-range commercial jet transports at cruise conditions. Such transports are powered by turbojet engines or by turbofan engines with by-pass air flows about 1.4 times primary air flows.

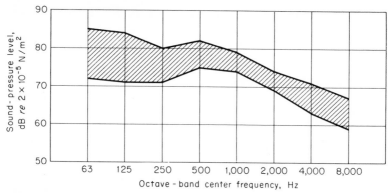

Fig. 18.20   Same as Fig. 18.19, except noise levels measured inside cockpit.

cockpit noise reduction is less than the cabin noise reduction because of additional flanking transmission paths (e.g., from the relatively larger area of windows) and because of the reduced space available within the cockpit to effectively install sound-absorptive treatment.

REFERENCES

1. L. L. Beranek, *Acoustics,* Chap. 13, McGraw-Hill, 1954.
2. L. L. Beranek, *Noise Reduction,* Chap. 20, McGraw-Hill, 1960.
3. L. L. Beranek, "Literature Survey for the FHA Contract on Urban Noise," *Bolt Beranek and Newman Inc. Rept.* 1460, January, 1967.
4. T. J. Schultz, "Technical Background for Noise Abatement in HUD's Operating Programs," *Bolt Beranek and Newman Inc. Rept.* 2005, September, 1970. Also, J. C. Webster, "SIL, Past, Present, and Future," *Sound and Vibration,* vol. 3, pp. 22–26, August, 1969.
5. K. D. Kryter, *The Effects of Noise on Man,* Academic, 1970.
6. S. S. Stevens, "Procedure for Predicting Loudness: Mark VI," *J. Acoust. Soc. Am.,* vol. 33, pp. 1577–1585, November, 1961.
7. "USA Standard Procedure for the Computation of Loudness of Noise," USAS 53.4, 1968, American National Standards Institute, Inc., 1430 Broadway, New York, N.Y. 10018.
8. ISO Recommendation R-532, 1966, "Method of Calculating Loudness Level."
9. L. L. Beranek, "Revised Criteria for Noise in Buildings," *Noise Control,* vol. 3, pp. 19–27, 1957.
10. T. J. Schultz, "Noise-criterion Curves for Use with the USASI Preferred Frequencies," *J. Acoust. Soc. Am.,* vol. 43, pp. 637–638, 1968.
11. R. W. Young, "Don't Forget the Simple Sound Level Meter," *Noise Control,* vol. 4, pp. 42–43, 1958. Also, J. H. Botsford, "Using Sound Levels to Gauge Human Response to Noise," *Sound and Vibration,* vol. 3, pp. 16–28, October, 1969.
12. R. W. Young, "Single Number Criteria for Room Noise," *J. Acoust. Soc. Am.,* vol. 36, pp. 289–295, 1964.
13. R. W. Young, "Measurement of Noise Level and Exposure," part II of Ref. 30.
14. R. W. Young and A. Peterson, "On Estimating Noisiness of Aircraft Sounds," *J. Acoust. Soc. Am.,* vol. 45, pp. 834–838, 1969.

15. R. W. Young, "Re-vision of the Speech-privacy Calculations," *J. Acoust. Soc. Am.,* vol. 38, pp. 524–530, 1965.
16. D. W. Robinson, "Towards a Unified System of Noise Assessment," *J. Sound Vibration,* vol. 14, 1971.
17. D. W. Robinson, "An Outline Guide to Criteria for the Limitation of Urban Noise," *Aeron. Res. Council Current Paper* 1112, H. M. Stationery Office, London, 1970.
18. ISO Recommendation R-266, 1962, "Preferred Frequencies for Acoustical Measurements," Ref. No. ISO/R-266-1962.
19. L. L. Beranek, "Specification of Acceptable Noise Levels," *Trans. ASME,* vol. 69, pp. 97–100, 1947.
20. L. L. Beranek, "Design of Speech Communication Systems," *Proc. Inst. Radio Engrs.,* vol. 35, pp. 880–890, 1947.
21. J. C. Webster, "Effects of Noise on Speech Intelligibility," pp. 49–73 of Ref. 29, 1969.
22. S. Fidell and K. S. Pearsons, "Audibility of Impulsive Sounds," *Bolt Beranek and Newman Inc. Rept.* 1851.
23. D. W. Robinson and L. S. Whittle, "The Loudness of Octave Bands of Noise," *Acustica,* vol. 14, p. 33, Fig. 14, 1964.
24. C. G. Bottom, "Progress toward the Unification of Noise Criteria," presented before British Acoustical Society, July 23, 1970.
25. W. E. Blazier, Jr., "A Field Measurement Study of the Sound Levels Produced Outdoors by Residential Air-conditioning Equipment," *ASHRAE J.,* pp. 36–39, May, 1967.
26. P. N. Borsky, "Community Reactions to Air Force Noise," *WADD Tech. Rept.* 60-689, Parts 1 and 2, Wright-Patterson Air Force Base, March, 1961.
27. P. A. Franken and G. Jones, "On Response to Community Noise," *Appl. Acoust. (Brit.),* vol. 2, pp. 241–246, 1969.
28. Cyril M. Harris, *Handbook of Noise Control,* Chaps. 35–40, McGraw-Hill, 1957.
29. "Noise as a Public Health Hazard," *Proc. Conf. Speech Hearing Assoc., Washington, D.C., February, 1969.*
30. J. D. Chalupnik, *Transportation Noises, A Symposium on Acceptability Criteria,* University of Washington Press, 1970.
31. K. D. Kryter, "Annoyance," pp. 69–84 of Ref. 30, 1970.
32. "Noise," final report prepared by the Committee on the Problem of Noise and presented to Parliament in July, 1963, H. M. Stationery Office, London, Cmnd. 2056.
33. "Urban Traffic Noise," DAS/CSI/68.47 (March, 1969) and (October, 1969); and 75.174 (April, 1970), Organization for Economic Co-operation and Development, Paris.
34. W. E. Scholes, "Traffic Noise Criteria," *Appl. Acoust.,* vol. 3, pp. 1–21, 1970.
35. W. J. Galloway and D. E. Bishop, "Noise Exposure Forecast: Evolution, Evaluation, Extensions and Land Use Interpretations," *FAA Rept.* No-70-9, August, 1970.
36. A. Cohen, "Noise Effects on Health, Productivity and Well-Being," *Trans. N.Y. Acad. Sci.,* ser. II, vol. 30, pp. 910–918, May, 1968. A. Cohen, J. Anticaglia, and H. H. Jones, "Sociocusis, Hearing Loss from Non-occupational Noise Exposure," *Sound and Vibration,* vol. 4, pp. 12–20, November, 1970.
37. M. Hatfield and J. J. Kaufman, "Compilation of State and Local Ordinances on Noise Control," *Congressional Record,* House of Representatives, vol. 115, pp. E9031–E9112, Oct. 29, 1969.
38. P. H. Parkin, H. J. Purkis, R. J. Stephenson, and B. Schlaffenberg, *London Noise Survey,* Building Research Station, Ministry of Public Building and Works, 1968, H. M. Stationery Office, London.
39. G. J. Thiessen, "Community Noise Levels," pp. 23–32 of Ref. 30, 1970.
40. L. L. Beranek, "Noise," *Sci. Am.,* vol. 215, pp. 66–76, December, 1966.

41. "Federal Aviation Regulations, Part 36: Noise Standards, Aircraft Type Certification," *Federal Register,* vol. 34, no. 226, pp. 18,815–18,878, Nov. 25, 1969.
42. W. C. Sperry, "Aircraft Noise Evaluation," *FAA Rept.* 68-34, September, 1958.
43. Department of California Highway Patrol, "Vehicle Noise Measurement Regulations" and "Vehicle Code Section 23, 130, as Amended in November 1970," January, 1968, P.O. Box 898, Sacramento, Calif. 95804.
44. "Research on Establishment of Standards for Highway Noise Levels," *Bolt Beranek and Newman Inc. Rept.* 1195, February, 1965.
45. "Objective Limits for Motor Vehicle Noise," *Bolt Beranek and Newman Inc. Rept.* 824, December, 1962.
46. D. E. Bishop and M. A. Simpson, "Noise Exposure Forecast Contours for 1967, 1970 and 1975 Operations at Selected Airports," *FAA Rept.* No-70-8, September, 1970.
47. W. J. Cavanaugh, W. R. Farrell, P. W. Hirtle, and B. G. Watters, "Speech Privacy in Buildings," *J. Acoust. Soc. Am.,* vol. 34, pp. 475–492, 1962.
48. R. D. Berendt, G. E. Winzer, and C. B. Burroughs, "A Guide to Airborne, Impact, and Structure-borne Noise Control in Multi-family Dwellings," *U.S. Department of Housing and Urban Development Rept.* FT/TS, January, 1968.   Order from Superintendent of Documents, U.S. Government Printing Office, Washington, D.C. 20402 ($2.50).
49. E. K. Bender and A. M. Collins, "Effects of Vibration on Human Performance: A Literature Review," *Bolt Beranek and Newman Inc. Rept.* 1767, Feb. 15, 1969.
50. *Evaluation Exposure of Humans to Whole Body Vibration, Intern. Organ. Standardization Rept.* 108/WG-7 (Secr. 8), February, 1967.
51. L. L. Beranek, W. E. Blazier, and J. J. Figwer, "Preferred Noise Criteria (PNC) Curves and Their Applications," presented at the 81st Meeting of the Acoustical Society of America, Washington, D.C., April 20, 1971.

Appendix to Chapter 18

E-WEIGHTED SOUND LEVEL

In a private communication of January 1971, S. S. Stevens described to the author the concept of a new "Ear-Weighted Sound-Level Meter." (Paper submitted to *J. Acoust. Soc. Am.* and titled "Perceived Level of Noise — Calculation Procedure Mark VII.")   If Stevens' Mark VII procedure for calculating "Perceived Level P in PLdB" were to be adopted by appropriate standards organizations, a unique opportunity would arise to add an "ear-weighting" to the sound-level meter that would yield sound-level meter readings in "dBE," which would approximately be equal numerically to the perceived levels calculated by Mark VII.   That is to say, the perceived level, which is an improved form of calculated loudness level, would be determined accurately from sound-pressure levels in third-octave bands, and it would be determined approximately by the levels read on an E-weighted meter.

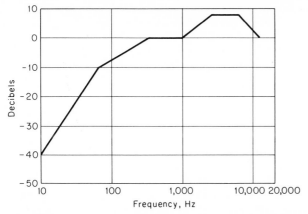

Fig. A.18.1   An electrical E-weighting characteristic for the sound-level meter that would yield readings in dBE approximately equal to S. S. Stevens' new perceived level in PLdB. The physical network for producing this characteristic would need to meet practical space and weight limitations. The electrical result would be a smoothed approximation to the characteristic which would give closely the same meter indications.

The practical result would be replacement of Stevens' loudness level, Zwicker's loudness level, and perceived noise level by the new perceived level, and replacement of A-weighted and N-weighted levels by E-weighted level. The new perceived level in PLdB would be *about* 13 dB lower than Zwicker's loudness level, 9 dB lower than perceived noise level, 8 dB lower than Stevens' loudness level, 2 dB lower than N-weighted level and 4 dB higher than A-weighted level.

The approximate shape of Stevens' E-weighted characteristic for the sound level meter is given in Fig. A.18.1. A practical network design would likely not be a six-segment curve with sharp transition points, but would probably be the best approximation to it whose physical embodiment could be fitted into the confines of a hand-held sound-level meter.

Appendix A
General References

R. A. Anderson, *Fundamentals of Vibrations,* Macmillan, 1967.

L. L. Beranek, *Acoustics,* McGraw-Hill, 1954.

L. L. Beranek, *Acoustic Measurements,* Wiley, 1949.

J. D. Chalupnik, (ed), *Transportation Noises, A Symposium on Acceptability Criteria,* University of Washington Press, 1970.

L. Cremer and M. Heckl, *Koerperschall,* Springer-Verlag, 1967.

W. Furrer, *Room and Building Acoustics and Noise Abatement,* Butterworth, 1964.

C. M. Harris, *Handbook of Noise Control,* McGraw-Hill, 1957.

D. N. Keast, *Measurements in Mechanical Dynamics,* McGraw-Hill, 1967.

K. D. Kryter, *The Effects of Noise on Man,* Academic, 1970.

G. Kurtze, *Physik und Technik der Laermbekaempfung,* Verlag G. Braun, 1964.

P. M. Morse and K. U. Ingard, *Theoretical Acoustics,* McGraw-Hill, 1968.

A. P. G. Peterson and E. E. Gross, Jr., *Handbook of Noise Measurement,* General Radio Company, 1967.

E. Skudrzyk, *Simple and Complex Vibratory Systems,* Pennsylvania State University Press, 1968.

H. Wiethaup, *Die Laermbekaempfung in der Bundesrepublik Deutschland,* C. Heymanns Verlag, 1961.

Appendix B
English System of Units

The English system of units is inherently confusing. In everyday American life, the *pound* (abbreviated lb) is used either as a force or as a weight, both having the units of force. To further complicate matters, some technical writers seek to provide a system parallel to the metric system by using either of two quantities for force and either of two corresponding quantities for mass. Thus, some adopt the pound as the unit of force and define a *slug* as the unit of mass. Others define a *poundal* as the unit of force and adopt the pound as the unit of mass. Neither the slug nor the poundal has found general acceptance in the literature, although we use the former in this text.

As one well-traveled acoustician said, "I have determined that 1 kg of butter bought in Zurich is exactly the same amount as 2.2 lb bought in New York. Whenever I wish to solve a technical problem in America without confusion, I immediately divide the number of pounds by 2.2 to obtain the equivalent number of kilograms. Then I work in the mks system, where force and mass are clearly distinguished."

Consistent Systems of Units Used in This Text

Two consistent systems of units are used in this text, the *mks* and the *fss* systems. To describe them, let us start with Newton's second law.

Force = mass × acceleration   (B.1)

In the *meter-kilogram-second* system
No. of newtons = no. of kg × no. of m/sec^2 (B.2)

In the *foot-slug-second* system
No. of pounds force (lb) = no. of slugs × no. of ft/sec^2 (B.3)

The relations among the magnitudes of the units are

$$1 \text{ kg} = 2.205 \text{ lb weight}$$
$$1 \text{ kg} = 0.0685 \text{ slug}$$
$$1 \text{ slug} = 14.59 \text{ kg}$$
$$1 \text{ newton} = 0.225 \text{ lb force}$$
$$1 \text{ lb force} = 4.448 \text{ newtons}$$
$$1 \text{ slug} = 32.17 \text{ lb weight}$$
$$1 \text{ lb weight} = 0.03108 \text{ slug} = 0.454 \text{ kg}$$

Example B.1: If 1 kg is to be accelerated 1 m/sec^2, we see, by Eq. (B.2), that a force of 1 newton is required. How many pounds (force) is required for the same result?

Solution: 1 kg equals (2.205/32.17) slug and 1 m/sec^2 = 3.28 ft/sec^2. Thus, by Eq. (B.3), 0.225 lb (force) is required.

Inconsistent Systems of English Units Used in the Literature

Two inconsistent systems of English units are commonly encountered, the *fps* and the *ips* systems.

In the *foot-pound-second* system (inconsistent system)

$$\text{No. of pounds force (lb)} = \frac{\text{no. of pounds weight (lb)}}{g} \times \text{no. of ft/sec}^2 \qquad (B.4)$$

where g is the acceleration due to gravity in units of ft/sec^2, that is, 32.17 ft/sec^2.

In the *inch-pound-second system* (inconsistent system)

$$\text{No. of pounds force (lb)} = \frac{\text{no. of pounds weight (lb)}}{g} \times \text{no. of in./sec}^2 \qquad (B.5)$$

where g is the acceleration due to gravity in units of in./sec^2, that is, 386 in./sec^2.

Mechanical engineers often use the *in.-lb-sec* system in the field of shock and vibration.

Example B.2: One kilogram is accelerated five meters per second^2. Find the force necessary to do this in newtons and pounds (force).

Solution:
$$1 \text{ kg} = 2.2 \text{ lb weight} = 0.0685 \text{ slug}$$
$$5 \text{ m/sec}^2 = 16.4 \text{ ft/sec}^2$$
$$F \text{ (newtons)} = 1 \times 5 = 5 \text{ newtons}$$
$$F \text{ (lb)} = 0.0685 \times 16.4 = 1.124 \text{ lb (force)}$$

Conversion Factors

The following values for the fundamental constants were used in the preparation of the factors:

$$1 \text{ m} = 39.37 \text{ in.} = 3.281 \text{ ft}$$
$$1 \text{ lb (weight)} = 0.4536 \text{ kg} = 0.03108 \text{ slug}$$
$$1 \text{ slug} = 14.594 \text{ kg}$$
$$1 \text{ lb (force)} = 4.448 \text{ newtons}$$
$$\text{Acceleration due to gravity} = 9.807 \text{ m/sec}^2$$
$$= 32.174 \text{ ft/sec}^2$$
$$\text{Density of } H_2O \text{ at } 4^\circ C = 10^3 \text{ kg/m}^3$$
$$\text{Density of Hg at } 0^\circ C = 1.3595 \times 10^4 \text{ kg/m}^3$$
$$1 \text{ U.S. lb} = 1 \text{ British lb}$$
$$1 \text{ U.S. gallon} = 0.83267 \text{ British gallon}$$

TABLE C.1   Conversion Factors

To convert	Into	Multiply by	Conversely, multiply by
acres	ft^2	4.356 x 10^4	2.296 x 10^{-5}
	miles^2 (statute)	1.562 x 10^{-3}	640
	m^2	4,047	2.471 x 10^{-4}
	hectare (10^4 m^2)	0.4047	2.471
atm	in. H$_2$O at 4°C	406.80	2.458 x 10^{-3}
	in. Hg at 0°C	29.92	3.342 x 10^{-2}
	ft H$_2$O at 4°C	33.90	2.950 x 10^{-2}
	mm Hg at 0°C	760	1.316 x 10^{-3}
	lb/in.^{2}	14.70	6.805 x 10^{-2}
	newtons/m^2	1.0132 x 10^5	9.872 x 10^{-6}
	kg/m^2	1.033 x 10^4	9.681 x 10^{-5}
°C	°F	(°C x 9/5) + 32	(°F − 32) x 5/9
cm	in.	0.3937	2.540
	ft	3.281 x 10^{-2}	30.48
	m	10^{-2}	10^2
circular mils	in.^{2}	7.85 x 10^{-7}	1.274 x 10^6
	cm^2	5.067 x 10^{-6}	1.974 x 10^5
cm^2	in.^{2}	0.1550	6.452
	ft^2	1.0764 x 10^{-3}	929
	m^2	10^{-4}	10^4
cm^3	in.^{3}	0.06102	16.387
	ft^3	3.531 x 10^{-5}	2.832 x 10^4
	m^3	10^{-6}	10^6
deg (angle)	radians	1.745 x 10^{-2}	57.30
dynes	lb (force)	2.248 x 10^{-6}	4.448 x 10^5
	newtons	10^{-5}	10^5
dynes/cm^2	lb/ft^2 (force)	2.090 x 10^{-3}	478.5
	newtons/m^2	10^{-1}	10
ergs	ft-lb (force)	7.376 x 10^{-8}	1.356 x 10^7
	joules	10^{-7}	10^7
ergs/cm^3	ft-lb/ft^3	2.089 x 10^{-3}	478.7
ergs/sec	watts	10^{-7}	10^7
	ft-lb/sec	7.376 x 10^{-8}	1.356 x 10^7
ergs/sec-cm^2	ft-lb/sec-ft^2	6.847 x 10^{-5}	1.4605 x 10^4
fathoms	ft	6	0.16667
ft	in.	12	0.08333
	cm	30.48	3.281 x 10^{-2}
	m	0.3048	3.281
ft^2	in.^{2}	144	6.945 x 10^{-3}
	cm^2	9.290 x 10^2	0.010764
	m^2	9.290 x 10^{-2}	10.764
ft^3	in.^{3}	1728	5.787 x 10^{-4}
	cm^3	2.832 x 10^4	3.531 x 10^{-5}
	m^3	2.832 x 10^{-2}	35.31
	liters	28.32	3.531 x 10^{-2}

TABLE C.1   Conversion Factors—Continued

To convert	Into	Multiply by	Conversely, multiply by
ft H_2O at $4°C$	in. Hg at $0°C$	0.8826	1.133
	lb/in.^{2}	0.4335	2.307
	lb/ft^2	62.43	1.602×10^{-2}
	newtons/m^2	2989	3.345×10^{-4}
gal (liquid U.S.)	gal (liquid Brit. Imp.)	0.8327	1.2010
	liters	3.785	0.2642
	m^3	3.785×10^{-3}	264.2
gm	oz (weight)	3.527×10^{-2}	28.35
	lb (weight)	2.205×10^{-3}	453.6
hp (550 ft-lb/sec)	ft-lb/min	3.3×10^4	3.030×10^{-5}
	watts	745.7	1.341×10^{-3}
	kw	0.7457	1.341
in.	ft	0.0833	12
	cm	2.540	0.3937
	m	0.0254	39.37
in.^{2}	ft^2	0.006945	144
	cm^2	6.452	0.1550
	m^2	6.452×10^{-4}	1550
in.^{3}	ft^3	5.787×10^{-4}	1.728×10^3
	cm^3	16.387	6.102×10^{-2}
	m^3	1.639×10^{-5}	6.102×10^4
kg	lb (weight)	2.2046	0.4536
	slug	0.06852	14.594
	gm	10^3	10^{-3}
kg/m^2	lb/in.^{2} (weight)	0.001422	703.0
	lb/ft^2 (weight)	0.2048	4.882
	gm/cm^2	10^{-1}	10
kg/m^3	lb/in.^{3} (weight)	3.613×10^{-5}	2.768×10^4
	lb/ft^3 (weight)	6.243×10^{-2}	16.02
liters	in.^{3}	61.03	1.639×10^{-2}
	ft^3	0.03532	28.32
	pints (liquid U.S.)	2.1134	0.47318
	quarts (liquid U.S.)	1.0567	0.94636
	gal (liquid U.S.)	0.2642	3.785
	cm^3	1000	0.001
	m^3	0.001	1000
$\log_e n$, or $\ln n$	$\log_{10} n$	0.4343	2.303
m	in.	39.371	0.02540
	ft	3.2808	0.30481
	yd	1.0936	0.9144
	cm	10^2	10^{-2}
m^2	in.^{2}	1550	6.452×10^{-4}
	ft^2	10.764	9.290×10^{-2}
	yd^2	1.196	0.8362
	cm^2	10^4	10^{-4}
m^3	in.^{3}	6.102×10^4	1.639×10^{-5}
	ft^3	35.31	2.832×10^{-2}

TABLE C.1 Conversion Factors—Continued

To convert	Into	Multiply by	Conversely, multiply by
m^3 *(Cont.)*	yd^3	1.3080	0.7646
	cm^3	10^6	10^{-6}
microbars (dynes/cm²)	lb/in.²	1.4513×10^{-5}	6.890×10^4
	lb/ft²	2.090×10^{-3}	478.5
	newtons/m²	10^{-1}	10
miles (nautical)	ft	6080	1.645×10^{-4}
	km	1.852	0.5400
miles (statute)	ft	5280	1.894×01^{-4}
	km	1.6093	0.6214
miles² (statute)	ft²	2.788×10^7	3.587×10^{-8}
	km²	2.590	0.3861
	acres	640	1.5625×10^{-3}
mph	ft/min	88	1.136×10^{-2}
	km/min	2.682×10^{-2}	37.28
	km/hr	1.6093	0.6214
nepers	db	8.686	0.1151
newtons	lb (force)	0.2248	4.448
	dynes	10^5	10^{-5}
newtons/m²	lb/in.² (force)	1.4513×10^{-2}	6.890×10^3
	lb/ft² (force)	2.090×10^{-2}	47.85
	dynes/cm²	10	10^{-1}
lb (force)	newtons	4.448	0.2248
lb (weight)	slugs	0.03108	32.17
	kg	0.4536	2.2046
lb H_2O (distilled)	ft³	1.602×10^{-2}	62.43
	gal (liquid U.S.)	0.1198	8.346
lb/in.² (weight)	lb/ft² (weight)	144	6.945×10^{-3}
	kg/m²	703	1.422×10^{-3}
lb/in.² (force)	lb/ft² (force)	144	6.945×10^{-3}
	N/m²	6894	1.4506×10^{-4}
lb/ft² (weight)	lb/in.² (weight)	6.945×10^{-3}	144
	gm/cm²	0.4882	2.0482
	kg/m²	4.882	0.2048
lb/ft² (force)	lb/in.² (force)	6.945×10^{-3}	144
	N/m²	47.85	2.090×10^{-2}
lb/ft³ (weight)	lb/in.³ (weight)	5.787×10^{-4}	1728
	kg/m³	16.02	6.243×10^{-2}
poundals	lb (force)	3.108×10^{-2}	32.17
	dynes	1.383×10^4	7.233×10^{-5}
	newtons	0.1382	7.232
slugs	lb (weight)	32.17	3.108×10^{-2}
	kg	14.594	0.06852
slugs/ft²	kg/m²	157.2	6.361×10^{-3}
tons, short (2,000 lb)	tonnes (1,000 kg)	0.9075	1.102
watts	ergs/sec	10^7	10^{-7}
	hp (550 ft-lb/sec)	1.341×10^{-3}	745.7

Appendix D
Decibel Conversion Tables*

It is convenient in measurements and calculations in electroacoustics to express the ratio between any two amounts of electric or acoustic power in units on a logarithmic scale. The *decibel* (1/10th of a *bel*) on the briggsian or base 10 scale is in almost universal use for this purpose.

Since voltage and sound pressure are related to power by impedance, the *decibel* can be used to express voltage and sound-pressure ratios, if care is taken to account for the impedances associated with them.

Tables D.1 and D.2 have been prepared to facilitate making conversions in either direction between the number of *decibels* and the corresponding power, voltage, and sound-pressure ratios. All numbers have been verified by digital computing machine.

To Find Values Outside the Range of Conversion Tables. Values outside the range of either Table D.1 or D.2 can be readily found with the help of the following simple rules:

TABLE D.1: *Decibels to Voltage, Sound-pressure, and Power Ratios*

Number of Decibels Positive (+). Subtract +20 db successively from the given number of decibels until the remainder falls within range of Table D.1. *To find the voltage (sound-pressure) ratio*, multiply the corresponding value from the

*Courtesy General Radio Company, West Concord, Mass.

right-hand voltage-ratio column by 10 for each time you subtracted 20 db. *To find the power ratio,* multiply the corresponding value from the right-hand power-ratio column by 100 for each time you subtracted 20 db.

Example.   *Given:* 49.2 db
49.2 db − 20 db − 20 db = 9.2 db

Voltage (sound-pressure) ratio: 9.2 db ⟶
2.884 x 10 x 10 = 288.4

Power ratio: 9.2 db ⟶
8.318 x 100 x 100 = 83,180

Number of Decibels Negative (−). Add +20 decibels successively to the given number of decibels until the sum falls within the range of Table D.1. *For the voltage (sound-pressure) ratio,* divide the value from the left-hand voltage-ratio column by 10 for each time you added 20 db. *For the power ratio,* divide the value from the left-hand power-ratio column by 100 for each time you added 20 db.

Example.   *Given:* −49.2 db
−49.2 db + 20 db + 20 db = −9.2 db

Voltage (sound-pressure) ratio: −9.2 db ⟶
0.3467 x 1/10 x 1/10 = 0.003467

Power ratio: −9.2 db ⟶
0.1202 x 1/100 x 1/100 = 0.00001202

TABLE D.2:   *Voltage and Sound-pressure Ratios to Decibels*

For Ratios Smaller than Those in Table. Multiply the given ratio by 10 successively until the product can be found in the table. From the number of decibels thus found, subtract +20 db for each time you multiplied by 10.

Example.   *Given:* Voltage (sound-pressure) ratio = 0.0131
0.0131 x 10 x 10 = 1.31

From Table D.2, 1.31 ⟶
2.345 db − 20 db − 20 db = −37.655 db

For Ratios Greater than Those in Table. Divide the given ratio by 10 successively until the quotient can be found in the table. To the number of decibels thus found, add +20 db for each time you divided by 10.

Example.   *Given:* Voltage (sound-pressure) ratio = 712
712 x 1/10 x 1/10 = 7.12

From Table D.2, 7.12 ⟶
17.05 db + 20 db + 20 db = 57.05 db

TABLE D.1   Decibels to Voltage (Sound-pressure) and Sound-power Ratios*

[*To account for the sign of the decibel.*   For positive (+) values of the decibel: Both voltage and power ratios are greater than unity. Use the two right-hand columns. For negative (−) values of the decibel: Both voltage and power ratios are less than unity. Use the two left-hand columns. Use the voltage columns for sound pressures.]

	Decibels	Power Ratio	Voltage Ratio
Example. *Given:* ±9.1 db. *Find:*	+9.1	8.128	2.851
	−9.1	0.1230	0.3508

Negative		Deci-bels	Positive		Negative		Deci-bels	Positive	
Voltage Ratio	Power Ratio		Voltage Ratio	Power Ratio	Voltage Ratio	Power Ratio		Voltage Ratio	Power Ratio
1.0000	1.0000	0	1.000	1.000	0.7079	0.5012	3.0	1.413	1.995
0.9886	0.9772	0.1	1.012	1.023	0 6998	0.4898	3.1	1.429	2.042
0.9772	0.9550	0.2	1.023	1.047	0.6918	0.4786	3.2	1.445	2.089
0.9661	0.9333	0.3	1.035	1.072	0.6839	0.4677	3.3	1.462	2.138
0.9550	0.9120	0.4	1.047	1.096	0.6761	0.4571	3.4	1.479	2.188
0.9441	0.8913	0.5	1.059	1.122	0.6683	0.4467	3.5	1.496	2.239
0.9333	0.8710	0.6	1.072	1.148	0.6607	0.4365	3.6	1.514	2.291
0.9226	0.8511	0.7	1.084	1.175	0.6531	0.4266	3.7	1.531	2.344
0.9120	0.8318	0.8	1.096	1.202	0.6457	0.4169	3.8	1.549	2.399
0.9016	0.8128	0.9	1.109	1.230	0.6383	0.4074	3.9	1.567	2.455
0.8913	0.7943	1.0	1.122	1.259	0.6310	0.3981	4.0	1.585	2.512
0.8810	0.7762	1.1	1.135	1.288	0.6237	0.3890	4.1	1.603	2.570
0.8710	0.7586	1.2	1.148	1.318	0.6166	0.3802	4.2	1.622	2.630
0.8610	0.7413	1.3	1.161	1.349	0.6095	0.3715	4.3	1.641	2.692
0.8511	0.7244	1.4	1.175	1.380	0.6026	0.3631	4.4	1.660	2.754
0.8414	0.7079	1.5	1.189	1.413	0.5957	0.3548	4.5	1.679	2.818
0.8318	0.6918	1.6	1.202	1.445	0.5888	0.3467	4.6	1.698	2.884
0.8222	0.6761	1.7	1.216	1.479	0.5821	0.3388	4.7	1.718	2.951
0.8128	0.6607	1.8	1.230	1.514	0.5754	0.3311	4.8	1.738	3.020
0.8035	0.6457	1.9	1.245	1.549	0.5689	0.3236	4.9	1.758	3.090
0.7943	0.6310	2.0	1.259	1.585	0.5623	0.3162	5.0	1.778	3.162
0.7852	0.6166	2.1	1.274	1.622	0.5559	0.3090	5.1	1.799	3.236
0.7762	0.6026	2.2	1.288	1.660	0.5495	0.3020	5.2	1.820	3.311
0.7674	0.5888	2.3	1.303	1.698	0.5433	0.2951	5.3	1.841	3.388
0.7586	0.5754	2.4	1.318	1.738	0.5370	0.2884	5.4	1.862	3.467
0.7499	0.5623	2.5	1.334	1.778	0.5309	0.2818	5.5	1.884	3.548
0.7413	0.5495	2.6	1.349	1.820	0.5248	0.2754	5.6	1.905	3.631
0.7328	0.5370	2.7	1.365	1.862	0.5188	0.2692	5.7	1.928	3.715
0.7244	0.5248	2.8	1.380	1.905	0.5129	0.2630	5.8	1.950	3.802
0.7161	0.5129	2.9	1.396	1.950	0.5070	0.2570	5.9	1.972	3.890

TABLE D.1   Decibels to Voltage (Sound-pressure) and Sound-power Ratios—Continued

Negative		Deci-bels	Positive		Negative		Deci-bels	Positive	
Voltage Ratio	Power Ratio		Voltage Ratio	Power Ratio	Voltage Ratio	Power Ratio		Voltage Ratio	Power Ratio
0.5012	0.2512	6.0	1.995	3.981	0.3350	0.1122	9.5	2.985	8.913
0.4955	0.2455	6.1	2.018	4.074	0.3311	0.1096	9.6	3.020	9.120
0.4898	0.2399	6.2	2.042	4.169	0.3273	0.1072	9.7	3.055	9.333
0.4842	0.2344	6.3	2.065	4.266	0.3236	0.1047	9.8	3.090	9.550
0.4786	0.2291	6.4	2.089	4.365	0.3199	0.1023	9.9	3.126	9.772
0.4732	0.2239	6.5	2.113	4.467	0.3162	0.1000	10.0	3.162	10.000
0.4677	0.2188	6.6	2.138	4.571	0.3126	0.09772	10.1	3.199	10.23
0.4624	0.2138	6.7	2.163	4.677	0.3090	0.09550	10.2	3.236	10.47
0.4571	0.2089	6.8	2.188	4.786	0.3055	0.09333	10.3	3.273	10.72
0.4519	0.2042	6.9	2.213	4.898	0.3020	0.09120	10.4	3.311	10.96
0.4467	0.1995	7.0	2.239	5.012	0.2985	0.08913	10.5	3.350	11.22
0.4416	0.1950	7.1	2.265	5.129	0.2951	0.08710	10.6	3.388	11.48
0.4365	0.1905	7.2	2.291	5.248	0.2917	0.08511	10.7	3.428	11.75
0.4315	0.1862	7.3	2.317	5.370	0.2884	0.08318	10.8	3.467	12.02
0.4266	0.1820	7.4	2.344	5.495	0.2851	0.08128	10.9	3.508	12.30
0.4217	0.1778	7.5	2.371	5.623	0.2818	0.07943	11.0	3.548	12.59
0.4169	0.1738	7.6	2.399	5.754	0.2786	0.07762	11.1	3.589	12.88
0.4121	0.1698	7.7	2.427	5.888	0.2754	0.07586	11.2	3.631	13.18
0.4074	0.1660	7.8	2.455	6.026	0.2723	0.07413	11.3	3.673	13.49
0.4027	0.1622	7.9	2.483	6.166	0.2692	0.07244	11.4	3.715	13.80
0.3981	0.1585	8.0	2.512	6.310	0.2661	0.07079	11.5	3.758	14.13
0.3936	0.1549	8.1	2.541	6.457	0.2630	0.06918	11.6	3.802	14.45
0.3890	0.1514	8.2	2.570	6.607	0.2600	0.06761	11.7	3.846	14.79
0.3846	0.1479	8.3	2.600	6.761	0.2570	0.06607	11.8	3.890	15.14
0.3802	0.1445	8.4	2.630	6.918	0.2541	0.06457	11.9	3.936	15.49
0.3758	0.1413	8.5	2.661	7.079	0.2512	0.06310	12.0	3.981	15.85
0.3715	0.1380	8.6	2.692	7.244	0.2483	0.06166	12.1	4.027	16.22
0.3673	0.1349	8.7	2.723	7.413	0.2455	0.06026	12.2	4.074	16.60
0.3631	0.1318	8.8	2.754	7.586	0.2427	0.05888	12.3	4.121	16.98
0.3589	0.1288	8.9	2.786	7.762	0.2399	0.05754	12.4	4.169	17.38
0.3548	0.1259	9.0	2.818	7.943	0.2371	0.05623	12.5	4.217	17.78
0.3508	0.1230	9.1	2.851	8.128	0.2344	0.05495	12.6	4.266	18.20
0.3467	0.1202	9.2	2.884	8.318	0.2317	0.05370	12.7	4.315	18.62
0.3428	0.1175	9.3	2.917	8.511	0.2291	0.05248	12.8	4.365	19.05
0.3388	0.1148	9.4	2.951	8.710	0.2265	0.05129	12.9	4.416	19.50

TABLE D.1   Decibels to Voltage (Sound-pressure) and Sound-power Ratios—Continued

Negative		Deci-bels	Positive		Negative		Deci-bels	Positive	
Voltage Ratio	Power Ratio		Voltage Ratio	Power Ratio	Voltage Ratio	Power Ratio		Voltage Ratio	Power Ratio
0.2239	0.05012	13.0	4.467	19.95	0.1496	0.02239	16.5	6.683	44.67
0.2213	0.04898	13.1	4.519	20.42	0.1479	0.02188	16.6	6.761	45.71
0.2188	0.04786	13.2	4.571	20.89	0.1462	0.02138	16.7	6.839	46.7⁻
0.2163	0.04677	13.3	4.624	21.38	0.1445	0.02089	16.8	6.918	47.86
0.2138	0.04571	13.4	4.677	21.88	0.1429	0.02042	16.9	6.998	48.98
0.2113	0.04467	13.5	4.732	22.39	0.1413	0.01995	17.0	7.079	50.12
0.2089	0.04365	13.6	4.786	22.91	0.1396	0.01950	17.1	7.161	51.29
0.2065	0.04266	13.7	4.842	23.44	0.1380	0.01905	17.2	7.244	52.48
0.2042	0.04169	13.8	4.898	23.99	0.1365	0.01862	17.3	7.328	53.70
0.2018	0.04074	13.9	4.955	24.55	0.1349	0.01820	17.4	7.413	54.95
0.1995	0.03981	14.0	5.012	25.12	0.1334	0.01778	17.5	7.499	56.23
0.1972	0.03890	14.1	5.070	25.70	0.1318	0.01738	17.6	7.586	57.54
0.1950	0.03802	14.2	5.129	26.30	0.1303	0.01698	17.7	7.674	58.88
0.1928	0.03715	14.3	5.188	26.92	0.1288	0.01660	17.8	7.762	60.26
0.1905	0.03631	14.4	5.248	27.54	0.1274	0.01622	17.9	7.852	61.66
0.1884	0.03548	14.5	5.309	28.18	0.1259	0.01585	18.0	7.943	63.10
0.1862	0.03467	14.6	5.370	28.84	0.1245	0.01549	18.1	8.035	64.57
0.1841	0.03388	14.7	5.433	29.51	0.1230	0.01514	18.2	8.128	66.07
0.1820	0.03311	14.8	5.495	30.20	0.1216	0.01479	18.3	8.222	67.61
0.1799	0.03236	14.9	5.559	30.90	0.1202	0.01445	18.4	8.318	69.18
0.1778	0.03162	15.0	5.623	31.62	0.1189	0.01413	18.5	8.414	70.79
0.1758	0.03090	15.1	5.689	32.36	0.1175	0.01380	18.6	8.511	72.44
0.1738	0.03020	15.2	5.754	33.11	0.1161	0.01349	18.7	8.610	74.13
0.1718	0.02951	15.3	5.821	33.88	0.1148	0.01318	18.8	8.710	75.86
0.1698	0.02884	15.4	5.888	34.67	0.1135	0.01288	18.9	8.810	77.62
0.1679	0.02818	15.5	5.957	35.48	0.1122	0.01259	19.0	8.913	79.43
0.1660	0.02754	15.6	6.026	36.31	0.1109	0.01230	19.1	9.016	81.28
0.1641	0.02692	15.7	6.095	37.15	0.1096	0.01202	19.2	9.120	83.18
0.1622	0.02630	15.8	6.166	38.02	0.1084	0.01175	19.3	9.226	85.11
0.1603	0.02570	15.9	6.237	38.90	0.1072	0.01148	19.4	9.333	87.10
0.1585	0.02512	16.0	6.310	39.81	0.1059	0.01122	19.5	9.441	89.13
0.1567	0.02455	16.1	6.383	40.74	0.1047	0.01096	19.6	9.550	91.20
0.1549	0.02399	16.2	6.457	41.69	0.1035	0.01072	19.7	9.661	93.33
0.1531	0.02344	16.3	6.531	42.66	0.1023	0.01047	19.8	9.772	95.50
0.1514	0.02291	16.4	6.607	43.65	0.1012	0.01023	19.9	9.886	97.72
					0.1000	0.01000	20.0	10.000	100.00

TABLE D.1   Decibels to Voltage (Sound-pressure) and Sound-power Ratios—Continued

Negative		Deci-bels	Positive	
Voltage Ratio	Power Ratio		Voltage Ratio	Power Ratio
3.162×10^{-1}	10^{-1}	10	3.162	10
10^{-1}	10^{-2}	20	10	10^{2}
3.162×10^{-2}	10^{-3}	30	3.162×10	10^{3}
10^{-2}	10^{-4}	40	10^{2}	10^{4}
3.162×10^{-3}	10^{-5}	50	3.162×10^{2}	10^{5}
10^{-3}	10^{-6}	60	10^{3}	10^{6}
3.162×10^{-4}	10^{-7}	70	3.162×10^{3}	10^{7}
10^{-4}	10^{-8}	80	10^{4}	10^{8}
3.162×10^{-5}	10^{-9}	90	3.162×10^{4}	10^{9}
10^{-5}	10^{-10}	100	10^{5}	10^{10}

*To find decibel values outside the range of this table, see pp. 613–614. Use the voltage-ratio columns for sound-pressure ratios.

TABLE D.2 Voltage (Sound-pressure) Ratios to Decibels*

(*Voltage and sound pressure ratios:* Use the table directly. *Power ratios:* To find the number of decibels corresponding to a given power ratio, assume the given power ratio to be a voltage ratio and find the corresponding number of decibels from the table. The desired result is exactly one-half of the number of decibels thus found.)

Example. *Given:* a power ratio of 3.41. *Find:* 3.41 in the table.

3.41 ⟶ 10.655 db x ½ = 5.328 db

Voltage Ratio	0.00	0.01	0.02	0.03	0.04	0.05	0.06	0.07	0.08	0.09
1.0	0.000	0.086	0.172	0.257	0.341	0.424	0.506	0.588	0.668	0.749
1.1	0.828	0.906	0.984	1.062	1.138	1.214	1.289	1.364	1.438	1.511
1.2	1.584	1.656	1.727	1.798	1.868	1.938	2.007	2.076	2.144	2.212
1.3	2.279	2.345	2.411	2.477	2.542	2.607	2.671	2.734	2.798	2.860
1.4	2.923	2.984	3.046	3.107	3.167	3.227	3.287	3.346	3.405	3.464
1.5	3.522	3.580	3.637	3.694	3.750	3.807	3.862	3.918	3.973	4.028
1.6	4.082	4.137	4.190	4.244	4.297	4.350	4.402	4.454	4.506	4.558
1.7	4.609	4.660	4.711	4.761	4.811	4.861	4.910	4.959	5.008	5.057
1.8	5.105	5.154	5.201	5.249	5.296	5.343	5.390	5.437	5.483	5.529
1.9	5.575	5.621	5.666	5.711	5.756	5.801	5.845	5.889	5.933	5.977
2.0	6.021	6.064	6.107	6.150	6.193	6.235	6.277	6.319	6.361	6.403
2.1	6.444	6.486	6.527	6.568	6.608	6.649	6.689	6.729	6.769	6.809
2.2	6.848	6.888	6.927	6.966	7.005	7.044	7.082	7.121	7.159	7.197
2.3	7.235	7.272	7.310	7.347	7.384	7.421	7.458	7.495	7.532	7.568
2.4	7.604	7.640	7.676	7.712	7.748	7.783	7.819	7.854	7.889	7.924
2.5	7.959	7.993	8.028	8.062	8.097	8.131	8.165	8.199	8.232	8.266
2.6	8.299	8.333	8.366	8.399	8.432	8.465	8.498	8.530	8.563	8.595
2.7	8.627	8.659	8.691	8.723	8.755	8.787	8.818	8.850	8.881	8.912
2.8	8.943	8.974	9.005	9.036	9.066	9.097	9.127	9.158	9.188	9.218
2.9	9.248	9.278	9.308	9.337	9.367	9.396	9.426	9.455	9.484	9.513
3.0	9.542	9.571	9.600	9.629	9.657	9.686	9.714	9.743	9.771	9.799
3.1	9.827	9.855	9.883	9.911	9.939	9.966	9.994	10.021	10.049	10.076
3.2	10.103	10.130	10.157	10.184	10.211	10.238	10.264	10.291	10.317	10.344
3.3	10.370	10.397	10.423	10.449	10.475	10.501	10.527	10.553	10.578	10.604
3.4	10.630	10.655	10.681	10.706	10.731	10.756	10.782	10.807	10.832	10.857
3.5	10.881	10.906	10.931	10.955	10.980	11.005	11.029	11.053	11.078	11.102
3.6	11.126	11.150	11.174	11.198	11.222	11.246	11.270	11.293	11.317	11.341
3.7	11.364	11.387	11.411	11.434	11.457	11.481	11.504	11.527	11.550	11.573
3.8	11.596	11.618	11.641	11.664	11.687	11.709	11.732	11.754	11.777	11.799
3.9	11.821	11.844	11.866	11.888	11.910	11.932	11.954	11.976	11.998	12.019
4.0	12.041	12.063	12.085	12.106	12.128	12.149	12.171	12.192	12.213	12.234
4.1	12.256	12.277	12.298	12.319	12.340	12.361	12.382	12.403	12.424	12.444
4.2	12.465	12.486	12.506	12.527	12.547	12.568	12.588	12.609	12.629	12.649
4.3	12.669	12.690	12.710	12.730	12.750	12.770	12.790	12.810	12.829	12.849
4.4	12.869	12.889	12.908	12.928	12.948	12.967	12.987	13.006	13.026	13.045

TABLE D.2   Voltage (Sound-pressure) Ratios to Decibels—Continued

Voltage Ratio	0.00	0.01	0.02	0.03	0.04	0.05	0.06	0.07	0.08	0.09
4.5	13.064	13.084	13.103	13.122	13.141	13.160	13.179	13.198	13.217	13.236
4.6	13.255	13.274	13.293	13.312	13.330	13.349	13.368	13.386	13.405	13.423
4.7	13.442	13.460	13.479	13.497	13.516	13.534	13.552	13.570	13.589	13.607
4.8	13.625	13.643	13.661	13.679	13.697	13.715	13.733	13.751	13.768	13.786
4.9	13.804	13.822	13.839	13.857	13.875	13.892	13.910	13.927	13.945	13.962
5.0	13.979	13.997	14.014	14.031	14.049	14.066	14.083	14.100	14.117	14.134
5.1	14.151	14.168	14.185	14.202	14.219	14.236	14.253	14.270	14.287	14.303
5.2	14.320	14.337	14.353	14.370	14.387	14.403	14.420	14.436	14.453	14.469
5.3	14.486	14.502	14.518	14.535	14.551	14.567	14.583	14.599	14.616	14.632
5.4	14.648	14.664	14.680	14.696	14.712	14.728	14.744	14.760	14.776	14.791
5.5	14.807	14.823	14.839	14.855	14.870	14.886	14.901	14.917	14.933	14.948
5.6	14.964	14.979	14.995	15.010	15.026	15.041	15.056	15.072	15.087	15.102
5.7	15.117	15.133	15.148	15.163	15.178	15.193	15.208	15.224	15.239	15.254
5.8	15.269	15.284	15.298	15.313	15.328	15.343	15.358	15.373	15.388	15.402
5.9	15.417	15.432	15.446	15.461	15.476	15.490	15.505	15.519	15.534	15.549
6.0	15.563	15.577	15.592	15.606	15.621	15.635	15.649	15.664	15.678	15.692
6.1	15.707	15.721	15.735	15.749	15.763	15.778	15.792	15.806	15.820	15.834
6.2	15.848	15.862	15.876	15.890	15.904	15.918	15.931	15.945	15.959	15.973
6.3	15.987	16.001	16.014	16.028	16.042	16.055	16.069	16.083	16.096	16.110
6.4	16.124	16.137	16.151	16.164	16.178	16.191	16.205	16.218	16.232	16.245
6.5	16.258	16.272	16.285	16.298	16.312	16.325	16.338	16.351	16.365	16.378
6.6	16.391	16.404	16.417	16.430	16.443	16.456	16.469	16.483	16.496	16.509
6.7	16.521	16.534	16.547	16.560	16.573	16.586	16.599	16.612	16.625	16.637
6.8	16.650	16.663	16.676	16.688	16.701	16.714	16.726	16.739	16.752	16.764
6.9	16.777	16.790	16.802	16.815	16.827	16.840	16.852	16.865	16.877	16.890
7.0	16.902	16.914	16.927	16.939	16.951	16.964	16.976	16.988	17.001	17.013
7.1	17.025	17.037	17.050	17.062	17.074	17.086	17.098	17.110	17.122	17.135
7.2	17.147	17.159	17.171	17.183	17.195	17.207	17.219	17.231	17.243	17.255
7.3	17.266	17.278	17.290	17.302	17.314	17.326	17.338	17.349	17.361	17.373
7.4	17.385	17.396	17.408	17.420	17.431	17.443	17.455	17.466	17.478	17.490
7.5	17.501	17.513	17.524	17.536	17.547	17.559	17.570	17.582	17.593	17.605
7.6	17.616	17.628	17.639	17.650	17.662	17.673	17.685	17.696	17.707	17.719
7.7	17.730	17.741	17.752	17.764	17.775	17.786	17.797	17.808	17.820	17.831
7.8	17.842	17.853	17.864	17.875	17.886	17.897	17.908	17.919	17.931	17.942
7.9	17.953	17.964	17.975	17.985	17.996	18.007	18.018	18.029	18.040	18.051
8.0	18.062	18.073	18.083	18.094	18.105	18.116	18.127	18.137	18.148	18.159
8.1	18.170	18.180	18.191	18.202	18.212	18.223	18.234	18.244	18.255	18.266
8.2	18.276	18.287	18.297	18.308	18.319	18.329	18.340	18.350	18.361	18.371
8.3	18.382	18.392	18.402	18.413	18.423	18.434	18.444	18.455	18.465	18.475
8.4	18.486	18.496	18.506	18.517	18.527	18.537	18.547	18.558	18.568	18.578

TABLE D.2   Voltage (Sound-pressure) Ratios to Decibels—Continued

Voltage Ratio	0.00	0.01	0.02	0.03	0.04	0.05	0.06	0.07	0.08	0.09
8.5	18.588	18.599	18.609	18.619	18.629	18.639	18.649	18.660	18.670	18.680
8.6	18.690	18.700	18.710	18.720	18.730	18.740	18.750	18.760	18.770	18.780
8.7	18.790	18.800	18.810	18.820	18.830	18.840	18.850	18.860	18.870	18.880
8.8	18.890	18.900	18.909	18.919	18.929	18.939	18.949	18.958	18.968	18.978
8.9	18.988	18.998	19.007	19.017	19.027	19.036	19.046	19.056	19.066	19.075
9.0	19.085	19.094	19.104	19.114	19.123	19.133	19.143	19.152	19.162	19.171
9.1	19.181	19.190	19.200	19.209	19.219	19.228	19.238	19.247	19.257	19.266
9.2	19.276	19.285	19.295	19.304	19.313	19.323	19.332	19.342	19.351	19.360
9.3	19.370	19.379	19.388	19.398	19.407	19.416	19.426	19.435	19.444	19.453
9.4	19.463	19.472	19.481	19.490	19.499	19.509	19.518	19.527	19.536	19.545
9.5	19.554	19.564	19.573	19.582	19.591	19.600	19.609	19.618	19.627	19.636
9.6	19.645	19.654	19.664	19.673	19.682	19.691	19.700	19.709	19.718	19.726
9.7	19.735	19.744	19.753	19.762	19.771	19.780	19.789	19.798	19.807	19.816
9.8	19.825	19.833	19.842	19.851	19.860	19.869	19.878	19.886	19.895	19.904
9.9	19.913	19.921	19.930	19.939	19.948	19.956	19.965	19.974	19.983	19.991
	0	1	2	3	4	5	6	7	8	9
10	20.000	20.828	21.584	22.279	22.923	23.522	24.082	24.609	25.105	25.575
20	26.021	26.444	26.848	27.235	27.604	27.959	28.299	28.627	28.943	29.248
30	29.542	29.827	30.103	30.370	30.630	30.881	31.126	31.364	31.596	31.821
40	32.041	32.256	32.465	32.669	32.869	33.064	33.255	33.442	33.625	33.804
50	33.979	34.151	34.320	34.486	34.648	34.807	34.964	35.117	35.269	35.417
60	35.563	35.707	35.848	35.987	36.124	36.258	36.391	36.521	36.650	36.777
70	36.902	37.025	37.147	37.266	37.385	37.501	37.616	37.730	37.842	37.953
80	38.062	38.170	38.276	38.382	38.486	38.588	38.690	38.790	38.890	38.988
90	39.085	39.181	39.276	39.370	39.463	39.554	39.645	39.735	39.825	39.913
100	40.000									

* To find ratios outside the range of this table, see p. 614.

Index

Index

A-weighted sound levels:
  in buildings, 584–585
  in communities, 579
  definition of, 78, 539, 555–557, 566
  permissible, 574–581
A-weighting network, 77
Absorption:
  in rooms: absorption coefficient, 223
    average for, 223
    noise level and, 226–230
    total, 226, 228, 241
  of sound in air: classical, 170
    molecular, 170
    outdoors, 170
    rooms, 240, 242
  by vibration on plates, 440
  (*See also* Attenuation, of sound)
Absorption coefficients:
  average of, room, 223
  beams on plates, 450–452
  bracing of plates, 450–452
  definition of, 220
  effect of air space on, 222, 223
  effect of thickness on, 222

Absorption coefficients (*Cont.*):
  graphs, 222, 223
  laboratory measured, 221, 222, 223
  noise reduction coefficient, 221
  normal incidence, 220
  plate discontinuity, 450–452
  porous materials (*see* Porous materials)
  random incidence, 221
  in rooms, 223
  Sabine, 221
  statistical, 221
Acceleration:
  of gravity, 46
  units of, 33
Acceleration level, 33
Accelerometers:
  acceleration range, 57
  acoustic sensitivity, 65
  analyzer, 80
  calibration, 81
  electrodynamic, 56
  frequency range, 54, 57
  frequency response, 54–56, 61, 62
  impedance, 55, 57